A Survey of the Forensic Sciences

Randall R. Skelton
2010

January 2011

Copyright © 2011 by Randall R. Skelton.

Back cover photograph © Ben Hutchings.

All rights reserved.

No part of this book may be reproduced or transmitted in any form except for educational purposes in the teaching of an actual course at an educational institution without the permission of the author. Inquiries may be directed to ASOTFS@yahoo.com.

Library of Congress Cataloging-in-Publication Data

Randall R. Skelton
  A Survey of the Forensic Sciences / Randall R. Skelton
  649 pp.
  Includes bibliographic references, glossary, and index
  ISBN: 978-1-4357-6762-1
  1. Forensic Science 2. Criminalistics I. Title

Available at www.lulu.com.

Printed in the United States of America.

ISBN: 978-1-4357-6762-1.

CONTENTS

List of Figures and Tables	vii
Preface	vii
Acknowledgements	ix
Chapter 1: Introduction to the Forensic Sciences	1
Chapter 2: Forensic Scientists: A Classification by Where You Find Them	11
Chapter 3: Basic Science: The Scientific Method	21
Chapter 4: The Criminal Investigation, Start to Finish	41
Chapter 5: Advanced Topics: Criminal Charges and the Court System	55
Chapter 6: The Crime Scene	67
Chapter 7: Forensic Photography and Forensic Art	79
Chapter 8: Basic Science: The Comparative Method	91
Chapter 9: Basic Science: Microscopes and the Physics of Light	101
Chapter 10: Crime Labs and Trace Evidence Examination	111
Chapter 11: Basic Science: The Chemistry of Oxidation Reactions	121
Chapter 12: Basic Science: Ballistics and the Physics of Projectiles	129
Chapter 13: Firearms Examination	141
Chapter 14: Toolmarks and Impressions Examination	153
Chapter 15: Basic Science: Genes and the Inheritance of Fingerprints	161
Chapter 16: Fingerprint Examination	173
Chapter 17: Questioned Documents Examination	183
Chapter 18: Basic Science: Analytical Methods and Instruments	195
Chapter 19: Serology of Blood	209
Chapter 20: Other Issues in Serology	221
Chapter 21: DNA Analysis	231
Chapter 22: Forensic Chemistry and Toxicology	243
Chapter 23: Advanced Topics: Drugs and Drug Testing	251
Chapter 24: Death Investigation	263
Chapter 25: Forensic Pathology	275
Chapter 26: Basic Science: Skeletal and Dental Anatomy	283
Chapter 27: Forensic Anthropology	297
Chapter 28: Forensic Odontology	311
Chapter 29: Forensic Entomology, Botany, and Geology	321
Chapter 30: Forensic Engineering	331
Chapter 31: Advanced Topics: Explosives	341
Chapter 32: Basic Science: Computers and Networks	349
Chapter 33: Forensic Computer Science	363
Chapter 34: Computer Network Security	375
Chapter 35: Forensic Image and Audio Enhancement	387
Chapter 36: Electronic Surveillance and Biometrics	399
Chapter 37: Forensic Databases	411
Chapter 38: Forensic Psychology	421
Chapter 39: Advanced Topics: Lie Detection and Hypnosis	433

Chapter 40: Psychological Profiling	441
Chapter 41: Statistical and Geographic Profiling	453
Chapter 42: Forensic Accounting	461
Chapter 43: Advanced Topics: Occupational Fraud	471
Chapter 44: Advanced Topics: Organized Crime	481
Chapter 45: Advanced Topics: Homeland Security	491
Chapter 46: Jurisprudence	503
Chapter 47: Other Forensic Sciences	517
Chapter 48: Advanced Topics: Forensic Science for Wildlife Protection	529
Chapter 49: Advanced Topics: Forensic Science Applied to History and Prehistory	539
Chapter 50: Advanced Topics: Pseudoscience	553
Chapter 51: Careers and Education in the Forensic Sciences	567
Chapter 52: What I Wish Defense Attorneys Knew About Forensic Science	575
Glossary:	583
Index:	626

LIST OF FIGURES AND TABLES

Figure 6.1:	Primary and Secondary Areas of a Crime Scene	69
Figure 6.2:	A Title Card Showing the Exhibit Number, Direction, and Investigator's Name.	74
Figure 7.1:	A Crime Scene Illustration	84
Figure 7.2:	Age Progression of a Suspect	85
Figure 9.1:	A Radiation Wave	102
Table 9.1:	Types of EMR by Wavelength	103
Figure 9.2:	Reflection & Refraction	104
Figure 9.3:	A Convex Lens	106
Figure 9.4:	A Concave Lens	107
Figure 9.5:	A Compound Microscope	108
Table 10.1:	Comparison of Montana's and Georgia's Crime Lab Sections	113
Figure 12.1:	A Ballistic Curve	131
Figure 12.2:	Shock Waves Produced by a Moving Bullet	133
Figure 12.3:	Yaw	134
Figure 12.4:	Entrance, Exit and "Keyhole" Bullet Wounds in Bone	138
Figure 13.1:	Handguns	143
Figure 13.2:	A Shotgun and a Rifle	144
Figure 13.3:	Various Types of Ammunition	146
Figure 13.4:	How Caliber is Determined	147
Figure 13.5:	A "Bullet's Eye" View Down the Barrel of a Rifle	148
Figure 15.1:	The Relationship Between Cell Nucleus, Chromosome, Loci, & Allele	161
Table 15.1:	Phenotypes and Genotypes for the Transfusion (ABO) Blood Group	165
Figure 16.1:	The Eight Basic Fingerprint Patterns (Class Characteristics)	175
Figure 16.2:	Fingerprint Minutia (Individual Characteristics)	177
Figure 17.1:	Two Checks Purportedly Written by John Doe	186
Figure 17.2:	John Doe's Known and Purported Signatures	188
Figure 18.1:	Paper Chromatography	200
Figure 18.2:	Diagrammatic Representation of a Gas Chromatograph	201
Figure 18.3:	A Electrophoresis Apparatus	202
Figure 18.4:	Diagrammatic Representation of a Spectrophotometer	204
Figure 18.5:	Mass Spectrum of Ethanol	205
Figure 19.1:	The Components of Blood When Centriguged	211
Figure 19.2:	A Simple Precipitin Test for Human Blood	214
Figure 19.3:	A Agglutination Reaction Between Antibodies and the Antigens on Cell Surfaces	215
Table 19.1:	Blood Typing for the Transfusion (ABO) Blood Group	217
Table 19.2:	Blood Typing for the ABO Blood Group Using the Absorption/Elution Method	218
Table 20.1:	Frequencies of PGM Electrophoresis Band Patterns	224
Table 20.2:	The Most Common Variant for 11 Genetic Marker Systems	225

Figure 21.1:	Diagrammatic Representation of a DNA Comparison Between Evidence DNA and the DNA of Two Suspects	234
Figure 24.1:	States with Medical Examiner Only, Coroner Only, and Mixed Systems of Death Investigation	264
Figure 26.1:	Lateral View of the Skull	286
Figure 26.2:	Frontal View of the Skull	287
Figure 26.3:	Inferior View of the Skull	288
Figure 26.4:	Superior View of the Skull	289
Figure 26.5:	Posterior View of the Upper Skeleton	289
Figure 26.6:	Frontal and Lateral Views of the Skeleton	291
Figure 26.7:	The Human Dentition: Tooth Names and Directions	292
Figure 26.8:	Cross Section of a Tooth	293
Figure 26.9:	The Army System of Tooth Designation	295
Figure 27.1:	The Process of Creating a Grave	299
Figure 27.2:	Dental Attrition	304
Figure 27.3:	Female and Male Pelves	304
Figure 27.4:	Postmortem and Premortem Trauma	307
Table 28.1:	Disasters that Could Happen in Spokane, WA In Order from Most Likely to Least Likely	315
Table 28.2:	DMORT Services	316
Table 30.1:	Additional Factors in a Traffic Accident that May be Considered by a Forensic Engineer	334
Figure 32.1:	The Data Movement and Storage Processes That Lead to the Creation of RAM Space and Slack Space	352
Figure 32.2:	The Relationship Between LAN's, Intranets, and the Internet	354
Figure 35.1:	Image Enhancement of a Digital Photograph Taken Under Low Light Conditions	388
Figure 35.2:	T-A, T-F, and T-F-A Spectrograms	392
Figure 35.3:	Voiceprint of a Japanese Woman Saying the Word "Minato"	395
Figure 36.1:	My Laptop Computer's Fingerprint Scanner	405
Figure 36.2:	Use of a Retina Scanner	406
Table 37.1:	Type of Items for Which the FBI Maintains a Database of Item Characteristics	413
Table 42.1:	Common Discrepancies in Financial Records	463
Figure 47.1:	An Example Terrorist Cell	521
Table 50.1:	Distinctions Between Science and Pseudoscience	561

PREFACE

This book stems from my inability, over the past decade, to find a textbook that works very well for my sequence of two classes on forensic science, which I teach at The University of Montana, Missoula. In these courses I attempt to survey the breadth of forensic science – not just a handful of the most familiar, most exciting, or most useful forensic sciences. Further, I approach the forensic sciences in an academic manner. In other words, I teach about forensic science – not how to actually do forensic science. I seek to teach students to be conversant in the forensic sciences, by which I mean that they could have a conversation with a forensic scientist about a topic in forensic science or about some evidence and be able to hold their own in the conversation. I have been unable to find within a single text both the great breadth and the focus on principle over practice that would best serve the needs of my students. Other texts on forensic science seem to me to be either too narrow in scope or are essentially laboratory manuals on forensic science methodologies.

Another of my goals with this book is to show that forensic science is, to some extent, a unified field, and not just a collection of bits and pieces from many fields that just happen to apply to civil or criminal legal processes. As such it has a coherent body of theory and principles that are shared by at least most of the individual forensic sciences. In this text I seek to highlight this body of theory by pointing out the ways in which the individual forensic sciences implement and express the principles of forensic science.

The first of the two classes this text seeks to serve is a lower division course that has been given a "natural sciences, non laboratory" designation with regard to The University of Montana's general education requirements. As such, I strive in this class to teach students a broad range of basic scientific principles, couched in the (hopefully) engaging subject matter of forensic science. This course focuses primarily on the forensic sciences practiced at a crime lab, and does not attempt to be especially cutting edge with regard to technological innovations in the forensic sciences.

The second of the two classes I teach is entirely different. It is an upper division course with two goals. The first goal is to be a thorough (though not exhaustive) survey of all the forensic sciences, both those practiced by the employees of a crime lab and those practiced by forensic scientists in other settings; and including forensic sciences that are more useful in civil cases as well as those that are more useful in criminal cases. Second, in this course I explicitly attempt to cover new and cutting edge forensic sciences and methodologies. Therefore, I cover such topics as biometrics, which includes technologies that are likely to become incorporated into more mainstream forensic science in the near future; and homeland security, which is an emerging application for the methods of forensic science.

The organization of this book arises from the differing natures of the two courses I teach and is likely to strike most people as odd (at best). The first unique feature of this text is that I recognize three types of subject matter: basic science, discussions of forensic sciences, and advanced topics. The core of the text is, of course, the discussions of the forensic sciences. The portions pertaining to basic science are probably optional for those readers who have a strong background in the sciences. Advanced topics refers to material that is also optional, depending on the nature of the course, and which either explores more deeply the science and technology behind some issue, or is a discussion of an application of the forensic sciences rather than of a particular forensic science. Portions of some chapters are labeled as basic science or advanced topics, and entire chapters are also so labeled. By this device I intend to enable users of the text to appropriately direct their readings and the readings of their students.

The chapter sequence roughly parallels that of my two classes. There are a lot of chapters. I intend that each chapter will correspond to an approximately one-hour lecture on that subject. I find that the vast majority of my students these days like their reading in small chunks, rather than in larger packages, so I have endeavored to keep the chapters short.

Especially odd to some people will be the fact that I am an anthropologist. Although some aspects of anthropology are not normally considered "scientific", I am a physical/biological anthropologist and share the broad education in the biological and physical sciences that is typical of most biological scientists. Further, in what could probably be considered a bout of mid-life foolishness, I recently completed a second Bachelor of Science degree, in Computer Science and Mathematical Sciences, with high honors.

In a more practical vein, I have been handling most of the forensic anthropology casework for the state of Montana for nearly two decades. In this capacity I am often directly involved with the forensic scientists who work at the Montana State Laboratory of Criminal Investigation. I am proud to note that several of the Crime Lab's employees are my former students, and three of them earned Master of Arts degrees with a concentration in forensic anthropology under my mentorship. These scientists and many others, as well as a host of current and former peace officers are personal friends of mine, and we discuss forensic science and forensic science education. Although I am proud of all these things, perhaps the most important advantage that an anthropologist brings to a writing project of this sort is that whatever training and specialization we have, we are all to one extent or another, ethnographers. Ethnography can be described as the art of observing a culture, while participating in it, then explaining the observations made and conclusions drawn to those who have not had the opportunity to participate in the culture. In this manner I can be considered an ethnographer for the culture of forensic science. I hope that I have successfully captured the material, social, cultural, and ideological elements of forensic science and that I can convey my observations and conclusions to my readers.

It is a certainty that I have made some errors in this book. I will try to make errata available on the internet if I can find a web site to host it. The easiest way to find the errata will be to use a search engine with the following text <"A Survey of the Forensic Sciences" Skelton errata> (include the quotes but not the <> brackets). I welcome you to point out any substantial errors you encounter, by sending me email at asotfs@yahoo.com with a subject line of "A SURVEY OF THE FORENSIC SCIENCES:

ERRATA" (capitalize, but don't include the quotes). An email with any other subject is likely to be filtered to my spam folder and deleted unread. I promise that I will research the issue you bring up and if warranted make an addition to the errata and a revision to the next version of the text (assuming there is one). By substantial errors, I mean something more important than punctuation, grammar, style, and the like.

I am not a consultant on the forensic sciences other than forensic anthropology and when I work for anyone other than the Montana State Laboratory of Criminal Investigation I expect to be paid. I am not certified as an expert witness outside of the state of Montana. I reserve the right to decline any requests for consultation or analysis for any reason. My caseload and general workload for my job are near maximum at present and I do not enjoy consulting on other than local cases, so expect that I will decline most of the time.

ACKNOWLEDGMENTS

My deepest thanks to Kelli Whitacker, Maia Hangas, James P. Cramer, Rod and Andi Skelton, Garry Kerr, Jennie Frankus, Greg Campbell, Tom Foor, Petey, Shadow, and several others who have all, in their own ways, contributed to keeping me from being just another statistic. Rod Skelton, who I am proud to call my father, was the primary proofreader for this text.

Without my friends and colleagues in forensic science and law enforcement, near and far, this book would not have been possible. Special thanks to Gary Dale, Bill Unger, Lynn Kurtz, Megan Ashton, Connie Muller, Willy Kemp, Jim Hutchison, Julie Long, and many others at the State Crime Lab. Special thanks also to Rick Browne and Larry Jackson who were often my sounding board for law enforcement issues.

Thank you to The University of Montana, College of Arts and Sciences for the time off to work on this book and to my department for their support. My younger colleagues, Noriko Seguchi and Ashley McKeown have been this tired old man's inspiration and have taught me the responsibilities of being a senior faculty member. To my past, present, and future students, this book is for you – literally. Thank you for all the times you have shown me by many methods of feedback, including your looks of understanding and blank stares of confusion, when I was getting it right and when I wasn't.

x

Chapter 1
What is Forensic Science

In this chapter we will struggle with defining forensic science. We will see that all definitions of forensic science have strengths and weaknesses, and that none is perfect. In discussing the strengths and weaknesses of each definition the issues in just what forensic science is and what it does will become apparent. Next, we will look at the history of forensic science and at the effect media portrayals of forensic science have had on the public's perception of it. The chapter will end with a discussion of four important principles of forensic science that are foundational for most of the forensic sciences.

1.1 Defining Forensic Science

Forensic science is very much a part of American popular culture. Almost everybody has heard of it, and has a firm idea of what it is. However, few people can give a precise definition of it. This is not unexpected, because even professionals and authorities in the field have trouble defining forensic science. Definitions have ranged from overly broad to overly narrow, and some have missed the point entirely. In my opinion, most existing definitions have at least partially missed the point of what forensic science is about.

First, forensic science is not forensics. **Forensics** is the sport of debate, in which teams of college students, travel to different colleges to engage other such teams in debates. We will see definitions of forensic science below, but in no way does it resemble forensics. Thus, the difference between forensic science and forensics is significant and obvious, and we should not use these terms interchangeably as is so often done.

Let's start with a broad definition – forensic science is the application of science to law enforcement. Although I normally prefer broad definitions, this one seems to be so broad as to include some people who we don't consider forensic scientists, and furthermore, they don't consider themselves forensic scientists either. For example, my friend Mike works for a county in the state of Maryland, and part of his job is to inspect restaurants to make sure that they keep and handle food under sanitary conditions as specified by those laws that regulate such things. Is Mike a forensic scientist? Most of us, including Mike, would agree that he is not. He does use principles of science in his work – for example the biological observation that bacteria grow best at a temperature above 40 degrees Fahrenheit. Further he is clearly enforcing laws. However, this broad definition has missed the point that forensic science involves the pursuit of evidence that might be used against a suspect in court. Since his job doesn't normally involve collection and analysis of evidence, it is actually more similar to the job of a traffic cop than to that of a forensic scientist.

Let's try a narrow definition. One way to narrow the definition of forensic science is to restrict it to those things having to do with criminal law. This gives us the definition – forensic science is the application of science to the enforcement of criminal law. Our narrower definition clearly rules out Mike's job as a restaurant inspector, because the laws he enforces are part of the body of law known as **regulatory law** or administrative law.

However, the narrow definition is still problematical, because it excludes the enormous number of people who consider themselves to be forensic scientists, but whose primary focus is on civil law and who testify in civil suits rather than at criminal trials. For example, a forensic economist may be asked to develop evidence and testify for one of the sides in a civil suit about the amount of income an injured person might expect to lose over their lifetime as a result of their injury. Forensic economists who do this type of work consider themselves to be forensic scientists, and so do we. Therefore, we don't want to exclude them in our definition.

Clearly, a compromise definition is in order. Richard Saferstein, the author of one of the most authoritative textbooks on forensic science offers this definition – forensic science is "... the application of science to those criminal and civil laws that are enforced by police agencies in a criminal justice system." This definition is better because it includes those forensic scientists who primarily practice in the civil arena. However, whether all civil laws (or even criminal laws) that forensic scientists become involved with are enforced by police agencies is debatable. The answer partially depends on how you define a police agency, which is a subject to be examined in another chapter.

Although Saferstein's definition captures part of the essence of forensic science, in my opinion it could be improved by changing the focus from police agencies to the courts, as the consumers of forensic science. After many years as a forensic scientist, after studying the field, and after talking with many forensic scientists and many other people whom we would not consider forensic scientists, I would suggest this definition – forensic science is the use of scientific methods to produce evidence that might be used in court. The main advantage of this definition is the focus on courtroom evidence. Developing evidence that might be used in court is central to the forensic sciences. In fact the word "forensic" means "having to do with the forum", with the forum in this case being a courtroom.

Like all other definitions, however, mine has a couple of weaknesses. First, this definition does not exclude many of the activities of police officers. From the traffic cop who measures skid marks to estimate the speed of vehicles involved in a collision, to the detective who draws conclusions from the evidence left at a crime scene, many police officers daily engage in activities that we would consider to be forensic science under this definition. Whether this is a good or bad result of the definition is arguable. In my opinion, although we may not consider police officers to be forensic scientists in the usual sense, they often do some forms of forensic science. It may actually be impossible to define forensic science in such a way that we exclude the activities of police officers, and I'm not sure that we really want to. Second, notice that I was careful to say that the evidence developed by forensic scientists might be used in court.

Although the goal of forensic science is to develop evidence that is strong enough to stand the scrutiny of the court, there are many situations in which the evidence is not actually used in a trial. There are many reasons for this. For example, it is often the case that the evidence demonstrates the innocence of a suspect, and therefore is not used in court. In other cases, when suspects are confronted with evidence of their guilt, they will plead guilty and the evidence does not need to be presented in court.

1.2 The Beginnings of Forensic Science

The use of science in the service of justice is undoubtedly ancient. Instances of the use of what could be considered forensic science are recorded for several early civilizations, such as the Ancient Greeks, the Romans, and the Ancient Chinese. It is safe to bet that the use of forensic science is even older, but not recorded because writing had not yet been invented.

As an organized field, however, forensic science had its beginnings in the 1800's as pioneering work in most of the branches of modern forensic science was carried out and published. More than occasional use of forensic science in criminal investigations began in the 1800's, and its use has become more pervasive through time. Here are a few of the beginnings of individual forensic sciences and a handful of the important people in the history of forensic science.

Toxicology was one of the first forensic sciences to become established. Mathieu Orfila (1787-1853), who was born in Spain but later moved to France, collaborated with police on several cases that involved poisonings. Orfila published the first forensically oriented book on poisons, their detection and their effects in 1814.

The foundation for identification of individuals, which is reflected today in forensic sciences such as fingerprinting, serology, DNA analysis, and biometrics, was provided by the work of Alphonse Bertillon (1853-1914), a citizen of France. In the late 1800's, Bertillon invented a method for distinguishing one individual from another by examination of their body measurements. **Bertillonage**, as it is often called, revolutionized the system of justice by providing a way to identify repeat offenders. Before Bertillonage came into widespread use it was impossible to determine definitely whether an offender before the court had been previously tried, perhaps with a different name, or whether this was their first offense. Bertillonage provided a way to establish a person's identity with reasonable certainty, regardless of what name they had provided the court. This gave judges the ability to impose more stringent punishments on repeat offenders.

Fingerprinting was developed during the 1800's, and several scientists contributed to it. Perhaps the greatest contribution was by Francis Galton (1822-1911). Galton, an English statistician, built on the work of previous researchers and systematized fingerprinting and the method by which fingerprints were classified for archiving. In 1892, he published a book that contained the first statistical proof of the uniqueness of fingerprints.

Hans Gross (1847-1915), an Austrian prosecutor and judge undertook the first attempt to organize the field of forensic science. In 1893 he published what could be considered the first textbook on forensic science, which included discussions of the insights that the physical and biological sciences could offer to investigations. He was the founder of the first forensic science journal, *Kriminologie*, which is still in publication today.

By the early 1900's research on blood and blood types had reached the point where what we today recognize as serology became a routine part of investigations that involved blood evidence. Leone Lattes (1887-1954), a professor at the Institute of Forensic Medicine at the University of Turin in Italy, invented the absorption-elution method for blood typing, which allowed the blood type of most people to be discovered from their dried blood or from other bodily fluids such as semen or saliva.

Firearms examination has been a part of forensic investigations since the invention of firearms. In the early 1900's, Calvin Goddard (1891-1955), an American Army colonel, began using the comparison microscope. A **comparison microscope** consists of two conventional microscopes, with their viewfields brought together using an optical bridge so that two items can be viewed at the same time. The comparison microscope enabled Goddard to compare bullets and other firearms evidence side by side, which allowed greater precision in matching, making the field of firearms examination more reliable.

Albert S. Osborn (1858-1946), an American forensic scientist, was instrumental in demonstrating that the results of document analysis were sufficiently precise that they could be used as courtroom evidence. In 1910, Osborn published the first significant text on the field of questioned document analysis.

The first crime lab in the world was established in 1910, in Lyons, France by Edmond Locard. Locard opened the lab with only two instruments, a microscope and a crude spectrometer, but added more instruments as time went on and was joined by other scientists. Locard's accomplishments were not restricted to the practical end of forensic science. He is renowned for his research on what we would now call trace evidence analysis, and he is credited with first proposing a fundamental precept of forensic science, the principle of cross transfer of evidence, which we will look at later in this chapter.

1.3 The History of Forensic Science in the United States

The first forensic science lab in the United States was opened in Los Angeles, California in 1923 by the Los Angeles Police Department. In 1932, the FBI organized a national forensic laboratory at Quantico, VA, which was intended to offer services to all the law enforcement agencies in the country. The FBI lab is now the largest in the world, performing over one million examinations every year. It has served as a model for other laboratories at the state, local, and international levels. In 1981 the FBI opened their Forensic Science Research and Training Center, which is dedicated to research in the forensic sciences and training of forensic scientists. In 1989, the U.S.

Fish and Wildlife Service opened a wildlife forensic laboratory in Ashland Oregon. This lab was created for the purpose of providing forensic support to wildlife law enforcement officers and agencies within the United States and throughout the world.

In the 1960's several events occurred that prompted the American justice system to place more value on forensic science. Prior to the1960's, police investigative methods focused on obtaining confessions from suspects. Examples of this approach can be seen in many classic films. The police would arrest a suspect and apply mild to severe forms of torture to get the suspect to confess to the crime. Then, the suspect would be tried on the basis of their confession and quite often found guilty. It is easy to see that this system could, and did, lead to abuses. Most people will confess to anything if subjected to sufficient persuasion, with the result that an innocent person is punished for something they did not do. Conviction of an innocent person is the worst possible miscarriage of justice. Not only is an innocent person unfairly punished, but the actual guilty party remains at large, free to commit other crimes.

Concerned about the state of the justice system, the U.S. Supreme Court handed down two important decisions during the 1960's, Escobedo v. Illinois (1964) and Miranda v. Arizona (1966), which demanded more thorough police investigations and put a premium on forensic evidence as opposed to confessions. The "reading of the rights" that police officers must now do while making arrests comes directly from Miranda v. Arizona, and the first of these is the right to remain silent. In other words these Supreme Court decisions gave suspects the right to not confess and to not be pressured into confessing.

The U.S. President (Lyndon Johnson, at that time) was also concerned about the state of the justice system, and in 1968 formed the President's Commission on Law Enforcement and Administration of Justice. This commission concluded that there was an urgent need for improved use of science in police work. Other governmental and civilian organization also joined in asking for increased reliance on forensic evidence and the forensic sciences during the 1960's and in following decades.

These governmental mandates caused a surge in the number of crime labs opened. In 1966 there were approximately 100 public crime laboratories operating at various levels of government – federal, state, county, and city. As I write now, about 40 years later, there are more than 350 such laboratories nationwide. These laboratories range in size from very small to very large, hiring anywhere from one to hundreds of employees. The Montana State Laboratory of Criminal Investigation, with which I most often work, is an example. It opened in the early 1970's and is a smaller laboratory, which currently employs about 35 people. Despite its small size, it provides a full range of forensic services, thanks to the quality and resourcefulness of its staff, and handles thousands of cases each year for law enforcement agencies throughout the state. In the year 2000, the Montana State Laboratory of Criminal Investigation receive evidence relating to more than 4,230 cases, despite Montana's relatively small population.

The use of forensic science increased in the U.S. through the 1970's and 1980's, as more crime labs opened and training of forensic scientists became more widely available. Many people point to the murder trial of O.J. Simpson in the middle 1990's as the single most important event for acquainting the public with forensic science. On

Friday, June 17, 1994, O.J. Simpson was arrested for the murder of his ex-wife Nicole Brown Simpson and her friend Ronald Lyle Goldman. Over the next 15 months many types of forensic evidence were discussed and presented at Simpson's trial, and the public received a fairly thorough education about forensic science. On October 3, 1995, Simpson received a verdict of not guilty, an outcome that conflicted with the opinion of the vast majority of Americans who were convinced of his guilt, primarily based on their understanding of the forensic evidence. Interest in forensic science, both among law enforcement professionals and the public, soared following the Simpson trial and has remained high ever since.

1.4 Forensic Science and the Media

Forensic science is one of those fields that has been blessed (or cursed) by considerable media attention. The emergence of forensic science into mainstream use during the late 1800's and early 1900's was no doubt catalyzed by the work of Sir Arthur Conan Doyle, who wrote many stories featuring the fictional master detective Sherlock Holmes over a period of time ranging from the 1880's through the 1920's. In these novels Holmes applies serology, fingerprinting, firearm identification, questioned document examination, and trace evidence analysis to the solution of several interesting mysteries. In 1894, Mark Twain's popular novel, *The Tragedy of Pudd'nhead Wilson and the Comedy of Those Extraordinary Twins*, was published. Part of this story was about the use of fingerprinting and it was instrumental in directing public attention toward this aspect of forensic science.

Although the middle 1900's abounded with mystery stories, featuring a variety of detectives, few of them focused on forensic science in a way that made a lasting impression on the public consciousness. Since the 1980's, various forms of forensic science have been featured in novels by several authors, such as Aaron Elkins and Patricia Cornwell, whose novels are very popular.

Television and the movies have also contributed to the public awareness of forensic science. The TV series *Quincy M.E.*, which ran between 1976 and 1983 with extensive reruns, featured Jack Klugman as the crusading Quincy, a medical examiner who took it upon himself to investigate countless crimes and social issues. More recent series that focus on forensic science, such as *The New Detectives*, *The FBI Files*, *Crime Scene Investigators*, *Bones*, *Numbers*, and numerous others have also captured the public imagination. Forensic science has also been a key element in a few movies, such as *The Bone Collector*, released in 1999.

1.5 Basic Science: First Principles of Forensic Science

There are four important principles that are basic to the forensic sciences. I will present them briefly here, and coming chapters will provide many varied examples of their application.

1.5.1 Principle #1: Locard's Exchange Principle

Locard's exchange principle, formulated in 1930, states, "whenever two objects come into contact, a transfer of material will occur." Since entering a location always involves two objects coming into contact, at least the soles of the shoes touching the ground or floor, this principle is extended to the idea that a person cannot be at a location without both leaving something there and taking something with them. What they leave may be as subtle as dust and dead skin cells, or as obvious as blood and bullets. What they take away may be as subtle as carpet fibers or as obvious as their victim's blood on their hands. The basic message of this principle is that evidence always exists which can be used to link the perpetrator of a crime to the scene of the crime. The evidence is always there, though it may be difficult to find.

1.5.2 Principle #2: Evidence Items and Standard Items

The items left by a perpetrator at a crime scene and the items taken away from the crime scene by the perpetrator are referred to as **evidence items**. When evidence items are discovered at a crime scene or in association with a suspect, their origin is initially unknown, and they must be matched with items known to originate with the suspect or the crime scene. This matching process demonstrates that the suspect was present at the crime scene. **Standard items** are items of known origin. They consist of the suspect's fingerprints taken directly from their fingers, the bullets collected after firing them from the suspect's gun, blood taken directly from the suspect's or the victim's veins, samples of fibers taken from the carpet at the crime scene, and similar items for which the origin is indisputable. The process of **matching** involves comparing the evidence items and standard items to show that they have the same origin. Other names for both evidence items and standard items exist. Evidence items are also referred to as unknown items or questioned items. Standard items are also referred to as known items or exemplars.

1.5.3 Principle #3: Class Characteristics and Individual Characteristics

Evidence and standard items are matched based on characteristics that they share. Characteristics, also called traits or characters, are either **class characteristics** or **individual characteristics**. Class characteristics are not unique to a single item, and other items of that type may share those characteristics. For example, I am typing on a Toshiba M400 computer. This means that my computer would not be matched with a Dell or Sony, or even with a Toshiba A7 computer. However, many other people have computers, which, like mine, belong to the class of Toshiba M400's. In contrast, individual characteristics are unique to one certain item of a class. For example, the serial number of my computer is 67099336H, and it was given this number during its manufacture. No other Toshiba M400 computer has this exact serial number – it is unique to this individual computer. All items have class characteristics, but not all items have detectable individual characteristics.

1.5.4 Principle #4: Exclusion, Consistent Match, Positive Match and Undetermined

When comparing an evidence item to a standard item, there are four results that may be obtained. First, it may be the case that neither the class characteristics nor the individual characteristics of the evidence item and the standard item are the same. In this case we have an **exclusion**. The evidence item is excluded from being from the same source as the standard item, and this normally means that you have the wrong suspect. Many forensic scientists and police officers do not appreciate exclusions, but in reality it is at least as important to demonstrate a suspect's innocence as it is to demonstrate their guilt. If even one class characteristic or individual characteristic differs between the evidence item and the standard item, which can not be explained as due to an event that occurred after the crime was committed, then the result is an exclusion.

If the class characteristics of the evidence item and the standard item are the same, and the individual characteristics have not yet been examined (or don't exist), then the result is a **consistent match**. In this case you can't exclude the possibility that the evidence and standard items are from the same source, but you can not say that they are for sure. Under these conditions, you might have the correct suspect, or you might not. For this reason, many forensic scientists use the term **indeterminate** or **inconclusive** in place of consistent match. Indeterminate should not be confused with the similar sounding term undetermined, explained below. Other forensic sciences more commonly use the terms circumstantial match or presumptive match. All of these terms carry exactly the same meaning as consistent match – that neither an exclusion nor a positive match is possible.

If both the class characteristics and the individual characteristics of the evidence and standard items are the same, then the result is a **positive match**, often simply called a "match". In this case you can say positively that the evidence and standard items are from the same source. In this situation it is likely that the suspect was present at the crime scene.

Situations do occur in which it is impossible to compare the characteristics of the evidence and standard items. For example, a fingerprint may not be distinct or a bullet may be too badly damaged for comparison. In this case the result is **undetermined** and no exclusion or any form of match is possible.

1.6 Questions for Review and Study

1. Forensic science is a surprisingly difficult field to define. Discuss some of the reasons for why this is true. Do you think that the way television and other media portray forensic science contributes to confusion over how to define it?
2. During what period in history did forensic science start to become scientific?
3. To which field of forensic science did each of the following scientists contribute: Orfila, Bertillon, Galton, Gross, Lattes, Goddard, Osborn, Locard?
4. During which period in history did crime labs appear?
5. Discuss how our justice system if different today from what it was like in the 1950's, and discuss some of the reasons for these differences.
6. How many television programs related to forensic science can you list? How many movies related to forensic science can you list? Now many novels that portray forensic science or a forensic scientist have you read?
7. Explain Locard's exchange principle in your own words.
8. In your own words, explain the difference between standard items and evidence items.
9. Choose some item that you own and list its class characteristics and its individual characteristics.
10. In your own words, explains the difference between a consistent match and a positive match. What other terms do some forensic scientists use to refer to a consistent match?
11. Explain the difference between the terms indeterminate and undetermined as they are applied to forensic matching.
12. Discuss why an exclusion is as important as a positive match in a forensic examination.
13. Look up the article on forensic science in Wikipedia (http://www.wikipedia.org/). In what ways do the definition of forensic science and the history of forensic science presented in the Wikipedia article differ from that presented in this chapter?

1.7 Sources

[Anonymous]. n.d. Pudd'nhead Wilson [Internet]. [cited 2007 Oct 16]. Available from: http://etext.lib.virginia.edu/railton/wilson/pwhompg.html.

American Academy of Forensic Sciences. 2007. So You Want to Be a Forensic Scientist! [Internet]. [cited 2007 Oct 16]. Available from: http://www.aafs.org/default.asp?section_id=resources&page_id=choosing_a_career.

Arneson E. 2000. Quincy, M.E.: The Crime-Fighting Coroner [Internet]. [cited 2007 Oct 16]. Available from: http://www.mysterynet.com/tv/profiles/quincy.

Arthur Conan Doyle Society. 2003. An Outline Bibliography of the Works of Sir Arthur Conan Doyle [Internet]. [cited 2007 Oct 16]. Available from: http://www.ash-tree.bc.ca/ac

Ball D, Ball V. 2007. Aaron J. Elkins [Internet]. [cited 2007 Oct 16]. Available from: http://www.booksnbytes.com/authors/elkins_aaronj.html.

Block EB. 1979. Science vs Crime. San Francisco: Cragmont Publications.

Eckert WG. 1996. Introduction to Forensic Sciences. Boca Raton (FL): CRC Press.

Hayes RA. 2000. Forensic Geologists Uncover Evidence In Soil And Water [Internet]. [cited 2007 Oct 16]. Available from: http://www.geoforensics.com/geoforensics/art-1101a.html.

Inman K, Rudin N. 2000. Principles and Practice of Criminalistics: The Profession of Forensic Science. Boca Raton (FL): CRC Press.

Jackson ARW, Jackson JM. 2004. Forensic Science: An Introduction. Upper Saddle River (NJ): Prentice Hall.

James SH and Nordby JJ. 2005. Forensic Science: An Introduction to Scientific and Investigative Techniques. Boca Raton (FL): CRC Press.

Kiely, TF. 2000. Forensic Evidence: Science and the Criminal Law. Boca Raton (FL): CRC Press.

Lee HC, Gaensslen RE, Harris H. 2007. Introduction to Forensic Science and Criminalistics. Columbus (OH): McGraw-Hill.

Meloan CE, James RE, Saferstein R. 1998. Lab Manual for Criminalistics: An Introduction to Forensic Science. 6th Edition. Upper Saddle River (NJ): Prentice Hall.

Nickel J, Fischer JF. 1998. Crime Science: Methods of Forensic Detection. Lexington (KY): University Press of Kentucky.

Rudin, N, Inman K. 2002. Forensic Science Timeline [Internet]. [cited 2007 Oct 16]. Available from: http://www.forensicdna.com/Timeline020702.pdf.

Saferstein R. 1998. Criminalistics: An Introduction to Forensic Science. 6th Edition. Upper Saddle River (NJ): Prentice Hall.

Siegel JA. 2006. Forensic Science: The Basics. Boca Raton (FL): CRC Press.

Siegel JA, Houck MM. 2006. Fundamentals of Forensic Science. Burlington (MA): Elsevier Science & Technology Books.

Tangled Web UK. 1998. Patricia Cornwell - Page 1 [Internet]. [cited 2007 Oct 16]. Available from: http://www.twbooks.co.uk/authors/pcornwell.html.

Wojtas O. 2007. TV crime rekindles interest in science. Times Higher Education Supplement 1794: 4(4).

Chapter 2
Forensic Scientists
A Classification by Where You Find them

In this chapter I will classify and organize the various forensic sciences and the forensic scientists who practice them. In the process we will take our first look at the broad spectrum of forensic scientists. There are several ways to classify forensic scientists. One example is provided by the American Academy of Forensic Sciences (**AAFS**), and is based loosely on subject matter. I will briefly introduce the AAFS sections below. In other systems forensic scientists are classified based on their goal; wherein some focus on linking a suspect to a crime scene, others focus on determining whether a crime has occurred, and others focus on determining how a crime was committed. I have chosen to adopt a classification for the purposes of this book based primarily on where the forensic scientist most commonly works – whether they are most commonly encountered at a crime lab, at a law enforcement agency other than a crime lab, in private practice, at a university, or at a hospital or similar treatment facility. In placing a particular type of forensic scientist into a particular workplace category I do not mean to imply that all of them work in that type of workplace, but that the indicated type of workplace is where they are most commonly encountered. The chapter will conclude with a consideration of attorneys and judges, who are not forensic scientists, but are intimately involved with the justice system.

2.1 The American Academy of Forensic Sciences (AAFS)

Although no organization exists that determines "officially" what is and is not a forensic science or into which category a forensic science falls, the AAFS is the closest we American forensic scientists have. The AAFS recognizes ten sections of the forensic sciences. These ten sections are membership categories, meaning that when you join the AAFS you must choose to affiliate with one of the ten sections. The first nine of these sections are: Criminalistics, Engineering Sciences, Jurisprudence, Odontology, Pathology/Biology, Physical Anthropology, Psychiatry & Behavioral Science, Questioned Documents, and Toxicology. The tenth section is General, and provides a membership category for those forensic scientists who do not fit into one of the nine other sections. As a forensic anthropologist, I am a member of the Physical Anthropology section.

The section to which a particular forensic scientist would belong is not always clear to someone not familiar with the AAFS. For example, where do serologists belong? The serologists I know are members of the Pathology/Biology Section. The largest section is Criminalistics. I will explain criminalistics further below, and the issue of exactly who is and is not a criminalist is somewhat complex. For now we can let a few examples of criminalists serve for our purposes. Familiar criminalists include fingerprint examiners, firearms examiners, and trace evidence examiners. Surprisingly to me (perhaps not to anyone else), the forensic chemists I know also belong to the Criminalistics Section, though I would not categorize them as criminalists.

Joining the AAFS is not so simple as just sending in money for a journal subscription. Each section defines its own criteria for membership. Everybody joins their section at the level of Associate Member, and over time may be promoted to Member, and finally to Fellow. Each section has their own criteria for promotion. For example, when I joined the AAFS I had to find at least one Fellow and one Member of my section who were willing to sponsor my application. I had to send them examples of case reports that I had done to demonstrate my active involvement in the field and at least minimal competence. After attending a certain number of meetings, I applied and was accepted for promotion to Member.

2.2 Forensic Scientists at the Crime Lab

We most often think of forensic scientists who work at a crime lab. Many texts and classes focus only on the crime lab forensic scientists. However, it is debatable whether it is accurate to state that most forensic scientists work at crime labs. The group of forensic scientist who have private practices seems to be at least nearly as large as the group of forensic scientists who work at a crime lab, and although I am unaware of any statistics on this issue, it would not surprise me to learn that the category of private practice forensic scientist is actually larger.

2.2.1 Criminalists

The largest category of scientists working at a crime lab are criminalists. The most common definition of **criminalist** is – a forensic scientist who uses the comparative method to match evidence recovered at a crime scene with evidence recovered from a suspect. Restated in terms from chapter 1, criminalists are those forensic scientists who do matching of evidence items with standard items with the goal of showing that a suspect was present at a crime scene. Although this definition is widely agreed upon, deciding exactly who is a criminalist is not simple because the term is also applied to several types of forensic scientists who do not use this method and have this goal. This reflects the fact that criminalist is to a certain extent a miscellaneous category in that forensic scientists who do not fit into any other category are often deemed to be a criminalist.

Here are some forensic scientists who are clearly criminalists: fingerprint examiners, firearms examiners, trace evidence examiners, toolmark examiners, and impressions examiners. These forensic scientists seek to perform matching, using fingerprints, bullets, hairs and other small items, toolmarks, and impressions, respectively.

Serologists and DNA analysts also fit this definition, because they seek to perform matching of blood and other body fluid evidence. Other forensic scientists who are considered criminalists as part of the "miscellaneous" category function of the term include evidence technicians and crime scene investigators, who specialize in collecting and handling evidence; and forensic technicians, who operate analytical instruments and handle other routine tasks at the crime lab.

2.2.2 Criminalists vs Criminologists

There is a great deal of confusion between the terms criminalist and criminologist, because they are such similar words. However, they describe very different types of people. A criminalist is a forensic scientist, whereas a **criminologist** is a social scientist, most often a type of sociologist. Criminologists study the justice system, why people become criminals, correctional practices, and other aspects of criminality as a social and societal phenomenon. One of the forensic sciences, jurisprudence, shares some of these concerns with criminology, but criminalists do not. In my career as a college professor of a forensic science, I have found that many students who seek my advice or take my classes are actually expecting to learn criminology. Similarly, I believe that many students who take criminology classes are expecting to learn forensic science. The two fields are mutually interfertile, and students of forensic science should take classes in criminology, and vice versa. Criminology classes are usually taught through sociology departments or criminal justice departments.

The American Board of Criminalistics defines criminalistics as "that profession and scientific discipline directed to the recognition, identification, individualization, and evaluation of physical evidence by application of the physical and natural sciences to law-science matters. In contrast, the American Society of Criminology defines criminology as "... scholarly, scientific, and professional knowledge concerning the etiology [origin], prevention, control and treatment of crime and delinquency. This includes the measurement and detection of crime, legislation and practice of criminal law, as well as the law enforcement, judicial and correctional systems."

2.2.3 Other Forensic Scientists at the Crime Lab

Several other forensic scientists besides criminalists work at crime labs. These forensic scientists have different goals from those of criminalists.

The main goal of forensic chemists, toxicologists, and blood alcohol analysts is to determine whether a crime has occurred or not. Forensic chemists analyze suspicious substances. If, for example, a white powder turns out to be sugar, then no crime has been committed. If it turns out to be cocaine, however, then a crime has been committed. Forensic toxicologists analyze blood and other body substances for the presence of poisons, alcohol, and drugs. If they find evidence of these substances in a person's body, then perhaps a crime has occurred. Blood alcohol analysts are most often involved in determining whether the amount of alcohol in a driver's blood was over the legal limit while they were driving. If so, then a crime has occurred.

Forensic pathologists have two goals – determining whether a crime has occurred and if so, determining how it occurred. They most often pursue these goals by doing autopsies of deceased people. Autopsies reveal the cause of death, which may provide evidence that a crime has occurred and may also provide evidence about how the crime occurred. During an autopsy, the pathologist will normally collect samples of body tissues and fluids for examination by toxicologists and other forensic scientists.

Questioned document examiners are similar to criminalists in that they also seek to perform matching. In many cases, however, they are not seeking to match a suspect to a crime scene, but rather to determine whether documents were produced at the time and place they are claimed to have been produced, and by the person claimed to have produced them. Questioned document examiners have other goals as well, including whether a crime has occurred. For example, if a check has been forged, then a crime has occurred. In some cases questioned document examiners seek to determine how the crime was committed. For example, did the person who forged the check use a copier, or did they copy the signature by hand onto a stolen check.

2.3 Forensic Scientists at a Law Enforcement Agency Other Than a Crime Lab

Forensic photographers, forensic artists, and forensic computer scientists most often work at a law enforcement agency that is not a crime lab, such as a police department. Although these forensic scientists may occasionally be employed by a crime lab, they most often work alongside police officers.

The most common task of forensic photographers is to document evidence at a crime scene. They may also do other jobs, such as photographing suspects or documenting the steps in an analysis or investigations. Nowadays there is considerable overlap with forensic image analysts, in that forensic photographers are usually trained in the methods of clarifying and enhancing photographs or other images.

Forensic artists make sketches of crime scenes and composite drawings of suspects, among other things. One of the newer trends is for forensic artists to undertake the task of reconstructing the facial features of a deceased person whose face has become disfigured or skeletonized.

Those forensic computer scientists who work with computers to recover evidence relating to crimes most often work for law enforcement agencies. This is also true of those who specialize in internet related crime. Forensic computer scientists who specialize in the prevention and investigation of hacking and other aspects of network security issues are most likely to be employed by a corporation or other organization that has sensitive or valuable data.

2.4 Forensic Scientists in Private Practice

A wide variety of forensic scientists work in private practice. By "private practice", I mean that they either own their own consulting business, or they work for a private company that provides some form of forensic services.

Some of the forensic scientists in private practice apply their expertise to solving crimes, just as would a forensic scientist who worked at a crime lab or a police station. Forensic accountants examine financial data to detect economic crimes, such as embezzlement or money laundering. Forensic engineers work on cases having to do with traffic accidents, bombs and explosives, fires, structural failure, and product failure. Forensic audio analysts specialize in improving the quality of audio evidence, especially recordings of conversations. Forensic image analysts specialize in enhancing the quality of images, usually digital images obtained from video recordings or digital cameras. Forensic odontologists most often seek to identify a deceased person from their dental records. In some cases forensic odontologists pursue the criminalist's goal of linking a suspect to a crime scene by matching a suspect's teeth to bite marks made on a person or a food item at the crime scene. Forensic knot examiners have expertise in knots and ligatures (the item in which a knot is tied), which can be helpful in cases where something or somebody has been tied up. Forensic podiatrists work with foot and shoe prints and impressions to help match a suspect to a crime scene.

Other forensic scientists in private practice provide expertise for civil cases. Forensic economists have the expertise to place a value on the result of an injury or some other event such as a breach of contract. There is a variety of forensic medical practitioners, such as forensic medical doctors, chiropractors, osteopaths, and naturopaths, who render opinions on the prognosis and required treatment for persons who have been injured. Forensic pharmacists share some of the same interests as toxicologists, especially with regard to prescription drugs and their effects, including the interactions between the effects of drugs that are taken together. Forensic meteorologists have expertise in weather and its effects on certain types of events such as traffic accidents.

2.5 Forensic Scientists at the University

There are several forensic sciences for which the number of cases that require their expertise is not large enough to support many (or even any) full time specialists, either at a crime lab, a law enforcement agency, or in private practice. Forensic scientists who practice these specialties most often work as a professor at a college, university, or museum, making their living by teaching classes, and consulting on cases as they come up.

My own specialty of forensic anthropology is one such forensic science. Forensic anthropologists handle locating and recovering buried evidence, quite often a buried body; and also specialize in estimating the age, sex, race, and other characteristics of a person from their skeletal remains. Perhaps it is good that these sorts of cases are relatively uncommon. In Montana, for example, we only have between 10 and 20 cases each year. As most of you know, there aren't all that many people here in Montana, but on the other hand it is such a vast state that it is easy to become lost or for a body to be hidden in a remote area. Regardless, these 10 to 20 cases per year do not justify having a full time forensic anthropologist in the state. Dr. Ashley McKeown, Garry Kerr, and I handle those that occur, with the help of our students, working them in around our other duties. Some larger jurisdictions have enough cases to hire a full time forensic anthropologist, and the U.S. military employs several of them. However, the vast majority of us work for a institution of higher learning or a museum. A similar situation holds true for the other forensic scientists who work at a university.

Forensic zoologists identify animal remains to help with enforcement of wildlife protection laws, or simply to make sure that they are not human. The Clark R. Bavin National Fish and Wildlife Forensics Laboratory in Ashland, Oregon, employs a few, and the rest work at a university or museum. One example is my colleague David Dyer, who is curator of the Philip L. Wright Zoological Museum of The University of Montana. Many of the cases submitted to us forensic anthropologists involve bones that are not human, and I pass them along to Mr. Dyer, who identifies which species they belong to.

Similarly, forensic botanists work with plant evidence, such as plant parts, pollen, and wood. Forensic entomologists work with insect evidence, often to estimate time since death, but also in cases involving insect damage to crops. Forensic geologists work with soils, dusts, rocks, and petrochemicals. Their work with soils and dusts overlaps with that of trace evidence examiners, but geologists have training in analysis of these substances that is beyond that of most trance evidence examiners, so they are consulted on complex cases. Similarly, the forensic geologist's expertise with petrochemicals partly overlaps the expertise of forensic chemists, but due to the long association between geologists and the oil industry the geologist's knowledge of these substances is often greater.

Forensic linguists have expertise in language and speech. Their areas of expertise include identification of a person who is speaking by the speaker's voice pattern, timing, and pronunciation; lie detection, and helping figure out the meaning of unusual terms used by criminals. Forensic mathematicians primarily work with DNA analysts to narrow down the probability of encountering another person with the same DNA profile as a suspect. They are also interested in logic errors, such as the prosecutor's fallacy and the defense attorney's fallacy that will be discussed in a coming chapter. Forensic mathematicians also model terrorist organizations as an aid to those agencies who combat terrorism. Forensic phylogenetics is the use of methods of evolutionary analysis to provide evidence about cases that involve rapidly evolving organisms, such as cases involving the transmission of the virus that causes AIDS.

2.6 Forensic Scientists in Hospitals and Other Treatment Facilities

Some forensic scientists work at hospitals or other treatment facilities. In this usage I am considering correctional institutions and some forms of government agencies as a type of treatment facility, because inmates and former inmates need treatment and services too, and those who provide them with the help they need can be considered forensic specialists.

The largest group of hospital based forensic scientists is forensic psychologists. Here I am using the term "psychologist" to refer to anybody who has special expertise on the mind and mental illness. Many forensic psychologists work at a hospital or correctional institution, where they provide treatment for offenders. However, forensic psychologists are not limited to this activity and their involvement in cases ranges from investigation to trial. One commonly encountered type of forensic psychologist is often called a "police psychologist". These scientists often work for a law enforcement agency, where they provide a variety of services including: counseling victims of crimes, counseling police officers who have had a traumatic experience, providing insight into the motivation and personality of criminals, and screening applicants applicants for police officer jobs. Other forensic psychologists do criminal profiling. Many lie detector operators and hypnotists are psychologists. Forensic psychologists are also in demand in court, where they render opinions on the competency of defendants to stand trial, participate in cases that invoke the insanity defense, and provide insights into the motivations and behaviors of witnesses and jurors.

Forensic nurses specialize in working with victims of violent crimes, particularly sexual assaults and domestic violence. Victims of violent crimes are often both physically and emotionally traumatized and forensic nurses have training in how to assist police officers with questioning and collection of evidence from the victim's body. Forensic nurses advocate for the victim, both while she or he is in the hospital and afterward.

Forensic social workers often pick up where forensic nurses leave off, to make sure that the victim of a crime receives necessary social services. Also, many forensic social workers work to provide social services to offenders.

2.7 Attorneys and Judges

Attorneys are not scientists. Their expertise lies elsewhere. However, being those people who specialize in the law, they are involved with nearly every criminal and civil case in some manner. Therefore, it is in our interest to learn something about them. Trials are adversarial by design. This means that there are two sides, the prosecution and the defense for a criminal trial, and the attorneys for the plaintiff and the defendant in a civil trial. In a trial the most convincing argument usually prevails, and the most convincing arguments are quite often those that are based on sound forensic evidence.

In a criminal trial the prosecuting attorney presents the government's case against the defendant, while the defense attorney works for the defendant to try to prove innocence. The two sides in a civil trial operate similarly. In most cases forensic scientists will be working with the prosecuting attorney, since there would likely be no trial without forensic evidence to back up the prosecution's claims. However, forensic scientists often work for the defense as well, and it is not uncommon for a case to involve evidence presented by both sides.

Defense attorneys are generally disliked by forensic scientists, and according to the defense attorneys I talk to, the feeling is mutual. This is a tragic situation. Although I have experienced my share of bullying by defense attorneys bent on discrediting my testimony, I strive to remember that this is their job. No matter how heinous the crime or how airtight the prosecution case seems, all defendants are entitled to the best defense possible. Wouldn't you want your attorney to work hard for you if you were on trial?

In most cases, though by no means all, judges are also attorneys. The most visible task of a judge is to preside at a trial to make sure that the trial occurs according to the rules and that the correct procedures are followed by both sides. In some cases it is the judge who decides the defendant's guilt or innocence after hearing both sides of the case. For serious crimes, however, guilt or innocence is usually decided by a jury. Finally, it is normally the judge's responsibility to impose a sentence upon someone whose guilt has been established.

2.8 Questions for Review and Study

1. In your own words, describe the nature and mission of the American Academy of Forensic Sciences (AAFS). If you wish, you may browse the AAFS website at http://www.aafs.org.
2. In your own words, define and describe criminalists and criminology. Describe the difference between criminalistics and criminology.
3. Make a list of the forensic scientists that you would expect to encounter during a tour of a crime lab.
4. Make a list of the forensic scientists that you might encounter during a visit to a police station.
5. Discuss what it means for a forensic scientist to work in "private practice".
6. Make a list of the forensic scientists that usually work in private practice.
7. Make a list of the forensic scientists that you might encounter at a university.
8. If your college or university lists the research and teaching interests of its faculty members, see how many you can find with an interesting in the forensic aspects of their discipline.
9. Make a list of the forensic scientists that you might encounter at a hospital or treatment facility.
10. Why are attorneys and judges often included in discussions of forensic scientists?

2.9 Sources

American Academy of Forensic Sciences. 2007. 2007 Membership Directory. Colorado Springs (CO): AAFS.

American Academy of Forensic Sciences. 2007. So You Want to Be a Forensic Scientist! [Internet]. [cited 2007 Oct 16]. Available from: http://www.aafs.org/default.asp?section_id=resources&page_id=choosing_a_career.

American Board of Criminalistics. 2004. American Board of Criminalistics By-Laws [Internet]. [cited 2007 Oct 16]. Available from: http://www.criminalistics.com/bylaws.cfm.

American Society of Criminology 2006. The American Society of Criminology [Internet]. [cited 2007 Oct 16]. Available from: http://asc41.com/.

Block E B. 1979. Science vs Crime. San Francisco: Cragmont Publications.

Eckert WG. 1996. Introduction to Forensic Sciences. Boca Raton (FL): CRC Press.

Inman K, Rudin N. 2000. Principles and Practice of Criminalistics: The Profession of Forensic Science. Boca Raton (FL): CRC Press.

Jackson ARW, Jackson JM. 2004. Forensic Science: An Introduction. Upper Saddle River (NJ): Prentice Hall.

James SH and Nordby JJ. 2005. Forensic Science: An Introduction to Scientific and Investigative Techniques. Boca Raton (FL): CRC Press.

Kiely, TF. 2000. Forensic Evidence: Science and the Criminal Law. Boca Raton (FL): CRC Press.

Lee HC, Gaensslen RE, Harris H. 2007. Introduction to Forensic Science and Criminalistics. Columbus (OH): McGraw-Hill.

Meloan CE, James RE, Saferstein R. 1998. Lab Manual for Criminalistics: An Introduction to Forensic Science. 6th Edition. Upper Saddle River (NJ): Prentice Hall.

Nickel J, Fischer JF. 1998. Crime Science: Methods of Forensic Detection. Lexington (KY): University Press of Kentucky.

Saferstein R. 1998. Criminalistics: An Introduction to Forensic Science. 6th Edition. Upper Saddle River (NJ): Prentice Hall.

Siegel JA. 2006. Forensic Science: The Basics. Boca Raton (FL): CRC Press.

Siegel JA, Houck MM. 2006. Fundamentals of Forensic Science. Burlington (MA): Elsevier Science & Technology Books.

Chapter 3
Basic Science
THE SCIENTIFIC METHOD

Many people are intimidated by science and the scientific method. Even more people don't really understand what it is. Much of the cause of this unfortunate situation is media portrayals of scientists as people who strut around in white lab coats and are the villains of many stories. It seems as if the "mad scientist" has taken the place of the "bad old witch" as the favored evil character against whom the hero must strive. Further, there is a seemingly increasing aspect of American culture that sees science and mathematics as difficult subjects to understand. Any schoolchild can tell you this – just ask one, especially a girl. The result of this has had a major and catastrophic impact on American society, both financially and practically. America has lost its edge in science. Nowhere is this more apparent than in the consumer electronics industry, where most of the early developments in the field of radio, television, telephones, and computers were made by Americans. Where is the center of innovation for these types of devices now? It's in Japan and other Asian countries. In new emerging fields such as genetic engineering and stem cell research, Americans stagger along, held back by the ignorance of religious conservatives in government. Therefore, the emergence of forensic scientists as heroes in modern American media is exciting to people like me who value and teach science. Perhaps, it will save our country from its status as a second rate nation in the field of science and technology.

It doesn't help that intelligible presentations of the scientific method are surprisingly rare. This stems from two factors. First, the scientific method is natural and easy. There really is not much to it. It is just a set of common sense steps that someone who wants to reliably acquire knowledge takes. To many scientists it seems such a basic thing that there is no need to formally discuss or present it. Second, science is broad, and each of its branches pursues what are fundamentally the same steps in different ways. So, it is easier to find a statement of the scientific method that is specific to a certain field, chemistry for example, but more difficult to find a general treatment of it.

In this chapter we will explore the scientific method on a very general level. We will start with a look at why we who work in the justice system must use the scientific method and be able to distinguish genuine science from non-science. We will look at what constitutes a scientific method and how all humans naturally use the scientific method to find out things about the world. We will look at an example of how most of us, with or without scientific training, would use the scientific method to solve a practical problem. The chapter will end with an advanced topics section on Ockham's Razor and biases in science.

3.1 Why the Justice System Must Use the Scientific Method

All those who work in the enforcement of justice in the United States – forensic scientists, police officers, attorneys, judges, and the rest, must be familiar with the distinction between science and non-science because the U.S. Supreme Court says so. In the case of **Daubert v. Merrell Dow Pharmaceuticals** (1993), the U.S. Supreme Court handed down a historic decision that requires judges to evaluate whether evidence presented in court was "derived by the scientific method". This decision places responsibility upon not only judges, but all people involved in an investigation to become familiar with the scientific method and how it can be used to produce evidence.

3.2 Defining the Scientific Method

What does it mean to be scientific? Basically, the **scientific method** is a systematic approach to exploring the world around us. That's all. You don't have to be wearing a white coat and work in a lab full of specimens and test tubes to be scientific. All you need is curiosity and a desire to find things out. Scientists find that many times they answer a question, only to find out that their answer has led me to even more questions. Thus, the quest for knowledge is never complete.

The scientific method can be defined as "the method by which replicable conclusions can be drawn about the nature of the world based on the results of experiments". This definition has three main components to it: replicable conclusions, the nature of the world, and experiments. Let's look at each of these three components in more detail.

3.2.1 Replicability

The first component of the definition of the scientific method is replicable conclusions. **Replicable** means that you get the same result (within limits) each time you repeat the procedure that produced the result. If the result is different each time, then it is likely that you are either not keeping the conditions the same each time or you are working in a realm that is not scientific. For example, the conclusions drawn by a firearms examiner are based on the expectation that the individual characteristics of a firearm are transferred to each and every bullet fired from it. Therefore, the transfer of the individual characteristics is replicable. If these details are transferred to the bullet replicably, (they are!) then a bullet fired by a certain firearm at a crime scene can be matched to a bullet fired by that same firearm in the crime lab, and a firearms examiner can match the evidence and standard items to place that firearm at the crime scene. If these detail were not transferred to bullets replicably, matching of firearms evidence would not be possible.

Replicability also means that different scientists should be able to obtain the same results (again, within limits). If two different scientists obtain different results that are not explainable by differences in the conditions under which they produced them, then something is wrong with those results and additional tests must be run until the two scientists can agree. Anything less is not scientific. I am not asserting, however, that scientists must always agree. Their interpretations of evidence may differ, but in order to be science the evidence upon which their interpretations are made must be replicable. For example, it can not be the case that Joe is the only firearms examiner at a certain crime lab who can fire a bullet from a weapon and obtain the same patterns of individual characteristics found on a bullet from the crime scene. In order to be replicable, the same details must be produced whenever Joe, any of his colleagues, or even you or I fire the weapon. However, it is possible for Joe and his colleagues to have a difference of opinion about whether the details of the evidence item and the standard item are sufficiently similar to produce a positive match or not.

3.2.2 Ordinary and Extraordinary Realities

The second component of the definition of the scientific method refers to the nature of the world. Scientists only work within what is called ordinary reality. **Ordinary reality** has the characteristic that it is the same for everybody. For example, everybody agrees that ocean water is salty, that fire produces heat and light, and that the human body has two arms. When applied to a murder, for example, everyone can agree that a person is dead, that their death was caused in a certain way, and that there must be a person responsible. Witnesses to the event should all be able to agree on the person responsible.

In contrast to ordinary reality, there may exist several realms of reality about which not all people agree. In many cases these arise from religious or philosophical views of the world and concern the underlying causes for events. For example, a devastating flood might be viewed by some people as the result of too much rain in the mountains and by others as due to the wrath of God upon people who are not living right. The rain in the mountains is a feature of ordinary reality, and could be measured to determine whether the amount was sufficient to cause the flood. Whether or not God caused the rain to fall is outside of ordinary reality, because scientists can not ask God whether he did or did not cause the flood. Realms of reality outside ordinary reality are called **extraordinary realities** or special realities. While there is only one ordinary reality there are many extraordinary realities. Since different religions have different views of God and the supernatural, they each create for themselves an extraordinary reality in which God and the supernatural work. It cannot be said whether extraordinary realities are in fact "real" or not. They are outside ordinary reality, therefore beyond what can be known reliably, and are matters of faith instead of science. Scientists work only within ordinary reality and typically remain steadfastly silent about extraordinary reality. A great deal of damage is caused by people who seek to mix ordinary with extraordinary reality, because this blurs the distinction between what can be known by everybody and what can be known only to those who are able to perceive the extraordinary reality involved.

Explanations of the world and its natural phenomena that are based on the actions of gods or spirits can not be replicated because the gods or spirits might do things differently each time. This is not to say that gods and spirits don't exist. Many scientists subscribe to faiths in which they do, but they are not a part of ordinary reality. Instead, they are a part of an extraordinary reality. Therefore, they, and the phenomena they might cause are not subject to scientific experimentation.

It is easy to see why evidence presented during a trial must be based in ordinary reality, and therefore scientific. Evidence presented against a person needs to be evidence that everybody can perceive. It is not fair to present something from an extraordinary reality, a divine revelation of the suspect's guilt, for example, as evidence against them. For an example of the abuses that can occur when features of extraordinary reality are allowed as evidence, read any account of the Salem witch trials that occurred in 1692 in the Massachusetts colony of the Eastern U.S..

3.2.3 Experiments

The third component of the definition of the scientific method is that the conclusions drawn are based upon data from some sort of experiment. An **experiment** is any procedure that produces an unambiguous result. An experiment can be performed in a variety of ways – it doesn't have to be something done in a lab with test tubes. It might also consist of an observation or collecting some form of specimen. So, an experiment might be mixing chemicals to see what happens, watching a bird or other animal to see how it behaves, or collecting a fossil to see what life was like in the past.

Investigation of a crime scene is a form of experiment, in that it consists of observing and collecting evidence. Actually, it is similar to the science of archaeology, in that the event of interest occurred in the past, and can not be directly observed. Therefore, the event is indirectly observed through a pattern of evidence produced during and because of the event. So, for example, if a murder occurs, there is no way to get into a time machine and go back in time to watch it. However, we know that the murder occurred because we find a corpse and we can tell cause of death from an autopsy. Examining the items at the murder scene may allow us to infer how the murder happened and perhaps who the murderer was.

3.3 A General Recipe for the Scientific Method

One thing that is not a part of the definition of the scientific method is a recipe for doing it. This is because the various branches of science are so diverse that it is difficult to come up with a step-by-step procedure that can cover all cases. There are, however, several parts of a scientific analysis that are widely shared among scientists. Here, I have combined them into a general recipe, which applies in general, but with considerable variation, to at least most sciences.

1.   Observe a phenomenon;

2. Ask a question about the phenomenon;
3. Make a hypothesis (preliminary answer) to your question based on all the evidence you have at the moment;
4. Make a prediction that follows from your hypothesis;
5. Gather data to test the prediction;
6. Analyze the data;
7. Draw a conclusion from the data as to whether the hypothesis is false;
8. If the hypothesis is false in some way, change it based on the new evidence you have and return to step 4 (*i.e.* make a new prediction).

In a typical investigation of a phenomenon a scientist may make several loops through steps 4 through 8 of this recipe, until he or she is no longer able to show that the hypothesis is false. Being unable to show that a hypothesis is false by no means proves that it is true. By continuously refining hypotheses we can get closer and closer to the truth.

Let's look at each of these steps in detail, framed within the simple example of trying to find out whether squirrels like potato chips.

3.3.1 Observing Phenomena

The world is full of things and events, each of which is a **phenomenon**. All phenomena are potentially the focus of scientific study. Let's take an example. Squirrels exist on the campus of The University of Montana. Students often feed the squirrels things, usually leftovers from their lunches or snacks.

3.3.2 Asking Questions

An important step in the scientific method is to formulate a question about a phenomenon that has been observed. Humans are naturally curious – just visit with a three-year-old sometime and you will see what I mean. Unfortunately parents and school teachers get tired of answering questions and the natural curiosity of children gets driven out of them. When I was in elementary school you had to raise you hand if you wanted to ask a question, and wait for the teacher to recognize you and call upon you. Any violation of this procedure resulted in some form of punishment. Being a teacher (of adults) myself, I recognize some need for this, but my best class sessions are always those in which I get to spend the entire period responding to student questions. Scientists must strive to recapture this natural human curiosity.

In asking questions, scientists must be careful to frame them in terms answerable in ordinary reality. Science can answer many questions about ordinary reality, but cannot answer any questions about extraordinary realities. An example question that can not be answered by science is: Why are we here? The word "why" implies a question about the purpose of humanity's existence. Most people who ask this question have in mind a question along the lines of: What cosmic forces have caused humanity to occur on this planet and what do these forces expect of us? This form of the question cannot be answered by science because the cosmic forces that may or may not control humans lives are part of an extraordinary reality and, therefore, not knowable by the means available to science. This question is one that can only be answered by faith.

Some rethinking might make a question answerable by science, however. For example if you change "Why are we here?" to "How did humanity come to exist on this planet?" it becomes answerable by science. Why humans are on the planet is a matter of faith, but how we evolved from the first forms of life that appeared on Earth has been studied and much is already known about it. Therefore, we scientists must be careful about what questions we ask, and the form of the answer we expect.

Definitional trivia cause problems in asking scientific questions. Sometimes a question seems to have no answer, but this is due to imprecise definition of the question or one of the terms in it. For example, the classic unanswerable question is: Which came first, the chicken or the egg? Many people think that this is a difficult question to answer, or even claim that it has no answer. They are mistaken; the problem is merely in the definitions of "chicken" and "egg". If we turn to the field of biology, we can find common definitions of a "chicken" as a member of the species *Gallus domesticus*, and an "egg" as a reproductive structure, encased in a membrane, produced by a female. When defined rigorously in this way, it is clear from the field of paleontology that eggs were produced by animals hundreds of millions of years before chickens evolved. At this point I am always interrupted by someone who believes that I have missed the point – that the question really is: Which came first, the chicken egg or the chicken. No, I haven't missed the point, and in fact the person who interrupted me has understood my point exactly – that the answerability of this question depends on how you define "chicken" and "egg". The interrupter now wants to redefine "egg". If I can be precise about what the interrupter means by "chicken egg", then the question is still answerable by reference to evolutionary theory. If you define "chicken egg" as an egg produced by a chicken, with the understanding that only a chicken can lay a chicken egg, then chickens came first, and the first chicken hatched out of an egg that was not a chicken egg because it was laid by a member of the species that evolutionarily preceded chickens. If you define "chicken egg" as an egg out of which a chicken hatches, then the egg came first – the first chicken egg was laid by a member of the species that preceded chickens and from it hatched the first chicken. Now red in the face, the interrupter will still maintain that I have missed the point and insist on defining "chicken egg" as an egg laid by a chicken and out of which a chicken hatches. Now, however, we have crossed the line into an extraordinary reality. To insist on this definition of "chicken egg" is to preclude the possibility that some earlier species evolved into chickens. As such it is either based in an extraordinary reality, such as creationism, or just plain wrong. However, the interrupter has finally succeeded at their goal of finding a way to define chicken and egg in such a way that the question is not answerable by science.

Here is an example question that arises from my earlier observation of the phenomenon that squirrels live on the UM campus and students often feed them. One day while walking to class, I observed a student trying to feed a squirrel some potato chips. The student wasn't having much luck. The squirrel appeared to be extremely nervous and hovered just beyond the range of the student's ability to throw a potato chip. Finally, the squirrel dashed over, grabbed a potato chip in its mouth and ran to the nearest tree, which it proceeded to scramble up, and finally disappeared into the leaves and branches. A question popped into my head. Do squirrels like potato chips? Not a very profound question, for sure, but one which interested me at the time. This question is one that can be asked in ordinary reality, so long as we make allowance for the fact that different squirrels may have different food preferences. Furthermore, the definitions of "squirrel" and "potato chip" seem obvious, so definitional issues should not make this question difficult.

3.3.3 Hypotheses

The next step in the general recipe for the scientific method is to form a hypothesis. The word hypothesis is a word that my granny would have described as "high falootin", but it is really a simple concept. A **hypothesis** is merely an educated guess as to the answer for the question, based on the evidence available at the moment. In many cases, the scientist will review the existing literature on the subject to see what hypotheses have been proposed in the past and the evidence presented to support them. Reading the literature is critical for all scientists, and as a forensic scientist you will be expected to keep abreast of new theory and practice in your specialty. After gathering as much information as possible, the scientist will present their hypothesis, which will be a tentative answer to some question, and which is compatible with as much of the currently known evidence as possible.

It is important to note that most hypotheses are false as originally stated. In fact most of science is spent trying to show that hypotheses are false. So, in formulating your hypothesis, you should not worry too much that you have come up with the best or the only possible hypothesis. The rest of the scientific method will test your hypothesis. What will be important is your decision at the end of the process.

In order to be useful, a hypothesis must be testable. There must be a way to test the possible answer to try to make it fail. If you design an untestable hypothesis, then you have a similar problem as occurs when you ask an unanswerable question, and science cannot be used to help you decide if it is right or not.

On that day that I formed the question "do squirrels like potato chips?", I stopped by the vending machine in my building and bought a small bag of them. The ingredients list included potatoes, salt, and a few things that I guessed were preservatives. In a horrible violation of what I would do if this were a more formal research project, I did not go to the library to look up the literature on squirrel food preferences. Instead I relied on what I already knew about squirrels, trusting that the scientific method reliably lead me to a correct conclusion. I knew that squirrels eat nuts, grains and other starchy foods in the wild, and potatoes are starchy. I also know that most animals enjoy the taste of salt. Given this, I figured that according to the best evidence I had at hand, squirrels should like potato chips. Therefore, my hypothesis was that squirrels like potato chips.

If a hypothesis has stood the test of time, in having survived numerous attempts to falsify it, then scientists give it the status of a **theory**. Examples of theories of this sort are Einstein's theory of relativity and Darwin's theory of evolution by natural selection. Sometimes a theory will stand as accepted knowledge for a time, then be falsified, such as when Bohr's theory of the atom was replaced by quantum theory. To attorneys, judges, and police officers, and most of the general population, however, the word theory has quite a different meaning. To these people, a theory is more like a hypothesis, perhaps even less than a hypothesis, and implies some speculative idea that has not yet been proven. One often hears the phrase, "That's just your theory." However, a scientific theory is as "proven" as it is possible to be at a given moment in time.

3.3.4 Making Predictions

Now that there is a hypothesis in hand, the process of testing it to see how well it stands up can begin. The procedure for testing a hypothesis starts with making a prediction about what should be observed under some set of experimental conditions, if the hypothesis is true. If the hypothesis is testable, then it will naturally suggest some sort of prediction.

Predictions can be stated as **"if...then" statements**, and this is one of the best ways to do it. A prediction framed as an "if...then" statement could have the following form.
1. If my hypothesis is true ...
2. then _____ should happen
3. when _____ is done.

The scientist making the prediction fills in the blanks with what they expect to happen and under what conditions it should happen. A prediction framed in this way provides an unambiguous way to test the hypothesis. The conditions called for are caused to occur (or allowed to occur naturally) and the scientist then observes what happens.

Applying this to the hypothesis that squirrels like potato chips yields the following "if...then" statement.
1. If squirrels like potato chips
2. then they should eat a potato chip
3. when I offer them one.

3.3.5 Experiments

Causing the conditions called for in the prediction to occur (or waiting for them to naturally occur) and observing the result is an experiment. The nature of the experiments that need to be conducted to test the hypothesis should now be clear. In our example, I now knew that I had to offer potato chips to squirrels and observe whether the squirrels ate the potato chips.

Note that the title of this section is experiments, plural, not experiment. Normally, a scientist can not make a decision about the reliability of the hypothesis with just one experiment, though there are exceptions to this rule. Performing the experiment several times give the scientist a way to know whether unusual or unanticipated circumstances are affecting the experiments. In our example, what if a squirrel is not hungry, or what if it is frightened away because somebody walked by. These circumstances could cause the squirrel to reject a potato chip even if squirrels really do like potato chips. Also, different squirrels may have different reactions to potato chips or to the person offering them, so we can't rely on experiments that don't include several different squirrels. The best studies use a variety of different conditions to elicit these types of unexpected results. Iterations of the experiment in which certain conditions are set up to test the reliability of the experimental procedure are called **control experiments**. For example, we may want to offer a squirrel something that we think squirrels don't like, say a piece of hot dog or even a non-food item, to see what a squirrel does when they don't like something. We may even follow up by offering squirrels something we know they do like. We could then have a better basis for interpreting whether the squirrels' reactions to potato chips were more similar to their reactions to things that they like or more similar to their reactions to things they do not like.

How many experiments is enough? There is no definite answer to this question, other than to say that more is better. A hypothesis that was tested by a few hundred experiments is more likely to be true than is one tested by only a handful of experiments. However, there are often practical limits on the number of repeated experiments. Cost, time, and other factors will usually have an impact on how many experiments a scientist actually performs. I decided to conduct my squirrel experiments as follows. Each morning when I was ready to go to class I bought a bag of potato chips from the vending machine. As I walked across campus I threw a potato chip to each squirrel I saw, and mentally noted whether the squirrel took the potato chip and did something with it or not. I did not feel that I had the time during my walk to class to actually wait to see whether the squirrel ate the chip or just handled it. This procedure allowed me to test the reaction of four or five different squirrels three days a week, under a variety of weather conditions. In all, I collected 34 observations over a time period of a little less than three weeks.

3.3.6 Data

The outcomes of the experiments performed is (or are) **data**. Data can also be referred to as "facts", "evidence", or "observations". Since types of experiments vary widely, types of data also vary widely and can come from a variety of sources. Data can be counts of something. For example, how many times the squirrel ate a potato chip when I offered it one. In other cases, data can be measurements of some item, descriptions of something, or any number of other things. The basic thing that qualifies something as data is that somebody observed something in some manner and recorded it.

Note that there is a debate among scientists and grammarians as to whether the word data is singular or plural. It is clear that "datum" is the singular of "data", and therefore "data" is to some extent plural. However, I follow the school of thought that sees the word "data" as similar to the word "herd" in that it implies a collection of an uncertain number of similar items. "Herd" is singular, in that we speak of "a herd" of elephants, and so too is "data" according to this school of thought. Therefore, I will consistently use data in the singular in this book. To those that disagree, I point out that language and word usage evolve through time and that this is what I observe to be the most common current usage.

Also note that "evidence" is another term for data. Those evidence items gathered at a crime scene are data. Standard items, obtained from a known source are also data, as are the results of comparisons between evidence items and standard items. What is not data is our interpretations of these comparisons. When we say we have an exclusion, a consistent match, or a positive match, we are drawing a conclusion based on the data and these conclusions are not data.

3.3.6.1 Anecdotes

Scientists have a view of data that differs from that of nonscientists, and which is very different from what some parts of the legal system consider good data. To most of us in our ordinary lives, and to a jury in a courtroom, the most reliable type of data is an eyewitness account. If I say that I saw a squirrel eating potato chips, then you either accept that as truth or consider me a liar. Sometimes in ordinary life, to even ask someone if they are sure about an observation is to imply they are a liar. To a scientist, however, an eyewitness account is the worst kind of data. Scientists call this type of evidence an "**anecdote**" and actually don't consider anecdotes to be scientific evidence at all. They are particularly suspicious of any interpretations that are made from anecdotal evidence. So, for example, many people claim to have seen Bigfoot, yet scientists remain unconvinced of the existence of Bigfoot. Non-scientists find this puzzling, since the total number of Bigfoot sightings is large, and all these people can't be mistaken, can they? To scientists, however, these sightings are all anecdotes, and are considered to have very little value as data. The justice system does, however, value what people say, and calls witness statements, suspect confessions, victims' descriptions, and all similar forms information provided by individuals, **testimonial evidence**. We will look at testimonial evidence in more detail in the next chapter.

3.3.7 Data Analysis

Data analysis refers to the process of preparing the data so it can be used to draw a conclusion. In some cases the number of successes out of all the experiments performed is counted. For example, we could count how many times a squirrel took the potato chip I offered it. In other cases the data are sorted into groups. For example, a forensic botanist might sort pollen grains by species. Data are quite often subjected to a mathematical procedure, such as calculating the average.

Data analysis also includes many of the examinations and tests that forensic scientists apply to evidence at the crime lab, or wherever they work. Some of these procedures will be explored more in coming chapters, and some are outside the scope of this book and to be found in those texts that cover laboratory procedures.

3.3.8 Drawing Conclusions

The final step in each circuit through the steps of the scientific method is to draw a conclusion as to whether the hypothesis tested is in some part false, or whether it can not be shown to be false with the current evidence. Notice here that I am careful to avoid the use of the word "true" in connection with a hypothesis, because a hypothesis can only be shown to be false – never can a hypothesis be shown to be true.

A hypothesis is considered **falsified** if any part of it conflicts with the data gathered. However, this evaluation is rarely so simple that one can look at the data and immediately make a decision as to the validity of the hypothesis. This is because If the scientist has done their test well, and collected data from many experiments, some of the experiments will have yielded conflicting results. Different branches of science differ in how often they expect conflicting results to be obtained. Within chemistry, for example, if hydrochloric acid is combined with sodium hydroxide, the result is always sodium chloride (table salt) and water, unless some extraordinarily rare set of conditions has occurred. However, in my field of anthropology, finding that two people act in exactly the same manner under a certain set of conditions is cause for astonishment. In most branches of the sciences, it is expected that there will be some variability in the results of the experiments.

The variability in experimental results is most often handled using statistics. Statistics provides methods for drawing reliable conclusions from real data of this sort. A typical use of statistics is to allow a hypothesis to be considered "not false" if the results obtained agree with the prediction made from the hypothesis a certain large percentage of the time. In this situation scientists express their conclusions in terms of the level of confidence they have in them. The 95% level of confidence is considered standard in statistics and in science, and is achieved when 95% of the experimental results agree with the prediction made from the hypothesis. In more complex types of experiments other statistical tests can provide an estimate of the level of confidence.

Let's now return to my hypothesis that squirrels like potato chips. Let's say that I did the experiment of offering a potato chip to a squirrel 100 times. If the squirrel ate the potato chip 10 of the 100 times, we would have to say that my hypothesis is false, especially if we observed that squirrels always eat sunflower seeds and never eat a gum wrapper. In this case the squirrels are treating a potato chip more like a gum wrapper than like a sunflower seed. If, however, we found that the squirrel ate the potato chip 97% of the time, we could claim that our hypothesis is not false, because we have obtained results better than the 95% level which is the standard in science and statistics. Intermediate results are more difficult to deal with. Let's say that the squirrel ate the potato chip 50% of the time. Clearly, we have to say that our hypothesis is false, but the squirrels did eat the potato chips some of the time, so perhaps some other factor is influencing the results. On the first day of my experiments I flipped a potato chip to 4 squirrels, but none of them showed the slightest interest in it. From this I was forced to conclude that my hypothesis was in some way false.

Notice that I have avoiding using the term "failure" in connection with a hypothesis that is shown to be false. This is not a failure – it is progress toward the correct answer or solution. The normal response to finding that one's hypothesis is false is to modify the hypothesis, make a new prediction, carry out experiments, and draw a new conclusion. This may occur several times as the scientist continually refines the hypothesis. Often, scientists expect to have to modify their hypothesis several times and have developed a way to avoid having to do huge numbers of experiments. If the first few experiments for testing a certain hypothesis do not yield results that agree with the predictions, then the scientist will stop experimenting at that point and proceed to modify the hypothesis and try again. Only when the results look promising under a certain hypothesis will the scientist go on to do a relatively large number of repeats of the experiment.

In the case of my squirrel experiments the problem was easy to see. I normally take my dog with me to class, and it was clear that the squirrels had their attention focused on my dog and not on the potato chip I was throwing toward them. This caused me to modify my hypothesis slightly, from "squirrels like potato chips", to "squirrels like potato chips, but are afraid of dogs". To accommodate this new hypothesis, I modified the conditions of my prediction, and I changed my if...then statement to this:

1. If squirrels like potato chips, but are afraid of dogs
2. then they should eat a potato chip
3. when I offer one and no dog is present.

After than first day, I left my dog in my office, much to his dismay, and found that when I was alone the squirrels did something with the potato chip 30 times out of 30 experiments – 100% of the time. Clearly 100% is greater than the standard of 95%, and based on these results, I can confidently say that squirrels like potato chips, but are afraid of dogs.

In my set of experiments I had to make two loops through the scientific method. This is pretty good – often many loops are required. It is often the case that one scientist will test a hypothesis and obtain satisfactory results, then another scientist will refine the hypothesis – by doing something like adding the proviso "but are afraid of dogs" to it. Often this refinement will lead to more satisfactory results. Yet another scientist (or one of the previous two) may further refine the hypothesis. Many interesting and exciting exchanges in scientific journals are of this form, with each new article presenting the results of the refinement of a hypothesis.

3.4 Knowing Truth In Science

Let me point out one more time that in the preceding discussion I was careful never to describe a hypothesis as "true". The closest I came to calling a hypothesis "true" was to say that it was "not false". Saying that something is "not false" implies a sort of truth, but it is a weak sort of truth. The best that a scientist can say about their hypothesis is that they have not discovered any data which shows that it is false. This recognizes that there may actually exist data that do show the hypothesis to be false, but which the scientist is not aware of. This is the most secure state that can be attained in science. There is always a possibility that someone will perform some experiments in a way that the scientist didn't think of, and which show that the scientist's hypothesis is false in some respect. To say that a hypothesis is "true" is to make the absurd and arrogant claim that there is no possibility that evidence exists or can be generated by any experiment in the future that shows the hypothesis to be false.

Because of this consideration, scientists never consider themselves to know actual truth. The truth they know at a certain moment in time is only a tentative truth that is based on the evidence available at that moment. Indeed, all scientists realize that their work will be superceded or built upon in new directions in the future. Just as Newton's theory of orbital motion was shown to be slightly inaccurate by Einstein's theory of relativity, the work of all scientists (including Einstein's) will one day be shown to be wrong in some respect. What we hope is that when the history of the subject is finally written in some distant time and place, our work is noted as having been on the right track.

Because scientists never really know truth, I am always dismayed when I hear the common phrase "scientific proof". This phrase is often used in public discourse, including TV advertisements. Unfortunately it is also used in the courtroom. Actually, scientific proof is an oxymoron – a contradiction in terms – and I hope that you can see why this is obviously so.

3.5 Humans Naturally Use the Scientific Method

I observe that humans naturally use the scientific method, often without being aware that they are doing so. Many people use the scientific method in a variety of different settings to learn something about the world. From the auto mechanic trying to figure out why a car won't start, to the medical doctor trying to discover the cause of a patient's cough, we all routinely use a form of the scientific method. I claim that as a human being, you are and have always been a scientist in the most practical of ways. I will now attempt to convince you of this by presenting a scenario, which I believe nearly everybody has encountered at one time or another and solved by the application of the process of hypothesis generation, prediction making, experimentation, and drawing conclusions from the experiments.

3.5.1 The Case of the Light that Wouldn't Come On

Let's say that you arrive home late at night, walk up to your front door, unlock the door, reach in to the light switch just inside the door, and flip it, but nothing happens. What do you do now? Here is a problem – the light won't come on. How do you fix the problem? I claim that as a human being, you will go through a mental and physical process of making and testing hypotheses. Further, I claim that you will test the various hypotheses in a logical order, based primarily on how easily the hypothesis can be tested and the likelihood that a certain phenomenon is the cause of the problem. The steps happen very rapidly in your mind, and prior to reading this chapter, you may not have had names for the various steps you go through. Nevertheless, I hope that you will now recognize what your mind is doing as you stand there in the darkness wondering why the light didn't come on.

The first thing you are likely to be doing while standing at your front door in the dark is running through a list of possible causes for the light not coming on. In scientific terms, you are generating hypotheses. Here is a list of some of them you might consider in this situation:
- The power is out;
- The bulb is burned out;
- The bulb is loose in its socket;
- The switch is broken;
- The wiring is faulty.

The next thing you will do is begin testing hypotheses. For those that are easiest to test, you may actually test them immediately. The easiest hypothesis to test is that the power is out. You can test this easily, and perhaps even without realizing that you are doing so by looking for evidence that the power is not out. Prediction -- if the power is out, then no electrical devices will be active at this moment. Is there a light burning in another room? Can you see the power indicator on your DVD player? Can you hear the refrigerator running? Are the neighbors' lights on? Checking these things is performing experiments. The first three of these would falsify the hypothesis that the power to your house is out. The last one would falsify the hypothesis that the power was out to the neighborhood.

Most people, after rejecting the hypothesis that the power is out, will turn to the mostly likely explanation for the light not coming on, which is that the bulb is burned out. This is an excellent next hypothesis to test, because in addition to being likely, it is also relatively easy to test, and furthermore you get to test the hypothesis that the bulb is loose in its socket at the same time. Here is your likely prediction – if the bulb is burned out, then the light will come on when I replace the bulb. So, off you go, to find a stepladder or a chair to stand on and a fresh bulb to try the experiment with.

If, after having replaced the bulb, the light still doesn't come on, the next most likely explanation is that the switch is broken. Although this is a rare occurrence, switches do have moving parts and do wear out. I'm told that if you flip a broken switch while looking at it in the dark you should see sparks jumping and be able to smell ozone. I'm not sure that this is always the case, but if I saw a switch sparking in the dark I would conclude that it was broken in some way. The definitive way to test whether the switch is broken is to buy a new one at a hardware store and use it to replace the old one. Be sure to turn off the circuit breaker first! People who wouldn't be comfortable replacing a light switch would call an electrician, which would be the person who could diagnose problems with the switch or wiring (using the scientific method) and carry out the repairs.

What do you think? Do you agree with me that humans naturally use the scientific method to learn about the world? If you can accept this, and realize that most of science is no more difficult than figuring out why a light won't come on, then science will seem less like some scary, esoteric thing that only a specialist can do, and more like something that you too can become proficient at. It is within your grasp to add the single most powerful tool ever discovered for learning about the nature of ordinary reality to your mental toolkit.

3.6 Advanced Topics: Ockham's Razor and the Experimenter Effect

In this section we will look at two advanced topics in science. The first is Ockham's Razor, which is typically used to justify why scientists typically test the simplest hypotheses first. Second, we will look at the issue of bias in science and at two common forms of scientific bias.

3.6.1 Ockham's Razor

Ockham's Razor ("Occam" is a commonly used Latinized form of Ockham) is the principle proposed by William of Ockham in the 1400's, who said that "Pluralitas non est ponenda sine neccesitate". For those who don't speak Latin, this translates as "entities should not be multiplied unnecessarily", and is most often expressed in the idea that the best answer is most often the simplest answer. Here again I claim that most people utilize the principle of Ockham's Razor naturally and intuitively, yet it is useful to state it explicitly. In more modern terms, we would express **Ockhams Razor** as the observation that if you have two hypotheses which both explain the observed facts then the simplest one is less likely to be false and you should test it first.

Obviously, the simplest hypothesis is not always the best one, and the principle of Ockham's Razor has often been criticized on these grounds. While it is definitely worth remembering that the simplest hypothesis is not always true, this does not invalidate the argument that you should test the simplest hypothesis first.

3.6.2 The Experimenter Effect

The **experimenter effect** refers to unconscious bias introduced into an experiment by the experimenter. The term **bias** here means manipulating the experiments or the data in such a way that the scientist's hypothesis is supported, when the hypothesis would have been shown to be false (or at least not as strong) without the manipulations. Bias introduced intentionally is simply fraud. However, it is possible for a scientist to introduce bias into their work without being aware of it, and this is what the experimenter effect refers to.

There are two recognized mechanisms by which bias can unconsciously be introduced into an experiment. The first is when the results of the experiment consist of small or difficult to detect effects, and the scientist wants very badly to observe them. An example is the "N-ray" experiments, of the early 1900's. N-rays do not actually exist. René Prosper Blondlot, a French physicist, "discovered" N-rays while working with X-rays. Many scientists were skeptical of the existence of N-rays because the properties they were supposed to exhibit were impossible according to the laws of physics. In 1903, Blondlot conducted a set of experiments in which he claimed to be able to detect N-rays using a relatively simple apparatus. These experiments were confirmed by some scientists in other laboratories, but many scientists were unable to detect N-rays in their labs. Detecting N-rays involved seeing a thread of calcium sulfide glow, ever so slightly, when in the presence of N-rays. We can imagine scientists huddled in the dark watching for the thread to glow slightly, and many of them believed that they actually saw it glow. N-rays were disproved by American physicist Robert W. Wood, during a visit to Blondlot's lab by pulling a trick on Blondlot's assistant. Wood secretly removed a critical piece of Blondlot's apparatus, but the assistant was still able to see the thread glow. Wood then replaced the critical component, but clumsily so that the assistant saw him and assumed that he was actually removing the component instead of replacing it. Believing this, the assistant was not able to see the thread glow, even though the device was functioning perfectly. Clearly, Blondlot, his assistant, and the other scientists who were able to detect N-rays were victims of their own wishful thinking. When looking for such subtle effects as a thread glowing slightly it is easy to let our expectations and our imagination lead us astray. Today, physicists design detection apparatus with electronic sensors to detect these types of subtle effects, which removes the fallible human element and makes the experiments much more reliable.

The second case in which the experimenter effect occurs is when many varied items must be treated in the same way. Under these conditions it is possible to unconsciously treat some differently from others. A classic example of this is provided by the detailed investigations into the relationship between race and skull capacity performed in the 1800's. Skull capacity was taken to be an estimate of brain size, and the underlying idea was that the larger the average brain size, the more intelligent people of a certain race were. These experiments were performed during the era of colonialism, in which Europeans and their descendants in North America and elsewhere believed that it was acceptable for them to exploit peoples of other countries because White peoples were superior. The skull capacity experiments were seen as a way to prove the superiority of White peoples by showing that their brains were larger. Skull capacity was measured by filling the empty skull with lead shot or mustard seed, and then pouring the shot or seed into a graduated cylinder (which is just a more accurate version of the common kitchen measuring cup) to measure its volume. It is thought that the experimenters, being White, desperately wanted to show White superiority, and so biased their experiments in subtle ways. It is easy, for example, to unconsciously pack the shot or seed more tightly into a White person's skull and less tightly into the skull of someone of another race. Not unexpectedly, the results showed that Whites did have (slightly) larger brains. These results were inaccurate, however, due to bias. In the natural, self-correcting progress of science, later researchers were able to convincingly demonstrate that there is no significant difference in the brain sizes of people of different races, once overall body size is accounted for.

The problem of bias seems to be just another historical curiosity, and in most cases it is, because science tends to be self-correcting. If one scientist gets it wrong, another will come along with data that sets the story right. However, in forensic science, bias is no mere academic trivia. In court, peoples lives and the conditions of their lives are at stake. Therefore, forensic scientists must be especially careful to be on the lookout for bias that might creep into their analyses. This is particularly problematical because law enforcement officers who bring evidence to the crime lab, and to whom the results of analyses will be given, are almost always convinced of the suspect's guilt and are looking to the forensic scientists to "prove" that guilt. There are many documented cases of forensic scientists introducing bias into their analyses, and it is thought that most of these cases are due to experimenter effect because of the pressure of law enforcement demands. It has been demonstrated that the results of fingerprint matching are influenced by what law enforcement authorities tell a fingerprint examiner about whether the evidence print is likely to be a match to a suspect or not. There have, in addition, been documented cases of fraud, which have resulted in egregious miscarriages of justice.

3.7 Questions for Study and Review

1. Discuss ways in which you agree and disagree with the following statement: the scientific method is a simple and natural way for humans to learn about the world.
2. In your own words, explain why judges, attorneys, police officers, forensic scientists, and all other persons involved in our modern justice system need to understand the scientific method.
3. Give a definition of the scientific method.
4. Discuss what it means for results to be "replicable" (there are at least two factors involved).
5. In your own words, explain the difference between ordinary reality and an extraordinary (special) reality.
6. In your own words, explain what an experiment is.
7. Make as long a list as you can of questions that can not be answered by the scientific method, and discuss why each of them are unanswerable questions for science. Could any of your questions be made answerable by carefully defining one or more of the terms in it?
8. What is a hypothesis? How does a hypothesis differ from a theory?
9. Many of the basic astronomical facts about our world have been historically difficult for people to perceive and scientists to demonstrate. Some examples of these "difficult" facts are: the world is round, the Earth orbits the Sun, and the Moon orbits the Earth. Choose one of these facts and treat it as a hypothesis. How would you test this hypothesis? Can you find a prediction, framed as an if-then statement that would allow you to test this hypothesis?
10. What is the relationship between "evidence" and "data"?
11. In your own words, define "level of confidence" as it applies to accepting or rejecting a hypothesis.
12. Discuss the concepts of "truth" and "proof" in science. If a scientist describes something as "true" or "proven" (actually, they should avoid these words), what does the scientist mean by this.
13. What does the principle of Ockham's razor suggest about how we should conduct a forensic investigation or examination?
14. Why is bias an issue in forensic science?
15. I am a dowser. I have the ability to locate water that is flowing beneath the surface of the earth. I have successfully located water for wells on a number of occasions. I locate underground water using two pieces of coat hanger wire, bent into an 'L' shape, with the short section of each piece stuck into the body of a Bic pen in such a way that it can rotate freely. I start out holding a Bic pen body in each hand, at arm's length from my body, with the long section of both 'L' shaped wires pointing in front of me. Then I begin walking. When I pass over underground water the wires will swivel right or left within the Bic pen bodies to align themselves parallel to the flow of the water. If I continue walking, the wires will rotate so that they point toward my chest. Because I am a scientist to the

core, I don't believe that I really have any supernatural gift for dowsing. In fact, there is no scientifically plausible mechanism by which the presence of underground water could be "communicated" to the coat hanger wires stuck into Bic pen bodies. Propose some hypotheses for how I can locate underground water successfully. Remember that I am fairly well educated in several sciences including geology and anthropology.

3.8 Sources

Berg RA, Moldwin M. 2005. What Defines Science? Skeptical Inquirer 29(2): p57(1).
Brown, JS. n.d. Problem Solving and the Scientific Method [Internet]. [cited 2007 Oct 17]. Available from: http://members.tripod.com/MrJBrown/sci.htm.
Carroll, RT. 2005. Blondlot and N-rays. In: The Skeptic's Dictionary [Internet]. [cited 2007 Oct 17]. Available from: http://skepdic.com/blondlot.html.
Dror IE, Charlton D, Peron AE. 2006. Contextual Information Renders Experts Vulnerable to Making Erroneous Identifications. Forensic Science International 156 (1): 74(5).
Ehrlich R. 2007. Science Will Never Explain Everything--that Is Why it Is So Useful! Skeptic 13(2): 17(3).
Ellis G. 2005. Are There Limitations to Science? Parabola 30(2): 63-67.
Gould SJ. 1996. The Mismeasure of Man. New York: W.W. Norton.
Kageyama Y. 2003. Openness to the Unknown: the Role of Falsifiability in Search of Better

Chapter 4
THE CRIMINAL INVESTIGATION, START TO FINISH

In this chapter we will examine the process of a criminal investigation from the discovery of a crime to the punishment of a criminal. The goal is to provide an overall picture of this process, which will serve as a foundation for further exploring the roles of forensic scientists in it. The ideas discussed in this chapter will also help in understanding why certain things we do simply have to be done in a certain way.

This chapter departs from the strict study of forensic science. Much of the material comes from the field of criminology, and serves as a brief introduction to the justice system as viewed by criminologists. I have not assigned this chapter an advanced topics (i.e. optional) designation because being aware of this process in its entirety is crucial for understanding how forensic science is used in investigations.

Let's start with the following model for the steps in a criminal investigation. I will assume that the investigation begins with the commission of a crime and ends with the sentencing of the criminal. The stages involved are then the following, in at least approximate order.

- Commission of the crime;
- discovery of the crime;
- police arrive at the crime scene;
- an investigation is conducted to obtain evidence;
- a suspect is arrested;
- the strength of the evidence is tested to determine whether the suspect will be tried;
- the pretrial phase begins, which includes several procedures such as sharing of evidence;
- trial;
- sentencing, if the defendant is found guilty;
- corrections.

4.1 Discovery of a Crime

Crimes are committed all around us all the time. I suspect that most of us commit crimes daily. The crime I commit multiple times each day is failing to keep my dog on a leash at all times as is required by city ordinance. Most crimes are not discovered or not reported. This is especially true for crimes that do not have a victim and crimes that are not very severe, such as not having my dog on a leash.

Let's say that a habitual criminal, Joe Baculum, is in downtown Missoula at the corner of Pine and Higgins. He wants to cross the street, but doesn't have the walk signal. He looks both ways. Nobody's coming. Nobody's looking. So, he crosses the street. The crime of jaywalking has been committed, but nobody knows it other than

Joe. Even if somebody saw Joe jaywalking, they are very unlikely to report it – it just isn't worth their time. Even a police officer would probably look the other way unless he or she was particularly zealous or having a bad day.

In contrast, let's say that Joe Baculum shoots and kills Bob Oosik at the corner of Pine and Higgins. It's quite likely that this crime will be discovered. For one thing, there is a body and other evidence present, such as blood and at least one bullet and firearm casing. Even if Joe can hide the body and clean up the blood without being seen, Bob's friends and family will miss him and eventually contact the police. This crime has a victim, and it's severe enough that anybody seeing the crime will report it.

The general principle is this: the more serious the crime the more likely it is to be discovered. For that matter, the more serious the crime the more likely that forensic scientists will become involved in the investigation of it.

4.2 Peace Officers and Law Enforcement Agencies

Citizens who want to report a crime do so by calling the police. A more general term for police is peace officer. A **peace officer** is someone with the authority to carry a weapon and to make arrests. In many states citizens share these powers in part, under certain conditions, but peace officers indisputably have greater powers in this arena than other citizens do.

An organization that employs peace officers is called a **law enforcement agency**. Most of the commonly encountered law enforcement agencies employ two types of peace officers. The first is the **patrol officer**, and the second is the **detective** (or investigator). Patrol officers are the ones we see throughout the community, wearing a uniform and driving a patrol car. The duty of patrol officers is to respond quickly to reports of crimes. Detectives are much less commonly seen and may not always be wearing a uniform and driving a marked car. The duty of detectives is to more thoroughly investigate crimes that need to be pursued beyond a simple response. Therefore, it is almost always a patrol officer who shows up first at the scene of a crime. Detectives may come later, or not at all, depending on the nature of the crime.

Peace officers exist at a variety of different levels of government. Each law enforcement agency has a certain set of laws it is charged with enforcing and a certain geographic area inside which it enforces those laws. This combination of laws and region is called a **jurisdiction**.

Different states have different systems of law enforcement agencies and peace officers. It is beyond the scope of this text to examine the system of every state, so I will choose Montana as an example, for the obvious reason that Montana is where I live and work. Montana's system of law enforcement is similar enough to that of most other states to serve as a good example. Montana recognizes the following agencies whose members are at least sometimes peace officers.

- City police. For example, the Missoula City Police. City police have jurisdiction over all crimes defined in state law, county law, and city ordinance within the city in which they are located.

- Sheriff's Departments. In most places in the United States Sheriffs Departments (or Offices) are the county police. For example, the Missoula County Sheriff's Department. Sheriff's departments usually have jurisdiction over all crimes defined in state or county law within a certain county. Sheriffs and their deputies are normally not expected to enforce city ordinances.
- State Police. In Montana we have several state police agencies.
    - Montana Highway Patrol: The Montana Highway Patrol is responsible for highway traffic safety management for the state of Montana including investigations, enforcement, and education. In addition, officers respond to requests for assistance from other city, county, state and federal law enforcement agencies.
    - Montana Department of Criminal Investigation (DCI). The DCI provides criminal investigative assistance to city, county, state and federal law enforcement agencies at their request. It also conducts investigations of allegations of misconduct within state government and is the enforcement arm for certain state agencies. These investigations include, but are not limited to, homicide and other violent crimes, organized crime activity, white collar crime, sex crimes, corruption, official misconduct, theft, complex financial crimes, and internal agency investigations. A branch of the DCI, the Narcotics Investigation Bureau (NIB) investigates dangerous drug violations and provides investigative assistance to city, county, state and federal law enforcement agencies as requested for drug related crimes.
- State Park Rangers. Their jurisdiction is crimes that occur within the boundaries of a state park.
- The State Fire Marshall Bureau. This bureau conducts investigations of arson.
- Prison guards in the course of their duties at the state's correctional facilities are considered peace officers.
- State Game Wardens. These peace officers have primary jurisdiction over wildlife-related crimes, such as poaching.
- Campus Security Officers. Such as the campus security officers here at U.M., who like to be called "public safety officers". Their jurisdiction is crimes that occur on the college campuses of the Montana University System.

The federal government has a huge number of agencies that have at least some agents who are peace officers, and which thus qualify as law enforcement agencies. Here is a partial list: the Federal Bureau of Investigation (FBI), Bureau of Alcohol, Tobacco, and Firearms (ATF), Coast Guard, Customs, Drug Enforcement Administration (DEA), Federal Emergency Management Agency (FEMA), Federal Protective Services, Fish and Wildlife Service, Internal Revenue Service, Military Police in all branches of the military, Postal Inspection Service, Secret Service, and the U.S. Treasury.

International law enforcement agencies and peace officers also exist. Every nation has its own system of law enforcement, and this applies as well to Native American nations in the U.S. Each reservation has a tribal police with jurisdiction over criminal and civil law on the reservation. The relationship between federal law, state law, and tribal law regarding Native Americans on and off reservations is difficult to sort out. Oversimplifying, tribal police are concerned with crimes where the alleged perpetrator is a Native American and the crime occurred on the reservation. An exception is if the crime is a "major crime", such as murder, in which case it is investigated by the FBI. Interpol is an international law enforcement agency recognized by most countries. It assists law enforcement agencies in member nations with crimes that cross national boundaries.

4.3 Crime Scenes, Searches, and Gathering Evidence

The location where a crime occurred is known as the **crime scene**. Crime scenes always contain some evidence, according to Locard's principle. We will look at crime scenes in more detail in another chapter.

For most crimes all necessary investigation is done by the responding patrol officer. If the case is sufficiently serious and complex, a detective will be assigned to investigate it. It is estimated that nationwide only about 25% of cases are investigated by a detective. Law enforcement agencies have to manage the time resources of their peace officers and unless the crime is very serious, cases won't be investigated unless the peace officers are fairly confident of actually determining who the perpetrator was.

Often there is evidence at a location other than the original crime scene. These locations are known as **secondary crime scenes**. Returning to the example with which we started this chapter, if Joe Baculum murdered Bob Oosik at the corner of Higgins and Pine, then tried to hide the evidence, it is likely that Joe put Bob's body in his car and drove to some location where he could hide the body. Joe's gun may also be hidden somewhere, or merely stored at Joe's house. Further, Joe may have left shoe prints, tire impressions, fingerprints, or other evidence at the crime scene. In a case such as this one the detective doing the investigation needs to collect evidence from Joe's car, the place Joe hid Bob's body, and Joe's house. Standard items for matching must also be obtained, including Joe's footwear, his car tires, and his fingerprints, at least. The corner of Higgins and Pine is the primary crime scene, and Joe's car, the place where Joe hid Bob's body, and Joe's house are secondary crime scenes. They are genuinely locations where a crime occurred, since it is illegal to transport or hide the body of a murder victim, and it is illegal to possess or hide a weapon that has been used in a murder.

The corner of Higgins and Pine is a public area, belonging to the City of Missoula. However, Joe's car and house are private property and in order to collect evidence or standard items there, peace officers must obtain a search warrant. It is likely that the peace officers will also need a search warrant to collect evidence at the place where Joe hid Bob's body. I order to obtain a search warrant a peace officer must convince a judge that a crime has been committed and that there is a strong likelihood that evidence of some sort exists at the location for which the warrant is being requested. There are a few specific situations in which a peace officer can justify searching without a warrant, but they almost always need to have "probable cause", which is some sort of indication that other evidence is present. We will examine warrants in more detail in the chapter on jurisprudence.

Prepared with search warrants, the detective and other peace officers will conduct searches of the primary and secondary crime scenes. They will gather any evidence they find. As this evidence is collected it enters the chain of custody. A **chain of custody** is an accounting of each item of evidence in a case from the time it is collected at a crime scene to the time it is presented in court. This accounting includes where the evidence has been stored, how it was stored, who analyzed it, and who had access to it. The goal is to prevent any tampering with the evidence by doing something to it which gives a stronger suggestion of the suspect's guilt or innocence. If the chain of custody is not secure, the evidence either will not be admitted into court, or the defense will be successful in having it suppressed or its impact minimized. In the O.J. Simpson trial, discussed in chapter 1, Simpson's defense attorneys were able to convince the jury of Simpson's innocence based primarily on allegations that the police had tampered with the evidence. Chain of custody is serious business, and some authorities believe that more prosecution cases are lost because of some irregularity in the chain of custody than for any other reason.

This point in the investigation provides the first opportunity for forensic scientist to become involved in the investigation. It is estimated that in 85% of cases where evidence is collected, it is collected by the responding patrol officer. In other cases a detective will collect evidence, and in some jurisdictions there are specialized crime scene investigators or even crime scene investigator teams. There are cases, however, in which a forensic scientist or a team of forensic scientists from the crime lab will assist in collecting evidence.

4.4 Physical and Testimonial Evidence

At this point in the investigation it is likely that the peace officers involved have a variety of evidence of two types. First, they have all the blood, fingerprints, tire tracks, hairs, bullets, etc. collected at a primary or secondary crime scene. This type of evidence is called **physical evidence**, and it consists of actual tangible items. In addition, it is likely that they have a body of testimonial evidence. **Testimonial evidence** consists of things that people say. Witness statements, confessions or explanations made by the suspect, descriptions given by a victim, and similar stories told by people to the peace officers constitute the testimonial evidence in the case. The body of testimonial evidence may be very large, especially if the peace officers have tried to put together the events leading up to the crime by interviewing many witnesses.

Testimonial evidence is the same thing as anecdotal evidence described in chapter 3, and scientists have no respect for it at all. However, not only do peace officer have respect for testimonial evidence, most of them prefer to work with it. It is estimated that physical evidence is collected in only 10% of all cases, though that figure is slowly improving. Also, in a large number of cases, peace officers only use physical evidence as leverage to obtain testimonial evidence. For example, if the firearms evidence links Joe Baculum to the crime scene at Higgins and Pine, peace officers are likely to use this fact to try to convince Joe to confess to the crime.

The problem with testimonial evidence is that it is widely known to be unreliable. Physical evidence is not completely reliable, but it is much more reliable than testimonial evidence. Most people believe that the best evidence is an eyewitness testimony, but this is not true for many reasons. The eyewitness may be mistaken, confused, lying, exaggerating, or in some other manner stating incorrect information. Confessions are widely thought to be reliable, but this is widely known in the justice system to be not true. For the most part, judges know about the unreliability of testimonial evidence, forensic scientists certainly do, and most attorneys probably do as well. The people who don't know how unreliable confessions are seem to be peace officers and those ordinary folks who sit on juries. However, this situation is improving too, beginning in the 1960's with the Supreme Court decisions described in chapter 1, and accelerated in recent times by the public's enthusiasm for forensic science shows on TV.

Physical evidence is usually taken to a crime lab, where it is examined or analyzed by forensic scientists. Physical evidence is much more reliable and impartial than testimonial evidence. Since it has to be analyzed and interpreted by human beings, however, it is not infallible.

4.5 Arrest

If successful, the investigation may result in an arrest. All arrests must ultimately be approved by a judge. In most cases a peace officer will present evidence to the judge showing that there has been a crime, that there is a high level of certainty that a certain suspect committed it, and that this suspect therefore needs to be arrested. If the judge considers this evidence to be sufficient, a warrant for the suspect's arrest will be issued.

There are situations when a peace officer may arrest a person on **probable cause**, such as when the officer observes the crime being committed or has good testimonial evidence that a certain person committed a crime. For example, if Bob Oosik is discovered lying at the corner of Pine and Higgins in a pool of his own blood, and with his last breath gasps to the police officer that Joe Baculum shot him, then the officer has probable cause to arrest Joe. In situations where the arrest is made on probable cause, the arrested person will appear before a judge in a hearing called a **first appearance**. At the first appearance hearing, the judge will hear evidence that the suspect really should have been arrested. If the evidence is not sufficient, the suspect will be released. When you hear of a suspect being released for lack of evidence, this is what has probably happened.

4.6 Testing the Evidence

When all the available physical and testimonial evidence has been gathered and digested, the detective will prepare a case and consult with the prosecuting attorney for the jurisdiction. Usually, it is the prosecuting attorney's decision whether or not to take the case to court, based on her or his perception of how strong a case they can make against the suspect based on the evidence at hand.

If the prosecuting attorney decides that the evidence is sufficient to warrant taking the case to court, then the strength of that evidence needs to be tested independently, either by a judge or a grand jury. In most misdemeanor cases there will be a **preliminary hearing** before a judge, in which the judge decides whether the evidence against the suspect is strong enough to try them. In most felony cases the evidence goes to a **grand jury**, which is a group of citizens who decide whether the evidence justifies taking the case to trial.

If the judge or grand jury decides that the evidence warrants a trial, then a trial will be scheduled. At this point the status of the suspect is changed to defendant.

4.7 Pretrial

The period of time between when it is decided that a suspect will be tried, and the trial actually begins is called **pretrial**. There are several things that happen during the pretrial phase. First, it must be decided whether to hold the defendant in jail until the trail, or not. Second, it is during this period that plea bargaining may occur. Third, it is during this period that the process known as discovery occurs, which may involve the forensic scientists who analyzed evidence for the case.

4.7.1 Forms of Pretrial Release

Most defendants are not held in jail before their trial. Almost all misdemeanor defendants are released, and most felony defendants are released as well. There are several reasons for releasing a suspect until their trial. First, under the American system of justice they are presumed innocent until proven guilty. Second, the suspect can usually prepare a better defense if he or she is free. Third, there often isn't enough jail space to hold pre-trial defendants. However, these things have to be weighed against the possibility that the defendant will commit another crime while free, or that the defendant will flee and not appear at their trial.

There are three types of pretrial release commonly used. The most familiar is **financial release**, also known as bail, where the defendant deposits a sum of money with the court, and gets it back if they appear for their trial. If the defendant fails to appear, they forfeit the money. Many defendants seek the help of a bail bond company, which will post their required bail for a fee. If the defendant appears for their trail, then the bail bond company gets their money back, and the defendant is only out the fee they paid to the bail bond company. If the defendant fails to show up, the bail bond company is out the money.

Conditional release is the situation wherein the defendant is released, but one or more limitations or conditions are placed upon them. For example, the defendant may be required to refrain from drinking alcohol, or to attend a safe driving school, or to be home by 10pm, or to check in with the police once a week, or any one of a number of such measures that are designed to keep the defendant available for trial and out of trouble. Nowadays it is common for conditional release to include the wearing of a tracking device tied in with the Global Positioning System (GPS) satellites. This device allows the defendant's location to be monitored at all times.

Recognizance release is simply letting the defendant go until their trial data. This is common when the defendant is considered to pose a minimal risk for fleeing or reoffending. Most misdemeanor defendants are released on their own recognizance.

4.7.2 Pleas and Plea Bargaining

There are three **pleas** that can be entered in court when the judge asks "How do you plead?" Guilty and innocent are self-explanatory. The third plea is no contest (nolo contendre). In most courts, **no contest** means the same as guilty, except that it does not imply the person is guilty for the purposes of any civil litigation that may follow the trial wherein the victim seeks a judgement against the defendant for damages caused as a result of the crime.

Many cases don't make it to trial because of **plea bargaining**. Plea bargaining is defined as the exchange of prosecutorial and judicial concessions for a plea of guilty. Usually this means that the defendant gives up their right to a trial in exchange for a lesser charge or a less severe sentence. Plea bargaining is an integral part of the process of administering justice. It has several important benefits. Most importantly, plea bargaining relieves caseload pressure on the courts. Plea bargaining also allows a defendant to avoid the publicity of a trial. There are some problems with plea bargaining, however. Especially troublesome is the possibility that some innocent persons may plead guilty in order to avoid the publicity of a trial. Plea bargaining may also result in excessive leniency.

4.7.3 Discovery Procedures

One important part of the modern pretrial process is the sharing of information between the prosecution and the defense. The days of the surprise witness, or of reserving some particularly damning piece of evidence to enter at a critical point as a surprise are pretty much gone. In general, all information to be presented at the trial has to be shared between the two sides. This insures fairness. The sharing of evidence, witness lists, and related information is called **discovery proceedings**. Discovery proceedings are important for forensic scientists because they can expect to be interviewed by the other side in the case. In many cases forensic scientists are expected to appear in court as expert witnesses. Most often they will be prosecution witnesses, but not always. This being true, the other side has the right to interview the forensic scientist and to depose them. **Deposing** a witness, or taking a deposition from a witness, means that the opposite side asks the witness what they are going to say, and writes it down. If the witness departs too radically in their actual testimony from what they said in the deposition, they can be held in contempt of court.

4.8 Trial

The defendant's trial is held in an appropriate court. There are many levels of court, which correspond roughly to the jurisdictions of peace officers. That is to say that there are city, county, state, federal, and international court systems. More about the court system will be discussed in the next chapter.

The conduct of the trial follows a set of rules designed to make the process fair (in theory). Trials for serious offenses are almost always jury trials. For misdemeanors, a judge may hear the case without a jury. Trials follow a certain format wherein the prosecution gets to speak first, and the defense gets the last word. Witnesses are questioned first by the side that subpoenaed them to appear. This is called **direct examination**. Then, the other side gets to question the witness in a step called **cross examination**. When all the evidence has been presented by both sides, and both sides rest their case, then the jury will deliberate about the evidence and determine guilt or innocence. If found guilty, the status of the defendant is changed to convict. More about the process of a trial will be discussed in the chapter on jurisprudence.

Forensic scientists are likely to be called to testify at a trial about the evidence they analyzed. Forensic scientists are considered to be expert witnesses, because they are testifying from the basis of some knowledge or training which is beyond that of most people. During direct examination, the attorney for their side will ask questions designed to bring out the results and conclusions of the analysis, including any evidence that links the defendant to the crime scene. During cross examination, the attorney for the other side will try to discredit or weaken their testimony by questioning their part in the chain of custody, their methods, and their interpretations. No forensic scientist that I have ever spoken with enjoys going to court. The experience of enduring a cross examination is unpleasant, to say the least. This is compounded by the fact that for most of us, including forensic scientists who work at a crime lab, and me, because I work for a state funded university, do not receive any money for testifying, other than a small fee to designed to be enough to cover expenses 20 years ago. Forensic scientists who work in private practice are, however, not so limited in the fees that they can ask for testifying.

4.9 Sentencing

The final phase in a trial is **sentencing**. Sentencing isn't usually done at the same session of the court during which guilt or innocence is determined. This is to allow for cooling off of emotions and for the judge to deliberate. In most jurisdictions, sentencing is done by a judge. Sometimes, the sentencing hearing will include witnesses for the prosecution and defense, chosen to present various points of view, such as the amount of suffering of the victim or their family, the stature of the defendant in the community, and other factors that may sway the judge to be more or less harsh or lenient.

In most cases the sentence, or at least the maximum sentence, is prescribed by law. However, there is still considerable latitude for a judge to impose greater or lesser penalties. Here are some of the penalties that can be imposed.
- Time in a correctional institution.
- Fines and other financial sanctions.
- Community service.
- Probation. This is a punishment in which the guilty person is released with some conditions, similar to the conditional release that may occur during pretrial.
- A sentence might be partly or totally suspended. This means that the guilty party does not have to fulfill the part of the sentence that is suspended.

4.10 Corrections

If a convict is sentenced to serve time in a jail, prison, or other facility, then we move into the area of **corrections**. There are correctional institutions at all levels of government. The most hardened and vicious criminals tend to move from the local level correctional institutions to the state and federal level ones. Therefore serving time in a more local institution is considered preferable to serving time in a federal institution. Federal correctional institutions are usually called penitentiaries. State level ones are usually called prisons. County or city level ones are normally called jails. A new level of correctional institution, the county or regional detention center, is becoming popular. Detention centers are often run by a private company that contracts with local jurisdictions to provide correctional services.

Most convicts do not actually serve their full sentence, but are paroled. **Parole** is setting convicts free before they have served their full sentence. Parole may come with significant restrictions, or not, depending on the severity of the offense, evidence that the convict has reformed, the convict's behavior in prison, and a number of other factors. When a convict has served their time, they are considered to have paid their debt to society. Felons may, however, forfeit certain rights, such as the right to vote and the right to bear arms.

4.11 The Roles of Forensic Scientists in a Criminal Investigation

In this chapter we have seen four places in the process of justice from discovery of a crime to serving time in a correctional institution where forensic scientists may become involved. They might become involved with collecting evidence at a crime scene. They will certainly become involved with analyzing physical evidence, if it is collected. They will probably be asked to give a deposition to the opposing side during the discovery process of pretrial. Finally, they will be expected to testify in court if the evidence they analyzed is presented against a defendant.

For some forensic scientists, however, this is only the beginning of their involvement. Those who practice forensic sciences useful in civil cases, such as forensic economists, may become involved in a civil trial following the criminal trial, in which the victim or their family attempt to sue the defendant for financial and other damages resulting from the crime. Those forensic scientists who primarily work in a hospital or treatment center, such as forensic psychologists, may find their work continuing during the convict's incarceration and even afterward.

4.12 Questions for Study and Review

1. What types of crimes are most likely to be discovered and/or reported?
2. What city, county, state, federal, and international law enforcement agencies are found where you live?
3. In your own words, define crime scene, secondary crime scene, and chain of custody.
4. In your own words, describe the difference between testimonial and physical evidence.
5. With regard to testimonial and physical evidence, who (forensic scientists, peace officers, attorneys, judges, etc.) prefers to work with which form?
6. In your own words, describe the term "probable cause".
7. Why do some cases have a preliminary hearing while others go to a grand jury?
8. In your own words, describe the various types of pretrial release.
9. What is the distinction between a plea of "guilty" and a plea of "no contest"?
10. In your own words, describe plea bargaining and the arguments for and against it.
11. In your own words, define "discovery proceedings" and "deposition". Explain why these are important for forensic scientists.
12. Explain the difference between direct examination and cross examination in a trial.
13. List the types of punishments that a judge may impose as part of a criminal's sentence.
14. In your own words, define the terms "corrections" and "parole".

4.13 Sources

Albritton WH, Thompson MH, De Ment, I. 1999. Standing Order on Criminal Discovery [Internet]. [cited 2007 Oct 17]. Available from: http://www.almd.uscourts.gov/rulesproc/docs/Criminal_Discover-534.pdf

American Bar Association. n.d.. How Courts Work [Internet]. [cited 2007 Oct 17]. Available from: http://www.abanet.org/publiced/courts/home.html.

Bulzomi MJ. 2001. Indian Tribal Sovereignty: Criminal Jurisdiction and Procedure [Internet]. FBI Law Enforcement Bulletin 70(6). [cited 2007 Oct 19]. Available from: http://www.fbi.gov/publications/leb/2001/june2001/june01leb.htm#page_25.

Clark MW. 2005. Enforcing Criminal Law on Native American Lands [Internet]. FBI Law Enforcement Bulletin 74(4). [cited 2007 Oct 19]. Available from: http://www.fbi.gov/publications/leb/2005/apr2005/april2005leb.htm.

Committee on Indian Affairs. 1995. The Tribal Nations of Montana: A Handbook for Legislators [Internet]. [cited 2007 Oct 17]. Available from: http://www.opi.state.mt.us/pdf/TitleI/MTTribal.pdf.

Cutler SP. 1998. Interpol [Internet]. FBI Law Enforcement Bulletin 67(12). [cited 2007 Oct 19]. Available from: http://www.fbi.gov/publications/leb/1998/dec98leb.pdf.

Department of the Army. 1985. Field Manual 19-20, Law Enforcement Investigations [Internet]. [cited 2007 Oct 19]. Available from: http://www.enlisted.info/field-manuals/fm-19-20-investigative-process.shtml.

Edwards JB. Homicide InvestigativeStrategies [Internet]. FBI Law Enforcement Bulletin 74(1). [cited 2007 Oct 19]. Available from: http://www.fbi.gov/publications/leb/2005/jan2005/jan2005.htm#page11.

Hendrie E. 2002. Inferring Probable Cause. FBI Law Enforcement Bulletin 71(2): 23-32.

Legal Information Institute. n.d. Criminal Law [Internet]. [cited 2007 Oct 17]. Available from: http://www.law.cornell.edu/wex/index.php/Criminal_law.

Mauriello TP, 1998. Criminal Investigation Handbook: Strategy, Law, and Science. New York: Matthew Bender and Company, Inc

Montana Code Annotated 2007 [Internet]. [cited 2007 Oct 17]. Available from: http://data.opi.state.mt.us/bills/mca_toc/index.htm.

Morley, Michael. n.d. Anatomy of a Murder: A Trip Through Our Nation's Legal Justice System [Internet]. [cited 2007 Oct 17]. Available from: http://library.thinkquest.org/2760/.

O'Connor T. 2004. Complaints, Indictments, Arraignments, Notifications, and the Pretrial Right to Counsel [Internet]. [cited 2007 Oct 17]. Available from: http://faculty.ncwc.edu/toconnor/arraignments.htm.

O'Connor T. 2004. Pretrial Procedures and Police Testimony [Internet]. [cited 2007 Oct 17]. Available from: http://faculty.ncwc.edu/toconnor/315/315lect08.htm.

O'Connor T. 2006. Format of a Criminal Trial [Internet]. [cited 2007 Oct 17]. Available from: http://faculty.ncwc.edu/toconnor/trialfrm.htm.

O'Connor T. 2006. Affidavits and Warrants [Internet]. [cited 2007 Oct 17]. Available from: http://www.apsu.edu/oconnort/3000/3000lect03c.htm.

Payton GT, Amaral M. 1996. Patrol Operations and Enforcement Tactics. San Jose (CA): Criminal Justice Services.

Ramage M. 1996. 96-02: Revised State Criminal Discovery Rules, Effective 10/1/96 [Internet]. [cited 2007 Oct 17]. Available from: http://www.fdle.state.fl.us/OGC/Legal_Bulletins/lb9602_9-24.html.

Sharps MJ, Hess AB, Casner H, Jones J. 2007. Eyewitness memory in context: toward a systematic understanding of eyewitness evidence. The Forensic Examiner 16(3): 20(8).

The Constitution of the State of Montana as Adopted by the Constitutional Convention March 22, 1972, and as Ratified by the People, June 6, 1972, Referendum No. 68 [Internet]. [cited 2007 Oct 17]. Available from: http://leg.mt.gov/css/mtcode_const/const.asp.

U.S. Government Printing Office. 2004. Federal Rules of Criminal Procedure [Internet]. [cited 2007 Oct 17]. Available from: http://judiciary.house.gov/media/pdfs/printers/108th/crim2004.pdf.

William J. Jameson Law Library. n.d. Indian Law and Tribal Law & Government Procedure [Internet]. [cited 2007 Oct 17]. Available from: http://www.umt.edu/LAW/library/Research%20Tools/Tribal%20Law.htm#Montana.

Knowledge. Philosophy of the Social Sciences 33(1): 100-121.

Latura B. n.d. The Scientific Method [Internet]. [cited 2007 Oct 17]. Available from: http://skepticx.myweb.uga.edu/skep_1.html#1.1.

Lectric Law Library. 1998. Daubert V. Merrell Dow Pharmaceuticals [Internet]. [cited 2007 Oct 17]. Available from: http://www.lectlaw.com/files/exp06.htm.

Legal Information Institute. n.d. Daubert v. Merrell Dow Pharmaceuticals, 509 U.S. 579 (1993) [Internet]. [cited 2007 Oct 17]. Available from http://supct.law.cornell.edu/supct/html/92-102.ZX.html.

Linder D. 2007. An Account of Events in Salem [Internet]. [cited 2007 Sep 25]. Available from: <http://www.law.umkc.edu/faculty/projects/ftrials/salem/SALEM.HTM>.

Nelson J. 2002. What is Science? The Science Teacher 69(3): p12(1).

Price DJD. 1975. Science Since Babylon, New Haven (CT): Yale Univ. Press.

Shrake DL, Elfner LE, Hummon W, Janson RW, Free M. 2006. What is Science? The Ohio Journal of Science 106(4): 130-135.

Thorburn WM. 1918. The Myth of Occam's Razor, Mind 27:345-353.

Chapter 5
Advanced Topics
CRIMINAL CHARGES AND THE COURT SYSTEM

This chapter is the first that I have designated as an advanced topic. It does not concern a forensic science, nor is it foundational for any other chapter. Instead, it examines two topics of indirect interest to forensic scientists: criminal charges and the court system. The section on criminal charges briefly explains some of the more commonly encountered offenses that often result in physical evidence to be examined. The section on the court system seeks to give a brief overview of the organization of federal, state, and international courts.

5.1 Criminal Charges

Although there is no need for forensic scientists to be experts on the various crimes with which a suspect may be charged, a general knowledge of them is useful. Detailed knowledge can be left to peace officers, attorneys, and judges. In this section I will present some of the more common criminal charges that may define the investigation for which a forensic scientist is analyzing evidence.

There are too many ways to break the law to go into all the possible crimes that a person can be accused of. Instead, however, let me list some of the more common charges that are brought against defendants and for which physical evidence is often collected. I will also not present here some crimes that will be defined in coming chapters, for example arson, which will be discussed in the chapter on forensic engineering. For each charge I will describe it as it is generally defined nationwide. In order to capture some of the variety from state to state I will describe Montana's version of the charge as appropriate.

5.1.1 Felonies and Misdemeanors

Criminal law makes a distinction between crimes based on how severe they are. The most serious crimes are called **felonies**. Most jurisdictions define a felony as a crime that is punishable by more than one year in a correctional institution or a more severe punishment such as execution. Crimes that carry a lesser sentence are **misdemeanors**. In most states the evidence for felonies is tested by a grand jury, and the evidence for misdemeanors is tested by a judge, as described in chapter 4. In most states felony trials are by jury, and misdemeanor trials are decided by a judge.

Although technically misdemeanors, many crimes such as parking violations that are normally not punished by a jail sentence are called **petty crimes**. Another category of crimes encountered in some states is **wobbly felonies**. Wobbly felonies can be charged either as a felony or a misdemeanor, and even if initially charged as a felony can be reclassified as a misdemeanor after the person who committed it has satisfactorily served their sentence.

5.1.2 Homicide

Homicide is the crime of causing the death of a human being. There are many forms of homicide. What makes causing a death to be **murder** is that there is malice involved. **First degree murder**, which is the most heavily punished type, occurs when the death is intentional and premeditated, or the death occurs in the course of a dangerous felony. In Montana this is called "deliberate homicide". **Second degree murder** is intentional, but not premeditated. For example if the perpetrator intended only to cause serious harm and not death, but death occurred. Second degree murder is also considered to have occurred if the death occurred due to an act that creates a grave risk of death, or if it occurs in the course of committing a non-dangerous felony. Montana does not recognize second degree murder – the conditions of second degree murder are included with those of first degree murder as deliberate homicide.

Manslaughter, sometimes referred to as third degree murder, does not require malice. It can be voluntary or involuntary. **Voluntary manslaughter** occurs when death occurs with intent to kill or inflict grievous bodily injury but under the influence of passion caused by sufficient provocation. Passions such as jealousy, anger, lust, and perhaps others can be cited to reduce the charge from murder to voluntary manslaughter. In Montana, this offense is known as "mitigated deliberate homicide". **Involuntary manslaughter** involves negligence or recklessness, or a death that occurs in the course of committing a dangerous misdemeanor. In Montana this offense is called "negligent homicide".

Some homicides are considered justifiable. **Justifiable homicides** are those that occur in self defense. in the defense of others, in the defense of property, during the prevention of a felony, or while arresting a felon.

5.1.3 Assault and Battery

Assault is any unlawful attempt to touch or strike a person without excuse or justification. Note that this does not require a significant blow that inflicts bodily injury. **Battery** is the result when the assault succeeds and the person is struck or touched. In Montana there is no distinction between assault and battery, it is all assault, whether you actually hit the person or just try to.

If a weapon capable of causing serious injury or death is used as the means of threat, the the assault is termed **assault with a deadly weapon**. **Aggravated assault and battery** occurs when the battery inflicts a serious bodily injury, or an assault and battery involves the use of a deadly weapon. In Montana, this is known as "aggravated assault".

5.1.4 Rape

Rape, is sexual intercourse achieved by force or threat of force against the will of the victim. In Montana this is called "sexual Intercourse without consent". The term "force" can include drugs administered by the rapist, and "without consent" can apply in cases where the victim is unable to understand what is happening. In Montana, if the sexual contact is something other than standard sexual intercourse, then it is called sexual assault, and it is common in Montana for the perpetrator to be charged both with rape and with sexual assault. In many jurisdictions the rapist must be a man. In Montana it can be a man or a woman. In most jurisdictions men don't rape men, they sodomize them instead. Montana is again gender neutral in that anybody can have sexual intercourse without consent with anybody else regardless of their gender.

Statutory rape involves sexual intercourse with a minor who is regarded by law as being so young that they are incapable of giving lawful consent to the act.

Spousal immunity, the idea that a man had the right to sex with his wife at any time, was present in some old laws but has been pretty much done away with in modern times.

5.1.5 Theft

Theft includes the various forms of stealing and is defined as taking another person's property. Montana refers to all forms of stealing as "theft", but other jurisdiction break down the type of theft. Many of the types of theft have both a **petit** (or "petty") form if the dollar value of the property is small and a **grand** form if the dollar value of the property is larger.

Larceny Is taking the property of another, knowing that it is another person's property, with the intention of depriving the person of their property permanently or infringing upon their rights of use of the property in a substantial way.

Embezzlement is the term applied when a person lawfully receives possession of the property of another, but then wrongfully converts that property to their own use. Embezzlement does not require intent to permanently deprive the legitimate owner of their property.

False Pretenses applies to persons who convince or trick others into giving them their property using any form of misrepresentation that the perpetrator knows is false.

Robbery is larceny that occurs when the legitimate owner of the property is present and the property is taken using force or the threat of force. If a weapon is used, then the term armed robbery applies.

Extortion is similar to robbery but uses the threat of harm in the future, rather than immediately.

Burglary is theft that involves breaking into a building (business, residence, etc.) in order to commit a felony theft. If the intent was not to commit a felony theft, then the charge is **breaking and entering**.

5.1.6 Forgery and Counterfeiting

Forgery is defined as creating, altering, signing, or in some other way being involved in the production of a document. This is a crime that requires intent. The purpose can be for the gain of the forger, or for the purpose of inconveniencing another person in some way. It does not apply to a situation in which nobody is financially damaged, as is the case when a celebrity hires someone to sign their photographs for them.

Counterfeiting is defined as producing an unauthorized copy of something that has value, such as currency, securities, software, or collectables. This is another crime that requires intent, and in most cases there must be an intent passing the copy off as if it were genuine. U.S. currency is an exception in that it is illegal to reproduce it for any purpose.

5.1.7 Alcohol-Related Crimes

Alcohol related crimes are extremely variable from state to state, so I may only speak in generalities. In general the following charges may apply, depending on the state.

Public intoxication refers to being visibly and obviously intoxicated while in a location that is not a private dwelling.

Driving under the influence (DUI) and driving while intoxicated (DWI) refer to operation of a motor vehicle with a blood alcohol over a certain amount. In most states the threshold amount is 0.08 percent. In addition, many states have laws prohibiting open containers of alcoholic beverages in a motor vehicle while it is being operated.

Alcohol in possession of a minor (or youth) and alcohol consumption by a minor (or youth) are charges that refer to a person under the age of 21 who has alcohol in their possession or has consumed it. Furnishing alcohol to a minor is a serious offence in which a person older than 21 provides alcohol to a person younger than 21.

Illegal distribution of alcohol refers to attempts to make alcohol available outside of the state's normal regulations and taxes.

5.1.8 Drug-Related Crimes

As we are all no doubt aware, it is illegal anywhere in the United States to manufacture, transport, sell, possess, or use drugs other than as ordered by a physician as treatment for a medical condition. The charges that describe these activities are self-explanatory – **possession of a controlled substance**, for example..

The various states all have laws that cover drug related crimes, and these laws are somewhat variable from state to state. Some nationwide standardization of treatment of drug related offenses is provided by federal law. In the Controlled Substances Act, federal law establishes a five-part classification of drugs. Each group of drugs, called a "schedule", is treated differently.

- Schedule I. Drugs in this category have a high potential for abuse and no recognized medical use in the U.S. Examples include heroin, marijuana, and LSD. Possession of these drugs for sale is punishable by up to 20 years in prison and up to $1 million dollars fine.
- Schedule II. Drugs in this category have a high potential for abuse, but do have a recognized medical use. Examples include morphine, methadone, PCP, most amphetamines, cocaine, and some barbiturates. Marijuana is transitional between schedule I and schedule II because its medical uses are being recognized. Like schedule I drugs, possession of these drugs for sale is punishable by up to 20 years in prison and up to $1 million dollars fine.
- Schedule III. This category includes drugs that have a lesser potential for abuse, do have a medical use, and have a low or moderate potential for dependance. The main examples are the barbiturates not included in schedule II, some codeine preparations, and anabolic steroids. Possession for sale can result in a jail term of up to five years and a fine of up to $250,000.
- Schedule IV. Drugs in this category have a low potential for abuse, have a medical use, and have a low potential for dependance. Examples include Darvon, and most tranquilizers (Vallium, Librium). Possession for sale can result in a jail term of up to three years and a fine of up to $250,000.
- Schedule V. This category includes drugs that have low abuse potential, have a medical use, and have very little potential for dependance. Examples include some of the opiate drug mixtures that also contain non-narcotic medicines, such as tylenol with codeine. Possession for sale can result in a jail term of up to one year and a fine of up to $100,000.

5.2 Courts

As we saw in chapter 4 for law enforcement agencies, there are several levels of courts. The county, state, and federal level are always important, no matter where you live, and if you live in a city, municipal courts may be important as well. In this section I will present the structure of the federal court (judiciary) system. I will again use Montana as an example state, and present the structure of the Montana court system as an example of a typical state system.

5.2.1 The Federal Judiciary System

The structure of the federal court system has changed and evolved throughout the history of the United States. The U.S. Constitution merely provides that the judicial power of the United States "be vested in one Supreme Court, and in such inferior courts as Congress may from time to time ordain and establish." Thus, the only court actually mandated is the Supreme Court. Congress has established and abolished other types of federal courts as national needs have changed over time.

At the present time, the structure of the federal court system is similar to a pyramid composed of three levels. At the top is the Supreme Court of the United States, the highest court. The middle level includes the U.S. Courts of Appeals. At the lowest level are the U.S. District Courts.

The **Supreme Court** of the United States consists of nine justices appointed for life by the President with the advice and consent of the Senate. One justice is appointed as the **Chief Justice** and has additional administrative duties related both to the Supreme Court and to the entire federal court system. The U.S. Supreme Court receives about 5,000 cases each year. Only a small fraction, about 150 each year, of the most interesting and important of these are actually considered by the court. In many of these the decision is published and has the force of law.

On the middle level are the 13 United States **Courts of Appeals**, also called the "circuit courts" in memory of an older system of courts out of which the Courts of Appeals evolved. Also a part of this level is the U.S. Court of Appeals for the Armed Forces. The judges who preside over the Courts of Appeals are appointed for life by the President with the advice and consent of the Senate. Each Court of Appeals has six or more assigned judges, depending on the caseload of the court. The judge who has served on the court the longest and who is under 65 years of age is designated as the **Chief Judge**, and performs administrative duties in addition to hearing cases. The Chief Judge serves for a maximum term of seven years. Since there are 13 Courts of Appeals, and 50 states, each Court of Appeals serves several states. For example, Montana is served by the 9th Court of Appeals, which is located in San Francisco, California. The 9th court also serves Idaho, Washington, Oregon, California, Nevada, Arizona, Alaska, Hawaii, Guam, and the Northern Mariana Islands.

On the lowest level of the federal court structure are the 94 **District Courts of the United States** and the specialized courts, such as the Tax Court, the Court of Federal Claims, the Court of Veterans Appeals, and the Court of International Trade. The 94 district courts each serve a district, which may itself be divided into divisions and may have several places where the court hears cases. Each District Court also has a bankruptcy unit.

District Court judges are appointed for life by the President with the advice and consent of the Senate, with certain exceptions beyond the scope of this book. Congress authorizes judgeships for each district based its caseload. In each district, the judge who has served on the court the longest and who is under 65 years of age is designated as the Chief Judge. The Chief Judge has administrative duties in addition to a caseload. The judges of each district may appoint one or more **Magistrate Judges**, who discharge many of the less important duties of district judges so that the judges can handle more trials. A full-time Magistrate Judge serves a term of eight years, and a part-time Magistrate Judge's term of office is four years. Magistrate Judges handle a variety of matters including civil consent cases, misdemeanor trials, preliminary hearings, and pretrial motions.

Most federal cases are initially tried and decided in the U.S. District Courts, and these courts are therefore called **courts of origination**, or courts of original jurisdiction. If one of the parties in the case is sufficiently dissatisfied with the result of the trial, they may appeal the case to the Court of Appeals that serves their region. The Courts of Appeals only hear cases that have been appealed after having been heard and decided in District Court. Similarly, the U.S. Supreme Court only accepts cases that have been appealed after having been heard and decided in a Court of Appeals. Courts that hear appeals of cases from lower courts are in general called **appellate courts** or "appeals courts".

Court of Appeals and District Court judges who are at least 65 years old and have served at least 15 years may choose to become Senior Judges. As Senior Judges, they may continue their service in several ways, including hearing cases, handling administrative matters, and handling various special tasks.

5.2.2 Montana's Court System as an Example State Court System

The Court System of the State of Montana is, like most state court systems, based roughly on the structure of the federal court system. Like the federal system, the structure is a pyramid. The tip of the pyramid is the Supreme Court of the State of Montana. The middle layer is the Montana State District Courts. The third layer is the county level Justices' Courts. In the case of cities, there are City Courts, which are considered third layer courts, and in the case of large cities there are Municipal Courts, which include the functions of both the second and third tiers.

Unlike the federal judges and justices, which are appointed by the President with advice and consent of the senate, almost all judges in the state court system in Montana are elected by the citizens of the jurisdiction they serve.

The Montana Supreme Court consists of one Chief Justice and six Justices. Each Justice is elected for an eight year term in a statewide nonpartisan election. Terms are staggered so that no more than two Justice positions are scheduled for election at the same time. A Justice of the Montana Supreme Court must be a citizen of the United States, must and have resided in Montana two years immediately before taking office, and must have been admitted to practice law in Montana for at least five years prior to the date of appointment or election. The Chief Justice is the administrative head of the Supreme Court, presides over Court conferences, and represents the Court at all official state functions. The Chief Justice presides at all oral argument sessions of the Supreme Court. In the event of the absence of the Chief Justice, the Justice having the shortest term remaining to be served presides as Acting Chief Justice.

The Montana Supreme Court is the highest Court of the State of Montana. Unlike the U.S. Supreme Court, which only hears appeals, the Montana Supreme Court functions both as an appellate court, and as a court of origination. The Montana Supreme Court does not conduct trials nor seat juries, and only considers legal arguments made by attorneys. The Montana Supreme Court has supervisory control over the other state courts and over the Montana State Bar Association, which certifies attorneys to practice in Montana. The Court has general authority to adopt rules of practice and procedure for courts, judges, and attorneys, which are subject to approval by the Legislature.

There are 21 Montana State District Courts, serving the 56 counties of Montana. Therefore, most counties share a district court, which is often located in one of the larger cities. For example, Missoula and Mineral Counties are served by District Court 4, located in Missoula, which is the fourth largest city in the state.

Montana District Court Judges must be Montana residents and must have practiced law in the state for 5 years. Each district court has an appropriate number of judges depending on their caseload. For example, District 4 is assigned 4 judges. The term of office of District Court Judges is 6 years. Like the state Supreme Court, the District Courts have both original and appellate jurisdiction. This means that some cases begin there, and that they also hear cases that have been appealed from lower courts.

At the level of the counties are the Justices' Courts. There is at least one Justices' Court in each county. For example, Missoula County has two of them, both located in Missoula. Judges of a Justices' Court are called Justices of the Peace (informally, JP's). JP's are elected for 4 year terms. Unlike District and Supreme Court Judges, a JP does not have to be a lawyer. JP's are paid by the county, whereas District and Supreme Court Judges are paid by the state.

The division of cases between Justices' Court and District Court is complicated. Here is a very oversimplified version. All civil cases where $5,000 or less is at stake are heard in Justices' Court. Civil cases involving larger sums are heard in District Court. All preliminary examinations both for misdemeanors and felonies, are heard in Justices' Court. You should recall from chapter 4 that preliminary examinations (hearings) are a court proceeding in which the strength of the evidence against the defendant is tested by a judge. In Montana, this is also commonly the hearing in which the defendant first makes their plea of guilty or innocent. Felonies that are to be prosecuted go to District Court. Lesser Misdemeanors, carrying a fine of $500 or less are tried in the Justices Court. More serious misdemeanors are shared by Justices' Court and District Court.

A City Court is established in every incorporated city in Montana. The jurisdiction of a City Court is similar to that of a Justices' Court, and they divide the caseload by the City Court hearing cases that occurred within the city limits and the Justices' Court hearing cases that occurred outside the city limits. For most practical purposes they are the same. In fact it is common for one judge to be both the Justice of the Peace and the City Court Judge. City courts are funded by the city, except that when combined with a Justices' Court the costs are shared between the city and the county.

City Court Judges have to meet the criteria of Justices of the Peace. They are elected in most cases, but there is a provision for appointing them.

In larger cities it is common to establish a Municipal Court. The act of creation of a Municipal Court abolishes the City Court, and the Municipal Court takes up the function of the City Court along with the functions of both Justices' Court and District Court for crimes that occur within the city limits. Municipal Courts divide jurisdiction with District Courts and Justices' Courts in a complex way. A Municipal Court is based within a certain city, and funded by that city. For example, Missoula has a Municipal Court. The Judge of a Municipal Court is elected to a 4 year term, and must be a lawyer for at least two years before the election. The decisions of a Municipal Court may be appealed to District Court.

Montana also has additional special purpose courts. One such is the system of Small Claims Courts. Small Claims Courts are civil courts that settle disputes where the amount of money involved is $3000 or less. They do not conduct jury trials – all proceedings are in front of a judge. Every Justices' Court has a Small Claims Court. Other Small Claims Courts may be established by District Courts. If called for by District Court, then they are funded by the state; if not, they are funded by the county. In all cases the judge of the Small Claims Court is appointed by the District Court Judges.

Another special purpose court in Montana is Water Court. Water Courts exist to adjudicate water rights. A Water Court Judge must be a current or retired District Court Judge, is appointed by the State Supreme Court, and serves for 4 years. Water Court is funded by the state.

The final special purpose court recognized in Montana is Worker's Compensation Court. The Workers' Compensation Court was created by the Legislature in 1975 to provide an efficient and effective forum for the resolution of disputes involving injured workers. It is attached to the Department of Labor and Industry (DLI) for administrative purposes. Workers' Compensation Court is the court of origination for most issues arising under the Workers' Compensation Act and some issues arising under the Occupational Disease Act. The Court also acts as an appellate court, conducting judicial review of DLI decisions. Appeals from decisions of the Workers' Compensation Court go directly to the Montana Supreme Court. The governor appoints the Workers' Compensation Judge for a six-year term of office.

5.2.3 International Courts

As was discussed in chapter 4, many Native American reservations are considered sovereign nations for matters that occur there. Just as tribal police exist, there also exist Tribal Courts to adjudicate disputes and crimes where the suspect is a Native American and the crime occurred on a reservation, and the crime is not a federally defined "major crime".

The International Court of Justice (ICJ), also known as the "World Court", is the official court of the United Nations (UN). It was established by the charter of the UN in

1945, and is located in The Hague, The Netherlands. The court hears cases related to legal disputes between members of the UN and gives opinion to member nations and other organizations about international law. The United States withdrew itself from compulsory jurisdiction of the ICJ in 1986 over questions of the impartiality of the ICJ, and now only recognizes the jurisdiction of the ICJ only on a case by case basis. The ICJ is composed of 15 judges elected by the general assembly and the security council of the UN.

The International Criminal Court (ICC) was established in 2002 by an international treaty called the Rome Statute of the International Criminal Court. Its jurisdiction is crimes against humanity, such as genocides and war crimes. There are currently 105 member nations (as of 2007), and a number of other countries are considering joining. Unfortunately, three of the largest countries, The People's Republic of China, India, and the United States, have not made any movement toward joining. Like the ICJ, the seat of the ICC is in The Hague. The proceedings of the ICC may take place in any of the treaty countries, however. The ICC consists of 18 judges, elected by delegates of the countries that are parties to the treaty.

5.3 Questions for Study and Review

1. Explain the differences between felonies, wobbly felonies, misdemeanors, and petty crimes.
2. What degrees of murder and/or manslaughter does your state recognize in law?
3. Explain the difference between assault and battery.
4. What forms and degrees of rape does your state recognize in law?
5. Explain the distinctions between theft, larceny, embezzlement, false pretenses, robbery, extortion, and burglary.
6. What is the difference between forgery and counterfeiting?
7. What forms of alcohol related crimes does your state recognize in law?
8. What forms of drug related crimes does your state recognize in law?
9. In your own words, describe the federal court system. Which federal courts are courts of origination, which are appellate courts, and which are both?
10. Describe how the state court system is organized in your state. Which of your state courts are courts of origination, which are appellate courts, and which are both?
11. What are the two international courts? How do the differ in terms of what types of cases they hear?

5.4 Sources

Administrative Office of the U.S. Courts. n.d. U.S. Courts [Internet]. [cited 2007 Oct 17]. Available from: http://www.uscourts.gov/.

[Anonymous]. n.d. Welcome to the Montana Workers' Compensation Court [Internet]. [cited 2007 Oct 17]. Available from: http://wcc.dli.mt.gov/.

[Anonymous]. n.d. Montana Counties: Justices of the Peace and City Judges [Internet]. [cited 2007 Oct 17]. Available from: http://www.statereporter.com/directory/mtCounty.html.

[Anonymous]. n.d. Brief History of the Montana Judicial Branch [Internet]. [cited 2007 Oct 17]. Available from: http://courts.mt.gov/history.asp/

[Anonymous]. n.d. The Montana Judicial System [Internet]. [cited 2007 Oct 17]. Available from: http://home.mcn.net/~montanabw/judsys.html.

Bulzomi MJ. 2001. Indian Tribal Sovereignty: Criminal Jurisdiction and Procedure [Internet]. FBI Law Enforcement Bulletin 70(6). [cited 2007 Oct 19]. Available from: http://www.fbi.gov/publications/leb/2001/june2001/june01leb.htm#page_25.

Committee on Indian Affairs. 1995. The Tribal Nations of Montana: A Handbook for Legislators [Internet]. [cited 2007 Oct 17]. Available from: http://www.opi.state.mt.us/pdf/TitleI/MTTribal.pdf.

Larson A. 2000. Criminal Charges [Internet]. [cited 2007 Oct 17]. Available from: http://www.expertlaw.com/library/criminal/criminal_charges.html.

LawInfo.com. 2007. Felonies & Misdemeanors. Available from: http://www.lawinfo.com/index.cfm/fuseaction/Client.lawarea/categoryid/137. Viewed September 27, 2007.

Montana Code Annotated 2007 [Internet]. [cited 2007 Oct 17]. Available from: http://data.opi.state.mt.us/bills/mca_toc/index.htm.

New Jersey Judiciary. 2001. Model Criminal Jury Charges [Internet]. [cited 2007 Oct 17]. Available from: http://www.judiciary.state.nj.us/charges/juryindx.htm.

Schubert FA. 2004. Criminal Law: The Basics. Los Angeles: Roxbury Publishing.

Wikipedia contributors. International Court of Justice [Internet]. Wikipedia, The Free Encyclopedia; 2007 Sep 4, 03:07 UTC [cited 2007 Sep 13]. Available from: http://en.wiklpedia.org/w/index.php?title=International_Court_of_Justice&oldid=155551862.

Wikipedia contributors. International Criminal Court [Internet]. Wikipedia, The Free Encyclopedia; 2007 Sep 13, 11:27 UTC [cited 2007 Sep 13]. Available from: http://en.wikipedia.org/w/index.php?title=International_Criminal_Court&oldid=157591307.

William J. Jameson Law Library. n.d. Indian Law and Tribal Law & Government Procedure [Internet]. [cited 2007 Oct 17]. Available from: http://www.umt.edu/LAW/library/Research%20Tools/Tribal%20Law.htm#Montana.

Chapter 6
THE CRIME SCENE

In this chapter we will explore crime scenes and crime scene processing. Although most forensic scientists only occasionally become involved in crime scene processing, this is another area where we can do a better job if we understand this process that is perhaps occurring outside of our field of view. There are many books on crime scene processing, and just as many approaches to it. In this chapter I can only scratch the surface of this enormous topic and present a general overview.

The chapter will begin with a discussion of crime scene investigators. From there it will move on to consider the steps of crime scene processing protocol. It will conclude with discussions of collecting standards and the requirement of crime scene investigators to testify in court.

6.1 Crime Scene Investigators

For the purpose of this book, a crime scene investigator is any person who is investigating the crime scene. As discussed in the previous chapter, this may be a patrol officer, a detective, an evidence technician, a certified CSI, someone from the crime lab, or a team composed of several of these people.

The term "crime scene investigator" evokes images of the several versions of the CSI television show. I find that nearly all my students watch it religiously, yet I, and many other forensic scientists, have never seen a single episode. In fact, most forensic scientists I speak with do not watch any of the TV shows related to forensic science. These shows tend to make us angry, even those with top quality technical advisors. There are several reasons why forensic scientists despise these shows. First, the CSI's on CSI and similar shows do more things than any forensic scientist. Not only are all of them adept at all branches of forensic science, but they also perform tasks that in the real world would be done by peace officers – not forensic scientists. Second, all the analyses of physical evidence are performed immediately and the results are available instantaneously. Of course this is a necessity for all the action to fit into a 60 minute time frame (minus time for commercials). This aspect of the shows is actually having a detrimental impact on the justice system. This impact is known as the "CSI effect" and it refers to the fact that jurors expect things to happen at the same pace as they do on their favorite shows. My apologies if I have gotten any part of the above wrong with respect to your favorite forensic science TV show. I refuse to watch any of them, even for the purpose of accuracy in this textbook. I hope that learning the reality of forensic science doesn't sour you on these programs as it has so many others in the past.

There do exist "real" **crime scene investigators**. These forensic scientists have specialized training in processing crime scenes. Evidence technicians are a form of CSI, though the heavyweight CSI's have earned certifications in crime scene processing from the International Association for Identification, which include 720 hours of training. This training includes a minimum of 80 hours training in latent fingerprint processing, 40 hours in major death investigation, 40 hours in advanced death investigations, 40 hours in photography, 40 hours in blood spatter interpretation and other training courses in arson investigation and forensic pathology. Considering that a typical three credit college class meets for between 40 and 45 hours each semester, the training required for CSI certification is the equivalent of about 18 college classes, or between two and three years of full-time college study.

What do "real" CSI's do? Surprisingly to many people, actually investigating crime scenes is only a relatively small part of the job. Much of the job consists of much less thrilling work, such as packaging evidence, maintaining equipment, preparing detailed reports, meeting with peace officers, sorting through photographs of crime scenes, pursuing continued education, and testifying in court. Would you find a TV show that focused on these realities of CSI work entertaining? Regardless, I heartily recommend pursuing CSI certification if you are interested in that line of work, with proper certification you won't have any problem whatsoever finding a job.

6.2 Layout of a Crime Scene

Let's start with the theory behind crime scene processing. This theory is reflected in how crime scene investigators think of the layout of a crime scene. Based on a combination of Locard's principle and the idea that humans have to move from one location to another by traveling across all of the intermediate distance, the layout of a crime scene is pictured roughly as a collection of concentric circles, like a target.

The "bullseye" of the target is the area where the most serious events happened. This is called the **primary area** of the crime scene. The surrounding area will constitute the **secondary area** of the crime scene, and includes all possible avenues leading into and out of the primary area. The primary area will have the most evidence, but the secondary area will also have some evidence because the perpetrator of the crime had to have moved through the secondary area to get to the primary area. Some authorities, in some cases, may also identify a tertiary area, which can be thought of as an even larger, roughly concentric circle that surrounds the secondary area.

For example, if a person was murdered in the bedroom of their house, that bedroom is the primary area. The primary area will contain the most evidence, including the victim's body, the victim's blood, possibly the suspect's blood, the suspect's fingerprints (unless they wore gloves), firearms evidence if a firearm was used, semen evidence if a sexual assault was involved, toolmarks if the bedroom door was forced open, and perhaps a variety of other evidence items. The rest of the house would be the secondary area, and although it probably doesn't contain as much evidence as the primary area, the perpetrator had to have moved through at least some of the other rooms of the house to get to the bedroom, and depending on exactly what he or she did, may have left evidence. An example of this is illustrated in Figure 6.1.
This is one situation where a tertiary area might be defined, which would consist of the

yard, sidewalk, and street surrounding the house. The perpetrator had to have moved through at least some of these areas as well, and may have left shoe prints in dirt, a cigarette butt on the sidewalk, or tire prints in the street.

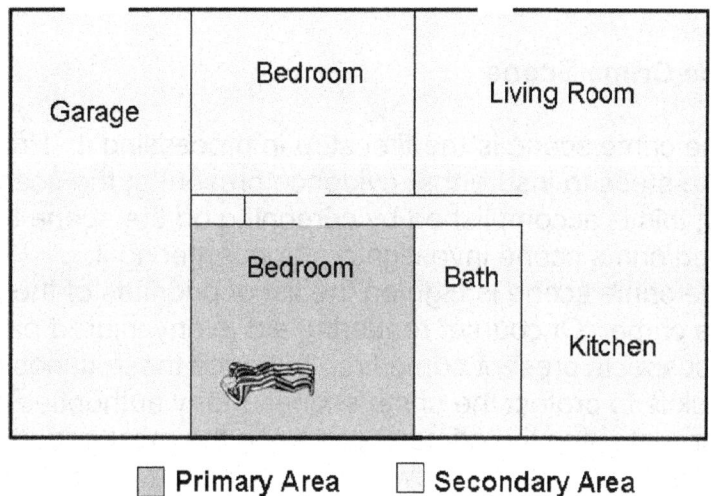

Figure 6.1: Primary and Secondary Areas of a Crime Scene

The most efficient crime scene investigations will utilize the fact that different areas of the crime scene will have different amounts of evidence. Naturally, the crime scene investigator will focus on the primary area, but also attempt to trace the perpetrator's movements through the secondary area as well, and possibly discover additional evidence. If a tertiary area is defined, there is probably little evidence present in it, but it's still worth a quick check.

6.3 Basic Crime Scene Processing Protocol

The task of gathering evidence at a crime scene sounds simple, but in fact crime scene processing is a very complex and intricate ballet of intertwining processes. As such, it is one of the most demanding yet creative tasks in forensic science. Each crime scene is different and will require a different approach to processing. However there is a basic crime scene protocol that should be followed in all crime scenes. These basic functions or tasks are as follows:
1. Protect the scene;
2. Interview witnesses and develop a theory of the case;
3. Preliminary examination to develop an evidence collection plan;
4. Documentation of evidence;
5. Evidence collection.

These five steps need not proceed exactly linearly, and they are in fact often intermingled with each other. However, there are certain things that must be done before others. For example, documentation of the evidence must occur before collection of the evidence, though documentation of some of the evidence may be performed before conducting a preliminary examination. I will briefly outline each of the steps listed above.

6.3.1 Protecting the Crime Scene

Protecting the crime scene is the first step in processing it. Protecting the crime scene means to take steps to insure that evidence present at the scene is not altered or destroyed. Usually, this is accomplished by cordoning off the scene to prevent anybody other than authorized crime scene investigators from entering it.

Protecting the crime scene is high on the list of priorities of the peace officer who first responds to the crime. Of course, rendering aid to any injured parties and apprehending any suspects present come first, but once these things are under control the officer's next task is to protect the crime scene. Many authorities on crime scene processing consider protecting the crime scene to be the most important aspect of the entire process. There is merit to this view, since it relates to maintaining a secure chain of custody, whereby it can be argued that any evidence recovered was not tampered with by anybody.

A crime scene is definitely a place where you don't want to have too many people around. Only those people responsible for the immediate investigation of the crime, the securing of the crime scene, and the processing of the crime scene should be present. Non-essential persons should never be allowed onto a crime scene unless they can add something other than contamination to the investigation. One way to discourage unnecessary people from entering the crime scene is to have only one entrance to the crime scene, which is guarded by an officer. The officer records names and asks each person what their reason for being there is. The officer can also explain that each person entering the crime scene will be required to submit fingerprints, shoe prints, and samples of blood, saliva, pulled head hair, and pulled pubic hair; so these items can be tested in cases of suspected contamination. This is not a bogus requirement, and most legitimate crime scene investigators will have such samples on file at the crime lab.

Protection of the crime scene also includes protection of the crime scene investigators. This is especially true if the suspect has not been apprehended. There are many cases of suspects still hiding at or near the crime scene. There should always be at least two people working the scene, at least one of whom should have a radio and a firearm. Investigator safety needs to be taken seriously. On two occasions, both recoveries of buried bodies, I have felt threatened. On one such occasion the suspect in the case had threatened to kill anyone he found poking around a certain area. This is the only case I have ever been on where I packed my .357 revolver in my backpack (concealing a weapon in this way is legal in Montana), figuring that if shooting started I would hide behind a rock or tree, and whoever came into my view had better be wearing a uniform. This was also the only case on which I did not take any students because I feared for their safety. On another occasion, the law enforcement agency I was working for (Montana DCI, described in chapter 4) took threats the suspects had made so seriously that they provided two SWAT teams to protect me and my students.

6.3.2 Interviewing Witnesses and Developing a "Theory" of the Case

The second step in processing a crime scene is to interview witnesses and develop a theory of the case. This is one situation in which even the most hard nosed forensic scientist would agree that it is necessary to collect testimonial evidence. The statements of witnesses and victims often provide the first idea of what happened.

A **theory of the case** is a hypothesis, based on the testimonial evidence obtained and any physical evidence immediately visible about the nature of the crime that occurred. A complete theory of the case would have a pattern similar to this – somebody did something to somebody somewhere at some time and in some manner – in which the underlined words will be substituted with actual names, places, times, and events. In victimless crimes, the "to somebody" would, naturally, not be a part of the theory of the case.

Note particularly that a theory of a case is a hypothesis. What do scientists do with a hypothesis? They try to show that it is false, and continually revise it to fit new evidence. Therefore, the forensic scientist will continue to process the crime scene, collecting evidence, which is constantly applied to the predictions that arise from the theory of the case, and which often lead to several revisions of the theory of the case. This approach is different from that adopted by many peace officers. We have all seen how peace officers treat a theory of the case on numerous TV shows and movies. They form a hunch, which nearly always turns out to be correct in the end, which they then pursue doggedly until the resolution of the story. The problem with following hunches is that it is tempting to make the evidence fit the hunch, as opposed to the forensic scientists' approach of making the hypothesis fit the evidence. Being too convinced of the accuracy of a theory of the case may also lead to overlooking important evidence. Clearly, the scientific approach is more fair, in that it lets the evidence tell its own story.

Many times the theory of the case is already known when the peace officer or forensic scientist arrives at the crime scene. This is because whoever reported the crime gave their theory of the case, which was relayed on to the responding officer. On most cases where forensic scientists become involved with processing a crime scene, the theory of the case will have been provided to them by the law enforcement agency that summoned them. In these situations it is likely that the theory of the case will be stated as if it is an established fact. Although being provided with the theory of the case removes the initial burden of having to form it from scratch, it does not mean that we should treat it as something more than just a hypothesis.

6.3.3 Preliminary Examination and Forming an Evidence Collection Plan

As is true of all complex tasks, evidence collection yields the best results if it is done according to a plan. Since every crime scene is different, each requires its own evidence collection plan. Sometimes the crime scene is small enough that a plan can be made simply by viewing it. In more complex crime scenes however, a preliminary examination is necessary. A preliminary examination is a slow, careful walk through the crime scene to determine the nature and location of evidence. The important items of evidence may be marked or flagged, and may be photographed, but should not be touched, disturbed, or collected. During this careful walk through the crime scene, the crime scene investigator gets their first chance to ascertain whether the theory of the case is consistent with the readily observed evidence. He or she can also locate points of entry and exit that might have been used by the perpetrator, and can determine the layout of the crime scene in terms of primary and secondary areas. Armed with this information, the crime scene investigator can form a plan for the best and most efficient way to collect the evidence.

The first rule when doing a preliminary examination is to look where you are going to step before you step there. You don't want to step on evidence. Another rule is to not touch anything. To help in remembering not to touch anything, crime scene investigators train themselves to always keep their hands either in their pockets or occupied by holding a notebook, camera, or flashlight. Finally, the crime scene investigators also needs to look up, over their heads, as bullet holes, blood spatter, and other evidence can often be located high on the walls or on the ceiling.

Some important types of evidence are small and difficult to see. Through the years, crime scene investigators have found that the lighting conditions are very important in determining whether evidence can be seen. One good technique to use indoors on hard floors is the oblique lighting technique (also known as side lighting). The room is darkened (after checking the light switch for fingerprints) and a strong flashlight is held about one inch from the floor while its beam is angled so that it just sweeps over the floor surface and is almost parallel to it. The light is then fanned back and forth. Evidence that is difficult to see under other lighting conditions, such as hairs and shoe prints, will usually be easily visible.

6.3.4 Documenting the Evidence

Documenting the evidence is the fourth step in crime scene processing protocol. This usually involves at least photography and sketching, and often videotaping. The general rule is that every bit of evidence that might be important in court needs to be documented. The goal is to record and preserve the nature and locations of items of evidence and the spacial relationships between them as they were originally encountered at the crime scene. This is part of establishing the chain of custody of evidence, and the standard to which crime scene investigators are held is that when they testify in court about how they collected the evidence they must be able to show the judge and jury the original nature and location of items as they existed at the crime scene. This is a high standard.

A video of the crime scene can provide a perspective on the general layout that cannot be as easily perceived in photographs and sketches. Video is a natural viewing medium to which people can readily relate. A plus is that most video recorders also record audio, so the investigator can make comments about the evidence being taped. However, the investigator and bystanders have to be careful about their language while taping is in process, as jurors may be easily shocked by any swearing that is recorded. Some crime scene investigators have told me that they always make video recordings with the audio off because the background noise picked up, such as the investigator's footsteps, are distracting. In any case it seems easy enough to keep the audio off most of the time and switch it on to record any comments that the investigator wants to make. Another advantage is that videotape can be easily copied, so investigators can work with copies and keep the original locked away to prevent damage.

The minimal standard for evidence documentation is photography. Photographs are made at the crime scene to record a pictorial view of what the scene looks like and to record items of possible evidence. Crime scene photographs are generally taken at three distances from the item being documents, a long range shot that shows the item in relation to the crime scene and the other evidence there, a medium range shot that allows the item to be located specifically among the items in the long range shot, and a closeup shot of the item.

Since photographs don't have audio recording capabilities, the investigator needs to keep detailed notes of each photograph taken. In most cases each photograph should include a title card, a direction indicator, and a scale. A **title card** is a piece of paper, index card, or other item upon which is recorded at least the case number, the item number of the evidence item being photographed, and the date. This insures that it is forever clear what the subject of the photograph is. A **direction indicator**, often as simple as a arrow pointing north, allows the photograph to be placed in proper orientation. A **scale** is some device of known size, which is included so that the correct perspective on the size of the evidence item can be determined later. For example, a photograph of blood that lacks a scale is difficult to interpret because it may not be obvious whether the blood is a single drop or a large pool. Clever investigators will use title cards of a known size and write notes about direction and location on the card so that the same item can serve simultaneously as a title card, a direction indicator, and a scale. This is illustrated in Figure 6.2.

For some types of evidence, such as fingerprints or shoe prints, the photograph is the means of collecting the evidence as well as documenting it. For this task a special photograph is taken that prints up at the actual size of the evidence item.

For all but the most simple crime scenes, sketching is also necessary. It is a property of the optics of cameras that they distort spacial relationships, and at best they are two dimensional representations of three dimensional objects. Anybody who has taken snapshots has probably experienced a picture not coming out exactly the way it looked through the viewfinder. The solution of this problem is to sketch the locations of the important items of evidence, giving measurements from these items to standard landmarks and to other evidence items. Nowadays, there are sophisticated mapping instruments available, such as laser transits, which can make this task quick and easy. Although forensic anthropologists routinely use these tools, their high cost is making them slow to be adopted by mainstream crime scene investigators.

A sketch can be made as if you are looking down from above. This is called an

Figure 6.2: A Title Card Showing the Exhibit Number, Direction, and Investigator's Name. Note that the card measures 4 X 6 inches.

overhead sketch. A sketch can also be made as if you are looking straight ahead at a wall, which is called an **elevation sketch**. Usually, the crime scene investigator will make a rough sketch, which will be redone more aesthetically later (perhaps by a forensic artist) if it is to be shown in court.

6.3.5 Collecting and Packaging the Evidence

After the evidence is documented, it can be collected. It is important to follow a plan for collecting the evidence in a logical and efficient manner. Normally this plan, or at least a preliminary version of it, will have been developed during the preliminary examination. The evidence collection process will usually start with the collection of the most fragile, or most easily lost evidence. In many cases the most fragile evidence is blood, which must be refrigerated within about two hours or bacteria will begin to grow in it, making its analysis more difficult. Investigators often find that some evidence is hidden underneath some other evidence. When this occurs, the evidence on top is collected first, then the previously hidden evidence is documented and finally collected.

Most items of evidence will be collected in paper containers such as packets, envelopes, and bags. Liquid items can be transported in non-breakable, leakproof containers. The rules for what types of items are packaged in what types of containers are very complex and beyond the scope of this class. Any items which may cross contaminate each other need to be packaged separately, such as blood from two separate pools. Information should be recorded on the container including the case number, the evidence item number, who collected the item, the time and date, and the nature of the evidence.

6.3.6 Interpreting the Crime Scene

Now that the evidence has been collected, packaged, and sent to the crime lab, it is time to spend a few moments reflecting on the evidence recovered and how it supports or falsifies the theory of the case. This process is known as interpreting the crime scene, and the best time to do it is while the investigator's memory of the evidence is fresh. Does the nature and location of the evidence support the hypothesis of what happened during this crime, or does it argue that some other series of events was more likely to have occurred?

6.4 Collecting Standards from Suspects and Victims

If the crime scene was at all complex, it is likely that processing it was a lengthy and tedious process. However, there is still more evidence collection to be done. At some point in the investigation standards items will usually have to be collected from a suspect and/or a victim. Once the crime scene has been processed it should be fairly clear what standard items need to be collected. For example, if hair is recovered at a crime scene, representative hairs for comparison will have to be obtained from the suspect and the victim.

6.5 Crime Scene Investigators and Courtroom Testimony

The crime scene investigator is usually expected to testify in court about the crime scene and any of the procedures they used to find, document, and collect the evidence. The chain of custody of the evidence is important because when the evidence gets to court the prosecution needs to be able to prove that the evidence was not tampered with. The procedures used will be thoroughly examined and challenged by a defense attorney, and many cases have been lost because of irregularities in the processing of the crime scene.

6.6 Questions for Review and Study

1. Discuss the various ideas concerning what a crime scene investigator (CSI) is and does.
2. In your own words, describe the difference between primary and secondary areas of a crime scene.
3. List the five steps of crime scene processing in the order in which they should ideally occur.
4. In your own words, describe what might be done to protect a crime scene.
5. What steps should be taken for the protection of crime scene investigators?
6. In your own words, describe what a "theory of the case" is, how it is developed, how it might be modified, and how it is used during crime scene investigation.
7. In your own words, describe the goals and process of a preliminary investigation of a crime scene.
8. What are some precautions that crime scene investigators will take to prevent damage to evidence during a preliminary examination of a crime scene?
9. Compare and contrast the three ways of documenting a crime scene in terms of how they are done and the type of information they can document. Which of these methods are always used?
10. All things being equal, what types of forensic evidence are collected first and what types are collected later?
11. In your own words, define crime scene interpretation.
12. How are standard items collected to be used for matching to the evidence items recovered from a crime scene.

6.7 Sources

Baldwin HB. 1998. Crime Scene Check List [Internet]. [cited 2007 Oct 17]. Available from: http://www.feinc.net/cklist.htm.

Baldwin HB. 1998. Crime Scene Interpretation [Internet]. [cited 2007 Oct 17]. Available from: http://www.feinc.net/cs-int.htm.

Baldwin HB. 1998. The Crime Scene Investigator [Internet]. [cited 2007 Oct 17]. Available from: http://www.feinc.net/csi.htm.

Baldwin HB. 1998. The Crime Scene Investigator (Job Description) [Internet]. [cited 2007 Oct 17]. Available from: http://www.feinc.net/csi-desc.htm.

Baldwin HB. 1998, Crime Scene Processing Protocol [Internet]. [cited 2007 Oct 17]. Available from: http://www.feinc.net/cs-proc.htm.

Behavioral Science Investigative Support Unit, FBI Academy. n.d. Crime Scene Photography Requirements of Criminal Investigative Analysis [Internet]. [cited 2007 Oct 17]. Available from: http://www.geocities.com/cfpdlab/csphoto.html.

Byrd M. n.d. Basic Concepts: Simple EZ Sketching by using all available resources [Internet]. [cited 2007 Oct 17]. Available from: http://www.crime-scene-investigator.net/sketch.html.

Byrd M. n.d. Hazards and a Crime Scene [Internet]. [cited 2007 Oct 23]. Available from: http://www.crime-scene-investigator.net/hazards.html.

Crawford KA. 1999. Crime Scene Searches [Internet]. FBI Law Enforcement Bulletin 68(1). [cited 2007 Oct 19]. Available from: http://www.fbi.gov/publications/leb/1999/jan99leb.pdf.

Dalrymple B, Shaw L, Woods K. 2007. Optimized Digital Recording of Crime Scene Impressions [Internet]. [cited 2007 Oct 23]. Available from: http://www.crime-scene-investigator.net/DigitalRecording.html.

Difonzo JH, Stern RC. 2007. Devil in a White Coat: the Temptation of Forensic Evidence in the Age of CSI. New England Law Review 41(3): 503-532.

Garrett RJ. 2003. A Primer on the Tools of Crime Scene Analysis [Internet]. Journal of Forensic Identification 53(6). [cited 2007 Oct 23]. Available from: http://www.crime-scene-investigator.net/ToolsofCrimeSceneAnalysis.html.

Ruslander HW. 2001. Searching and Examining a Major Case Crime Scene [Internet]. [cited 2007 Oct 23]. Available from: http://www.crime-scene-investigator.net/searchingandexamining.html.

Schiro G. n.d. Protecting the Crime Scene [Internet]. [cited 2007 Oct 17]. Available from: http://www.crime-scene-investigator.net/evidenc1.html.

Schiro G. n.d. Examination and Documentation of the Crime Scene [Internet]. [cited 2007 Oct 17]. Available from: http://www.crime-scene-investigator.net/evidenc2.html.

Schiro G. n.d. Collection and Preservation of Evidence [Internet]. [cited 2007 Oct 17]. Available from: http://www.crime-scene-investigator.net/evidenc3.html.

Schiro G. n.d. Special Considerations for Sexual Assault Evidence [Internet]. [cited 2007 Oct 17]. Available from: http://www.crime-scene-investigator.net/evidenc4.html.

Staggs S. n.d. Forensic Photography for the Crime Scene Technician [Internet]. [cited 2007 Oct 17]. Available from: http://www.crime-scene-investigator.net/fet-ol.html.

Staggs S. 1997. Using Video to Record the Crime Scene [Internet]. [cited 2007 Oct 23]. Available from: http://www.crime-scene-investigator.net/video.html.

Thomas PW. 2001. Video Guide Lines for Evidence Scenes [Internet]. [cited 2007 Oct 17]. Available from: http://www.crime-scene-investigator.net/videoguidelines.html.

Chapter 7
FORENSIC PHOTOGRAPHY AND FORENSIC ART

In this chapter we will consider two forensic sciences – forensic photography and forensic art. As discussed in chapter 4, these are two forensic sciences for which those who practice them are most often employed at a law enforcement agency. Forensic photography and forensic art are closely connected to crime scene processing, which is why I have chose to place the chapter pertaining to them here. Both fields, however, are more extensive than just crime scene related activities.

Forensic photography and forensic art share many similarities, in that many things that can be illustrated by sketching or drawing can also be illustrated by photography, and vice versa. The main difference in the way these forensic disciplines are used is that forensic art is primarily used to illustrate unknown people, items, and events, including hypotheses about how a crime was committed. In contrast, forensic photography focuses on documenting things that exist and are in the possession of the law enforcement community. Therefore, forensic photography is used to document the nature and condition of evidence items such as fingerprints, impressions, blood spatter, and other items found at the crime scene or being processed in the crime lab. A forensic artist may be called upon to generate a composite drawing of a suspect not in custody, whereas a forensic photographer will document what a suspect looks like once they have been arrested.

7.1 Photography and Art in Crime Scene Documentation

From chapter 6 it should be fairly clear that forensic photographers and artists often play a role in crime scene processing. As with most things having to do with crime scenes, the necessary photography and sketching is most often accomplished by a peace officer. However, in complex crime scenes the forensic photographer and artist will be asked to contribute their expertise.

Each crime scene requires a slightly different procedure. Certain types of crime scenes, such as domestic violence and sexual assault scenes, require special procedures. For example, it is common for a victim of domestic violence to later deny that the incident occurred because they are dependent upon their abuser for basic support, such as food and shelter, or are intimidated by their abuser into recanting their testimony. In these cases, successful prosecution relies on photographic documentation of the victim's injuries, especially if it can be done in such a way that they are linked to the abuser or can at least be shown to be of human origin. If the shape of a certain bruise can be shown to be in the shape of a human hand, then the abuser can't believably claim that the victim simply fell down.

Crime scene sketches are usually drawn in overhead perspective, as if looking down from above. The physical details of the crime scene, such as walls, furniture, doorways, etc., are sketched in, along with the locations of the various items of evidence. Standard technique includes measuring the distances from the items of evidence to permanent landmarks within the crime scene and using this information to plot the location of the evidence on the sketch. Using a sketch of this sort, investigators and forensic scientists can make and test hypotheses about the relationships between the items of evidence using what archaeologists call the law of association. The **law of association** says that the closer together two items are the more likely it is that there is some sort of relationship between the items. Though not formally given a name by non-archaeologists, this logical principle is as fundamental in investigating a crime scene as it is in the practice of archaeology. For example, if there is a screwdriver lying next to a windowsill upon which are scratches, it is likely that this was the screwdriver used in the activity that resulted in scratching the windowsill.

Each separate type of evidence has a detailed set of methodologies for most effectively capturing it by photography or sketching. In the case of photography, the differences may include preferred type of film, pre-treatment of the evidence to be photographed, different methods of lighting, or the use of alternative forms of light source such as ultraviolet light. In the case of sketching, deciding whether to use an overhead view or a plan view, whether to sketch in pencil or some other medium, and the level of detail required in the sketch are concerns.

7.2 Forensic Photography

Forensic photography is the application of the skill of photography to investigations. The main features of this forensic science will be discussed below. On one level it is simple to say that photographers take pictures. However, the things they take pictures of are many and varied.

7.2.1 The History of Forensic Photography

The first use of forensic photography dates to the 1840's in Belgium. Most of the early use of photography in the justice system was to photograph suspects and inmates for purposes of identification. By the late 1800's this practice was common and law enforcement agencies began to experience the problem of having so many photographs that it was difficult to organize them in such a way that a useful collection of them could be retrieved quickly. One common solution was to organize photos of criminals by type of crime, so that a victim of, for example, a robbery, could look at the photos of known robbers to see whether they recognized the person who had robbed them.

Alphonse Bertillon, discussed in chapter 1, was a forensic photographer, and one of the first to use photography to document a crime scene. He developed his system of identification based on body measurements because he became convinced that photographs were not reliable for identification unless they were taken under strictly specified conditions of lighting and angle.

By the early 1900's forensic photography had emerged much as we know it today. Its use in crime scene documentation and its other uses became routine. Much of the advancement in forensic photography since has been due to changes in the technology of cameras.

7.2.2 The Practice of Forensic Photography

Forensic photography is practiced in a variety of locations. Forensic photographers may work for law enforcement agencies, a crime lab, or for a private company. Their tasks are varied and some degree of specialization exists within their ranks.

Certification is important for forensic photographers and is the criteria by which professional forensic photographers are distinguished from amateurs. There are two important boards that oversee certification of forensic photographers. One is the International Association for Identification's (IAI) Forensic Photography Certification Board. IAI certification requires forty hours of training in photography and imaging techniques, three years of supervised experience as a forensic photographer, and a set of tests. The Evidence Photographers International Council (EPIC) also offers certification of forensic photographers. EPIC certification is a process that takes up to three years. First, a candidate for certification must have been a member of EPIC for at least a year and have taken at least one of its schools, which consists of three days of presentations and workshops. In addition there is a battery of tests that may take up to two years to complete.

7.2.3 Tasks of Forensic Photographers

Those forensic photographers who work for a law enforcement agency are often called **police photographers**. Usually, their job consists of photographically documenting evidence at crime scenes, preparing photographic exhibits of evidence for court, and in some jurisdictions, photographing suspects. **Crime lab photographers**, like their police photographer colleagues, photographically document crime scenes and prepare exhibits for court. In addition, they are often called upon to work with evidence that consists of photographs or other likenesses. They may work with forensic artists to do age progression of suspects whose photographs are out of date. Crime lab photographers also work closely with the other forensic scientists at the crime lab to photographically document steps in the analysis of the evidence. Many private forensic consulting companies offer forensic photography services. Photographers who work for such companies are sometimes asked to assist with documentation of criminal evidence, but are more often hired by one side or another in a civil case.

7.2.4 Challenges: Why a Forensic Photographer Can Usually Take Better Pictures Than a Cop

Although almost anybody can use a "point and shoot" camera to take photos that are acceptable for everyday use, photography involves a large and exacting set of skills, which are only mastered by a specialist who has had considerable training and experience. This is why certification of forensic photographers is such a seemingly long and rigorous process. Photography seeks to capture light that is reflected from the subject and captured by the optics of the camera. This reflected light is then captured digitally or on film. Thus, all details involved with the light, including the nature, location, and intensity of its source; the angles between the light source, the subject, and the camera; and the amount of light gathered by the camera are critical if the details of the subject are to be captured as clearly as possible. Also, the time period over which light is gathered is important. The shorter the time interval during which light is allowed to enter the camera, the less likely that movement of the subject will cause distortion of the image. Certain films and digital imaging devices require longer periods of light gathering.

Although much of the information required to take high quality photographs can be learned in classes, the actual details of how all these factors work together, particularly with respect to a certain camera, have to be worked out through experience.

Peace officers have a large number of areas of expertise with which they are expected to be familiar, and the basics of photography is one of them. However, for the best results, a forensic photographer with their greater knowledge of the specific details of photography and cameras can generate higher quality photograph. The higher quality of photographs can easily translate into doing a better job of convincing a judge or jury of the truth of the information obtained in the photograph.

7.2.5 Forensic Photography and Digital Image Analysis

Increasingly, forensic photographers are embracing the techniques of digital image analysis and enhancement. This is natural, because a photograph is an image, and if it originally exists on film rather than in digital form it is usually simple to convert it to a digital image. Once in digital form a variety of manipulations can be made to the image to enhance the viewer's ability to extract information from it. Most people I talk with who do this use Adobe Photoshop software, which is certainly a leader in the field. Using Photoshop or similar software, portions of the image can be magnified, arrows and other markers can be added to direct the viewer's eye to a certain item in the image, and various filters and color changes can be applied to bring out details. It is important to be sure that this manipulation of the image only clarifies existing details and does not actually add information that was not originally present in the photograph.

Training in digital image analysis is now a required part of the training that forensic photographers must undergo to obtain certification. At this point in time there still exists a cadre of forensic digital image analysts who are not photographers, or at least are not primarily photographers. It is unclear whether this forensic specialty will be subsumed under forensic photography in the future.

7.3 Forensic Art

Forensic art also has applications beyond the crime scene. Forensic artists consider their discipline to include crime scene sketching and illustration, composite drawing, and facial reconstruction. We have already discussed crime scene sketching in chapter 6.

There is a certification process for forensic artists, administered by the Art Certification Board of the International Association for Identification. The certification focuses on composite drawing and the lowest level of certification requires 40 hours of training in composite drawing plus one year of experience. Higher levels of certification require additional training and experience.

7.3.1 The History of Forensic Art

The use of art in solving crimes must date back nearly to the times that human beings first learned to draw, which we know to be more than 30,000 years. The use of an artist to portray the likeness of a suspect, which we now call composite drawing, first appeared systematically in the 1800's, and Bertillon, whose name you should now recognize, was instrumental in its beginnings. Crime scene sketching is even older, and recorded instances date back to the ancient Greeks.

7.3.2 Crime Scene Illustration

As previously discussed, crime scene sketching is most commonly done by hand, at the crime scene, by whoever is processing it. Although an adequate sketch can be made by most people who have had a little training, the skills possessed by a technical artist can result in better and more easily interpretable sketches. Crime scene sketches made by a non-artist are rarely pretty, and they often have small discrepancies due to the fact that they are drawn quickly with more attention to recording information than to being neat. Often, especially if a sketch is intended to be entered as evidence in court, a forensic artist, working in the comfort of their workshop or studio, will clean up the crime scene sketch and combine it with information from photographs and other data to produce a more visually appealing diagram, called a **crime scene illustration**. A crime scene illustration will often be used in court as a visual aid for testifying investigators to refer to as they present their evidence and theory of the crime. Figure 7.1 shows a crime scene illustration and the rough sketch upon which it is based.

Figure 7.1: A Crime Scene Illustration. The left panel shows a rough sketch made by an investigator at the crime scene and the right panel shows the sketch professionally redone, including flipping the directions so that north is to the top as is conventional.

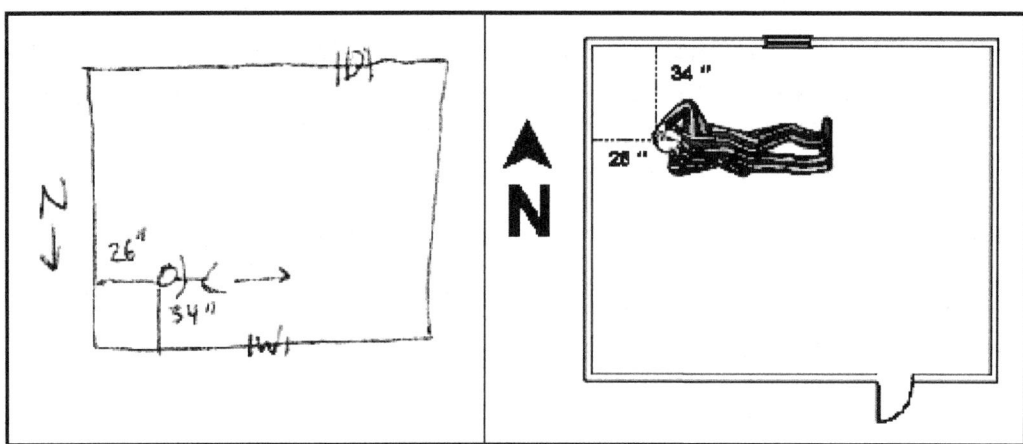

Crime scene illustrations are an accepted aid in presenting a case in court. What is not generally accepted, however, is a **crime scene reconstruction**. A reconstruction is a visual recreation of the theory of the crime, either using a series of still frames or a computer-generated animation. Crime scene reconstructions do an excellent job of demonstrating the theory of the crime – perhaps too good. The practice of using crime scene reconstructions has been criticized by some experts in jurisprudence as being so convincing that a jury may think of it as being a factual account rather than an artistic interpretation of a hypothesis about what might have happened. Therefore, crime scene reconstructions are rarely allowed in court.

As is true for forensic photography, forensic art has become highly computerized. In recent times, quite a lot of the work of producing crime scene illustrations and reconstructions is done using computerized methods. A wide spectrum of drawing software exists. These programs facilitate drawing shapes and lines, and usually allow for scanned photographs to be imported and integrated into the illustration being produced. Using these techniques allows better illustrations to be produced in less time.

7.3.3 Composite Drawing

Composite drawing is perhaps the task for which forensic artists are best known. A composite drawing consists of the face or body of a suspect, composed from witness or victim descriptions. Composite drawings were traditionally made by hand, but in recent times there are tools and software packages available to help the forensic artist.

One such is the Identikit, which consists of standard face outlines, hair styles, noses, mouths, eyes, etc. which can be combined like the parts of a Mr. Potato Head to produce a face. A composite sketch usually takes one and a half to three hours to complete by hand, but a fraction of that time using software.

A completed composite drawing is considered an evidence item, to which a suspects appearance can be matched. A composite drawing can be circulated to law enforcement agencies, published in a newspaper, or posted as a wanted poster, in the hope that someone will be able to identify the suspect. It is often accompanied by a description of the suspect that includes details of the appearance of the suspect, such as sex, age, race, height, weight, hair color, etc. Occasionally, a composite drawing of the face of a victim is made for inclusion in a missing person poster. Suspects have been located in tens of thousands of cases using composite drawings.

7.3.4 Age Progression and Updating

Another set of tasks carried out by a forensic artist is age progression and updating. If a significant amount of time has passed since a composite drawing or photograph has been made, it is likely that the subject's appearance has changed. Age progression consists of changing the details of the drawing or photograph in such a way

Figure 7.2: Age Progression of a Suspect

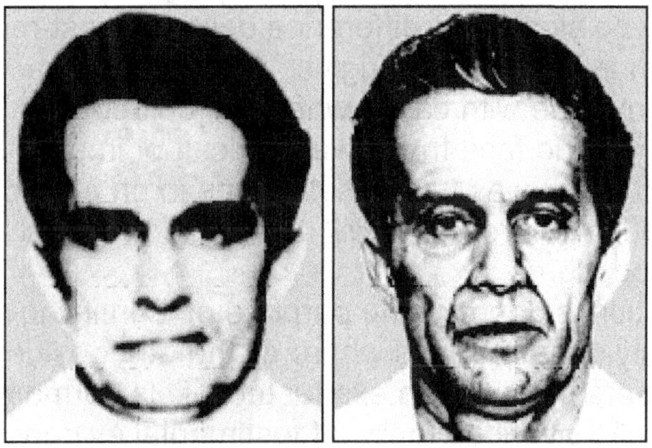

Photograph taken in 1979   Age Enhanced Photograph

as to be consistent with the current age of the suspect or missing person. This is especially important in cases that involve missing children, since children's faces often change dramatically as they mature. The likenesses of adults need age progression less often, but it may still be necessary after the passage of several years. Figure 7.2 shows an age progression of a wanted person, posted on the FBI website.

Updating of a composite drawing or photograph is similar, and may involve making age-related changes. However, it is more often done in cases where the police have received information that the fugitive has changed their appearance by adopting a new hair style, growing or shaving facial hair, or wearing glasses. These updates can help make sure that the image reflects the current appearance of the subject.

7.3.5 Facial and Postmortem Reconstruction

Post-mortem reconstruction and facial reconstruction are methods for reconstructing the face of a deceased person whose actual face has been damaged or disfigured beyond recognition, perhaps due to decay or burning of the body. Post-mortem reconstruction was developed by and traditionally considered part of the expertise of forensic pathologists. Similarly, facial reconstruction was developed by forensic anthropologists who continue to practice this technique in many places. Forensic art has claimed these procedures as its own, however, and I believe that this is appropriate because the the result is better reconstructions. Anybody can learn the science involved in doing the reconstructions, but art involves a talent not available to everybody. On the couple of occasions in which I have attempted a facial reconstruction, the result scarcely looked human. One of my colleagues, Garry Kerr, has done many of them and still teaches our forensic anthropology students how to do it, but the result sometimes looks more like Garry than like the photograph of the person taken in life! Clearly, the degree of artistic training and skill required to produce a successful likeness of the face is beyond the capabilities of most forensic pathologists and anthropologists.

There is probably no significant difference between post-mortem and facial reconstruction other than their historical origins. Post-mortem reconstruction was traditionally done in the morgue with cases where an unidentified body has sufficient trauma or decomposition of the face that it was difficult or impossible to recognize facial features. Facial reconstruction was traditionally done in an anthropology lab in cases where only a skull remained. Hereafter, I will use the term "facial reconstruction" to refer to both of these procedures.

Facial reconstruction is done for the purpose of identifying a deceased person. Further, it is normally only done in cases where all the other evidence that might lead to the identification of the person has been exhausted. This normally includes fingerprints and DNA analysis, as well as various forms of testimonial evidence. In this situation, the best that can be done is to generate an approximate likeness of the person and circulate it to law enforcement and the media in the hope that someone will recognize the likeness.

Facial reconstructions can be two dimensional or three dimensional. Both techniques involve using tissue depth markers, which are approximations of the amount of tissue (flesh and skin) overlying the skull at certain standard locations. The tissue depth varies over the surface of the skull - for example it is usually thin on the forehead and thicker in the regions that lie on either side of the nose between the eyes and the corners of the mouth. Tissue depth varies between people of different ages, sexes, races and nutritional statuses, and standard tables of tissue depths have been compiled for a reasonable variety of people. The tissue depth markers are usually made of pencil eraser material cut to appropriate length.

A two dimensional reconstruction is done using a radiograph (x-ray picture) of the skull or a tracing of the skull from a radiograph. Tissue depth markers are attached to the radiograph and photographed. A drawing is made from the resulting photograph using the tissue depth markers as guides to outline the face and fill in the features. Two dimensional reconstruction is relatively quick to do compared to three dimensional reconstruction, but probably less accurate.

A three dimensional reconstruction is made using the skull, or preferably a plaster cast of the skull, to which tissue depth markers are attached. The areas between the tissue depth markers are then filled in using modeling clay. The facial features are then added using a mixture of scientific knowledge about such things as the relationship between the length of the anterior nasal spine and the length of the nose, and artistic skill. A wig completes the reconstruction, and sometimes other accouterments are added, such as eyeglasses.

With either the two or three dimensional method, a photograph is made of the resulting reconstruction for circulation to law enforcement and the media. The success of facial reconstruction, both in terms of generating leads about the person's identity and in terms of being able to match the reconstruction with a photograph of the person in life, is remarkably low. When a facial reconstruction is compared with a "lineup" of several photographs of people of similar age, sex, and race, people are able to match the reconstruction with the photograph between about 25% and about 70% of the time. This amounts to between about two and six times the number of correct matches expected from chance alone, so there is merit to the method. However, with such relatively low percentages, one wonders whether facial reconstruction offers any improvement over simply publishing the age, sex, and race of the deceased person. In relatively few cases, has circulating a photograph of a facial reconstruction led to the identification of the deceased person, but it needs to be kept in mind that facial reconstruction is used only in the most difficult cases, when all other leads have not produced results.

7.4 Questions for Study and Review

1. List the tasks that a forensic photographer might perform.
2. Since almost anyone can use a modern camera, why is it still important to have trained forensic photographers?
3. How do forensic photographers use digital image analysis?
4. List the tasks that a forensic artist might perform.
5. What is the difference between crime scene illustration and crime scene reconstruction?
6. Why would a composite drawing be made of a suspect, given that cameras are so plentiful?

7. In your own words, describe the difference between age progression and updating of images of suspects.
8. Describe the process of facial reconstruction.
9. What is the difference between a two-dimensional and a three-dimensional facial reconstruction?

7.5 Sources

[Anonymous]. n.d. Forensic Art [Internet]. [cited 2007 Aug 10]. Available from: http://www.forensicartist.com/.

[Anonymous]. n.d. History of Forensic Art [Internet]. [cited 2007 Aug 10]. Available from: http://www.forensicartist.com/history/index.htm.

Baldwin H. n.d. Photographic Techniques for the Laser or Alternate Light Source [Internet]. [cited 2007 Oct 17]. Available from: http://www.feinc.net/als-phot.htm.

Baldwin HB. n.d. Forensic Computer Composites [Internet]. [cited 2007 Oct 23]. Available from: http://www.feinc.net/composit.htm.

Brown RE, Kelliher TP, Tu PH, Turner WD, Taister MA, Miller KWP. 2004. A Survey of Tissue-Depth Landmarks for Facial Approximation [Internet]. Forensic Science Communications 6(1). [cited 2007 Oct 21]. Available from: http://www.fbi.gov/hq/lab/fsc/backissu/jan2004/research/2004_01_research02.htm.

Evidence Photographers International Council, 2000. Certification [Internet]. [cited 2007 Oct 17]. Available from: http://www.evidencephotographers.com/EPICcertification.html.

Evison MP. 2001. Modeling Age, Obesity, and Ethnicity in a Computerized 3-D Facial Reconstruction [Internet]. Forensic Science Communications 3(2). [cited 2007 Oct 21]. Available from: http://www.fbi.gov/hq/lab/fsc/backissu/april2001/evison.htm.

Grover CC. 1980. Selection and application guide to police photographic equipment. Washington (DC): U.S. Dept. of Commerce.

Hart R. 1998. Evidence Photography. Popular Photography 62(4): 44-50.

International Association for Identification. n.d. Forensic Artist Certification Process [Internet]. [cited 2007 Oct 17]. Available from: http://www.theiai.org/certifications/artist/index.php.

International Association for Identification. n.d. Forensic Photography Certification [Internet]. [cited 2007 Oct 17]. Available from: http://www.theiai.org/certifications/imaging/index.php.

LeMay J. 2003. Using scales in photography (1): there are many things a crime scene technician needs to know. The Forensic Scientist 1(1): 87-90.

Miyasaka S, Mineo Y, Kazuhiko I, Sueshige S. 1995. The computer-aided facial reconstruction system. Forensic Science International 74(1-2): 155-165.

Moenssens AA. 2004. The Origin of Legal Photography [Internet]. [cited 2007 Nov 1]. Available from: http://forensic-evidence.com/site/EVID/LegalPhotog.html.

Pex JO. n.d. Domestic Violence Photography [Internet]. [cited 2007 Oct 17]. Available from: http://www.crime-scene-investigator.net/dv-photo.html.

Phillips SS, Haworth-Booth M, Squires, C. 1997. Police pictures : the photograph as evidence. San Francisco : San Francisco Museum of Modern Art : Chronicle Books.

Staggs S. n.d. Forensic Photography for the Crime Scene Technician [Internet]. [cited 2007 Oct 17]. Available from: http://www.crime-scene-investigator.net/fet-ol.html.

Taister MA, Holliday SD, Borrman HIM. 2000. Comments on Facial Aging in Law Enforcement Investigation [Internet]. Forensic Science Communications 2(2). [cited 2007 Oct 21]. Available from: http://www.fbi.gov/hq/lab/fsc/backissu/april2000/taister.htm.

Taylor K. 2001. Forensic Art and Illustration. Boca Raton, FL: CRC Press.

Wikipedia contributors. 2007. History of forensic photography [Internet]. Wikipedia, The Free Encyclopedia; 2007 Apr 1, 22:21 UTC [cited 2007 Aug 10]. Available from: http://en.wikipedia.org/w/index.php?title=History_of_forensic_photography&oldid=119584417.

Wikipedia contributors. 2007. Forensic photography [Internet]. Wikipedia, The Free Encyclopedia; 2007 Oct 8, 12:34 UTC [cited 2007 Oct 17]. Available from: http://en.wikipedia.org/w/index.php?title=Forensic_photography&oldid=163075808.

Wikipedia contributors. 2007. Forensic facial reconstruction [Internet]. Wikipedia, The Free Encyclopedia; 2007 Oct 5, 06:37 UTC [cited 2007 Oct 17]. Available from: http://en.wikipedia.org/w/index.php?title=Forensic_facial_reconstruction&oldid=162410364.

Zimmerman MR. 1992. PCs help U.S. agency find missing children; system allows artists to 'age' photos. PC Week 9(1): 19 21.

7.6 Acknowledgements

The wanted poster from which Figure 17.2 was created was posted on the FBI's website, http://www.fbi.gov/wanted/fugitives/vc/murders/webb_de.htm. I accessed it on November 1, 2007.

Chapter 8
Basic Science
THE COMPARATIVE METHOD

In this chapter we will look at record keeping and the use of the comparative method. Both of these are topics basic to the forensic sciences, especially to criminalistics. The chapter will begin with a consideration of record keeping in science and in the process of an investigation, including the analyses of forensic scientists. It will continue with an examination of the comparative method, and conclude with a look at some terminology issues.

8.1 Record Keeping

Scientists never rely on their memory. As we will explore further in the chapter on lie detection and hypnosis, memory is a tricky business. Memories fade and become confused with other events and even dreams. Therefore, scientists always keep detailed records. In this section we will explore the very important process of record keeping. If you observe scientists at work you will find a great deal of attention is paid to keeping records of their experiments and analyses. The requirement to keep detailed records is universal in science from the lab of the chemist to the excavation pit of the archaeologist. Nowhere is keeping good records as important as in forensic science, where concerns over chain of custody and handling of evidence demand a detailed accounting of everything that was done with an item of evidence. When forensic scientists testify in court as expert witnesses, they are expected to rely upon their written notes in making their testimony and not to rely upon their memory.

As a student in several chemistry, biology, and anthropology lab classes, I absolutely loathed taking notes. The note-taking process seemed to me to be tedious and completely unnecessary. "Who cares whether the solution turns blue?", I thought. Having gained some experience in the decades since then I now realize that knowing the color of the solution helped the person grading my laboratory work to draw a conclusion about whether the correct proportions of reagents had been added to the solution. Lacking that information, the necessary conclusion could not be drawn, and this fact was represented in the score I received for that laboratory session and the class overall. Now that I have matured quite a bit, I see not only the desirability of keeping good notes, but its necessity. In my late teens and twenties my memory was sharp enough to get me through writing up my lab report without too many significant errors, but now that I am in my 50's, such is not the case. Even as a young man I misremembered some details, and occasionally the score on my lab report would reflect this. Today it would be impossible for me to write a forensic anthropology report on a case without referring to the notes I made during my inspection of the evidence. Further, the only outcome that rested on my memory as a student was my grade.

However, as a forensic scientist, much more than a grade depends on the accuracy of my analyses, reports, and testimony. A person's lifestyle, or even their very life, may be riding on my testimony, so I better get the details exactly, precisely right.

Here's a grim scenario. What if, just after performing an analysis, a forensic scientist were to suddenly become incapacitated? Will the case then wait for the forensic scientist to recover, or will the case be dismissed? What if the forensic scientist had died? These things do happen, and cases must continue regardless. Therefore, it is important that notes be kept to a sufficiently high standard of detail that someone else could take over that part of the case using the original forensic scientist's notes.

8.1.1 Keeping Records

The process of developing detailed records for a case begins with the patrol officer who responds to a crime scene. The officer will carry a notebook into which she or he will write many things. One type of information an officer will be careful to record is the significant times associated with their response to the report of the crime, including the time they were dispatched to the scene, the time they left for the scene, the time they arrived at the scene and the time they left the scene. The officer will also record notes about any testimonial evidence taken from witnesses, victim, and suspects. These notes will consist of names, times at which things happened, and the relevant details of the testimonies of persons interviewed. The officer will also record details about the victims, suspects, and witnesses, such as whether they are injured, what they look like, their mental state, whether they appear intoxicated, and their attitude. Other things the officer may note include whether the doors and windows are open or closed, whether the lights are on or off, and whether any form of short-lived evidence is present (such as the odor of cigarette smoke). If the evidence must be moved, disturbed, or altered in any way by emergency medical personnel, fire fighters, the victim, witnesses, or by the officer themself, then the officer will record this information as well. The notes taken by the officer will be kept in case he or she needs to testify, and will form the body of information the officer will include in their report about their response to the crime.

This process of record keeping continues with detectives and crime scene investigators making detailed notes of what they did and what they discovered. As discussed in chapter 6, documenting the crime scene is a type of record keeping that forms an essential part of a criminal investigation. The video recordings, photographs, and sketches made by the crime scene investigators, along with their notes and lists of evidence, create the foundation for interpreting the results of forensic analyses.

When physical evidence arrives at a crime lab, an exacting process of recording everything about the evidence and what is done with it is begun, and continues throughout the analyses performed by the forensic scientists who work with the evidence.

8.1.2 Using Records

The most important use of records is in interpreting the evidence. While it might be possible to draw some conclusions about the crime without documentation, the conclusions drawn are usually relatively weak. For example, finding a suspect's fingerprints anywhere at the scene is sufficient to establish that they were there, but it creates a better and tighter case if it is recorded that the fingerprints were on the murder weapon.

Certain types of evidence are worthless without documentation. For example, the knowledge that there was blood spatter present tells us very little other than the fact that a violent crime occurred. If we know the location of the blood spatter, however, we can draw a conclusion about where the event that caused the victim to bleed occurred. Furthermore, if the investigator has recorded what the blood droplets look like and the pattern of the blood droplets on a surface, conclusions might be drawn about the nature of the event that caused the victim to bleed, the amount of force applied, whether the victim was moving at the time, and where the criminal and the victim were positioned relative to each other.

Another use for records is as a source of information to rely on in interviews with law enforcement agents or attorneys, and in court. For most witnesses, referring to notes while giving testimony would be regarded as suspicious. Expert witnesses such as forensic scientists, however, are not only allowed to refer to their notes during testimony, they are expected to. It is unprofessional for a forensic scientist to speak entirely from memory without consulting their notes for exact details.

8.1.3 How to Keep Good Records

Although taking lecture notes is one of the central experiences of being a college student, I have found that many students balk at the idea of taking lab or field notes. I believe that much of this reluctance stems simply from being confused about what sorts of things to record. When starting out in the sciences, many students are unsure about what information should be recorded in lab or field notes. The principle that it is better to record information that later turns out to be unneeded than to fail to record critical information suggests that we should record as much as possible. However, there is never enough time to jot down everything, so strategic decisions have to be made about what to record. I suggest that the situation is very similar to taking lecture notes in a classroom. While listening to a lecture, most students write down those things the professor says which are not generally known or obvious. A student writes these things down because they want to remember them later for study, and because ultimately the professor will ask questions on a test that will require knowledge of this material in order to earn a good score. Similarly, the crime scene investigator takes notes about things that are not generally known or obvious because they want to remember them later when trying to interpret evidence, and ultimately an attorney or judge may ask about these details in a courtroom. Forensic scientists working at a crime lab take similar notes for similar reasons. So, just as the question "What lecture notes should I take while listening to this lecture?" is answered by asking yourself what you think the professor is going to ask on the test, the question "What should I record in my notebook?" can be answered by asking yourself what things about the evidence will you need to be able to tell the attorneys, judge, and jury if you testify in court. Also realize that it is most important to simply do your best. All forensic scientists have on occasion failed to record some vital item of evidence. We are human, after all, but as long as we do our very best, embarrassing omissions can at least be kept to a minimum.

8.2 The Comparative Method

The comparative method is one of the basic tools shared by all scientific fields. It is also a general method used by non-scientists and in everyday life. An astronomer may use the comparative method to determine the nature of a certain star. A philosopher may compare and contrast the ideas of Hobbs with those of Locke. In everyday life we ask and answer many questions using the comparative method. For example, the question "Is that dog a dachshund or a golden retriever?" is answered by comparing the pooch in question with what we know of the characteristics of these very different breeds. When we ask these sorts of questions, we are comparing what we observe about the item of interest (star, philosophical writing, dog, etc.) with some standard (the characteristics of stars, another philosophical writing, the characteristics of the various breeds of dog, etc.). To summarize, the comparative method usually involves comparing observations about some item to known characteristics of that type of item.

8.2.1 Compare and Contrast

One essential part of the comparative method is captured in the phrase "compare and contrast". When making a comparison it is as important to note the differences between the items being compared as it is to note the similarities. It is the differences that allow us to distinguish between items. When comparing a motorcycle to a sedan, for example, there are many similarities, both have wheels and an engine, but the distinction between them is the difference in the number of wheels.

As discussed in chapters 1 and 2, one very important task in forensic science is matching items. Items are said to match (*i.e.* a positive match is obtained) if all of their characteristics are the same and are said to not match (*i.e.* an exclusion is optained) if they are different in even one characteristic. For example, one difference between the genetic markers of blood recovered at the crime scene and the blood of a suspect is sufficient to show that the blood at the crime scene did not come from the suspect. Remember that in forensic science we seek not just to put people in jail but to put the correct person in jail. Not only do we not want to harm an innocent person, but we don't want to close a case with the guilty person still at large in the community. Therefore, it is as important, perhaps more so, to find those differences that exonerate a suspect as to find the similarities that incriminate them.

Given the importance of differences in the comparative method, it is surprising that some criminalistic specialties only focus on similarities. Fingerprint matching, for example, is based on finding enough "points of similarity" in the fine details of the prints to convince the examiner that the two prints were made by the same person. Fingerprint matching is considered highly reliable, yet one wonders whether not paying attention to the differences between prints has perhaps led to some cases in which the result was a positive match when it should have been an exclusion.

8.2.2 Differences that Do and Don't Exclude a Match

Although finding differences between items often means that there is no link between the crime and the suspect, this is not always true. There are certain types of differences that can be produced after the suspect left the crime scene. For example, after leaving the crime scene a suspect might step on a sharp object that produces a cut in the bottom of one of their shoes. The fact that the shoe prints left at the crime scene lack this cut does not mean that the suspect did not leave them. However, if the shoe prints at the crime scene have a cut mark that is not present in the suspect's shoes, then the suspect could not have made the shoe prints unless they were in possession of a time machine. Therefore, the nature of differences needs to be evaluated logically before drawing conclusions. The training and experience of forensic scientists prepares them to make these types of determinations.

8.2.3 Complete Comparisons

It is critically important for a comparison to be complete. An incomplete comparison can be misleading. If we were comparing cattle with goats, we would find a large number of easily observed similarities. Both have hooves, four legs, horns, a tail, eat grass, and live on a farm. The differences, such as size and stoutness, are visible but fewer in number than the similarities. Unless we set a goal of being complete, it would be possible to produce a long list of similarities before finding a difference. When applied to the forensic sciences, this means that a "quick and dirty" analysis may reveal only similaries, thus incriminating a suspect, whereas a more through analysis would yield one or more differences that would exonerate the suspect.

Another important aspect of the comparative method is that comparisons need to be detailed. A classic example from evolutionary biology illustrates this point. Birds and bats both have wings which function very similarly and have a lot of visual similarities. It takes detailed observation to be able to find that there are some differences between them. Therefore, many folk taxonomies, such as that presented in the Bible (Leviticus 11:13-19; Deuteronomy 14: 12-18) classify bats as a kind of bird. However, a detailed comparison reveals that the wings of birds and bats are constructed out of slightly different arrangements of bones, and therefore are not the same type of wing. Of course, to get this far you would have to ignore the most obvious difference between birds and bats -- having feathers versus having fur. The ancient Hebrew writers of the Bible must have observed this difference but thought it to be less important than having wings. In contrast, modern biologists consider having fur to indisputably link bats with the four-footed mammals despite the fact that they have wings. This difference does not mean that the ancient Hebrews were stupid. It merely means that experts in a particular field are better able to notice and interpret the details of similarities and differences.

When applying this idea to the forensic sciences, we again note that the training and experience of the forensic scientist performing the analysis are important. This is why training standards for forensic scientists, especially those who will testify in court, are very high. Most forensic specialties from forensic art to fingerprint examination require certification, which involves specialized training and examinations designed to make sure that the forensic scientist in question has an appropriate level of expertise.

8.2.4 Where do Individual Characters Come From?

As discussed in chapter 1, the characteristics of an item fall into two types, class characteristics and individual characteristics. Class characteristics are those that apply to a group of item of the type being examine. Individual characteristics are those that can be used to distinguish one individual example of that type of item from the group. Let's take the example of a shoeprint. The tread pattern of the sole of a shoe is specific to each model and brand. This tread pattern will be found on the shoe prints of everybody who wears that model and brand of shoe. Therefore, tread pattern is a class character. Size is another class character of shoe prints.

Individual characteristics are those that are unique to one individual item - not to a group of items. They arise out of the observation that no two items are ever exactly the same. We all know that fingerprints, faces, and snowflakes are unique, but we seldom consider the fact that no two shoes, or guns, or computer printers is exactly the same. Individual characteristics are those features that distinguish one particular unique item from all other items of the same type that possess the same class characteristics. For a shoeprint, the individual characters are features of use and abuse of the shoe, such as wear patterns, cuts, or scratches. These traits will be unique to each shoe, because each shoe has an individual history of being worn by its owner and has a unique history of random events that cause some minor change to the sole. These characteristics allow, for example, Joe's shoes to be distinguished from all other shoes of the same brand, style, and size.

We can benefit from considering where individual characteristics come from. Individual characteristics may occur on an item due to uniqueness of construction, uniqueness of random occurrences, or uniqueness of usage. **Uniqueness of construction** refers to the fact that no process can produce exactly identical items. In particular, this is easy to observe for things produced by hand by human beings. You can easily verify this by trying to write your signature the same way twice. When you compare two of your signatures, even if produced only seconds apart, you will undoubtedly find subtle differences. The truth of uniqueness of manufacture is probably least obvious for items that are mass produced by machines, yet in this case it is still true because the machinery that produces the item is in a constant state of change. Machinery can usually produce items that are more similar than is possible for a human being to produce, yet the various parts of a machine are in a constant state of wear, variations in temperature cause parts to shrink or contract, and other tiny variations occur in the environment causing each consecutive item produced by a machine to be slightly different. Although differences in manufacture must always exist, it is sometimes devilishly difficult to find them.

Uniqueness of random occurrences usually refers to forces that cause deterioration of or damage to the item. For example, no two items ever break in exactly the same way. As a experiment, buy a dozen eggs and drop them one by one onto a hard surface. Then examine the broken egg shells and you will find that each egg broke in a slightly different pattern. This principle is true of tears, stains, cuts, scratches, and other defects that an item will acquire over time due to random accidents.

Uniqueness of usage refers to the fact that no two items of the same type are ever used in exactly the same way. For example, no two people ever walk in exactly the same places for exactly the same amount of time. Therefore, the wear on the soles of their shoes must be different. No two screwdrivers are ever used on exactly the same number and type of screws, so their blades must have slightly different wear patterns.

Put together, these three factors ensure that, with very few exceptions, all items of a given brand, model, and type are unique in some way. The task of the criminalist is to discover the unique features and find a way to match these unique features to some feature of the evidence. It is possible, however, for a limited number of the random features of two or more items of the same type to look the same given the ability of the human examiner to detect all possible differences.

8.3 Terminology Issues: Identification and Individualization

There exists a terminology issue with respect to matching that you should be made aware of. Many authorities speak of consistent and positive matching, and in my observations this is the majority usage and I have adopted it for this textbook. However, other authorities prefer to use the term individualization to refer to the process of working toward a positive identification. Similarly, if an exact match is made between the class and individual characteristics of an evidence item and a standard item, then the evidence item may be said to have been individualized. Those who follow this terminology restrict the use of the term "identification" to placing an item in its proper class or type. For example, it's a bone, a piece of glass, or a drop of blood.

Terminology issues of this sort are expected within a field as large and diverse as the forensic sciences. Now that you are aware of this issue, hearing or reading someone using the term "individualize" should not confuse you.

8.4 Questions for Study and Review

1. Why do forensic scientists never rely on their memories of the procedures they used and the results of their examinations?
2. List some of the things that a crime scene investigator would keep a record of in their notebook.
3. List some of the things a forensic scientists at a crime lab would keep a record of in their notebook.
4. In your own words, describe how a forensic scientist's notes are used during interpretation of evidence and in giving courtroom testimony.
5. In your own words, describe the comparative method.
6. Describe a situation in which the individual characteristics of an evidence item and a standard item differ, but for which you could still conclude that you have a positive match.
7. In your own words, explain why comparisons must be complete.
8. Describe some of the processes that lead to the production of individual characteristics in an item.

8.5 Sources

Byrd M. n.d. Written Documentation at a Crime Scene [Internet]. [cited 2007 Oct 23]. Available from: http://www.crime-scene-investigator.net/document.html.

Discovery Education. 2007. Scientific Method [Internet]. [cited 2007 Oct 18]. Available from: http://school.discoveryeducation.com/sciencefaircentral/scifairstudio/handbook/scientificmethod.html.

Fox Chase Cancer Center. 2005. Ethics Policies and Procedures: Scientific Recordkeeping [Internet]. [cited 2007 Oct 18]. Available from: http://www.fccc.edu/ethics/RecordKeeping.html.

Inman K, Rudin N. 2002. The origin of evidence. Forensic Science International 126 (1): 11-16.

Peterson JL, Mihajlovic S, Gilliland, M. 1984. Forensic Evidence and the Police : the Effects of Scientific Evidence on Criminal Investigations. Washington, D.C. : U.S. Dept. of Justice, National Institute of Justice

Saferstein R. 1998. Criminalistics: An Introduction to Forensic Science. 6th Edition. Upper Saddle River (NJ): Prentice Hall.

Schafer JR. 2000. Color of Law Investigations [Internet]. FBI Law Enforcement Bulletin 69(8). [cited 2007 Oct 19]. Available from: http://www.fbi.gov/publications/leb/2000/aug00leb.pdf.

University of Florida, Office of Technology Licensing. 2002. Good Record Keeping: Procedures for Academic Laboratory Settings [Internet]. [cited 2007 Oct 18]. Available from: http://rgp.ufl.edu/otl/goodrecords.html.

Wolfs F. n.d. Appendix E: Introduction to the Scientific Method [Internet]. [cited 2007 Oct 18]. Available from: http://teacher.pas.rochester.edu/phy_labs/AppendixE/AppendixE.html.

Chapter 9
Basic Science
MICROSCOPES AND THE PHYSICS OF LIGHT

In this chapter we will look at one of the tools commonly used by criminalists and other forensic scientists – microscopes. Microscopes enlarge or magnify an image, allowing details to be viewed that are not visible otherwise. To understand how microscopes work, we will first need to explore the physics of electromagnetic radiation and the science of optics.

9.1 Electromagnetic Radiation (EMR)

Light, radio waves, microwave radiation, infrared (heat) radiation, ultraviolet radiation, x-rays, and gamma radiation seem to be quite different things. They are certainly produced in different ways and used for different purposes. Despite these differences, all of these phenomena are versions of the same thing – **electromagnetic radiation**. As implied by this name there is a deep physical relationship between electricity, magnetism, and these various forms of radiation. Electromagnetic radiation is one of the most common things in the universe. Most objects emit some form of electromagnetic radiation, and those that don't interact with it in complex, though predictable, ways. Optics is the branch of physics that studies the properties of EMR and the interaction of EMR with various materials.

9.1.1 Ways of Conceptualizing EMR: Particles vs Waves

There are two valid, though completely different ways of looking at electromagnetic radiation: as a stream of particles called photons, or as a wave. Both of these views are true. Electromagnetic radiation is both a stream of photons and a wave of radiation. One way to think of this duality is that the radiation exists in a state in which it has the ability to manifest itself as either a stream of photons or as a wave, and the choice of which form it manifests in is made by the person detecting or using the radiation. If the person detecting the radiation designs an apparatus that detects waves, such as a radio receiver, then the radiation will be received as a wave. If, on the other hand, the person detecting the radiation uses a scintillation counter, which is a device that measures flashes of light emitted as particles interact with matter, then the radiation will obligingly behave as particles. I prefer to view radiation as a wave, and so the remainder of this section will primarily take that approach.

Figure 9.1: A Radiation Wave

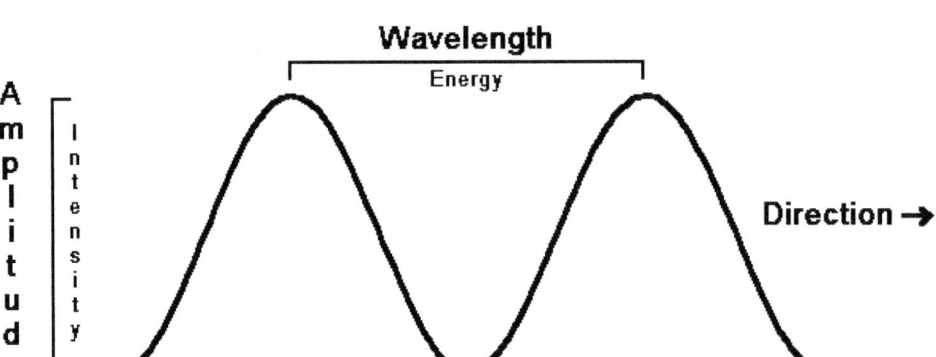

9.1.2 Properties of EMR

A unit of electromagnetic radiation has three important properties: **intensity**, **energy**, and **direction**. If the radiation is viewed as a stream of photons, then the intensity is the number of photons in the stream, the energy is the amount of energy inherent to each particle, and the direction is the direction in which the photons are moving. If the radiation is viewed as a wave, as illustrated in Figure 9.1, then the intensity is known as amplitude and is the height of each crest of the wave. The energy of a wave is reflected in its frequency, the number of crests and troughs per second, and is expressed in hertz. The number of hertz is the number of cycles from crest to trough and back to crest, per second. The energy of a wave can also be expressed as its **wavelength**, where the wavelength is the distance from one crest (or trough) to the next. The direction of a wave is the direction in which the wave is traveling, which is perpendicular to the dimension in which the crests and troughs occur. Note that the speed of the photons or waves of all forms of electromagnetic radiation is constant and equal to the speed of light, about 300,000,000 meters per second (186,000 miles per second).

9.1.3 Types of EMR

The various forms of electromagnetic radiation differ in their energy (their frequency or wavelength). Table 9.1 below shows the relationship between these forms of radiation and wavelength. Radio is the lowest frequency electromagnetic radiation, followed in order by microwave, infrared, visible light, ultraviolet, x-rays, and gamma rays, which are the highest frequency radiation.

Table 9.1: Types of EMR by Wavelength

Type of Electromagnetic Radiation	Wavelength
Radio	Longer than 10cm
Microwave	10cm to 1mm
Infrared	1mm to 700nm
Visible Light	700nm to 400nm
Ultraviolet	400nm to 1nm
X-ray	1nm to 0.01nm
Gamma Rays	Shorter than 0.01mn

9.1.4 Interactions Between EMR and Matter

One of the most useful things about the electromagnetic spectrum is that radiation of different wavelengths interacts with matter in different ways. For example, visible light easily penetrates the earth's atmosphere. It is no accident that the human eye has evolved to detect this form of radiation, since it is the most common form available at the earth's surface. The eye detects visible light using two types of cells in the retina of the eye, both of which act more like particle detectors than wave detectors. Rod cells respond well to the intensity of light, but are not sensitive to the frequency of light. What this means is that rod cells only perceive shades of gray – not color. Cone cells, in contrast, are sensitive to the frequency of light, perceiving the lowest visible frequency of light as red, and the highest as violet. Yellow, green and other colors of light are found between these extremes. The familiar separation of light into its component colors by a prism, or by water droplets in the atmosphere to form a rainbow, illustrates the range of colors visible to the human eye.

Although visible light can penetrate the atmosphere, it cannot effectively penetrate human skin and tissue. In contrast, x-rays are very effectively blocked by the atmosphere, so that only a few photons of x-ray radiation ever reach the surface of the earth. However, x-rays can penetrate skin and flesh to produce the radiographs so commonly used as diagnostic tools in medicine.

EMR can interact with matter in a number of ways. It can pass through the matter, it can be reflected by the matter, it can be absorbed by the matter, it can be refracted by the matter, and in some cases it can cause the matter to luminesce. During **luminescence** the matter is stimulated by the energy of the EMR striking it to itself radiate EMR, but of an energy different from that of the radiation striking it. Each type of matter interacts with radiation of a certain energy differently from every other type of matter. This property makes electromagnetic radiation extremely useful. Figure

9.2 illustrates reflection and refraction which are two of the ways that EMR interacts with matter.

Figure 9.2: Reflection & Refraction

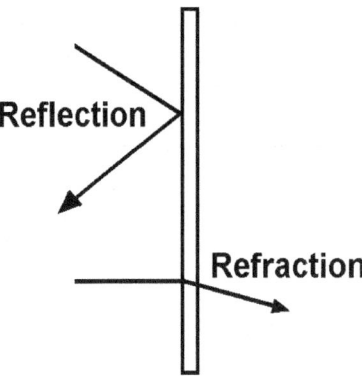

In the following discussion I will focus on light as an example of EMR because it is the most familiar form of EMR. Any property of the interaction of light with matter is shared by other forms of EMR as well. Perhaps the most well known way in which light interacts with matter is **reflection**, wherein the light waves bounce off the surface of the matter they strike. All materials that are not perfectly transparent reflect light. Our ability to see depends on this property. Light striking an object is reflected and picked up by our eyes. Many of us are most familiar with reflection as seen in a mirror, which is a sort of double reflection. When we see ourselves in the mirror, some of the light striking our bodies is reflected onto the mirror, which then reflects them back to us so our eyes can pick them up.

Refraction occurs when light passes from one material to another, for example from air into glass. At the intersection of the two materials the rays of light are bent (more precisely their direction is changed). This bending is known as **refraction**. Each type of material has a characteristic refractive index, which is the amount of refraction it produces when light waves enter it. Therefore, the amount of refraction, and thus the direction of the light waves can be controlled by choosing a substance with an appropriate refractive index.

Absorption occurs when the EMR neither reflects from a material nor passes through it. The energy of the EMR is at least partly absorbed by the material. Physical or chemical changes may occur in the molecular structure of the material as a result, but from our point of view it seems as if the material simply captures the EMR. The opposite of absorption is **transmission**, wherein the light passes through the material without any decrease in its intensity.

9.1.5 How Forensic Scientists Harness the Interactions Between EMR and Matter

Forensic scientists utilize the electromagnetic spectrum in a variety of ways. In most of the forensic applications, electromagnetic radiation is viewed as a wave and in most cases the energy property of the wave is expressed as a wavelength. One way in which forensic scientists use the electromagnetic spectrum, visible light in this case, is photography. Black and white photography records the intensity of the light striking the photographic film. Areas of more intense light are whiter than areas of less intense light. Color photography captures not only the intensity of light, but its energy as expressed in its color. The direction of light can be captured by a hologram, which takes advantage of the way in which the right and left eyes see things slightly differently to produce a three-dimensional image.

Refraction is one of the most useful interactions between EMR and matter for forensic scientists because this is the interaction upon which microscopes are based. Refraction also forms the basis of other types of instruments. For example, the refractive index of a solution of sugar in water is a function of the amount of sugar in the solution. Therefore, a refractometer can be used to measure the refractive index of a sugar solution to estimate the sugar content fairly precisely.

Absorption forms the basis of several analytical instruments or their parts. For example, most substances absorb certain wavelengths of EMR. A spectrophotometer is an instrument that measures how much EMR of a certain wavelength is absorbed during passage through a solution, allowing the quantity of the substance of interest can be estimated fairly precisely.

9.2 Lenses

A **lens** is a piece of material that has a certain shape to it and is designed to cause EMR to refract in a controlled manner. In most common applications the material of which the lens is made is glass, plastic, or some similar transparent material, and the EMR being refracted is visible light. Lenses are used not only in microscopes, but also in telescopes, eyeglasses, binoculars, magnifying glasses, and cameras.

As discussed above, one way to control refraction is by choosing a material with an appropriate index of refraction. However, it is much more common to control refraction by the shape of the lens. Since refraction occurs at the surface of the refractive material, the curvature of the surface where a light ray strikes influences the direction of the refracted light waves. Utilizing this principle, waves of light can be manipulated in a variety of useful ways using lenses made of chunks of glass or plastic molded or ground to selected shapes.

Lenses that are convex on at least one surface refract light waves in such a way that the waves of light converge toward a single point. Commonly available magnifying glasses have convex lenses. Most of us have had the experience of using a magnifying glass to concentrate the sun's rays to a concentrated point, thereby producing a charred spot on a piece of wood or paper. When this happens, light waves arriving from the sun arrive at the back surface of the lens (the back surface being the one farthest from the charred spot). At this surface they are refracted in such a manner that they converge at the charred spot. The point at which light waves converge, the charred spot in our example, is called the focal point of the lens. The distance between the focal point and the front surface of the lens (the surface closest to the focal point) is called the focal distance. A **convex lens** is illustrated in Figure 9.3.

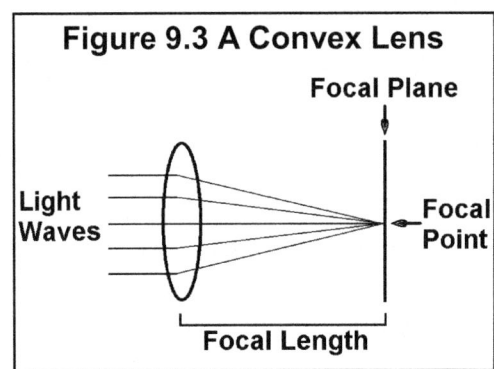

Using a magnifying glass to magnify something reverses this process. During magnification, light reflecting from the item being viewed strikes the front surface of the lens (the surface closest to it) and emerges from the back surface spread out in such a way as to form an image that is larger than the actual object being viewed. The image produced by a lens is called a virtual image.

Lenses that are concave on at least one surface refract light in the opposite manner from a convex lens, causing the light waves to diverge from each other. This pattern of refraction doesn't produce useful magnification, but concave lenses are sometimes combined with convex lenses in a microscope to reduce certain forms of distortion. The main use of concave lenses is in eyeglasses that correct the vision of near-sighted people. The refraction of a **concave lens** is illustrated in Figure 9.4.

9.3 Microscopes

Microscopes are used to magnify the image of a small item so that the item or its details are more visible. Since identifications are made on the basis of matching details, microscopes are widely used by forensic scientists, especially criminalists.

Microscopes have the advantage that, unlike some other types of instruments, the microscope gives a direct image of the item being viewed, not some abstract pattern that has to be interpreted. Most of the professional quality microscopes allow a camera to be attached so that the image being viewed can be photographically documented.

The methodology of using microscopes, called **microscopy**, appears straightforward at first -- you stick something under a microscope, peer into the eyepiece, twiddle a few knobs, and an image of the item you want to see comes into view. Although this procedure works for some types of items, other items require more elaborate procedures and microscopy is yet another skill that requires training and experience to be applied well.

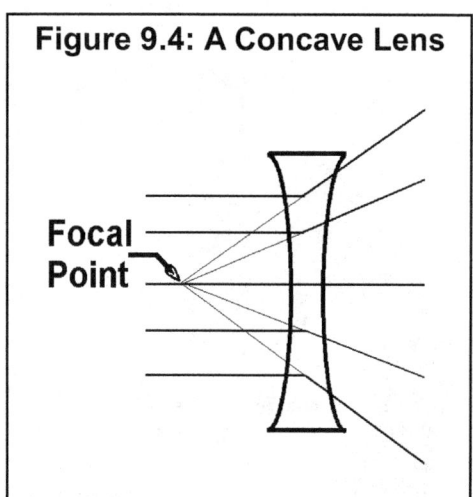

9.3.1 Types of Microscopes

A single lens can be used as a microscope, as in a magnifying glass. A microscope consisting of a single lens, *i.e.* a magnifying glass, is called a simple microscope. Most modern microscopes are **compound microscopes**, meaning that they are constructed of two or more lenses held in a tube in such a relationship as to produce higher powers of magnification. Most modern microscopes are constructed as shown in Figure 9.5, below, and usually contain three main lenses. Two of the lenses are visible in Figure 9.5, the ocular and the objective lenses. The third lens, if present, is called the field lens, and is hidden inside the body tube. Often, a few different objective lenses of differing power are attached to a turret, which can be rotated to bring any of the objective lenses into use. This allows for different levels of magnification. Some compound microscopes also have interchangeable ocular lenses of various magnifications.

The specimen to be viewed is placed on the stage of the microscope, directly under the objective lens, and the focus knobs are adjusted until a clear virtual image is obtained. Certain types of specimens are placed on microscope slides and covered with a cover plate. A microscope slide is simply a rectangular piece of thin glass, usually about one inch wide and two inches long. A cover plate is smaller and thinner piece of glass that is used to cover the specimen. Thus, the specimen is sandwiched between the slide and the cover plate. A variety of different types of small-sized specimens are prepared and viewed using microscope slides, ranging from blood to hairs. Specimens placed on microscope slides are often stained with a dye that enhances the details of the specimen.

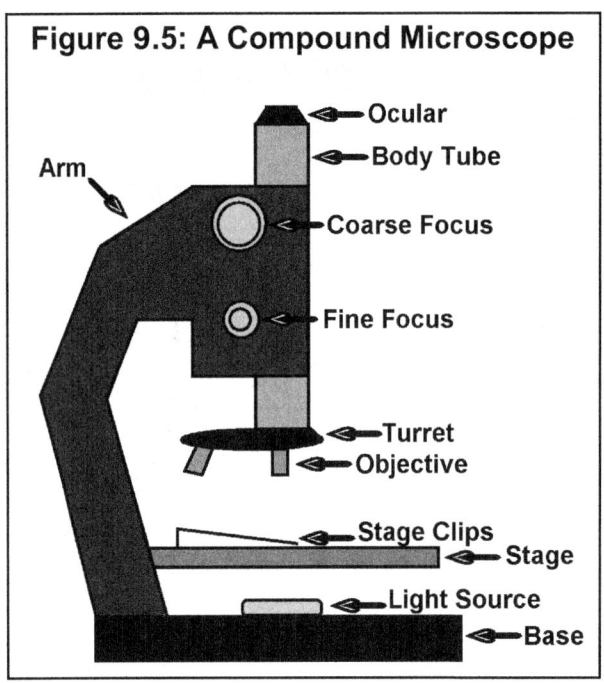

9.3.2 Illumination

The specimen to be viewed must be illuminated by providing a source of light. In general there are two types of illumination schemes, reflective and transmissive. In **reflective illumination** an external source of light is shined onto the specimen, or reflected onto it using a mirror. Some of the light waves reflected by the specimen enter the objective lens and are magnified. In **transmissive illumination** the specimen is illuminated by a light source below the stage, and the light passes through a hole in the stage, through the specimen, and into the objective lens. Transmissive illumination is usually used with microscope slides and reflective illumination is usually used with larger specimens, such as an insect or a fragment of bone.

9.3.3 Comparison Microscopes

Criminalists often use a special type of microscope called a comparison microscope, which was introduced in chapter 1. A comparison microscope consists of two compound microscopes joined together by an optical bridge. An optical bridge is a device that replaces the objective lens on both microscopes and uses a system of lenses, prisms, and mirrors to form a composite virtual image wherein the specimens magnified by both microscopes can be seen at the same time when looking into the occular of the bridge. This allows two specimens to be directly compared, side by side, under magnification. Using a comparison microscope, a criminalist can compare two fingerprints, two hairs, two bullets, a tool with a tool mark, and many other such items that can potentially be matched by comparison of minute details.

9.3.4 Electron Microscopes

A newer type of microscope uses a stream of electrons instead of visible light waves as an illumination source. This type of microscope is known as an electron microscope. Electron microscopes are constructed very differently from light microscopes, with the most obvious difference being that the image is viewed on a computer monitor rather than by looking into an ocular lens. Electron microscopes are capable of higher magnification, higher resolution, and greater depth of field than light microscopes. Electron microscopes are rarely used by forensic scientists because the items they examine can usually be adequately magnified by good quality light microscopes, with no need to resort to the extreme magnification ability of an electron microscope.

9.4 Questions for Study and Review

1. In your own words, define the following terms: microscope, EMR, optics, photon, microscopy.
2. In your own words, describe the energy, intensity, and direction properties of EMR, both when the EMR is viewed as a particle and when it is viewed as a wave.
3. Which of the properties of EMR is reflected by the categorization of the electromagnetic spectrum into radio, visible light, x-rays, etc.?
4. In your own words, describe the following interactions between EMR and matter: luminescence, reflection, refraction, absorption, and transmission.
5. Which of the interactions between EMR and matter is harnessed to construct a microscope lens?
6. Which of the interactions between EMR and matter is used as the basis for a spectrophotometer's ability to detect the presence and quantity of a substance?
7. In your own words, define the following terms that apply to lenses: concave, convex, focal point.
8. List the main parts of a typical compound microscope and describe what each part does.
9. What is the difference between reflective and transmissive illumination?
10. How does a comparison microscope compare to other types of compound microscopes?

9.5 Sources

[Anonymous]. n.d. The Electromagnetic Spectrum [Internet]. [cited 2007 Oct 18]. Available from: http://csep10.phys.utk.edu/astr162/lect/light/spectrum.html.

Bagnell CR. 2000. Class Notes for Pathology 164 - Light Microscopy [Internet]. [cited 2007 Oct 18]. Available from: http://www.med.unc.edu/microscopy/path164.htm.

Beeson S. 1995. Submodule 2: Lenses, Mirrors, and Prisms: Objective 2 [Internet]. [cited 2007 Oct 18]. Available from: http://acept.la.asu.edu/PiN/mod/light/lenses/pattLight2Obj2.html.

Holophile Inc. 2002. History [Internet]. [cited 2007 Oct 18]. Available from: http://www.holophile.com/history.htm.

Jones TH. n.d. History of the Light Microscope [Internet]. [cited 2007 Oct 18]. Available from: http://web.archive.org/web/20021001072245/www.utmem.edu/~thjones/hist/hist_mic.htm.

Karl S. 1998. Introduction to Light Microscopy [Internet]. [cited 2007 Oct 18]. Available from: http://nsm1.fullerton.edu/~skarl/EM/Microscopy/IntroLightMicro.html.

Laboratory for High Energy Astrophysics at NASA. n.d. Electromagnetic Spectrum: Measuring the Electromagnetic Spectrum [Internet]. [cited 2007 Oct 18]. Available from: http://imagine.gsfc.nasa.gov/docs/science/know_l1/emspectrum.html.

Laboratory for High Energy Astrophysics at NASA. n.d. The Electromagnetic Spectrum: More about the Electromagnetic Spectrum [Internet]. [cited 2007 Oct 18]. Available from: http://imagine.gsfc.nasa.gov/docs/science/know_l2/emspectrum.html.

The Paper Project. 2000. History of the Microscope [Internet]. [cited 2007 Oct 18]. Available from: http://paperproject.org/microscopehistory/index.html.

Thomas DJ. 2001. Exercise 1: Introduction to the oil immersion compound microscope [Internet]. [cited 2007 Oct 18]. Available from: http://www.lyon.edu/webdata/users/dthomas/microbiology/labweb/lab01.html.

Wikipedia contributors. 2007. Comparison Microscope [Internet]. Wikipedia, The Free Encyclopedia; 2007 Oct 18, 15:58 UTC [cited 2007 Oct 18]. Available from: http://en.wikipedia.org/w/index.php?title=Comparison_Microscope&oldid=165427598.

Chapter 10
CRIME LABS AND TRACE EVIDENCE EXAMINATION

Although I have shown in previous chapters that many forensic scientists work outside a crime lab, crime labs still remain the most familiar venue for forensic scientists. Crime labs have several important functions, and in this chapter we will look at the function and organization of a crime lab. Following this discussion, we will take our first thorough look at a branch of criminalistics commonly practiced at a crime lab, i.e. trace evidence analysis. Trace evidence analysis was the specialty of Locard, the historical figure discussed in chapter 1, and is a quintessential example of a criminalistic forensic science.

10.1 Crime Labs

A **crime lab** is a laboratory dedicated to the processing, by forensic scientists, of physical evidence relating to crimes. It brings together scientific instruments and the trained forensic scientists who use them in a location where they can collaborate in analyzing evidence and developing presentations for court. The work of crime labs is directed toward criminal cases, not civil cases, and given perpetually tight government budgets and high crime rates, they are almost always busy places. Similar forensic scientists who work for private companies can provide forensic analyses for civil cases.

As discussed in a chapter 1, the world's first crime labs appeared in the 1910's through 1930's, and crime labs became mainstream in the 1930's through 1960's. Especially since the 1960's, and the landmark Supreme Court decisions discussed in chapter 1, the number of crime labs at all levels of jurisdictions has increased dramatically.

10.1.1 Crime Lab Sections

Crime labs are organized into a variety of sections (or divisions or departments), each devoted to the analysis of a certain type of evidence. The Montana State Laboratory of Criminal Investigation (hereafter MSLCI) can serve as an example of a smaller sized crime lab. The MSLCI is divided into nine sections: Breath Alcohol Program, Crime Scene Investigation, Drug Chemistry, Firearms and Toolmarks, Latent Prints and Impressions, Pathology, Serology/DNA, Toxicology, and Trace Evidence. The MSLCI also has arrangements with other forensic scientists for handling specialized types of evidence that are encountered less frequently. For example, when the MSLCI has a case that involves skeletal remains, they send this evidence to The

University of Montana, where our Forensic Anthropology Team handles the examination and analysis.

Here are some thumbnail descriptions of the main purpose of each of these nine sections of the MSLCI.

- The Breath Alcohol Program provides maintenance of breath alcohol analyzers used by law enforcement agencies throughout Montana and trains officers in their use. Toxicologists and technicians skilled in maintenance of breath alcohol analyzers work in this section.
- The Crime Scene Investigation Section supplies specialized equipment and personnel for collecting evidence at a variety of serious crime scenes throughout Montana. Evidence technicians work in this section.
- The Drug Chemistry Section provides analysis of drugs submitted by law enforcement agencies. The type of drug, as well as its quality and quantity, can be determined. Forensic chemists work in this section.
- The Firearms and Toolmarks Section analyzes guns, bullets, bullet casings, other firearms evidence, and other weapons evidence, often for the purpose of matching bullets or casings to the firearm that fired them. This section also handles the comparison of tools to marks left by tools on surfaces in order to determine whether or not an individual tool make a certain mark. Firearms and toolmarks examiners work in this section.
- The Latent Prints and Impressions Section identifies develops, preserves, compares, and matches skin ridge patterns from finger, palm, and bare foot prints left on a variety of surfaces. This section also handles impressions evidence such as shoe or tire impressions. Fingerprint examiners and impressions examiners work in this section.
- The Pathology Section provides autopsies in which a variety of samples and evidence are collected to help in determining time, cause, and manner of death. Forensic pathologists, including the Montana State Medical Examiner and the Associate State Medical Examiner work in this section.
- The Serology/DNA Section identifies body fluids such as blood, semen, and saliva, and identifies genetic markers in these samples that can be used to match them to a particular individual. Serologists and DNA analysts work in this section.
- The Toxicology Section determines the presence and quantity of alcohol, drugs, and poisons in blood or biological tissues. Toxicologists work in this section.
- The Trace Evidence Section examines small items, such as hairs, fibers, paint flakes, gunshot residues and glass flakes in order to match evidence collected from suspects with evidence collected at crime scenes. Trace evidence examiners work in this section.

Now, let's compare the MSLCI with a larger crime lab, which I visited about 20 years ago, but never actually handled a case for, the Georgia Bureau of Investigation's Crime Lab in Decatur, GA (hereafter the GBICL). The GBICL has the following sections: Administrative, Alcohol Testing, Drug Identification, Firearms, Forensic Biology, Latent Prints, Pathology, Photography, Questioned Documents, Toxicology, and Trace Evidence. The GBICL's Photography Section corresponds to the MSLCI's Crime Scene Investigation Section. With the exception of the GBICL's Administrative and Questioned Documents section, the correspondences between the organization of the two crime labs are clear, as shown in Table 10.1. The administrative section of the GBICL handles record keeping and evidence routing, which are handled by evidence technicians at the MSLCI who are not assigned to a specific section. The questioned documents section handles many forms of questioned documents. Montana does not have enough questioned document cases to justify hiring a full time questioned documents examiner, so they send their questioned documents evidence to the FBI Lab in Quantico, VA for analysis.

Table 10.1: Comparison of Montana's and Georgia's Crime Lab Sections

MSLCI Section	GBICL Section
	Administrative
Breath Alcohol Program	Alcohol Testing
Drug Chemistry	Drug Identification
Firearms and Toolmarks	Firearms
Serology/DNA	Forensic Biology
Latent Prints and Impressions	Latent Prints
Pathology	Pathology
Crime Scene Investigation	Photography
Toxicology	Toxicology
Trace Evidence	Trace Evidence
	Questioned Documents

Some interesting differences exist between the MSLCI and some other crime labs, however, in who handles impressions evidence and who handles edge matching of torn or broken items. At many crime labs impressions, evidence is handled by Firearms and Toolmarks Section because the nature of the individual characteristics are similar for all three types of evidence. At the MSLCI, however, impressions evidence is handled by the Latent prints and Impressions Section, and I see good reasons for this too. Some types of fingerprints are an impression and the methods used to examine them are similar to those using to examine other types of impressions, such as footwear and tire tread impressions. At most crime labs, edge matching is handled by the Trace Evidence Section, since matching the edges of broken or torn large items is essentially similar to matching the broken edges of paint and glass chips, which are in the domain of trace evidence examiners. At the MSLCI, however, edge matching is handled by the firearms and toolmarks examiners, for reasons that escape me. As we say here in Montana, there are three ways to do everything: the right way, the wrong way, and the Montana way.

10.1.2 Functions of a Crime Lab

One of the most important functions of a crime lab is to facilitate collaboration between the forensic scientists who work there. Although a crime lab may be organized into sections, the forensic scientists assigned to different sections routinely collaborate in analyzing evidence from a case. For example, if the body if a murder victim is autopsied by a forensic scientist who works in the Pathology Section, tissue samples will be collected for analysis by scientists in the Toxicology Section, any bullets recovered will be passed along to the Firearms Section, gunpowder residues will be collected and assigned to the Trace Evidence Section, and any blood or other body fluids suspected to belong to the murderer rather than the victim will be handed off to the Serology Section. Since all of the forensic scientists involved work in the same building, the distribution of evidence to the appropriate specialists can be accomplished easily.

Another important function of a crime lab is to maintain a secure chain of custody by making sure that evidence doesn't leave the building, and that good records are kept of who had access to the evidence and what analyses and procedures they performed. In many crime labs the evidence is stored in a central evidence locker and each forensic scientist who analyzes some of it will go there to get it. When fetching evidence for analysis, the forensic scientist will record the evidence taken, time, date, the purpose of the analysis, and his or her name in a log book designated for the case. When returning the evidence after analyzing it, a similar notation is made. When the case is taken to court, the logbook may be used to document the chain of custody.

The third function of the crime lab is to efficiently provide resources for the forensic scientists who work there. For example, a GCMS instrument costs many thousands of dollars. It is beyond the resources of an individual forensic scientist to own a GCMS of their own, so the goverment provides these instruments for them. Laboratory space, other instruments, computers, internet connections, and various disposable items are also resources, which are provided by a government and efficiently allocated by a crime lab to the scientists who work there.

10.2 Trace Evidence Examination

The term **trace evidence** refers to evidence that is small in size. Trace evidence includes hairs, fibers, paint and glass chips, stains, and dusts that are picked up or left at a crime scene. Trace evidence examiners compare the characteristics of items in order to find a match. In most crime labs it is also trace evidence examiners who match items based on the patterns of tear and breakage found on their edges.

Trace evidence is often collected at a crime scene by vacuuming the area. Other trace evidence can also be recovered by combing, sweeping, or beating items; or by simple visual inspection. Another technique for collecting trace evidence is "lifting" it with tape. In this procedure tape is pressed against the evidence then pulled free in such a way that some of the evidence adheres to the tape. Trace evidence can often be found clinging to the clothing or body of a suspect or victim.

10.2.1 The History of Trace Evidence Examination

The technique of matching the pattern of a torn edge with the larger item from which it is torn dates back to the late 1700's when a murderer was convicted based on matching the torn edge of a piece of newspaper used as wadding in his pistol with a piece found in his pocket. In 1910, Victor Balthazard and Marcelle Lambert published the first comprehensive study of hair, and Balthazard used his knowledge to solve a murder case based on hair evidence. In 1916, Albert Schnieder pioneered the technique of collecting trace evidence using a vacuum cleaner. In 1920 Edmond Locard published a treatise on evidence that included a discussion of his work with dusts. In 1950, Max Frei-Sulzer developed the tape lift method of collecting trace evidence.

10.2.2 Hairs

The color, thickness, and detailed composition of hairs can be compared using a comparison microscope. A hair collected as evidence is considered to match that of a hair sample of known origin if all the features are the same. Hair evidence is not as useful as some other types of evidence because hair has no individual characteristics – only class characteristics. Therefore, it can be used to exclude a suspect, but not to positively identify a subject. This doesn't mean that hair has no value as evidence or that it can't help to link a suspect to a crime. When collected properly, documented thoroughly, and interpreted cautiously, hair evidence can provide important corroborating evidence placing a suspect at a scene. An experiment conducted in Canada showed that the odds of the hairs from two different people matching is about 1 in 4500. This is not good enough for positive identification, but is still reasonably accurate evidence.

Hair may also contain DNA, especially in the root or any adhering follicular tissue. Therefore, DNA testing may be used to identify the person who was the source of the hair. DNA analysis will be discussed in a coming chapter. We will see when in that chapter, however, that DNA also has only class characteristics.

Hair is an appendage of the skin, which grows out of a hair follicle. Hair has a root, which is embedded in the follicle, a shaft, and a tip. The hair shaft is composed of 3 layers. The cuticle forms the outermost layer and the medulla forms the innermost portion, with the cortex comprising the layer between the cuticle and the medulla. The cuticle is formed by overlapping scales that always point toward the tip end of the hair. Although the scale pattern doesn't vary much from one person to another, it does vary from one species to another. So, you can determine whether a hair is human or not by examining it under a microscope. The cortex of hair is the middle layer. It contains elongated cells and granules of pigment. The color and pigment granule pattern of a hair is a class characteristic that can be compared between individuals. The medulla is a collection of cells forming the innermost layer of the hair. The medulla varies in size between different species, allowing the hair of non-humans to be distinguished from that of humans..

Experts in hair analysis can usually determine the person's race from their hair. This is done by looking at thickness, color, and curliness. As an anthropologist I must point out, however, that race is more a social and cultural phenomenon than a biological phenomenon and hair form only approximately correlates with race. Hair examiners can also tell which part of the body -- scalp, beard, pubic area, armpit, or general body, that a hair came from. As you know, the hairs from these regions vary in coarseness and curliness. Contrary to what most people would expect, it is usually not possible to estimate the age or sex of a person reliably from their hair, but the length of the hair and its grayness may offer some clues.

It is also possible to determine whether a hair fell out, was pulled out, or was cut. Cut hairs will have an obvious cut end and will lack a root. Whether a hair was shed naturally or pulled out is distinguished by examination of the root. Hairs go through a life cycle beginning with the anagen stage, in which the hair is actively growing; a catogen or intermediate stage; and a telogen stage, in which hair growth has ceased and the hair is ready to be shed. The root of the hair has a different shape in all of these stages, with the telogen stage characterized by a club-shaped root. If a hair has a club-shaped root, it is presumed to have fallen out naturally, but if the root has a shape characteristic of the anagen or catogen phase it is presumed to have been pulled, especially if there is any adhering follicular tissue.

Standard hairs for comparison need to be collected from the victim and from any potential suspects. About 50 to 100 hairs from each region should be collected. The regions that need to be sampled include at least the scalp, and may include the pubic or other areas, depending on the case.

10.2.3 Fibers

Fibers can be shed by clothing, carpets, or other fabrics or textiles. Like hairs, fibers only have class characteristics. The utility of fibers lies, however, in the fact that there are so many different kinds of them. The task of the forensic scientist is to narrow down the source of the fiber to a few possibilities, preferably one. For example, it is sometimes possible to determine that a certain fiber is of a certain composition, and is from a certain type of product, manufactured by a certain company, between a certain range of dates.

Fibers are classified into two broad groups – natural and human produced. **Human produced fibers** are also referred to as artificial or man-made. The term "man-made" is more common than "human produced", but I prefer to avoid the sexism of the former term. Before the 20th century there were only natural fibers. The first human produced type of fiber, rayon, was introduced in 1911.

Natural fibers are obtained from animals or plants. Animal fibers from mammals (such as wool) are hairs, and are compared using methods described above. Silk is another animal fiber, which is produced by a caterpillar. The most common plant fiber is cotton, though many others exist, such as linen, jute, and hemp. Human produced fibers are marketed under hundreds of trade names, even though there are relatively few actual chemically distinct substances out of which they are made. There is, however, an enormous amount of variation in the characteristics of the fiber of a certain chemical type as manufactured by different companies.

Among the human produced fibers, there are some that are derived from natural plant or animal products. An example is rayon, made from cellulose derived from cotton or wood pulp. Fibers of this sort are called **regenerated fibers**. Other human produced fibers are produced entirely from chemicals, mostly petroleum products. Polyester is an example of a type of fiber produced from chemicals refined from petroleum. Fibers of this type are called **synthetic fibers**.

The most obvious class characteristics of fibers are their diameter and their color. A comparison microscope is used to match fibers with respect to these characteristics. If the diameter and color of two fibers match, the examiner will look at more detailed comparisons including: the cross-sectional shape of the fiber, the chemical composition of the fiber and any dye used to color it, the presence and composition of additives, and striations or other markings on the fiber. The cross-sectional shape of the fiber refers to the shape of the fiber as viewed from a cut end. Some fibers are round in cross section, some are oval, some are ribbon-like, and others are triangular. The chemical composition of the fiber, whether it is rayon, nylon, acetate, or another material, can be determined from the degree the material refracts light (its refractive index), by which frequencies of light it absorbs (spectrophotometry), what it smells like when burned, whether it dissolves in certain acids, and similar tests. The composition of the dye used to color the fiber is usually determined by extracting it with a solvent using a process called thin layer chromatography that will be discussed in a later chapter. Many synthetic fibers contain additives, such as titianium dioxide powder, to reduce their shininess. The presence and composition of these additives constitute a data item that can be used for matching. Many fibers contain striations or other markings made by the machines used to produce them, and these can be used for matching. These features are all class characteristics and can not be used to positively link a person to a crime scene.

10.2.4 Paint and Glass Chips

When paint chips flake off a painted surface, or are broken or scraped off, the fragments break away from the painted surface in a manner that leaves a pattern of breakage on the edges of the paint chip and the original painted surface. This edge breakage pattern can be matched. Similarly, fragments of broken glass have edges that can be pattern matched to the original piece of glass from which they were broken. The pattern of breakage of paint or glass is an individual characteristic which can be used to positively link the glass or paint fragment to a crime scene. Paint or glass fragments can adhere to a vehicle, clothing, or a person's skin in certain types of situations.

Paints and glasses can also have distinctive chemical compositions and other characteristics. These features can be determined, using an appropriate method and used in matching, although they are class characteristics rather than individual characteristics.

10.2.5 Dusts

Dusts are composed of fine particles of material. A particular dust often consists of several different types of particles, such as soil, sand, animal and human dander, plant and insect fragments, and fragments of industrial materials such as steel or plastic. Because of the diversity of particles, and the fact that certain types of particles are found only in certain locations, dusts from many locations are unique and can be matched to tie a suspect to a crime scene. Although dust is everywhere and often adheres to a suspect's shoes or clothing, it is very difficult to detect, collect, and analyze. Therefore, dust is used less often in matching than other types of trace evidence.

10.2.6 Torn and Broken Edge Matching

Of the several types of examination that a fiber examiner can make, one of the most important is matching a piece of fabric from a crime scene with a larger piece of fabric from which it was torn, usually some item of clothing. As with the broken edges of paint or glass chips, the shape of a torn edge is at least partly random in nature, so this is an individual characteristic, which can be used to make a positive match. Similarly, almost any fragment of a broken item can be matched with other contiguous fragments or with the original item from which it was broken. A vast array of items from a paper match torn from a matchbook to the fragments of a broken condom can be positively matched in this way.

10.3 Questions for Study and Review

1. Does a crime lab normally process evidence for civil cases?
2. Where is the closest crime lab to where you live? Into what sections are the forensic scientists who work there organized?
3. List the functions of a crime lab.
4. In your own words, define trace evidence.
5. List the types of evidence that are examined by trace evidence examiners.
6. Is it possible to make a positive match between two hairs or two fibers? Why or why not?
7. Describe the structure of a hair.
8. List the class characteristics of hairs.
9. What sorts of information can be recovered from analysis of hairs?
10. What is the difference between a natural fiber and a human-produced fiber?
11. What is the difference between a regenerated fiber and a synthetic fiber?
12. List the class characteristics of fibers.
13. Is it possible to make a positive match between two paint chips, or between a paint chip and the painted item from which it came? Why or why not?
14. In your own words, describe the analysis of dusts.
15. In your own words, describe the process of edge matching.

10.4 Sources

Deedrick DW. 2000. Hairs, Fibers, Crime, and Evidence [Internet]. Forensic Science Communications 2(3). [cited 2007 Oct 21]. Available from: http://www.fbi.gov/hq/lab/fsc/backissu/july2000/deedrick.htm.

Deedrick DW, Koch SL. 2004. Microscopy of Hair Part 1: A Practical Guide and Manual for Human Hairs [Internet]. Forensic Science Communications 6(1). [cited 2007 Aug 12]. Available from: http://www.fbi.gov/hq/lab/fsc/backissu/jan2004/research/2004_01_research01b.htm.

Georgia Bureau of Investigation. n.d. DOFS Division of Forensic Sciences [Internet]. [cited 2007 Oct 18]. Available from: http://www.state.ga.us/gbi/fordiv.html.

Kramer RE. n.d. Case Solved With Aid of Fracture Match [Internet]. [cited 2007 Oct 23]. Available from: http://www.geocities.com/cfpdlab/fracture.htm.

Koons RD, Buscaglia J, Bottrell M, Miller ET. 2002. Forensic glass comparisons. In: Forensic Science Handbook. 2nd ed. R. Saferstein (Editor), pp. 161–213. Upper Saddle River (NJ): Prentice-Hall.

McJunkins SP, Thornton JI. 1973. Glass fracture analysis: A review. Forensic Science 2(1): 1–27.

Meloan CE, James RE, Saferstein R. Lab Manual for Criminalistics: An Introduction to Forensic Science. 6th Edition. Upper Saddle River (NJ): Prentice Hall.

Nickell J, Fischer JF. 1998. Crime Science: Methods of Forensic Detection. Lexington (KY): University of Kentucky Press.

Rudin, N, Inman K. 2002. Forensic Science Timeline. [Online]. Available from: http://www.forensicdna.com/Timeline020702.pdf. Accessed 2007 Oct 16.

Saferstein R. 1998. Criminalistics: An Introduction to Forensic Science. 6th Edition. Upper Saddle River (NJ): Prentice Hall.

Scientific Working Group on Materials Analysis (SWGMAT) Evidence Committee. 1999. Trace Evidence Recovery Guidelines [Internet]. Forensic Science Communications 1(3). [cited 2007 Oct 18]. Available from: http://www.fbi.gov/hq/lab/fsc/backissu/oct1999/trace.htm.

State of Montana Department of Justice. n.d. Forensic Science Division [Internet]. [cited 2007 Oct 18]. Available from: http://www.doj.mt.gov/enforcement/crimelab/.

Scientific Working Group on Materials Analysis. 2005. Forensic Human Hair Examination Guidelines [Internet]. Forensic Science Communications 7(2). [cited 2007 Oct 21]. Available from: http://www.fbi.gov/hq/lab/fsc/backissu/april2005/standards/2005_04_standards02.htm.

Scientific Working Group on Materials Analysis. 2005. Forensic Fiber Examiner Training Program [Internet]. Forensic Science Communications 7(2). [cited 2007 Oct 21]. Available from: http://www.fbi.gov/hq/lab/fsc/backissu/april2005/standards/SWGMAT_fiber_training_program.pdf.

Working Group for Materials Analysis. 2004. Introduction to Forensic Glass Examination [Internet]. Forensic Science Communications 7(1). [cited 2007 Oct 21]. Available from: http://www.fbi.gov/hq/lab/fsc/backissu/jan2005/standards/2005standards4.htm.

Chapter 11
Basic Science
THE CHEMISTRY OF OXIDATION REACTIONS

In this chapter we will explore some of the basics of chemistry. Chemistry is a large subject, to which justice can not be done in a single chapter. I hope, however, that by focusing on the centrality of chemistry to the forensic sciences, I will encourage students to take classes in this fascinating subject. Since I only have time to focus on a tiny bit of chemistry, oxidation reactions seem the most relevant, and in this chapter we will explore the chemistry of oxidation reactions. The most interesting of the oxidation reactions for forensic scientists is the oxidation of gunpowder, because this is the reaction that occurs when a firearm is discharged, causing the release of energy that propels a bullet down the barrel and out toward its target.. Fires and explosions are also oxidation reactions.

The chapter will begin with a discussion of chemistry in the forensic sciences, then continue to present some of the very basic concepts in chemistry such as atoms, molecules, and chemical reactions. Next, we will go into some depth in an exploration of oxidation reactions. The chapter will conclude with a look at the oxidation reactions that happen when ammunition is fired.

11.1 Chemistry in the Forensic Sciences

Chemistry is a subject of prime importance for the forensic sciences. Most forensic scientists either have a degree in chemistry, or at least have a large amount of classwork in chemistry. The Montana State Laboratory of Criminal Investigation has joined the majority of crime labs in the U.S. that will no longer consider students for internships or volunteer positions unless they have had at least the year-long basic college general chemistry sequence, with the laboratory component. This applies not only to the Drug Chemistry, Toxicology, and DNA sections, as you would expect, but also to the Latent Prints and Impressions and the Firearms and Toolmarks sections. This is because the preparation of the basic dyes, solutions, and other materials used in analyses in these sections requires basic chemistry laboratory skills. Also, chemistry classes usually include instruction in basic laboratory safety, which is as important at a crime lab as it is at any lab.

Most advertisements for entry level criminalist positions that I have seen require at least 20 college semester credits in chemistry. This usually means the basic general chemistry sequence, which is usually 10 credits spread over a year, a class or two in organic chemistry or biochemistry, and quantitative analysis and/or instrumental analysis. The latter course is most important, because many things done at a crime lab require quantitation of substances using instruments. Many years ago I applied for a job as a criminalist at a crime lab in California. I did not even make the eligibility list for the position because the quantitative analysis class I took while at a branch of the University of California was offered by the Department of Agriculture, rather than the Department of Chemistry. The classes offered by the two departments were virtually the same, and the term that I wanted to enroll in a quantitative analysis class, the one offered by Chemistry was already full. So I took the agricultural version. I learned a lot of quantitative analysis, but when it came to the State of California's requirements for criminalists, agriculture was not chemistry.

I advise any students who really want a job working in a crime lab to major in chemistry. If not chemistry, then biology, but make sure that you have taken a heavy dose of chemistry classes.

11.2 Atoms, Elements, and Molecules

I'm sure most of you know that matter is composed of atoms of chemical elements. There are 116 elements currently known to science (as of 2007). Only the first 92 of these occur naturally (uranium is number 92) and the others only exist when produced in a laboratory or as the result of a nuclear reaction. An **atom** is a particle, and is the smallest indivisible bit of an element. Each element has a name and a **chemical symbol** or abbreviation. For example, the chemical symbol for uranium is U. The symbol for oxygen is O, carbon is C, and hydrogen is H. Some elements have a two letter symbol, such as He for helium. Often the symbol is a short form of the English name of the element, but some are abbreviations of their name in Latin. For example, Pb is the symbol for lead, plumbum in Latin.

Atoms are composed of combinations of three types of elementary particles, **protons**, **neutrons**, and **electrons**. Protons and neutrons are found in the nucleus of the atom, and the electrons exist in a probability cloud around the nucleus, with indeterminate location until detected. You may find it helpful to think of an atom in terms of the Bohr model, in which the atom resembles a solar system, with the nucleus taking the place of the sun in the center and the electrons orbiting around it. This model is inaccurate, but it has the virtue of being easily visualized. Now, to visualize the true state of affair, the quantum model, don't see the electrons as discrete particles orbiting the nucleus, but instead visualize a hazy cloud, encircling the nucleus on all sides. Out of this cloud, an electron will appear if somebody seeks to detect it. However, until detected the electrons actually exist everywhere in the cloud simultaneously. This sounds too metaphysical to be true, but I am assured that it is the actual reality of an atom, and that it has been verified by numerous experiments.

The number of protons in the nucleus determine the element of which the atom is a particle. For example, carbon has six protons, oxygen has eight, and hydrogen has one. Protons carry a positive electrical charge, and electrons carry a negative electrical charge. Atoms are neutral in charge overall, meaning that there is normally one electron for each proton, which balances the charges of these two types of particles. Neutrons are neutral in charge and are located in the nucleus of the atom along with the protons. The number of neutrons in an atom is usually about the same as the number of protons, but can vary slightly. In most cases the "normal" variety of an element encountered has the same number of neutrons as protons, but different isotopes of an element have more neutrons than the "standard" count. If there are sufficiently many of these "extra" neutrons, the isotope will be radioactive. For example, normal carbon is C^{12}, with the superscripted 12 denoting the total number of protons plus neutrons in the nucleus of the atom. We know that carbon has six protons (one less and it would be boron, one more and it would be nitrogen), therefore the other six of the 12 particles must be neutrons. C^{13} has seven neutrons and exists in nature, though it is much less common than C^{12}. C^{14} is also found naturally on earth, though it is radioactive, and is used in the widely known radiocarbon dating method for determining the age of ancient items.

Molecules are combinations of atoms. For example, an oxygen atom is extremely reactive and will immediately combine with another nearby atom. Often, this other atom is another oxygen atom, so a molecule of oxygen is produced that is symbolized as O_2. This form of notation, O with a subscripted 2, means that this molecule is formed of two oxygen atoms. When one atom joins with another, this is called **bonding**, or forming a bond, between them. So, O_2 is composed of two atoms of oxygen bonded to each other. The subscripted number refers to how many atoms of that element are part of the molecule. Sometimes you will find three oxygen atoms bonded together to give O_3, which is called ozone.

Atoms of different elements can also bond together. For example, water is H_2O. This means that each water molecule is composed of two atoms of hydrogen and one atom of oxygen. Similarly, carbon dioxide is CO_2, composed of one atom of carbon bonded with two atoms of oxygen.

11.3 Chemical Reactions

Under certain conditions molecules can react with each other. When molecules react, the atoms of the elements in the molecules recombine to produce one or more different molecules. In a chemical reaction the atoms are neither created nor destroyed, just recombined to form new molecules. Therefore, the number of atoms of each element has to be the same after the reaction as it was before, though at least some of the atoms will be bound to different atoms than they were originally bonded to. We will see some examples below. In order to actually change an atom into one of another element you need to have a nuclear reaction – not a simple chemical reaction.

11.4 The Oxidation Reactions

The type of reaction we are concerned with in this chapter is called oxidation. **Oxidation** is a type of reaction in which oxygen combines with some other element or molecule. The most familiar form of oxidation reaction for most people is **combustion** (burning), but oxidation is a very common type of reaction and is responsible for the rusting of iron among other things. Let's start our examination of oxidation by focusing on combustion.

11.4.1 Fuels

Many types of materials will burn. One group of molecules that combusts particularly well is the hydrocarbon group. The **hydrocarbons** are a group of molecules that contain hydrogen and carbon. Many hydrocarbons are familiar to us as flammable gasses or liquids. For example, methane is the primary component of natural gas and also a large component of animal flatulence. So, if any of you have ever lit your farts on fire, you were burning methane. The methane in farts is produced by bacteria in an animal's gut that break down indigestible parts of food, such as fiber, and produce methane as a byproduct. Another familiar hydrocarbon is propane gas, which is used to power gas barbecues, gas torches, and the stoves in campers and travel trailers. Octane, a component of gasoline, is another hydrocarbon. Other than methane, most hydrocarbons are produced by refining them from petroleum (crude oil).

Methane is the simplest hydrocarbon and its molecule is symbolized as CH_4. Thus, it is composed of one carbon atom to which four hydrogen atoms are bonded. Methane burns to produce carbon dioxide and water. The oxidation of methane can be written as follows:

$$CH_4 + 2O_2 \rightarrow CO_2 + 2H_2O$$

This means that one molecule of methane (CH_4) reacts with two molecules of oxygen ($2O_2$). Notice that this is two molecules of oxygen – not two atoms – there a total of 4 oxygen atoms here. The reaction produces one molecule of carbon dioxide (CO_2) and two molecules of water ($2H_2O$). Notice that the methane and oxygen molecules have been ripped apart and their atoms have been recombined to form new molecules. The carbon atom originally part of the methane molecule is recombined with two of the oxygen atoms present to form a molecule of carbon dioxide, and the four hydrogen molecules of the original methane are recombined with the two remaining oxygen atoms to produce two water molecules.

Methane is a simple molecule, so its oxidation reaction is simple. Other flammable substances, such as paper or gunpowder, are more complex mixtures of large and small molecules so their oxidation reactions are more complex, but the basic nature of the reaction remains the same.

11.4.2 Energy and Chemical Reactions

Some chemical reactions are used to produce chemicals of interest, and others are used to produce energy. Oxidation is a type of reaction in which the desired product is usually the energy it produces. For example, we burn propane to generate heat to cook food, burn methane to keep warm, burn octane to propel a vehicle, and burn gunpowder to propel a bullet. Reactions that consume energy to produce new molecules out of some original molecules are called **endothermic**. Reactions that produce energy when they occur are called **exothermic**. Oxidation is an exothermic reaction.

Reactions that release energy do so because the products of the reaction are more energetically efficient than the original components of the reaction. In our example of burning methane, the bonds between the carbon atom and the hydrogen atoms in the methane molecule require more energy to establish and maintain than the bonds between carbon and oxygen or hydrogen and oxygen do. This difference in energies is released as heat when the reaction occurs. So, the amount of energy released during an oxidation reaction is the difference between the energy stored in the methane and oxygen molecules and the energy stored in the carbon dioxide and water molecules.

There is one other important energy consideration. Hydrocarbons are fairly stable. They don't burn spontaneously, except in very rare and exceptional conditions. If this weren't true they would be very dangerous to have around. What you have to do to get them to burn is to strike a match or in some other way give them a little kick of energy. Once a hydrocarbon starts to burn, the energy that the oxidation reaction liberates keeps it burning until either the fuel is exhausted or the oxygen is exhausted. This small bit of starting energy is referred to in general as the energy of activation. When referring to oxidation reactions, the energy of activation is usually called **ignition energy**, and the device or mechanism that supplies the ignition energy is called an **ignitor**.

The ignition energy is needed because in order for the oxidation reaction to proceed at least one of the bonds between a hydrogen and a carbon in the hydrocarbon must break. That frees a hydrogen to react with oxygen, energy is liberated, and the reaction commences. So, the activation or ignition energy is the amount of energy required to break one of the hydrogen-carbon bonds.

11.4.3 Making a Fire

If you have read this far you now know how to make a fire. You need three things – a fuel, oxygen, and an ignitor to supply ignition energy. However, we have not yet considered the fact that in order for oxidation to occur there must be a proper mixture of fuel and oxygen.

For each fuel there is a certain range of mixtures of the fuel with oxygen that will allow combustion to occur. In a car engine, for example, the mixture of gasoline with air is critical. If there is too much gasoline the mixture is too rich and the car won't run well. Similarly, if there is not enough gasoline the mixture is too lean and the car won't run well. Cars have components, such as fuel injectors that are designed to keep the gasoline to air mixture within the range that allows good burning, which in this case is between 1.3 and 6 parts per hundred of gasoline to air.

11.4.4 Spontaneous Combustion

Spontaneous combustion occurs when something starts burning without being ignited. Arson suspects often rely on the excuse of spontaneous combustion, but rarely get away with it. The fact that there are specific conditions under which a fuel will burn, makes spontaneous combustion a rare phenomenon.

Spontaneous combustion can occur when some natural heat producing process occurs in poorly ventilated containers or areas. For example, improperly dried hay stored in barns provides an excellent situation for the growth of bacteria whose activities will generate heat. If the hay is not properly ventilated, the heat will build to a level that will support other types of heat-producing chemical reactions in the hay. Eventually, as the heat rises, the ignition temperature of hay is reached, spontaneously setting off a fire. Another situation in which spontaneous combustion is known to occur is when rags soaked with certain types of highly unsaturated oils, such as linseed oil, are stored in improperly ventilated containers. Heat builds up to the point of ignition as a result of a slow, heat-producing chemical reaction between the air and the oil. However, spontaneous combustion will not occur with hydrocarbon lubricating oils, and it is not expected to occur with most of the fats and oils that are found in a household.

11.4.5 Sources of Oxygen

For many oxidation reactions, air is the source of oxygen. About 20% of the air in the atmosphere at sea level is oxygen. However, there are other sources of oxygen that can be used. In particular, air does not exist within firearm cartridges or bombs. Therefore, the gunpowder inside a cartridge and the explosives in a bomb do not rely on oxygen from the atmosphere, and get it from some other chemical. Chemicals that supply oxygen are known as **oxidizing agents**.

11.4.6 An Example: The Oxidation of Black Powder

Let's look at the oxidation of black powder as an example of these more complex, but highly relevant, oxidation reactions. Black powder is a form of gunpowder, normally used in muzzle-loading rifles. Most firearms cartridges use another form of gunpowder, yet black powder can serve as an example for the types of explosions of relevance to forensic scientists. Black powder, and other gunpowders are classified as low explosives, following a classification we will explore further in the chapter on explosives. It is composed of a mixture of the following chemical ingredients: 75% potassium nitrate (KNO_3, also known as niter or saltpeter); 15% charcoal, which is nearly pure carbon (C); and 10% sulfur (S). In this combination the potassium nitrate acts as an oxidizing agent, and the carbon in the charcoal is the fuel. As ignition energy is applied to black powder, oxygen is liberated from the potassium nitrate and combines with the carbon in the charcoal to produce heat and gases. The chemical reaction of gunpowder can be written like this:

$$2KNO_3 + 3C + S \rightarrow N_2 + 3CO_2 + K_2S + heat$$

N_2 is nitrogen gas and K_2S is potassium sulphide, a solid. These molecules are produced by the nitrogen (N) and potassium (K) liberated from the KNO_3 when the oxygen is released. In fact the purpose of mixing sulphur into black powder is to give something other than oxygen for the potassium to combine with. Without the sulphur, the potassium would tend to react with some of the oxygen, making that oxygen unavailable for oxidizing the carbon, which would cause the explosion produced to be weaker.

11.5 Explosives in Ammunition

Gunpowder is an explosive. What makes something an explosive as opposed to a simple fuel is the speed of the oxidation reaction. Explosives burn much more rapidly than a normal fuel such as a hydrocarbon. A detailed examination of explosives will follow in a later chapter.

Most firearm cartridges have two explosives in them. One is the gunpowder that explodes and propels the bullet down the barrel of the firearm and toward a target. The second explosive is found in the primer of the cartridge. The explosive in the primer is chosen for its low ignition energy. What you want in a primer is an explosive with an ignition energy low enough that the small amount of heat and shock caused by the firing pin or hammer striking it is enough to ignite it. This causes a flash of energy which ignites the gunpowder, thus firing the bullet.

11.6 Questions for Study and Review

1. Most entry-level positions at a crime lab require applicants to have college level classes in what field?
2. Define the following terms: atom, element, chemical symbol, proton, neutron, electron, isotope, molecule, chemical bonding.
3. Define the following terms: oxidation, combustion, fuel, hydrocarbons, exothermic, endothermic, ignition energy, ignitor, oxidizing agent.
4. What are the chemical formulas for molecular oxygen, methane, carbon dioxide, water?
5. What chemical substances are used in making black powder, and what does each substance do in the mixture?
6. In your own words, describe the two types of explosives used in ammunition.

11.7 Sources

Bauer RC, Birk J, Marks PS. 2006. A Conceptual Introduction to Chemistry. Columbus (OH): McGraw-Hill Companies.

Klatt EC. 2007. Firearms Tutorial [Internet]. [cited 2007 Oct 18]. Available from: http://library.med.utah.edu/WebPath/TUTORIAL/GUNS/GUNINTRO.html.

Michon, GP. 2007. Black Powder / Blackpowder / Gunpowder [Internet]. [cited 2007 Oct 18]. Available from:
http://home.att.net/~numericana/answer/chemistry.htm#blackpowder.

Wikipedia contributors. 2007 Cartridge (firearms) [Internet]. Wikipedia, The Free Encyclopedia; 2007 Oct 15, 17:15 UTC [cited 2007 Oct 18]. Available from: http://en.wikipedia.org/w/index.php?title=Cartridge_%28firearms%29&oldid=164752164.

Wikipedia contributors. 2007. Combustion [Internet]. Wikipedia, The Free Encyclopedia; 2007 Oct 18, 18:49 UTC [cited 2007 Oct 18]. Available from: http://en.wikipedia.org/w/index.php?title=Combustion&oldid=165462945.

Wikipedia contributors. 2007. Spontaneous combustion [Internet]. Wikipedia, The Free Encyclopedia; 2007 Sep 11, 07:44 UTC [cited 2007 Oct 18]. Available from:
http://en.wikipedia.org/w/index.php?title=Spontaneous_combustion&oldid=157109004.

Chapter 12
Basic Science
BALLISTICS AND THE PHYSICS OF PROJECTILES

In this chapter we will examine the physics of motion, with particular attention to the motion and energy of a projectile, a topic referred to as ballistics. We will apply ballistics to the subject of firearms. A solid foundation in ballistics helps the firearms examiner with many tasks, probably most obviously the determination of bullet pathways through space.

The chapter will start with a consideration of ballistics in general, then will proceed to investigate, in turn, each of the phases of ballistics as related to firing a firearm.

12.1 Ballistics

Ballistics is the study of the motion of projectiles, such as bullets. There are three branches of ballistics, when it is applied to firearms. **Internal ballistics** is the study of projectiles inside the firearm. **External ballistics**, is the study of the behavior of the projectile as it travels through the air between the muzzle of the firearm and the target. **Terminal ballistics** is the study of the interaction between the projectile and the material of the target it strikes.

12.2 Internal Ballistics

Internal ballistics is very important in firearms examination because it involves things that happen to the cartridge and bullet while they are still inside the firearm. Many of the details that a firearms examiner looks at when examining bullets or casings involve marks or damage to the bullet or casing that were received while they were inside the firearm.

The process of firing a bullet is fairly straightforward. A cartridge, which consists of a primer, a chamber filled with gunpowder, and a bullet, is loaded into the firing chamber of the firearm. When the trigger of the firearm is pulled, its firing pin or hammer strikes the primer, igniting the small amount of explosive therein, which produces a burst of heat and energy. The primer's burst of energy causes the gunpowder in the cartridge to ignite, creating energy and gasses that propel the bullet through the barrel of the firearm and toward the target. Finally, the spent cartridge casing may be ejected from the firing chamber, depending on the type of firearm.

During this process there are several opportunities for the characteristics of the firearm to be transferred to the casing or the bullet. Marks may be left on the casing as it is moved into and out of the firing chamber. Some sort of mark will almost always be left on the casing by the firing pin or hammer. Marks will almost always be transferred to the bullet from imperfections or rifling in the barrel of the firearm. These marks will contain a mixture of class and individual characteristics that may allow a firearms examiner to match bullets or casings recovered at a crime scene to one specific firearm.

12.3 External Ballistics

External ballistics is the movement of a projectile, such as a bullet, through the air toward a target. The path that the bullet follows is called its trajectory. The trajectory of a projectile may be important to a forensic scientist as he or she attempts to interpret or reconstruct a crime scene in terms of what events actually happened. In particular, figuring out a bullet's trajectory can help in determining where the shooter and the victim were located and the spatial relationships between them.

12.3.1 Ballistic Curves

The trajectory that a projectile follows has one of a family of shapes called ballistic curves, often referred to as simply "the" ballistic curve. The trajectory is always curved. Gravity acts to draw the projectile toward the ground. Therefore, the firearm must fire the bullet slightly upward, so that by the time it reaches the target it has dropped sufficiently to strike the desired point. See Figure 12.1 for an example ballistic curve. Ballistic curves are well known and mathematically describable using formulas. One fairly simple set of formulas describes the X and Y positions of the projectile on a ballistic curve at a certain amount of time after it was propelled toward the target. These equations may be written like this.

$X = \frac{1}{2}a_x t^2 + v_x t + x_0$, where X = the distance traveled horizontally at time t, a_x = the acceleration of the projectile in the horizontal direction, v_x = the initial velocity of the projectile in the horizontal direction, x_0 = the initial horizontal position from which the projectile was propelled.

$Y = \frac{1}{2}a_y t^2 + v_y t + y_0$, Where Y = vertical height relative to the starting position at time t, a_y = the acceleration of the projectile in the vertical direction, v_y = the initial velocity of the projectile in the vertical direction, y_0 = the initial vertical position from which the projectile was propelled.

Figure 12.1: A Ballistic Curve

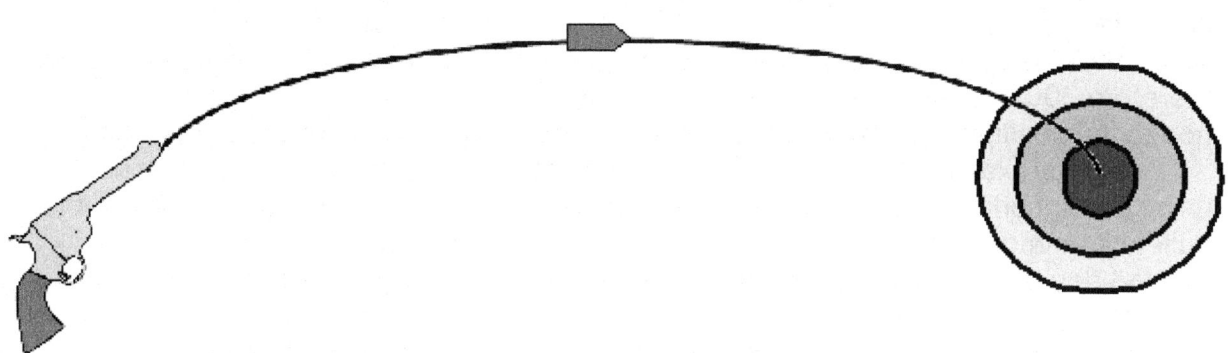

12.3.2 Factors that Affect the Shape of a Ballistic Curve

We can learn a great deal by looking at the various factors that determine the shape of the ballistic curve. The following factors all have some impact on the exact shape of a ballistic curve and are important for us to understand.

1. Location of the firearm and target. It is important to know the starting and ending points of the trajectory.
2. Projectile velocity. Velocity is the speed at which the projectile travels through the air. This is usually expressed as muzzle velocity, which is the speed of the bullet as it leaves the barrel of the firearm. The greater the muzzle velocity the "flatter" the curve.
3. Atmospheric drag or air resistance. This is the friction of the projectile against the molecules of air, which tends to slow the projectile down.
4. Weight (mass) of the projectile. The heavier the projectile, the larger the effect of gravity pulling it toward the ground and the less flat the ballistic curve.
5. Firing angle. The firing angle is the angle between the barrel of the firearm, and a level line toward the target. When a firearm is aimed, the firing angle is almost always at least a little upward, in order to counteract the effect of gravity dragging the projectile toward the ground. Most sighting mechanisms on firearms take this into account.

12.3.3 An Example: Was it Murder or an Accident

Let's look at an example of using the ballistic curve equations. A suspect says she was shooting her pistol at a target 20 feet away, missed, and accidentally shot her husband whose body was discovered about 660 feet (about 200 meters) away from where she claims she was standing. Is this possible? A firearms examiner determines the muzzle velocity of her pistol to be exactly 984 feet per second (300 meters per second) with the ammunition she was using. She says that she was shooting from the hip, which puts the initial height of the pistol about 1 meter (m) above the ground, and that the target was about level with the pistol. We can assume that the acceleration in the horizontal direction is 0, and that the only acceleration in the vertical direction is the acceleration due to gravity, equal to –9.8m/s. We can assume that the muzzle velocity of the bullet is all in the horizontal direction (since the target was level with the firearm), therefore 300 m/s; and that the vertical velocity is 0. We can arbitrarily set the initial horizontal position, x_0, at 0, and we know that the initial vertical position, y_0, is 1m.

In our example, how long will the bullet stay in the air? We can determine this using the second of the equations above and solving for elapsed time, t, at which the bullet hits the ground, *i.e.* its height, Y, becomes 0. Therefore, we can set up the second equation as follows. In writing these calculations I use the symbol '➜' to mean "implying". In this context, implying means that the previous equation in the sequence can be rewritten in the way shown without violating any principles of mathematics. Note that in the calculations below I have converted everything to metric units for the calculations. This is actually not necessary – the formulas work fine using feet and feet per second.

$$Y = 0 = \tfrac{1}{2}a_y t^2 + v_y t + y_0 = \tfrac{1}{2}(-9.8)t^2 + 0 + 1 = -4.9t^2 + 1 \;\rightarrow\; -4.9t^2 = -1$$
$$\rightarrow\; 4.9t^2 = 1 \;\rightarrow\; t^2 = 1/4.9 \;\rightarrow\; t^2 = .20 \;\rightarrow\; t = \sqrt{(.20)} = .45 \text{ seconds.}$$

Therefore, the bullet will stay in the air for 0.45 second, which is a little less than half a second.

In our example, how far does the bullet travel in the 0.45 seconds it is in the air, ignoring air resistance? Now we can use the first of the formulas above. We will set t = 0.45, which allows us to set up the equation as follows, then plug and chug to find the distance traveled, X.

$$X = \tfrac{1}{2}a_x t^2 + v_x t + x_0 = (\tfrac{1}{2})(0)(0.45^2) + (300)(0.45) + 0 = 0 + 300(0.45) + 0 = 135m$$

We now know that in the 0.45 seconds the bullet was in the air it traveled about 135 meters. 135 meters converts to about 443 feet.

The suspect's story can not be true as told. She could not have shot her husband who was 660 feet away while aiming at a target 20 feet away because the bullet would only travel 443 feet under these conditions. Further, the calculations we made ignored air resistance, which slows a bullet and prevents it from traveling quite as far as the calculations suggest. Either her husband was much closer to her than the 660 feet she claims (she was actually standing closer to him or she moved his body), or the pistol was not aimed level with the target, or some other factor in her story is incorrect. Therefore, we conclude that this incident should probably be investigated as a possible murder.

But wait! Let's be thorough investigators and not stop at the quick solution. We know that the firing angle has an impact on how far the bullet travels. One very likely explanation for how the woman might have hit her husband 200 meters away is that she was actually, perhaps unintentionally, aiming somewhat upward rather than level with the target. In fact, the firing angle must have been higher than she intended if she missed the target and fired over the top of it. She says that she was shooting from the hip, which is done by holding the firearm near the hip, rather than raising it to use the sighting mechanism to aim. A person's control of aim and firing angle is poor under these conditions. When the firearm is not aimed perfectly level, some of the bullet's velocity is in the vertical direction, which means that it will stay aloft longer and travel farther. When a projectile is fired at some velocity V, at an angle θ to the horizontal, the amount of velocity in the horizontal and vertical directions can be calculated using the pair of formulas below, in which cosθ is the cosine of θ and sinθ is the sine of θ. Cosines and sines of any angle can be looked up in a table or calculated by pressing a button on a scientific calculator.

$$\text{Horizontal velocity} = v_x = V(\cos\theta)^2$$
$$\text{Vertical velocity} = v_y = V(\sin\theta)^2$$

Let's assume that due to her poor aim when shooting from the hip, the woman was actually aiming very slightly upward, at an angle of 5° from perfectly horizontal. Using the same values as in our previous example this gives us the following velocities.

$$v_x = V(\cos\theta)^2 = 300(\cos 5)^2 = 300(0.9962)^2 = 300(0.9924) = 297.72 \text{ m/s}$$
$$v_y = V(\sin\theta)^2 = 300(\sin 5)^2 = 300(0.0872)^2 = 300(0.0076) = 2.28 \text{ m/s}$$

Recalculating the time the bullet stays in the air using a non-zero velocity in the vertical direction gives a quadratic equation, which may be solved using the methods in any algebra textbook. With a vertical velocity of 2.28 m/s the bullet now stays in the air for 0.741 seconds. In this time it travels at 297.72 m/s for about 221 meters. Since the husband was 200 meters away, the woman could have accidentally shot him if her firing angle was only 5° upward from level. Therefore, this incident was probably an accident after all.

12.3.4 Other Aspects of Projectile Behavior

In addition to the factors that affect a projectile's trajectory, there are some characteristics of projectile behavior that have an effect on how well it travels through the air and on its accuracy.

As a projectile travels through the atmosphere it generates **shock waves**, which are caused by the air being compressed as the projectile moves through it. Most projectiles will be accompanied by at least two shock waves, the strongest and most important of which emerges from the bullet's nose and is called the Mach cone. A projectile also generates a wake, which is a highly turbulent flow of air behind the base. All traveling projectiles have a wake. Figure 12.2 illustrates these two shock waves.

Figure 12.2: Shock Waves Produced by a Moving Bullet

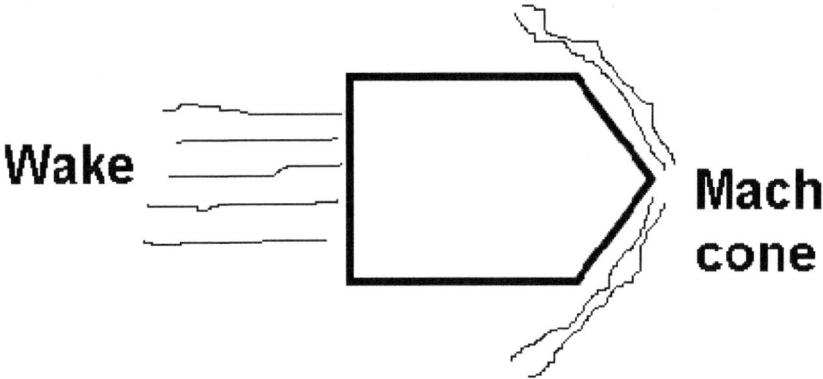

Yaw is the difference between the direction the projectile is traveling and the direction its axis is pointing. Inevitably, the point of the projectile will point slightly up, down, or to one side or the other, compared to the direction it is traveling in. The main effect of yaw is that it increases the amount of atmospheric drag, but it also decreases accuracy. Figure 12.3 illustrates yaw.

Figure 12.3: Yaw

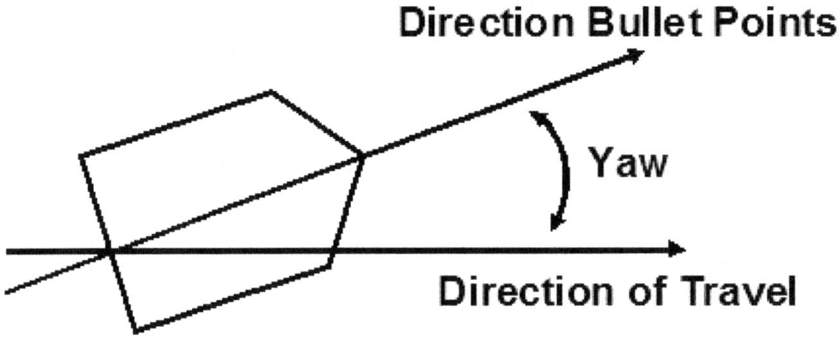

Most firearms have rifling in their barrel which consists of ridges that twist in a certain direction and impart a **spin** to the projectile as it passes through. Spin is very important. Without spin, the effect of yaw causing atmospheric drag will cause the bullet to begin tumbling. A tumbling bullet can behave unpredictably and is very inaccurate. Spin prevents tumbling by stabilizing the motion of the bullet. The overall effect of spin is to increase accuracy.

12.4 Terminal Ballistics

Terminal ballistics is the aspect of ballistics concerned with what happens when a projectile strikes its target. As forensic scientists we are most concerned with a subfield of terminal ballistics called "wound ballistics" which is concerned with the effect of a projectile on animal (human) tissue.

12.4.1 Kinetic Energy

In general, the amount of damage to the target is determined by the amount of kinetic energy transferred by the projectile to the tissue it strikes. The kinetic energy of a projectile, abbreviated as K, is a function of the mass of the projectile, the velocity of the projectile, and the acceleration due to gravity (in a downward direction). These factors relate together as shown in this formula:

$$K = \frac{MV^2}{2g}$$

where: M = projectile mass (weight), V = projectile velocity, g = gravitational acceleration.

Gravitational acceleration is a constant (on Earth), thus projectile weight and velocity determine the kinetic energy possessed by a projectile. In order to cause the maximum damage to the target you want to have as heavy a projectile as possible, moving as fast as possible. Bullets are often made of lead, which is one of the heaviest metals.

12.4.2 The Mass vs Velocity Debate

People who are interested in causing as much damage as possible, or even death, to the target often debate the merits of a heavier bullet vs a faster moving bullet. In terms of kinetic energy the speed of the bullet is clearly most important. Kinetic energy increases in linear proportion to the bullet weight, so if you double the weight of the bullet while keeping the speed constant you double the kinetic energy. However, the kinetic energy increases proportionally to the velocity squared, so if you double the speed of the bullet while keeping its weight constant you increase the kinetic energy by a factor of $2^2 = 4$.

Some, authorities, however, believe that kinetic energy is not the primary concern in wounding capability. This is partly true because what is important in wounding capability is how much of the kinetic energy is transferred to the target. I think you can easily convince yourself that you shouldn't overlook the importance of projectile mass in causing damage by considering the case of a particle of zero mass traveling at infinite speed (theoretically, the speed of light is the fastest anything can travel). What you are describing is a photon of light, and we are being bombarded by billions of them at this very moment without damage (so long as it's just visible light), and even those forms of EMR that cause damage do so via interactions with molecular bonds and not by imparting kinetic energy to what they strike.

Also, the nature of the wound is important in that a small cut to a major artery can kill more effectively than a large amount of damage to some non-critical tissue.

This is a complex issue to which no definitive answer can be found, despite a large amount of theoretical and practical research. The debate continues whenever hunters, peace officers, or some types of criminals meet.

12.4.3 Transfer of Kinetic Energy to the Target

A variety of factors determine the actual amount of kinetic energy transferred to a target, and these are important considerations in terminal and wound ballistics.

The yaw of the projectile has an effect. If the projectile doesn't strike exactly point forward it causes more damage, but doesn't penetrate as far. In the worst cases of yaw, if the projectile is actually tumbling end over end, a maximum amount of damage is caused, but the penetration is minimal.

The size or caliber of a bullet is important in that a larger one will cause more damage, even if it is the same weight as a smaller one. A projectile that is round will cause more damage than a pointed one. If a bullet deforms inside the body, especially if it become flattened, it will cause more damage. In general, there is a tradeoff between penetration and damage. If a bullet stops inside the target, then all the kinetic energy of its motion has been transferred to the target in the form of damage. If a bullet passes completely through the target, however, it has failed to transfer all of its kinetic energy to the target. Different types of bullets are designed for a specific penetration vs damage strategy. For maximum penetration, some bullets have pointed tips and are coated with a jacket made of a harder metal, such as copper, brass, or steel. Some may even add a teflon coating on top of this. These bullets will penetrate well, perhaps even penetrating body armor or through wood, glass, or metal. However, they minimize the actual amount of energy transferred to the target, thus minimizing some aspects of the damage caused. Bullets designed to maximize damage have rounded or flat tips and are hollowed out to encourage the bullet to deform. A hollow point bullet will deform to a mushroom shape and cause a maximal amount of damage, trading off penetration.

12.4.4 Wound Ballistics and Tissue Damage

There are three primary mechanisms of by which a projectile causes tissue damage. These are laceration and crushing, shock waves, and cavitation.

Laceration and crushing are caused by the projectile ripping through, smashing, and displacing the tissues in its track. This mechanism is thought to be the most important one for handguns.

Shock waves, similar to those produced as the projectile flies through the air can occur in tissue if the projectile velocity is greater than 2,500 feet per second. The shock waves cause compression of tissues that lie ahead of and around the bullet. This is thought to be the most important wounding mechanism in rifles, which often produce projectile velocities in this range. The velocity of bullets fired from handguns is usually too slow to cause damage by this mechanism.

Cavitation occurs when the projectile pushes tissue aside, creating a temporary cavity that collapses once the projectile has passed. Even though the tissue is merely pushed aside it is damaged in the process. Cavitation occurs at projectile velocities over 1,000 feet per second, which can be found in many handguns and all modern rifles. Cavitation is a significant wounding mechanism and can even cause bone fracturing when the bullet passes close to the bone but doesn't actually touch it.

The examination of bullet paths and wounds in a body is usually made by a forensic pathologist rather than a firearms examiner, who examines the bullets, firearms, and casings. By looking at a bullet's path through the body and the wounds it has caused, a forensic pathologist can determine whether the wound was likely to have caused death or not.

12.4.5 Range of Fire

One of the most common determinations made by a firearms examiner, is the range of fire. This classification is made on the basis of the appearance of the wound and gunpowder residues. Whenever a gun is fired, there is a considerable discharge of powder, in addition to the bullet. This powder consists of unburned gunpowder and some of the products of the oxidation of the gunpowder that did burn. The powder travels a certain distance, normally about three feet, more or less. Therefore, the amount and distribution of powder relative to the gunshot wound can give information about how far away the gun was from the victim.

Gunshot wounds are typically classified as: contact, intermediate range, or distant range. Contact wounds occur when the muzzle of the firearm is in contact with the target. Distant range is far enough away that no powder is transferred to the target. Intermediate range (also called close range) is everything in between contact and distant, and is diagnosed by the fact that some powder is transferred to the target. When powder contacts skin it makes small marks called **powder stippling**, that result from slight penetration of the powder and slight burning due to the fact that the powder is hot.

Examination and evaluation of wounds usually brings a forensic pathologist into the picture. Forensic pathologists often collaborate with firearms examiners on cases that involve a shooting.

Contact entrance wounds characteristically have soot on the outside of the skin, and muzzle imprint, or laceration of the skin from the effects of expanding gases. Intermediate (close-range) wounds will show powder stippling, but lack a muzzle imprint and laceration. The area of powder stippling will depend upon the distance from the muzzle. Distant range wounds lack powder stippling and usually exhibit a hole roughly the caliber of the projectile fired.

12.4.6 Entrance and Exit Wounds

Entrance wounds into bone or other hard material typically exhibit beveling of the surface in a direction away from the weapon (the direction in which the projectile is traveling). This results from bone being chipped off by the passage of the projectile. The chips of bone are blown in the direction the projectile is traveling and are removed from the surface farther along the projectile's trajectory. Figure 12.4A illustrates an entrance wound.

As discussed above, some bullets are designed to stop inside the target without exiting. In many situations, however, an exit wound will be present because the bullet was fired from a firearm powerful enough to propel the bullet completely through the target, or the target may be thin in the location the bullet strikes.

Exit wounds are generally larger than entrance wounds, due to the fact that the

Figure 12.4: Entrance, Exit and "Keyhole" Bullet Wounds in Bone

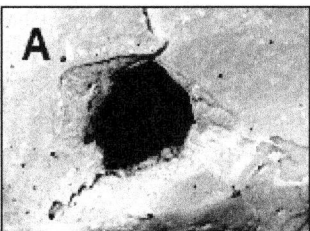
A typical entrance wound in bone. The arrow indicates where bone has been bent in the direction of the projectile's travel (away from us).

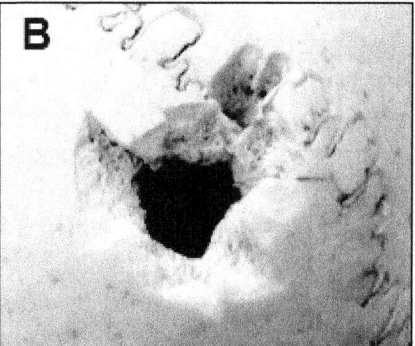
A typical exit wound in bone. Note the material has been blasted off in the direction of the projectile's travel (toward us).

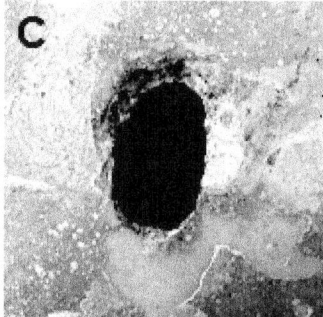
A typical "keyhole" wound in bone. Note that the upper portion resembles an exit wound, and the lower portion resembles an entrance wound.

bullet has expanded inside the target. Exit wounds usually do not exhibit powder stippling, even at close range. In bone, beveling may be present on the side away from the entrance wound. In some cases there are multiple exit wounds because the bullet strikes something hard and fragments inside the target. If the exit wound is "shored" by the area of exit being in contact with a firm support, then the exit wound may take on appearances of an entrance wound because the bullet may ricochet back into the target. An exit wound in bone is shown in Figure 12.4B.

If the projectile strikes the bone at an angle, it may produce an effect wherein the wound takes on characteristics of both an entrance and exit wound. This produces a pattern where one part of the wound will be larger than another and the overall shape of the wound is like a keyhole. A keyhole wound in bone is shown in Figure 12.4C.

12.4.7 Conclusions from Wound Ballistics

By putting together all the information about entrance and exit wounds, projectile trajectory, and range of fire, a forensic investigator can often draw a scientifically supportable conclusion about the events that surrounded a shooting. Often the person who does this is a forensic pathologist, sometimes working in collaboration with a firearms examiner.

12.5 Questions for Study and Review

1. In your own words, define ballistics and its three branches as applied to firearms.
2. In your own words, describe what happens to ammunition and all its components between the time it is loaded into a firearm and the time a bullet hits a target.
3. Discuss how the nature of the motion of a rifle bullet through space determines how the rifle is aimed.
4. A rifle that fires a bullet with a velocity of 500 meters per second is fired by a tall man. The man's shoulder is 2 meters above the ground (I said he was tall!) and the rifle is aimed exactly level with the ground. How long will the bullet stay in the air, and how far will it travel before hitting the ground?
5. How far would the bullet travel if it was aimed at an angle of 15 degrees upward from the level, i.e. $\theta = 15$. A calculator says that $(\sin 15)^2 = 0.07$ and $(\cos 15)^2 = 0.93$, approximately.
6. In your own words, define the following terms as they relate to the motions of projectiles: shock waves, mach cone, wake, spin, yaw.
7. In your own words, summarize the debate over whether the velocity of a projectile or the mass of a projectile is more important in causing damage to a target. Which of these does the formula for kinetic energy suggest is more important?
8. Discuss the factors that have an impact on how much of a projectile's energy is transferred to a target.
9. In your own words, define the following terms as they apply to wound ballistics: laceration and crushing, shock waves, cavitation.
10. What are the three ranges of fire? How is range of fire determined?
11. How are entrance wounds distinguished from exit wounds in a hard material?

12.6 Sources

Fr. Frog. 2007. A (Very) Short Course in Internal Ballistics [Internet]. [cited 2007 Oct 18]. Available from: http://www.frfrogspad.com/intballi.htm.

Intermedia Outdoors, Inc. 2007. Ballistics For Dummies [Internet]. [cited 2007 Oct 18]. Available from: http://www.rifleshootermag.com/shooting_tips/ballistics_0303/index.html.

Jason A. n.d. Crime Scene Analysis & shooting incident reconstruction [Internet]. [cited 2007 Oct 18]. Available from: http://www.projectile.com/AJpro.htm.

Nennstiel R. n.d, How Do Bullets Fly [Internet]. [cited 2007 Oct 18]. Available from: http://www.fulton-armory.com/fly/index.htm.

Ross AH. 1995. Gunshot Wounds: a Summary [Internet]. [cited 2007 Oct 18]. Available from: http://www.soton.ac.uk/~jb3/bullet/gsw.html.

Wilson JDD, Buffa AJ, Lou B. 2006. College Physics. Menlo Park (CA) Benjamin Cummings.

Chapter 13
FIREARMS EXAMINATION

In this chapter we will investigate firearms examination. As discussed in previous chapters, firearms examination is one of the criminalistic specialties. The primary concern of firearms examination is to determine if a bullet, cartridge case, or other ammunition component was fired from a specific firearm. However, firearms examiners also perform other types of analyses, such as firearms function examination and range of fire determinations. We will explore these types of analyses in this chapter and also look at some of the characteristics of firearms and ammunition.

The chapter will begin with a brief history of firearms examination and a consideration of some of the statistics related to firearms fatalities in the United States. It will continue with an introduction to the types of firearms and ammunition. Next it will consider the tasks of firearms examiners. The chapter will conclude with an introduction to databases of firearms related information.

13.1 The History of Firearms Examination

The history of firearms examination goes back at least to the 1700's, when examiners used fairly crude methods to match a bullet (usually a ball) to the gun from which it was fired. In the early 1800's Henry Goddard, matched a bullet to the mold used to make it and solved a famous murder case.

The scientific matching of firearms evidence is a product of the late 1800's and 1900's. For example, Professor Alexandre Lacassagne, working in the late 1800's and early 1900's in France, used rifling marks on bullets to match them to the guns that fired them. By the 1920's the field of firearms examination was revolutionized by the development of the comparison microscope, as described in chapter 1. This revolutionary instrument was the product of the work of four men. Charles E. Waite compiled a massive amount of data from gun manufacturers. John H. Fisher, a physicist and Howard Gravelle, a chemist, developed the actual instrument. Calvin Goddard (so far as is known, no relation to Henry Goddard), a U.S. Army Colonel, worked out the actual process of doing the comparisons to make an identification. These specialists established what is essentially the procedure for matching firearms evidence that is used today.

13.2 Some Statistics on Firearms Related Deaths

The statistics on gun related deaths are appalling. According to statistics compiled in the year 2000, the U.S. leads the world in firearms related deaths due to homicide and accident, though there are countries that may have higher firearms related suicide rates. In particular, the U.S. stands out among the more industrialized countries as having approximately three times as many firearms related deaths as our nearest competitor (Finland). When firearms related deaths involving children are examined, the death rate in the U.S. is four times greater than that of our nearest competitor (Scotland). This is a national embarrassment. Better training in firearm use and safety could help alleviate this problem. Switzerland provides a good example. Boys and girls are taught firearm use and safety beginning at a young age, and military training that includes use of firearms is mandatory for young men. The number of homicides or attempted homicides involving firearms, reported by Swiss police in 2006, was 34 in the entire country – a tiny fraction of the figure for the U.S.

13.3 Firearms

Modern firearms are manufactured in a variety of shapes and sizes to fit multiple purposes. There was a time when the forensic scientist was faced with a less complex situation, with fewer types, models, and mechanisms available for use. Unfortunately, the proliferation of firearms in the U.S. includes not only sheer numbers but also a staggering array of models. Included in this array are semi-automatic and even automatic weapons built primarily for military usage. Virtually any type of gun can be found on the streets in use by youth gangs, persons involved in drug trafficking, paramilitary "survivalists" and even what we would consider "ordinary" citizens.

13.3.1 Handguns

From the very start, handguns (pistols) were conceived as compact weapons for self defense. Even though today there are handguns made specifically for target competition or hunting, most are still designed with self defense in mind. Thus, handguns are compact for concealability and ease of carrying. This becomes a part of the legal definition of a handgun, as they are considered "concealable" and therefore deemed dangerous and are controlled by law in most states. The two most common defensive handguns are the double action revolver and the semiautomatic pistol. These two types of handguns are shown in Figure 13.1.

Revolvers are simpler in design, and are usually considered more reliable and safer than semiautomatics. Revolver are also easier to master. Revolvers are thought to be more accurate than semiautomatics, thought this is a matter of intense debate. The limitations of revolvers are that they are limited to six shots or fewer, are relatively slow to reload, the gap between barrel and cylinder makes them less efficient, and the trigger pull is greater. Revolvers come in a variety of barrel lengths. Some have short barrels for concealability and others have longer barrels for accuracy or greater punch. There are two types of action for revolvers: single-action and double-action. Single-action revolvers require the shooter to cock the hammer before the trigger is pulled. Double action revolvers cock and fire with a single pull of the trigger.

Semiautomatic pistols use the energy generated by the fired cartridge to eject the empty cartridge case, load the next cartridge, and ready the firing pin for the next shot. This means that semiautomatics are faster to shoot, since they can be fired as fast as one can pull the trigger. There are some revolver enthusiasts, however, who believe that they can fire a revolver as quickly as a semiautomatic pistol by "fanning" (holding back the trigger while using the heel of the other hand to rapidly pull the hammer back and let it go), though studies have shown that it simply isn't possible. Instead of being limited to the five or six rounds of the revolver, many semiautomatic pistols are designed to carry 15 to 19 rounds or more. The main disadvantage of the semiautomatic pistol is that it is more complex, therefore less reliable and requiring more practice to use well. The final disadvantage of a semiautomatic pistol is that the cartridges need to be fairly short in order to pass through the more complex pathway from the magazine to the firing chamber. Therefore, they are limited in length and in the amount of powder that can be put in them. So, some revolver cartridges can be more powerful.

Figure 13.1: Handguns

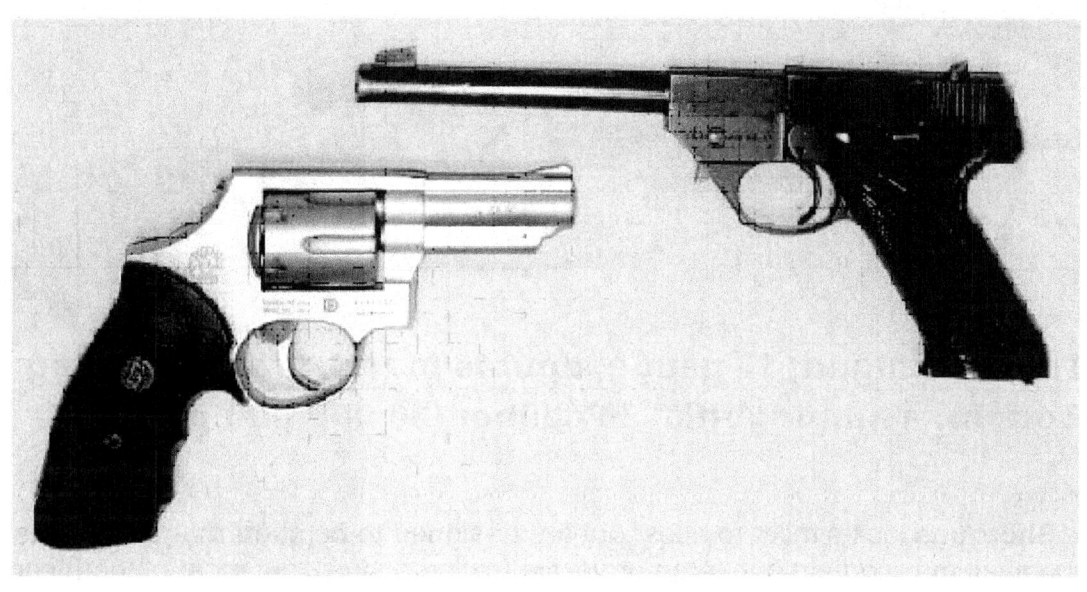

Left: Revolver, double action, 357 caliber
Right: Semiautomatic pistol, 22 caliber

13.3.2 Rifles and Shotguns

Rifles and shotguns are referred to as long guns. They have longer barrels and a stock. They are designed to be fired using both hands. For these reasons they are harder to carry, conceal, and use. Therefore they are less commonly used in the commission of crimes and are more loosely regulated than handguns. A rifle and a shotgun are shown in Figure 13.2.

Rifles were designed for hunting and war, and therefore designed to be accurate at long distances. Rifles get their name from the fact that they have **rifling** in the barrel, which consists of grooves designed to add spin to a bullet in order to increase accuracy. There are several mechanisms used to get the cartridge into the firing chamber of a rifle, called **actions**. One is simply to stick it in by hand each time. A rifle of this sort is called a single shot rifle. Bolt action rifles have a handle, called the bolt, that is manipulated to eject the old cartridge and feed in a new one from a magazine. These are still common for hunting rifles. Military rifles are usually semiautomatic or automatic, having a detachable magazine holding 5 to 50 cartridges. Pump action and lever action rifles also exist, usually in smaller calibers, and have magazines below the barrel.

Figure 13.2: A Shotgun and a Rifle

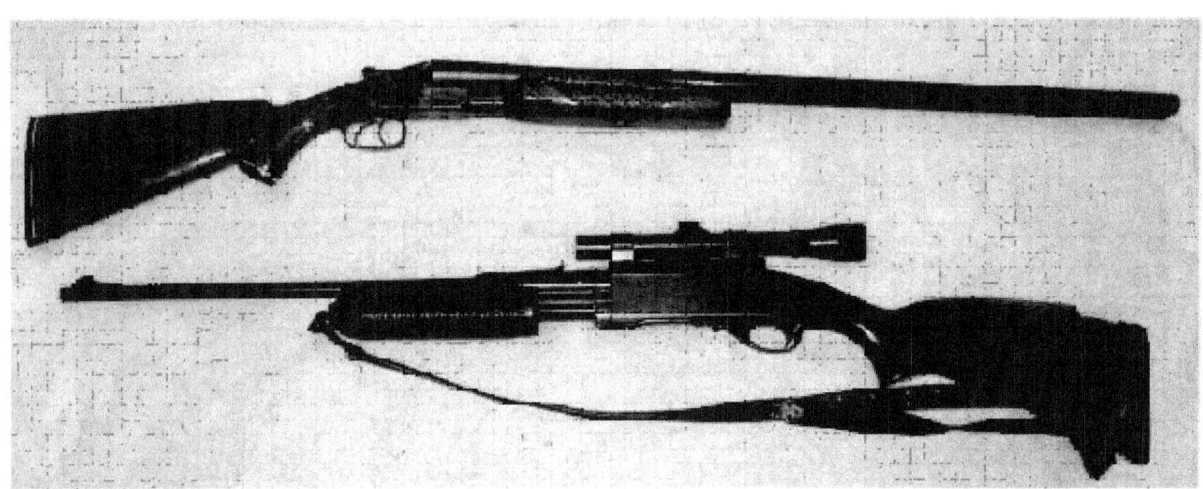

Top: a shotgun, 12 gauge, double barrel, breech loading
Bottom: a hunting rifle, 30 caliber (30'06), pump action

Shotguns look similar to rifles, but are designed to be short range weapons. A shotgun shell may contain one large projectile (called a slug), but most of them contain many small, usually spherical, projectiles called "shot". There may be a few pellets of large shot, or many tiny pellets of smaller shot. When a shotgun is fired the shot spread out over a relatively large area, making it more likely that at least some of the shot will hit the target. This has several consequences. First, shotguns are the preferred hunting weapons for things like birds that are fairly delicate creatures but which move fast. Second, the effect of a shotgun blast has a very limited range before the shot become so widely scattered and so low in energy that they have little effect. The effect of a shotgun blast at short range can be devastating, but at long range may cause no damage at all. Third, the simplicity of shotguns makes them easy to use by a beginner.

Shotguns are used more often in committing crimes than rifles, though not as often as handguns. Criminals who use a shotgun are most often aiming for an intimidation effect. People know that if someone shoots a shotgun at them they are not likely to miss and it's really going to hurt bad. Many criminals cut the barrel of the shotgun shorter, to produce a "sawed-off" shotgun. These modified shotguns are more concealable and have a wider pattern of shot dispersal, but have a shorter range.

Shotguns are available as single shot (break action), double barrel, pump action, and semiautomatic.

13.3.3 Other Types of Firearms

Semiautomatic versions of submachine guns (such as the Uzi) are classed as pistols for legal reasons. They usually hold 20 to 30 rounds, but in other ways are similar to conventional handguns of equivalent caliber. The are so expensive and difficult to obtain that they are seldom used except by members of organized crime (including drug-dealers). However, some of them are showing up in the hands of inner-city gangs in recent time, probably through the drug dealer connection.

Air guns use the pressure of a compressed gas, such as air or carbon dioxide, to propel a small projectile. Generally known as "BB guns" or "pellet guns" these firearms have been around for over 200 years. They fire small projectiles with only a weak amount of force, so are rarely used in the commission of a crime, and are unlikely to kill or seriously harm anybody.

Criminals on a tight budget, such as young criminals, may try to build their own firearms. These are usually crude and almost as dangerous to the person firing them as to the intended target because they have a tendency to discharge hot gasses and powder out the back, and may even explode. Most of them are designed to fire cheap small caliber ammunition.

13.4 Ammunition

Firearms must fire some sort of ammunition. Air guns fire small projectiles that are round (BB's) or oval (pellets). These are single items that are not propelled by gunpowder. They are often poured into a tube that allows them to be fed into the firing chamber one at a time. Shotguns fire **shells**. Shells are usually a plastic or cardboard casing into which is packed powder, a "wad" or disk of material that separates the powder from the shot, and the shot. The end is crimped to hold everything in place. A primer is placed in the base of the shell to ignite the gunpowder when struck by the shotgun's firing pin or hammer.

Rifles and pistols fire **cartridges** (also called rounds). A cartridge is composed of a casing, usually made of brass, that holds gunpowder. A bullet, usually made of lead, is then forced into the neck of the cartridge, sealing in the powder. There is a primer in the base of the cartridge. Figure 13.3 shows a handgun cartridge, a rifle cartridge, and a shotgun shell.

13.5 Firearms Matching

The most common examination performed by firearms examiners is firearms identification, which consists of comparative analysis for the purpose of matching ammunition components to the firearm that was used to fire them. The firearm thought to have been used to fire the evidence bullets or casings is usually referred to as the "suspect" firearms – not necessarily because it belongs to a certain suspect, but because it is suspected to be the one used to produce the evidence. The firearms examiner will compare evidence bullets and casing to the suspect firearm and to standard bullets and casing fired from the suspect firearm to identify similarities and differences. When making these comparisons the firearms examiner will look for both class characteristics and individual characteristics.

Figure 13.3: Various Types of Ammunition

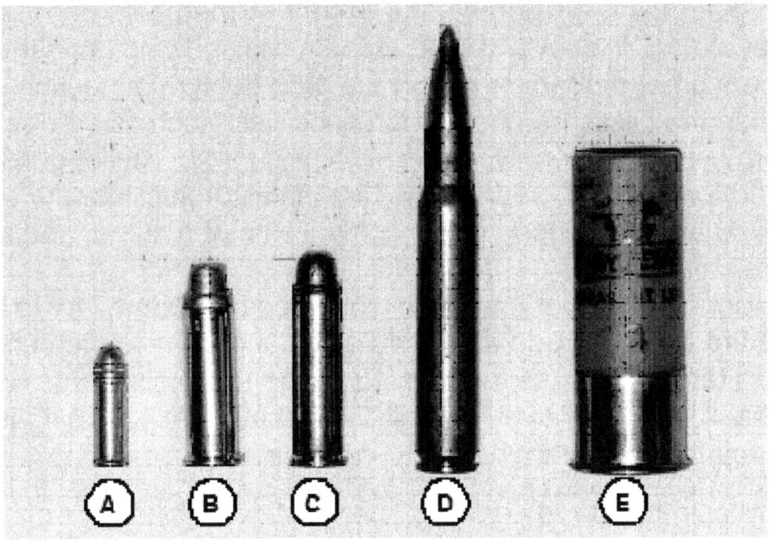

A: Pistol Cartridge, 22 long caliber
B: Pistol Cartridge, 38 caliber
C: Pistol Cartridge, 357 caliber,
D: Rifle Cartridge, 30'06 caliber
E: Shotgun Shell, 12 gauge

It is the ability for one item to transfer characteristics to another (Locard's principle again) that makes firearms identification possible. Both bullets and cartridges are softer than the material from which a firearm is made, so both pick up marks from the firearm as they travel from magazine to firing chamber and out during the internal phase of ballistics, as described in chapter 12. These marks are the class and individual characteristics that a firearms examiner will look for to make an identification.

13.5.1 Class Characteristics of Firearms and Ammunition

The first step in any comparison is to determine whether the items you are comparing have the same class characteristics. As you should remember, class characteristics are those that would be common to a group of items of a certain type. Class characteristics of firearms that relate to the bullets fired from them include the caliber of the firearm, the cartridge type, and the rifling pattern of the barrel of the firearm.

Caliber is the size of a bullet, measured as its diameter, and is a class characteristic. In the U.S., calibers are usually measured and expressed in hundredths of an inch, and occasionally in thousandths of an inch. A bullet that is 22 hundredths of an inch (.22) in diameter is called a 22 caliber bullet, and a bullet that is 357 thousandths of an inch in diameter is called a 357 caliber bullet. Outside of the U.S., firearms manufacturers use the metric system and would refer to a 22 caliber bullet as an 5.6mm caliber bullet. The caliber of a bullet corresponds to the inner diameter of the barrel of the firearm that fires it. Bullet caliber is illustrated in Figure 13.4. Shotgun shells are not designate by caliber, but instead by **gauge**, where the smaller the gauge the larger the diameter of the shell and the shotgun barrel. For example, a 12 gauge shotgun has a larger diameter barrel than a 20 gauge shotgun, but a smaller diameter barrel than a 10 gauge shotgun.

Figure 13.4: How Caliber is Determined

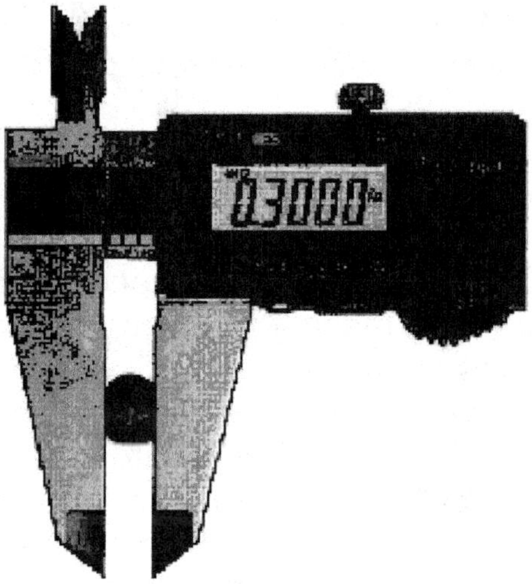

Bullets of the same caliber can have a variety of different shapes, ranging from round to pointed. The shape of a bullet is also a class characteristic.

Bullets are attached to a casing to form a cartridge. The casing holds the powder that propels the bullet. The resulting cartridges can vary in shape, length, and the amount of powder. For example, 22 caliber ammunition comes in three common types: short, long, and long rifle. To a certain extent these are specialized for the type of firearm that shoots them, and a type of ammunition must fit into the firing chamber of the firearm that will shoot it. My 22 semiautomatic pistol will take 22 shorts or longs, but not long rifle. My father's 22 rifle will take any of these three types of 22 ammunition.

Some cartridges are rim fire cartridges and some are center fire. Rim fire cartridges have primer distributed around the rim of the base of the cartridge and are designed to be hit by a firing hammer that strikes the cartridge in this location. Center fire cartridges are designed to be hit by a firing pin that strikes a primer device in the center of the base of the cartridge.

Rifling is a pattern of grooves and lands on the inside of the barrel of all rifles and most handguns. **Grooves** are depressions cut down the length of the barrel. The spaces between the grooves are called **lands**. The grooves are usually spiral in nature, so that they twist as they progress down the barrel. This is what imparts a spin to the bullet and improves its accuracy as described in chapter 12. Firearms of differing manufacturers, and even different models made by the same manufacturer, have

**Figure 13.5: A "Bullet's Eye"
View Down the Barrel of a Rifle**

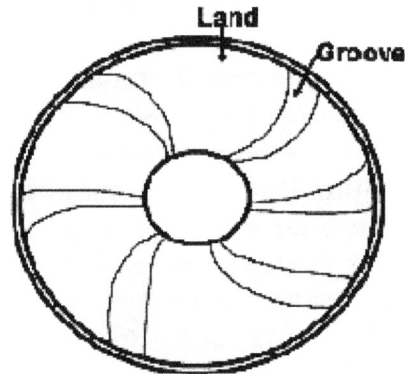

different patterns to their rifling. First, the number of grooves will vary. This often has to do with the quality of the firearm, since it saves money to make fewer grooves. However, you don't want to have too many grooves, or the effect of the rifling will be lost. A typical number of grooves is around 6 to 8. Second, the rifling can twist clockwise (right) or counterclockwise (left). There is no advantage to either direction. Third, the degree of twist can vary. Each manufacturer has their own idea about how severe the amount of twisting should be. Figure 13.5 illustrates rifling.

As we know from chapter 1, class characteristics can only be used to rule out an unknown being of the same source as a standard. A positive identification or match can't be made using class characteristics. However, if the class characteristics match, the firearm examiner will look for individual characteristics that can be used to make a positive match.

13.5.2 Individual Characteristics of Firearms and Ammunition

As we remember from chapter 1, Individual characteristics are those characteristics that are unique to an item and which can be used to make a positive match by showing that an evidence item must have come from the same source as a standard.

The internal surfaces of the magazine, firing chamber, and barrel may look smooth (not counting the rifling in the barrel), but in reality they have small imperfections. These imperfections mark the bullets and casings. In addition, the firing pin or hammer will have imperfections on the surface that strikes the cartridge to fire it, and these imperfections will likewise leave individualizing marks on the casing. These marks transferred to the bullet and casing are called a **mechanical fingerprint**.

These individual characteristics of a firearm arise in the manner described in chapter 8. Initially they come from the process of a firearm being manufactured by machines that are in a constant process of change, actually in the process of wearing out. Because every machine wears out a little more each time it is used, it is slightly different each time and it will transfer these slight differences to the surfaces of each firearm made with it. So, even a brand new firearm will contain individual characteristics that can be transferred to a bullet or casing. Individual differences also arise during use and abuse of the firearm. Every firearm has a unique history of ammunition fired, cleaning and maintenance, corrosion, etc. These produce additional individual characteristics that can be transferred to bullets or casing.

13.5.3 Collecting Standard Items

The standard bullets and casings used for comparison are obtained by firing the suspect firearm. Bullets pose an obvious problem for collection. They are sometimes fired into a target, but this may cause damage to the bullet. Most often bullets are collected by firing them into a water tank. Water is much denser than air and does a good job of slowing and stopping a bullet without adding any additional marks to it. A typical water tank used for this purpose is about 3 feet wide, 3 feet deep, and 10 feet long. The bullet is fired from one end and recovered from the bottom of the tank. A typical handgun bullet may travel two to four feet, and a typical rifle bullet may travel 4 to 6 feet. The casing discharged during the firing of the firearm is usually allowed to fall onto a padded surface, which will not leave marks on it.

13.5.4 Matching Ammunition to Firearms

Typically, the firearms examiner will make sure that the class characteristics of the evidence items and the suspect firearm agree before firing the weapon to collect standard bullets and casings. Having evidence items and standard items in hand, the firearms examiner will then examine the standard items to see if a distinctive pattern of individualizing marks can be found. If so, then he or she will look for this pattern of marks on the evidence items. If the pattern of marks on the evidence items are similar, then the firearms examiner will use a comparison microscope to match them.

When this examination is concluded, the firearms examiner can come to one of three possible decisions.
1. The evidence was identified as having been fired from the suspect firearm (a positive match);
2. the evidence was determined not to have been fired from the suspect firearm (an exclusion);
3. indeterminate, the evidence could not be identified or eliminated as having been fired from a suspect firearm (a consistent match or undetermined).

Why is there some evidence for which it can not be determined whether it was fired from a certain firearm or not? Usually it is because the evidence consists of a bullet that is damaged to the point that few individualizing marks can be detected, or perhaps none at all. Also, some firearms simply don't leave good individualizing marks on the ammunition fired from them. This may imply a consistent match, in that the caliber and other class characteristics match, or it may imply that the evidence was too damaged to obtain any type of match at all.

13.6 Unknown Firearm Examination

Sometimes bullets and/or casings are recovered, but no suspect firearm is discovered. In this case a firearms examiner may be able to narrow down the list of possible firearms that produced the evidence to a few out of the hundreds of possibilities. This process is called **unknown firearm examination**, because the firearm that produced the evidence is unknown at the time.

The firearms examiner will almost always be able to determine the caliber of the firearm, from the size of the bullets or casings recovered. However, if the bullet is fragmented and the person who did the shooting picked up their casings, then this may not be possible.

If a casing is recovered, it may be possible to determine whether the cartridge was fired from a revolver or a semiautomatic, because most revolvers have a hammer that hits the cartridge to fire it, while most semiautomatics have a firing pin that strikes the cartridge in the center of its base.

There are a number of different rifling patterns used in the firearms industry, though not every manufacturer's rifling pattern is unique. If a bullet is recovered, the firearm that fired it can often be narrowed down to a small list of manufacturers that use the rifling pattern observed on the bullet. The FBI maintains a database of all firearms manufactured, by manufacturer, caliber, type of firearm, and rifling pattern. As each manufacturer produces a new style of firearm or updates an old one, they submit the information to the database. The FBI then makes the database available to firearms examiners around the country.

13.7 Firearm Function Examination

A **firearm function examination** is performed to determine whether a particular firearm can be fired accidentally, and if so can how can it happen. In many shooting cases a suspect will claim that a firearm was fired unintentionally, during the process of cleaning it, picking it up, putting it down, dropping it, or some other activity other than pointing it at the victim and pulling the trigger. A firearm function examination can determine whether the activity claimed by the suspect could have led to an accidental firing of the weapon. Many stories given by suspects about how a firearm was accidentally fired and another person was killed have been shown to be either true or false by firearm function examination.

Accidental discharge of a weapon is common. Hopefully, most of the time everyone is following good safety practices and not pointing the business end of the firearm at someone they don't want to shoot. For example, my grandfather gave me an old 30-'06 hunting rifle for Christmas one year (shown in Figure 13.2), but neglected to warn me that when unloading it the last cartridge didn't eject from the chamber. So, the first time I went hunting I came home, ejected all the cartridges from the rifle, and, like I had been taught as a basic safety practice, pointed it up at the ceiling and pulled the trigger. Bang! I put a nice hole through the ceiling and roof of my parents' bedroom, but nobody was hurt. A firearms examiner could have examined my Christmas rifle and determined that there indeed was a problem with ejecting the final cartridge from the chamber and that, therefore, my explanation for the hole in the roof might be true.

A firearm function examination usually begins with a number of tests to make sure that the firearm is functioning properly. The firearm is inspected to determine whether it has a safety, which is a mechanism that prevents it from being fired until the safety is taken off. Is so, it is determined whether the safety is functioning properly or not. Finally, specific detail of the workings of the firearm are examined to determine whether the firearm could be discharged in the manner claimed.

13.8 Firearms Related Databases

There are several databases of information of interest to firearms examiners. The FBI maintains databases of characteristics of firearms and ammunition. The NIBIN/IBIS system of databases of characteristics of firearms and ammunition evidence from crimes is administered by the Bureau of Alcohol, Tobacco, and Firearms. The NIBIN/IBIS system will be described in more detail in the chapter on forensic databases.

13.9 Questions for Study and Review

1. In your own words, describe the similarities and differences between revolvers, semiautomatic pistols, rifles, and shotguns. What other types of firearms also exist?
2. Describe the types of ammunition used in the various types of firearms.
3. List the class characteristics of firearms and ammunition.
4. List the individual characteristics of ammunition components recovered at a crime scene and discuss the mechanisms that produce these characteristics.
5. Describe rifling, including it's nature, its purpose, and the differences that exist between the rifling of firearms made by different manufacturers.
6. How are standard ammunition components generated and collection for comparison to evidence.
7. In your own words describe unknown firearm examination.
8. In your own words describe the nature and goal of firearm function examination.

13.10 Sources

[Anonymous]. n.d. NIBIN [Internet]. [cited 2007 Oct 18]. Available from: http://www.nibin.gov/.

[Anonymous]. 2003. Types of Firearms [Internet]. [cited 2007 Oct 18]. Available from: http://www.gunsandcrime.org/guntypes.html.

Block EB. 1979. Science vs Crime. San Francisco: Cragmont Publications.

Doyle JS. 2007. Fundamentals of Firearms ID [Internet]. [cited 2007 Oct 18]. Available from: http://www.firearmsid.com/A_FirearmsID.htm.

Fingerhut LA, Cox CS, Warner M. 1998. International Comparative Analysis of Injury Mortality: Findings From the ICE on Injury Statistics. Advance Data 303: 1-20.

Hamby JE. 1999. The History of Firearm and Toolmark Identification [Internet]. [cited 2007 Oct 18]. Available from: http://www.firearmsid.com/A_historyoffirearmsID.htm.

Jason A. n.d. Crime Scene Analysis & shooting incident reconstruction [Internet]. [cited 2007 Oct 18]. Available from: http://www.projectile.com/AJpro.htm.

Klatt EC. 2007. Firearms Tutorial [Internet]. [cited 2007 Oct 18]. Available from: http://library.med.utah.edu/WebPath/TUTORIAL/GUNS/GUNINTRO.html.

Schehl SA. 2000. Firearms and Toolmarks in the FBI Laboratory, Part 1 [Internet]. Forensic Science Communications 2(2). [cited 2007 Oct 21]. Available from: http://www.fbi.gov/hq/lab/fsc/backissu/april2000/schehl1.htm.

Vizzard WJ. 1999. Reexamining the Importance of Firearms Investigations [Internet]. FBI Law Enforcement Bulletin 68(5). [cited 2007 Oct 19]. Available from: http://www.fbi.gov/publications/leb/1999/may99leb.pdf.

Wikipedia contributors. 2007. Gun politics in Switzerland [Internet]. Wikipedia, The Free Encyclopedia; 2007 Dec 3, 23:40 UTC [cited 2007 Dec 18]. Available from: http://en.wikipedia.org/w/index.php?title=Gun_politics_in_Switzerland&oldid=175598978.

Chapter 14
TOOLMARKS AND IMPRESSIONS EXAMINATION

The main part of this chapter will be devoted to two related topics – toolmarks examination, and impressions examination. Since these types of examinations are done using the comparative method to link evidence recovered at a crime scene with evidence recovered from a suspect, they are within the field of criminalistics. Most often, matching of this type of evidence is done by firearms examiners because the marks that constitute individual characteristics of toolmarks and impressions are often similar to those used in matching firearms evidence. As discussed in the chapter on crime labs, however, impressions evidence is also commonly handled by fingerprint examiners. After considering these forensic sciences the chapter will continue with a brief exploration of how forensic podiatry can contribute to matching footprint and shoeprint evidence, and conclude with an advanced topics section on casting technology.

14.1 The History of Toolmarks and Impressions Examination

Like many forensic sciences, the use of impression evidence, especially of footprints, is likely to be ancient. The first recorded instance of impressions examination dates to the early 1800's in England. In this case impressions of footprints and of corduroy cloth with a visible patch were used to identify a murderer. Impressions evidence has been used routinely since the late 1800's and this field has developed along with firearms and toolmark examination and fingerprint examination.

Toolmark examination is also likely to be ancient. It first appears in the early 1900's as a companion forensic science to firearms examination, and has been linked with firearms examination ever since. In 1969 this relationship was further formalized in the title of the Association of Firearms and Toolmark Examiners (AFTE), the largest professional society of these forensic scientists.

14.2 Toolmarks Examination

Toolmarks consist of scrapes, gouges, and impressions left in a material during the process of using a tool on that material in some way. If the material to which the tool is applied is softer than the material of which the tool is made, then the class and individual characteristics of the tool can be transferred to the toolmark in a manner similar to how the characteristics of a firearm are transferred to the casings and bullets fired through it. For example, if a criminal used a screwdriver to pry open a window in order to gain entry into a house, it is likely that marks made by the screwdriver would be left in the wood of the windowsill, and that these mark record some of the class and individual characteristics of the screwdriver. If the characteristics of the toolmark can be matched to a particular tool, then it can be shown that the tool was present at the crime scene. Figure 14.1 shows a tool mark next to the tool that made it.

In some situations the tool marks can be compared directly with the tool, simply by placing the tool so that it is obvious that the nicks, scratches, and projections on the tool line up with corresponding features of the toolmark. In many cases, however, it is preferable to make a cast of the tool mark, which should reconstruct the portion of the tool that left the mark. The cast can then be compared to the tool using a comparison microscope. Yet another method is to make a cast of the toolmark, make a standard toolmark using the suspect tool in a similar material as the evidence toolmark, and make a cast of the standard toolmark. The two casts are then compared using a comparison microscope.

14.2.1 Class Characteristics of Tools

Possibly the most important class characteristic of a tool is what type of tool it is. A hammer will leave different types of toolmarks than will a screwdriver. Within a certain type of tool, class characteristics include the size and shape of the tool, as well as other characteristics shared by some, but not all, tools of that type. Let's say for example, that an examiner is investigating marks left on a windowsill by a screwdriver that was used to pry the window open. First, the examiner will determine that the marks were made by a screwdriver, as opposed to a pry bar or some other tool. Screwdrivers come in a large assortment of widths, lengths, and types of point. Let's say that the marks being examined are consistent with a standard screwdriver that has a blade 1/4 inch wide. If the suspect screwdriver is a #2 Phillips head, then it did not make the marks in question and an exclusion exists. If the class characteristics match, the examiner can move ahead to look at the individual characteristics.

14.2.2 Individual Characteristics of Tools

As is true for many other types of evidence, individual characteristics of tools include features of manufacture and features that arise due to the use and abuse of the tool. They usually consist of nicks, scratches, areas of corrosion, and other defects that are unique to the individual tool. Sometimes individual characteristics are missing or difficult to see in the toolmark, preventing a positive identification from being made.

14.2.3 Results of Matching Toolmarks with Tools

Our familiar categories for results of matching attempts apply here. An examination can return a positive match, a consistent match (usually described as inconclusive), an exclusion, or a result of undetermined. A result of inconclusive usually occurs if the individual characteristics present in the toolmark are too indistinct to adequately compare with standard items.

14.3 Impressions Examination

Various types of impressions, such as shoe impressions, tire impressions, etc. can also be examined and matched using methods similar to those for toolmarks. As with firearms or toolmarks, comparisons are made, often using the comparison microscope, to match class characteristics and individual characteristics of impressions. Depending on the type of impression, the matching can be performed using the impression itself and the item suspected to have made the impression, or by using casts. A cast can be made of the evidence impression, or casts can be made of both the evidence impression and a standard impression made by pressing the suspect item into an appropriate material.

14.3.1 Class and Individual Characteristics of Impressions: Tire Impressions as an Example

Impressions usually have both class and individual characteristics. The class characteristics will include the type of item that made the impression, along with the size of the item. Other class characteristics may exist depending on the type of item. Individual characteristics may exist due to uniqueness of manufacture or to use and abuse of the item that made the impression.

Let's take automobile tire impressions as one example of the types of impressions examined by an impressions examiner. Let's say that the crime scene is a muddy field and tire tracks are visible at the edge of the field. This is one situation in which a cast will probably be made of the impressions, since it is difficult to remove blocks of mud that contain the impressions, though I know of this having happened on some occasions. The class characteristics of tires include the size (especially their breadth or "profile" as tire sellers call it) and the tread pattern.

The tread pattern can often be used to narrow down the tire that produced the impression in a manner similar to unknown firearms examination. Tire manufacturers have more or less unique tread patterns that they use for their tires. Different tires produced by the same manufacturer will also have different tread patterns.

These class characteristics may allow the range of possibilities for the vehicle that made the impressions to be narrowed down some. For example, certain tire sizes are only used on pickups and SUV's, and never on sedans. If a certain tread pattern is found to be from a "performance tire" or "touring tire" then it is not likely that the vehicle that made it was an inexpensive model.

Most of the useful individual characteristics of a tire are nicks, scratches, gouges, and similar defects produced by use and abuse of the tire. One feature of use that is not considered an individual characteristics is the depth of the tread. As the tire is driven on, the tread wears and the depths of the grooves of the tread become more shallow. This is considered a class characteristic, since it is possible for two tires to have been driven on in a manner similar enough that their treads are worn the same amount. However, in the course of doing this, these two tires probably picked up different nicks and gouges as one of them ran over a broken bottle that the other avoided, and similar events happened.

There are many types of impressions and all of them are examined using similar methods.

14.4 Forensic Podiatry

Forensic podiatry is a relatively new field that overlaps to a certain extent with impressions examination in that the objects of study by forensic podiatrists are foot impressions or shoe impressions. Podiatrists are medical doctors who specialize in disorders of the feet, and they can bring their expertise in foot anatomy and foot abnormalities to the study of evidence involving feet.

14.4.1 The History of Forensic Podiatry

The first recorded case of the use of forensic podiatry dates to 1935, and involved a chiropodist (like a chiropractor for the feet) rather than a podiatrist. Originally, forensic podiatry focused on matching characteristics of the external surfaces of feet and shoes, to impressions and similar forms of evidence. In more recent times forensic podiatry has incorporated new knowledge concerning the effects of internal forces, such as biomechanics and foot anatomy into their analyses.

14.4.2 The Practice of Forensic Podiatry

Forensic podiatrists work with types of evidence that we would consider impressions and prints. The distinction between an impression and a print is normally whether the evidence is two dimensional, consisting of some type of image of the foot or shoe on a surface; or whether the evidence is three dimensional, having be pressed into a surface. A two dimensional representation is a **print**, i.e. a shoeprint or a footprint, and a three dimensional representation is an **impression**. Forensic podiatrists also examine the insides of shoes, where the impressions of feet have been made during the wearing of the shoe. Standard items that forensic podiatrists work with include feet, shoes, insoles, and other items related to the foot.

14.4.3 Class and Individual Characteristics of Feet

Although forensic podiatrists work with foot and shoe prints and impressions, they do not normally match skin ridge patterns as would a fingerprint examiner, nor do they normally match the features of use and abuse of a shoe as would an impressions examiner. Instead, the contribute their expertise on features of people's feet.

Most of the characteristics examined by forensic podiatrists are class characteristics, such as the size of the foot or shoe. Forensic podiatrists apply their knowledge of biomechanics, foot anatomy, and abnormalities of the foot to identify, for example, features of a footprint that are consistent with a person suffering from club foot or clawing of the toes. Since there are many people who suffer from these ailments, these are primarily class characteristics. However, some disease processes are random. For example, no two people have the same pattern of arthritic spurs of the toe joints. To the extent that these types of features can be distinguished, they constitute individual characteristics.

14.4.4 Matching in Forensic Podiatry

Since forensic podiatrists work primarily with class characteristics, they are familiar with the procedures that need to be used to narrow down the range of possible suspects based on class characteristics. A similar problem is encountered in working with genetic markers, and will be discussed in more detail in the chapter titled "other issues in serology". The result is framed in terms of the probability that the suspect left the evidence. More precisely, it is framed in terms of the probability that an individual other than the suspect would have the same combination of class characteristics seen in the evidence. This is a form of consistent match, but a form that is more certain that other types of consistent matching. Some authorities use the term "**strong consistent match**" to refer to this level of matching. In exceptional circumstances, individual characteristics can be used to make a positive match, though this is rare. Of course, the matching of evidence by a forensic podiatrist may result in an exclusion or an inconclusive (consistent, but not positive) match.

14.5 Advanced Topics: Casting Technology

Casting can be defined as reproducing some form of evidence using a durable material, in such a manner as to preserve the important class and individual characteristics of the evidence. All those forensic scientists who work with toolmarks and impressions make casts of the evidence, at least occasionally. Other pursuits, such as facial reconstruction, also commonly utilize casting. Therefore, we will take a brief look at the process of casting and its technology.

14.5.1 Stabilizing the Evidence

The first task in collecting some forms of impressions evidence is stabilizing it, meaning to prevent it from deteriorating during the process of collecting the evidence. This is a particular concern with impressions found in loose material such as sand or snow. Stabilization is not as often a concern for toolmarks.

The best way to stabilize impressions in a material such as sand is to apply some form of glue to stick the sand grains together until a cast can be made. There are products available to do this, but one of the most common is cheap hair spray. Many of the less upscale hair sprays are simply a plastic dissolved in a solvent such as alcohol.

When sprayed onto a surface, be it hair or sand, the plastic sticks the individual hairs or sand grains together. Forensic anthropologists, also find cheap hair spray to be useful for stabilizing bones that are in poor condition.

Impressions in snow present a nearly insurmountable problem for stabilization. One of the most highly recommended methods is to spray the impression gently with aerosol automobile primer, the substance sprayed onto the bare metal of the automobile to prepare the surface for painting. The directions on every can of primer say to spray on many very light coats, and this is particularly critical for impressions in snow, as a single full blast from the aerosol can create a hole in the impression causing loss of detail. There is also a specialized product, called snowprint wax that can be used.

Need I say it? I guess I better. Photograph any such fragile impression before beginning any form of stabilization, during the stabilization process, and after the stabilization process but before any cast is made.

14.5.2 Casting Materials

There is a wide range of casting materials available. By **casting material** I mean the substance that is poured, squeezed, sprayed, or otherwise placed into the impression in a liquid state and which will harden to form the cast.

The simplest and most common casting materials are forms of plaster, of which there are several. Plasters are mixed with water to form a liquid, which is then poured carefully and gently into the impression and allowed to harden. Plaster of Paris is common and often used. However, there are a variety of specialized casting plasters, with slightly differing characteristics such as hardness and setup time, that allow the cast to be as good as it can be under a particular set of conditions. Most forms of plaster are just fine for many types of impression, such as footwear and tire impression, under a wide variety of conditions. However, most plasters generate heat as they set up, which makes them unusable for casting impressions in snow. One form, dental stone, generates less heat, especially when mixed with cold water, and this is the form of plaster recommended for casting impressions in snow. One should always follow the instructions on the bag of plaster for mixing and for time to allow the product to harden.

In addition to plasters there are several rubber, plastic, and resin materials available for casting. These are preferred for toolmarks because they are not as liquid and therefore stay in an oddly shaped or oddly placed mark better than a plaster product. Also, these materials are known for accurately reproducing even the most minute details. Some of them are simply squeezed out of a tube and pressed into the impression with a wooden popsicle stick or similar device. Some are two-part materials, including a product and a hardener that must be mixed with it. These are also pressed into an impression. One of the most widely used is a silicone product called Mikrosil. Some of these products are designed to be used with a releasing agent, which is a substance applied to the impression before the casting material and which prevents the casting material from sticking so tightly to the material in which the impression resides that it can not be removed without damage.

Other casting materials exist as well, including foams and gels. These materials are used in special situations and do a fine job of reproducing the details needed by toolmarks or impressions examiners to do matching.

14.5.3 The Casting Process Using a Footwear Impression Example

Let me describe a typical casting process, using the example of a footwear impression in dried mud. Under these conditions no special precautions need be taken to stabilize the impression. Many crime scene investigators will have in their vehicle one or more pre-measured ziplock baggies with about 2 pounds of plaster. It is important to follow the directions for each specific material, but most plasters will require about 266 milliliters of water for two pounds of plaster. 266 milliliters is about 9 ounces, or about 3/4 of a 12 ounce beverage can. The water is poured into the plastic bag containing the plaster, and the bag is reclosed. Massaging and squeezing the bag will mix the plaster and water. Once the plaster and water have been mixed, and documentary photographs have been taken of the footwear impression, the liquid plaster can be gently poured into the impression. The plaster should harden in about one hour, at which point it can be lifted from the impression. The cast is then cleaned of adhering dirt and carefully packaged for transport to the crime lab. Careful records are kept of how the cast was made, when it was made, who made it, the case number, and other relevant details. Often, these details are written directly onto the cast using a felt tip marker.

14.6 Questions for Study and Review

1. Discuss the class characteristics of hammers, crowbars, screwdrivers, and saws.
2. Discuss the individual characteristics of tool marks and the processes that lead to their production.
3. Discuss the class characteristics that would be observed in footwear impressions.
4. Discuss the individual characteristics that might be observed in footwear impressions and the processes that lead to their production.
5. In your own words, describe the goals and methods of forensic podiatry.
6. What sorts of evidence might a forensic podiatrist be able to infer from impressions of feet that an impressions examiner might not?
7. Under what conditions might a cast be made of an impression?
8. In your own words, describe the process of making a cast of an impression. In your description be sure to consider the differing nature of the surfaces in which the impressions exist.

14.7 Sources

[Anonymous]. 2004. Tool Marks Fight Crime: Identification System Recognizes Manufacturing Processes. Industrial Engineer 36(6): p17(1).

Block EB. 1979. Science vs Crime. San Francisco: Cragmont Publications.

Bodziak WJ. 2000. Footwear Impression Evidence: Detection, Recovery and Examination. Boca Raton (FL): CRC Press.

Byrd M. n.d. Simple Tire Standards Collection [Internet]. [cited 2007 Oct 23]. Available from: http://www.crime-scene-investigator.net/tirestandards.html.

Byrd M. n.d. Other Impression Evidence [Internet]. [cited 2007 Oct 23]. Available from: http://www.crime-scene-investigator.net/otherimpressionevidence.html.

C.A.S.T. 2006. Track Impressions in Snow [Internet]. [cited 2007 Aug 20]. Available from: http://members.aol.com/varfee/mastssite/snow.html.

C.A.S.T. 2006. Casting Impression Evidence [Internet]. [cited 2007 Aug 20]. Available from: http://members.aol.com/varfee/mastssite/casting.html.

DuPasquier E, Hebrard J, Margot P, Ineichen M. 1996. Evaluation and Comparison of Casting Materials in Forensic Sciences - Applications to Tool Marks and Foot/shoe Impressions. Forensic Science International 82 (1): 33-43.

Evident, Inc. 2007. Product Catalog, Casting and Impression Materials [Internet]. [cited 2007 Aug 20]. Available from: http://www.Evidentcrimescene.com.

Hamby JE. 1999. The History of Firearm and Toolmark Identification [Internet]. [cited 2007 Oct 18]. Available from: http://www.firearmsid.com/A_historyoffirearmsID.htm.

Held DE. 2001. Handwriting, Typewriting, Shoeprints, and Tire Treads: FBI Laboratory's Questioned Documents Unit [Internet]. Forensic Science Communications 3(2). [cited 2007 Oct 21]. Available from: http://www.fbi.gov/hq/lab/fsc/backissu/april2001/held.htm.

La Trobe University. 2001. Forensic Podiatry [Internet]. [cited 2002 Feb 2]. Available from: http://www.latrobe.edu.au/podiatry/forensicpod.htm.

Liukkonen M, Majamaa H, Virtanen J. 1996. The Role and Duties of the Shoeprint/toolmark Examiner in Forensic Laboratories. Forensic Science International 82 (1): 99-108.

Nichols RG. 1997. Firearm and Toolmark Identification Criteria: a Review of the Literature. Journal of Forensic Sciences 42(3): 466-474.

Saferstein R. 1998. Criminalistics: An Introduction to Forensic Science. 6th Edition. Upper Saddle River (NJ): Prentice Hall.

Schehl SA. 2000. Firearms and Toolmarks in the FBI Laboratory [Internet]. Forensic Science Communications 2 (2). [cited 2007 Oct 18]. Available from: http://www.fbi.gov/hq/lab/fsc/backissu/april2000/schehl2.htm#Toolmark.

Scientific Working Group on Imaging Technologies (SWGIT). 2003. General Guidelines for Photographing Footwear Impressions [Internet]. Forensic Science Communications 5(4). [cited 2007 Oct 21]. Available from: http://www.fbi.gov/hq/lab/fsc/backissu/oct2003/2003_10_guide03.htm.

Scientific Working Group on Imaging Technologies (SWGIT). 2003. General Guidelines for Photographing Tire Impressions [Internet]. Forensic Science Communications 5(4). [cited 2007 Oct 21]. Available from: http://www.fbi.gov/hq/lab/fsc/backissu/oct2003/2003_10_guide04.htm.

Chapter 15
Basic Science
GENES AND THE INHERITANCE OF FINGERPRINTS

In this chapter we will investigate elementary genetics and inheritance. These subjects establish a background for fingerprint examination, and for serology and DNA analysis which come later.

The chapter will begin with a consideration of how genetic material is stored, retrieved, and used as a blueprint for making a protein product. Next the focus will shift to inheritance – first inheritance of simple traits, then inheritance of complex traits. The chapter will conclude with a look at the inheritance of fingerprints.

15.1 Storage of Genetic Information: Chromosomes and DNA

Inherited information for how to construct and maintain the human body and all its metabolic processes is stored in packages called **chromosomes**. Chromosomes reside within a structure called the cell nucleus, which is found inside almost every type of cell of the body. Chromosomes are small structures, visible under a microscope with the proper staining, but not to the naked eye. Chromosomes are constructed of a framework that contains **DNA**, a type of molecule that is used for genetic encoding because it has chemical properties that make it easy to copy. New copies of the chromosomes need to be made whenever a cell is going to reproduce itself by cell division, or whenever the cells used in sexual reproduction are created.

Most organisms have several chromosomes. For example, human cells have 46 chromosomes. Each chromosome is actually two strands of DNA twisted around each other in a spiral pattern called a double helix. The two strands are not random, and they have a very specific relationship to each other.

Figure 15.1: The Relationship Between Cell Nucleus, Chromosome, Loci, and Allele

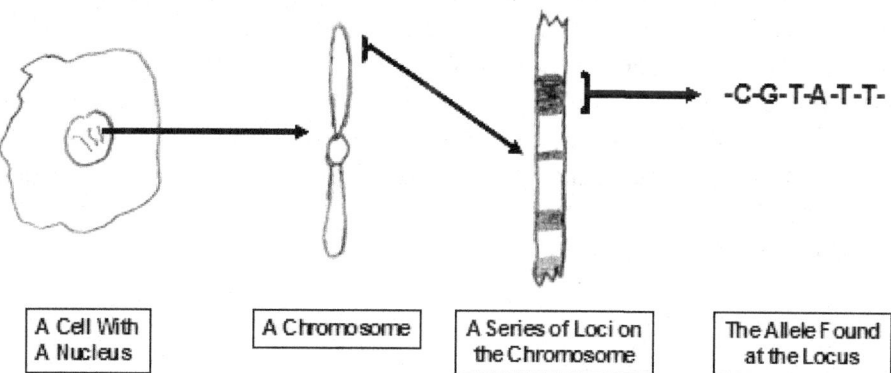

15.1.1 Loci and Alleles

Genetic information is encoded in the DNA that makes up the strands of the chromosomes. A location on a chromosome where the information for some particular feature of the human body is stored is called a **locus** (plural loci). For example, a person's blood is categorized into one of four **blood types** (A, B, AB, or O) for the purpose of blood transfusions. The genetic instructions for which of these blood types a person will have is stored at a specific location on chromosome 9. Due to the efforts of the human genome project, the exact locations of most loci are known.

At each locus there is a sequence of DNA that is the code or blueprint for a particular product, usually a protein. These actual genetic instructions are called **alleles**. For the transfusion blood group locus there are three possible alleles that may be found there: A, B, and O. Note that there is no AB allele, and how this blood type is coded for will be discussed in section 15.2.1.

15.1.2 The Genetic Code

There is a straightforward relationship between the DNA of an allele and the product it produces. Genetic information is stored in DNA using an 'alphabet' of four different molecules, called **bases**, that are chained together to form the DNA strand. The four bases are each represented by a letter: A, C, G, or T. For example, a small portion of a DNA strand may consist of -A-C-G-T-C-C-T-G-A-.

Remember that the two strands of a chromosome have a specific relationship to each other. Each of the DNA bases on a strand has an affinity for being paired with a certain other base on the other strand. A and T like to pair up, and C and G like to pair up. In our example, a small section of a chromosome would look like this.

```
-A-C-G-T-C-C-T-G-A-
-T-G-C-T-G-G-A-C-T-
```

Every A on one strand is paired with a T on the other strand, and every C on one strand is paired with a G on the other strand (and T's with A's, and G's with C's). Therefore, each strand has the information necessary to exactly specify the other strand. Only one strand of the pair will have the actual genetic sequence for making a protein, but the other strand has the information in a complementary form.

The pairing of bases on the two DNA strands of a chromosome is what gives chromosomes the power to exactly copy themselves. When a chromosome is replicated, the two strands separate, and each attracts the DNA bases necessary to form its complementary strand. This is exactly what happens when chromosomes are copied just before cell division.

15.1.3 How the Genetic Code Specifies the Construction of Proteins

The proteins coded for by an allele are also molecules that consist of chains of smaller molecules. The subunits of protein are called **amino acids**. There are about 20 amino acids used in the human body, each with a specific set of chemical properties. Proteins are constructed using these building blocks in such a way that the protein can do its job in the body.

For example, the collagen of skin, bone, and connective tissue needs to be tough and stretchy, so it has a repetitive structure that utilizes the amino acids glycine and hydroxyproline. Glycine is a simple amino acid with a shape that is straight, but hydroxyproline has a kink to it. The typical sequence of collagen is -glycine-glycine-hydroxyproline-glycine-glycine-hydroxyproline- so that each third amino acid is hydroxyproline. This gives the collagen molecule a kink every third amino acid and results in a structure that is spirally shaped, like a spring, capable of being stretched without breaking. Likewise, all proteins are constructed to have specific functions by using specific amino acid building blocks that have the desired properties.

Since there are about 20 amino acids but only 4 DNA bases, the DNA bases have to be taken in groups of three, called a **codon**, to specify an amino acid. Taking the DNA bases one at a time you could only code for 4 amino acids. Taking them two at a time you could code for 4 X 4 = 16 amino acids, which is not quite enough. Taking them three at a time allows you to code for 4 X 4 X 4 = 64 amino acids, which is more than enough. Because there are "extra" codons, the genetic code is said to be redundant. This means that it is common for a certain amino acid to be coded for by more than one codon. Some amino acids have as many as five codons that specify them.

To summarize: the genetic information for the human body and its systems is stored in chromosomes, which consist of two strands of DNA that are chains of bases which occur in an order that encodes the instructions for making a protein. Each chromosome has many loci, at each of which is found at least one allele (possibly more) that contains the instructions for making a certain protein. Proteins are composed of amino acids, which are specified by codons in the DNA sequence.

15.2 Inheritance of Simple Traits

Next, we need to explore inheritance. The most important things to accomplish are to see how genes are inherited and to examine how genes and the environment interact. We will start with simple inheritance, such as the inheritance of the transfusion blood groups. Later in the chapter we will look at more complex forms of inheritance.

It is usually no surprise to most people that parents pass on their genetic characteristics to their offspring. However, human parents do not pass on all of their genetic information to their children – only half of it. Each child is a unique combination of the genes of both its parents, and there are always a few things about a child that are not predictable from the genes of both parents. So, each individual is genetically unique, except for identical (monozygotic) twins that occur when a single embryo splits early in development and two individuals develop from the result of one conception event. Even so, identical twins are never completely identical. They always have some subtle differences. One of these differences is that the details of their fingerprints differ.

In order to understand inheritance, we need to look at (1) how genetic information is passed from parent to child, (2) how combinations of inherited genetic information are expressed in a person, and (3) where the unique differences come from. Looking forward toward the use of this information in forensic science, genes passed from parent to offspring result in class characteristics, and the unique differences result in individual characteristics.

Remember that humans have 46 chromosomes. Twenty-three of the 46 chromosomes were inherited from our mother, and were passed on to us through her ovum. The other 23 chromosomes were inherited from our father, through his sperm. Therefore, ova and sperm break the general rule that human cells have 46 chromosomes, and only have 23. This allows a new individual with the correct number of chromosomes, 46, to be created from two cells that have 23 chromosomes each.

Therefore, with the exception of those chromosomes involved in determining which sex we are, we have two of each chromosomes, so two loci for each trait (one on each of the two chromosomes) and therefore two possibilities for alleles we could inherit for each trait. In simpler words, we have two alleles for each trait – one that came from Mom on a chromosome we inherited from her, and one that came from Dad on a chromosome we inherited from him.

15.2.1 Dominant and Recessive Alleles

One factor that makes inheritance interesting is that some alleles are always expressed if present, whether on Mom's original chromosome, or Dad's original chromosome, or both. Others are only expressed if the child has two copies of it, one inherited from Mom and one from Dad. Those alleles that are always expressed when present are called **dominant alleles**, and those that are only expressed when no dominant allele is present are called **recessive alleles**. For example, alleles A and B of the transfusion blood group system are dominant, and allele O is recessive. Therefore, if someone inherits an A allele from one parent and either another A or an O from the other parent, they will express the feature coded for by the A allele, and we say that they have blood type A. If an O allele is inherited, it is not expressed if an A allele is also inherited. Similarly inheriting either two B alleles or a B and an O allele will result in blood type B. Blood type O results only when an O allele is inherited from both parents. If a person inherits an A allele from one parent and a B allele from the other, they will express the characteristics of both A and B, because both of these alleles are dominant. Therefore, this person's blood type will be AB.

15.2.2 Genotypes Haplotypes, and Phenotypes

In the language of genetics, each of us has a genotype and a phenotype. The **genotype** is a notation of what alleles we have inherited. The **phenotype** is the condition we express. It is important to keep the distinction between these two things in mind.

The form of notation for genotypes, used most commonly by geneticists, uses abbreviations for alleles. For the transfusion blood group, A, B, and O are used as natural abbreviations for allele A, allele B, and allele O. Using this notation, the genotype of an individual can be expressed as combinations of these abbreviations. For example, in the paragraph above we talked about people who inherited an A allele from one parent and an O allele from the other. We say that these people have genotype AO. The order of the letters doesn't matter, and AO means exactly the same thing as OA.

A **haplotype** is also a notation of what genetic information is present at some region, but for a single chromosome. Further, when using the term haplotype, it does not matter whether the region is a locus, with one or more alleles that code for something, or whether it is a region that does not contain instructions for building a protein. If the region is a locus, then a haplotype is simply half of a genotype. Therefore, a person with an AO genotype has both an A haplotype and an O haplotype. If the region of the chromosome is not an actual locus (i.e. it does not contain instructions for making some protein), there may still be a DNA sequence of interest there. In fact the region may have several alternative DNA sequences, each of which may be named or notated, and used in DNA matching of suspects. In this situation each of the alternative sequences is a haplotype. Because many of the chromosome regions with which DNA analysts work are not loci, they find it more convenient to think in terms of haplotypes than genotypes.

Those genotypes for which the two alleles are the same: AA, BB, and OO for the transfusion blood group, are called **homozygous**. AA and BB are called homozygous dominant genotypes, because they are homozygous genotypes in which both alleles are the dominant ones. OO is a homozygous recessive genotype for similar reasons. AO, BO, and AB are called **heterozygous** genotypes because two different alleles were inherited.

The phenotype is the actual expression of the trait in the body. So, for our example, there are four possible phenotypes: the person is either blood type A, blood type B, blood type O, or blood type AB.

The phenotype and the genotype relate together in a specific way that depends on which alleles are dominant and which are recessive. Table 13.1 shows the relationship between genotypes and phenotypes for the transfusion blood group system.

Table 15.1: Phenotypes and Genotypes for the Transfusion (ABO) Blood Group

Phenotype (Blood Type)	Corresponding Genotypes
A	AA or AO
B	BB or BO
AB	AB
O	OO

15.3 Inheritance of Complex Traits

Some characteristics are inherited in a more complex way than simple systems like the ABO blood group. Some examples are stature (height while standing), skin color, and fingerprint patterns. Complexity is introduced whenever more than one locus is involved in the expression of a trait. For example, there are thought to be between seven and nine loci involved in the expression of stature. Geneticists are reluctant to speculate about exactly how many loci are involved in fingerprint patterns, but it is thought to be several.

Complex characteristics can take on a wide spectrum of values. Unlike the ABO blood group, for which there are only four phenotypes, complex characteristics can be expressed as a wide variety of phenotypes. For example, consider the wide range of human heights or human skin colors.

15.3.1 Effects of Several Loci

Complex characteristics arise from the combined effects of alleles at several loci. One reason for the large spectrum of phenotypes comes partly from the fact that there are many loci. Let's assume for the sake of illustration that there are nine loci for stature, each with two alleles: a short version and a tall version. Let's further assume that all the alleles have equal effects (all are dominant). Given this situation there are 19 possible phenotypes. The best way to see this is to count the number of tall alleles, which can range from 0, if there are two short alleles at all nine loci; to a total of 18 tall alleles, if there are two tall alleles at all nine loci. In reality, there are many more than 19 phenotypes, because we have not accounted for dominance effects, or for the fact that all 9 loci may not have equal effects. Theoretically, there could be up to $2^9 = 512$ distinct phenotypes that arise from the nine loci if each has just two alleles. However, for all we know each, of these loci has 10 or 20 alleles – the human genome project identifies loci, not alleles, so the number and nature of alleles for each locus remain mostly unknown (as of 2007).

15.3.2 Effects of the Environment

The second important factor affecting the expression of complex characters is not their genetics, but the effect of non-genetic factors. Geneticists refer to everything that is not genetic as "the **environment**", a usage that is somewhat different from the meaning of "environment" in ordinary English. In the case of stature, it is known that nutrition plays a role, although nobody is entirely sure what aspects of nutrition may cause one person to be shorter or taller than another person with an identical genotype. The effect of nutrition on stature may come from the fact that periods of starvation are rare (in the U.S.), it may come from total calories, or it may come from the amounts of certain nutrients in the diet. The research I have seen points in so many different directions that it seems likely that all of these factors, and more, are environmental effects on stature. The effects of the environment are what make the distribution of phenotypes in complex characteristics such as stature continuous and the number of phenotypes extremely large. Each person's individual nutritional history will be unique and thus even identical twins will have eaten a different selection of foods over their lifetimes and thus have slightly different statures.

15.3.3 Expressivity

A third phenomenon that affects complex traits more than simple traits, though simple traits are not immune to it, is called expressivity. **Expressivity** refers to the situation in which a person's phenotype does not reflect their actual genotype. This often surprising situation can occur in a variety of ways, all related to the fact that our metabolisms and developmental pathways from embryo to adult are complex systems with many interactions. We each have a personal biochemistry that is unique and within which our alleles express themselves. Therefore, people who are genetically the same, such as identical twins, may still differ in several ways.

One profound difference in personal biochemical environment occurs in the hormonal differences between females and males. For example, male hormones, such as testosterone, have an adverse effect on scalp hair follicles. Therefore, males are more commonly bald and more profoundly bald than females. If a man and a woman have identical alleles for hair retention or loss, the man will experience more actual loss of hair than the woman. Stature also varies a slight amount by sex, apparently due to differences in the expression of alleles within female and male hormonal biochemical systems.

Within the complex biochemical systems of our bodies, the products of certain loci have the ability to mask, alter, or enhance the phenotypic effect of the products of other loci. This interaction between loci is often poorly understood and can take several forms that are interesting but beyond the scope of this simple introduction to genetics and inheritance.

15.3.4 Blended Inheritance

Complex traits exhibit a form of inheritance that appears to be blended, meaning that the phenotype of offspring is usually in some way intermediate between that of the parents. This is often seen for inheritance of skin color. If parents of differing skin colors have a child, the child's skin color is usually lighter than that of the dark skinned parent and darker than that of the light skinned parent. Blending inheritance is not so clearly seen with stature because of the environmental effects. One phenomenon that many people have observed is that the children of immigrants are often taller than their parents. This suggests that the parents were short because of the environment in which they were raised, which often had limited food resources. When their children are born in the U.S., they are not as nutritionally limited and may be taller than either of their parents. The pattern of alleles these children have, say the number of tall alleles for the nine loci, are still probably intermediate between those of their parents, but their better nutrition makes a difference.

Also, it is not impossible, merely unusual, for a child to inherit, for example, more tall alleles, or more short alleles, than either parent has. For example, let's say that both Mom and Dad have one tall allele and one short allele for each of the 9 loci. The exact allele that a parent passes on for each locus is random and not affected at all by other alleles that are passed on. Therefore, the child of these two parents could inherit all of Mom's tall alleles and all of Dad's tall alleles, for a total of 18 tall alleles. It is equally likely that the child could inherit all Mom's short alleles and all Dad's short alleles. In either case the child would not have a genotype or a phenotype that is intermediate between that of Mom and Dad and may differ drastically in stature from either parent.

Blended inheritance is, in actuality, only an illusion. On the level of alleles, inheritance follows the same rules as for simple inheritance. Complex traits sometimes appear to have blended inheritance because of the complexities of having several loci, environmental effect, and expressivity effects.

15.3.5 Heritability

Since the phenotype, or expression, of a complex trait can often be influenced both by genetics and the environment, geneticists and other scientists have developed methods for estimating how much of the expression of a trait is determined by the underlying genetics and how much by environmental factors. The results are sometimes surprising.

Geneticists use the idea of **heritability**, which is defined as the proportion of the variability in the phenotype that is controlled by genetic factors. Using this idea, geneticists divide variability in phenotypes into two parts, variability due to genetic factors and variability due to environmental factors. Variability here refers to statistical variance, which can be calculated using a formula that can be found in any statistics book. The variance calculated this way is called **phenotypic variance**, and is symbolized as V_p. Phenotypic variance is seen as being composed of two parts – variance due to genotype, V_g, and variance due to the environment, V_e. This gives us the formula:

$$V_p = V_g + V_e$$

Some geneticists also add on a third term, the variance due to interaction between genotype and environment, V_{ge}. Adding this term makes the model more accurate, because there are some cases where the genotype and the environment interact. Smoking and the genetic predisposition to lung cancer is an example of genotype-environment interaction. Some people are genetically disposed to develop lung cancer and will get it even if they don't smoke. Others will never get lung cancer no matter how much they smoke. However, there are some people who will not get lung cancer if they don't smoke, but will get it if they do. For the most part this last term can be ignored, however, and we will do so here.

Ignoring V_{ge}, heritability can be calculated using this equation.

$$\text{Heritability} = \frac{V_g}{V_p} = \frac{V_g}{V_g + V_e}$$

When defined this way, a heritability score of 1 (100%) means that the trait is totally determined by genetics. A heritability score of 0 means that the trait is completely uninfluenced by genetics.

Heritability is notoriously difficult to measure in humans, and most heritability estimates are regarded as imprecise unless replicated several times. The main problem in measuring heritability is in separating the effects of genetics from the effects of the environment. The classic scientific approach would be to hold the environment constant while varying the genetics and then hold the genetics constants while varying the environment. This is fairly easy to do with nonhumans, like bacteria, fruit flies, or even mice, but very difficult to do with humans. However, there is one situation in which we have human beings who are genetically identical, and that is identical twins. Geneticists refer to identical twins as **monozygotic**, or MZ twins. This is different from normal **dizygotic** (DZ) twinning, where the twins develop from two completely different fertilized ova. DZ twins are no more closely related than normal brothers and sisters. Interestingly, DZ twinning appears to be a very heritable trait, although MZ twinning does not. Heritability studies using MZ twins, especially MZ twins raised separately so that they experience a full range of environmental difference, are the best we can do for heritability estimation in humans.

Heritability studies have shown that very few things are purely determined by genetics and very few things are completely uninfluenced by genetics. There seem to be three broad categories of traits.

1. Traits that are highly heritable. There are several traits for which the heritability is about 80% to 90%. These tend to be things having to do with anatomy, such as stature and the class characteristics of fingerprints. Interestingly, this category also includes some talents and abilities, such as musical talent and mathematical ability.
2. Traits that are moderately heritable. These traits are about equally influenced by genetic and non-genetic factors. The heritability is about 50%. These include things like basic personality types (introverted vs extroverted, etc.). The individual characteristics of fingerprints probably fall into this category.
3. Traits that are slightly heritable. The heritability for these traits is usually about 10%. This group of traits includes some surprising things that can be considered minor aspects of human behavior. For example, whether a person always flushes the toilet twice, whether a person hold a cup or beer can with their pinky finger sticking out, and the types of foods, beer, etc. people enjoy.

15.4 The Inheritance of Fingerprints

Skin ridge patterns are called **dermatoglyphics** by geneticists, and are complex genetic traits with moderate to high heritability. The underlying genetics of dermatoglyphics is so complex that it is still incompletely understood, and geneticists are reluctant even to speculate about how many loci might eventually be found to be involved. Nor do we have a good idea of the expressivity issues that impact fingerprints, though they undoubtedly exist.

One of the more recent theories, which seems to make sense, though making sense is certainly no form of proof, relates the genetics of dermatoglyphics to some of the most fundamental and ancient loci in the human genome, called homeobox genes. Homeobox genes are loci that form part of the system of controlling the process of the development of an embryo and fetus. They help in orchestrating the complex series of events that lead to the formation of a human body plan. Whatever loci are responsible, the skin ridge patterns begin forming before the third month of gestation. The genetics of dermatoglyphics seem to at least partly control the class characteristics of fingerprints, which we will look at more in a coming chapter and consist of the overall pattern of arches, loops, and whorls. The heritability of these basic fingerprint patterns has been explored, and seems to be around 80%.

The individual characteristics of a fingerprint, which will also be discussed in a coming chapter, seem to be more influenced by environmental differences. A former graduate student of mine, who now works in the Latent Prints and Impressions section of the Montana State Laboratory of Criminal Investigation, did her thesis on the heritability of fingerprint characteristics. She found that many individual characteristics have heritabilities between about 10% and 60%.

For individual characteristics of fingerprints, some of the environmental differences seem to be extremely small variations in the environment within the uterus, which influence the division of cells as the tissues destined to become skin ridges are being formed. Skin ridge patterns are so sensitive to these differences that even identical twins, who have the same alleles and who share the same uterus during gestation have different individual characteristics of their fingerprints.

Let me clear up a potential point of misconception. Doctors have been known to state that identical twins have the same fingerprints, but they are referring to the class characteristics of the fingerprint, not to the individual characteristics of the fingerprint. Identical twins do most often have exactly the same class characteristics of their fingerprints, but the individual characteristics, being more subject to environmental effects, always differ.

15.5 Questions for Study and Review

1. Discuss the relationships between alleles, loci, DNA, chromosomes, and the cell nucleus.
2. Discuss how the information for a biological feature of an individual, such as blood type O, is encoded in DNA.

3. Discuss how the two strands of a DNA molecule pair with each other and the properties that this pairing behavior gives to DNA.
4. Discuss the relationship between the bases of DNA and the amino acids of proteins.
5. For the purpose of blood transfusions, a person's blood type is often expressed as a combination of their ABO type and their Rh type. For example, O$^+$ or AB$^-$, where the + and − signify Rh positive and Rh negative, respectively. The Rh blood group system has two alleles, positive and negative. The positive allele is dominant and the negative allele is recessive. List the possible genotypes a person could have for the Rh blood group, and the phenotype corresponding to each genotype.
6. In your own words, explain the difference between a genotype and a phenotype.
7. In your own words, explain the difference between a genotype and a haplotype.
8. In your own words, explain the difference between simple genetic traits and complex genetic traits.
9. Discuss the genetic concept of the environment and how it differs from the common understanding of this term.
10. In your own words, explain why some characteristics seem to exhibit blended inheritance, given that we know that all things are inherited as distinct alleles.
11. In your own words, explain the concept of heritability and how it is estimated in humans.
12. Discuss the inheritance of dermatoglyphics.

15.6 Sources

Boaz NT, Almquist AJ. 1999. Essentials of Biological Anthropology. Upper Saddle River (NJ): Prentice Hall.

Brues AM. 1990. People and Races. Long Grove (IL): Waveland Press.

Goldsmith TH. 1994. The Biological Roots of Human Nature. London: Oxford University Press.

Mange AP, Mange EJ. 1997. Genetics: Human Aspects. Sunderland (MA): Sinauer Associates.

Molnar S. 1998. Human Variation: Races, Types, and Ethnic Groups. 4th Edition. Upper Saddle River (JN): Prentice-Hall.

Muller-Ford CS. 2004. Analysis of Dermatoglyphic Heritability: a Study of Phenotypic Relationships. Thesis submitted in partial satisfaction of the requirements for the M.A. degree in Anthropology, The University of Montana – Missoula.

Wikipedia contributors. 2007 Haplotype [Internet]. Wikipedia, The Free Encyclopedia; 2007 Oct 13, 16:41 UTC [cited 2007 Oct 19]. Available from: http://en.wikipedia.org/w/index.php?title=Haplotype&oldid=164306192.

Chapter 16
FINGERPRINT EXAMINATION

In this chapter we will take a look at the examination of fingerprints and similar skin ridge prints, such as palm and sole prints. We will look at how they arise, how they are described, how they are detected and collected, and how they are used to link a suspect to a location.

The chapter will begin with a brief history of fingerprint examination. From there it will turn to a consideration of the theory underlying fingerprint expression and matching, and continue with an investigation of characteristics of fingerprints and matching. Next, we will examine how prints are collected and made visible, and the chapter will conclude with an introduction to databases of fingerprints.

16.1 The History of Fingerprint Examination

The history of fingerprint examination is especially well documented. The knowledge that the fingerprints of each person are unique can be traced back to the Early Babylonians, who pressed their fingers against soft clay to guard against the forgery of important documents.

During the late 1600's through the 1800's several scientists worked to classify the basic fingerprint patterns, based on whorls, loops, and arches. By late 1800's, fingerprints had become established as an accepted part of forensic work in the United Kingdom. Other countries, including the U.S., were slower to adopt fingerprints as evidence and many people continued to be skeptical of fingerprint evidence into the 20th century.

Two pivotal events for the acceptance of fingerprint evidence in the U. S. occurred in 1903. First, the New York Police Department began routinely collecting the fingerprints of people they arrested. Second, two prisoners at Leavenworth State Prison in Kansas who were both named Will West, and who looked identical (it was discovered later that they were identical twins) were shown to have different fingerprints.

By the 1910's it became apparent that the U.S. needed a nationwide fingerprint registration system. The U.S. Department of Justice decided to set up a fingerprint bureau at Leavenworth Prison and gave the authorities there $60 with which to do it. So, the Leavenworth officials set up a system using convicts to do the work. The convicts didn't do a very good job, and in fact sabotaged the system. In 1924, after a period of intense controversy, the fingerprint bureau was taken over by the FBI, at that time under the famous J. Edgar Hoover. The FBI got the mess straightened out, and many thousands of cases have been solved by drawing on the fingerprint bank.

Since the 1990's, computerized databanks of fingerprints have been developed called Automated Fingerprint Identification Systems (AFIS). These systems have dramatically reduced the effort required to retrieve and match fingerprints.

16.2 Dermatoglyphics

Dermatoglyphics are skin ridge patterns. The term "dermatoglyphics" is also used to refer to the study of skin ridge patterns. Skin ridges are also termed dermal ridges or friction ridges. They occur on the fingers and palms of the hand, and on the toes and soles of the feet. The skin ridges have associated oil glands that cause the ridges to be coated with oil, various types of dirt, other substances that are trapped by the oil. Dermatoglyphic prints are made when a surface is touched and the skin oils and other substances on the skin ridges are transferred to the surface, thus leaving a print of the ridge patterns on that surface.

We most often think of fingerprints. However, the palms of the hands, soles of the feet, and toes have skin ridge patterns that are as unique as fingerprints and which can also be transferred to surfaces. Fingerprint examiners work with all of these types of skin ridge patterns and prints – not just with fingerprints.

16.3 Theory of Fingerprints: Permanence and Individuality

The use of fingerprints as evidence is based on two theoretical principles, which also apply to palm and foot prints. These are the principle of permanence and the principle of individuality.

The **principle of permanence** states that a person's fingerprints are unchanging throughout their life. The skin ridges on the fingers are formed during the third to fourth month of fetal development and do not change significantly thereafter. As a person grows from infant to adult, their fingerprints grow in size as their fingers grow, but the pattern remains the same. Imagine drawing a picture on an uninflated balloon, then blowing it up. The picture will increase in size due to even stretching in all directions, but the picture on the inflated balloon remains recognizable as the picture drawn on it when uninflated. The skin ridge patterns on a child's fingers grow in a similar manner. Of course scars will change a person's fingerprints, but are easily recognized and their effects compensated for. Warts push apart an area of skin ridges and are equally easy to compensate for in a comparison. Advanced age causes the appearance of creases and a flattening of the skin ridges, which make comparison difficult, but do not alter the basic skin ridge pattern.

The **principle of individuality** states that no two individuals have the same fingerprints. Indeed each finger of each individual has a unique skin ridge pattern. Print examiners believe that a fingerprint will be unique to a certain individual and can be used to positively identify that person and exclude every other person on earth. The principle of individuality has a considerable amount of theoretical, practical, and experimental evidence that supports it. In the more than a century that fingerprints have been routinely compared world wide, no two areas of skin with friction ridges on any two persons, including identical twins, have been found to contain the same individual characteristics in the same relationship. With the advent of modern computerized databases of prints, millions of fingerprints have been compared and no two have been found to be identical. Recent studies comparing the fingerprints of cloned monkeys show that they, just like identical twin humans, have different fingerprints.

16.4 Class Characteristics of Fingerprints

Like most types of evidence, fingerprints have class characteristics and individual characteristics. As with other forms of evidence class characteristics can be used to exclude a suspect or achieve a consistent match, but a positive identification of a suspect requires matching individual characteristics.

In the case of fingerprints the class characteristics are the fingerprint patterns. **Fingerprint patterns** are divided into **arches**, **loops**, and **whorls**. Further, most fingerprint experts recognize two types of arches, plain arch and tented arch; two types of loops, ulnar loop (toward the little finger) and radial loop (toward the thumb); and four types of whorls, plain whorl, double loop whorl (also called double loop or double whorl), central pocket whorl, and accidental whorl. This gives a total of 8 possible fingerprint patterns. These patterns are illustrated in figure 16.1.

Figure 16.1: The Eight Basic Fingerprint Patterns (Class Characteristics)

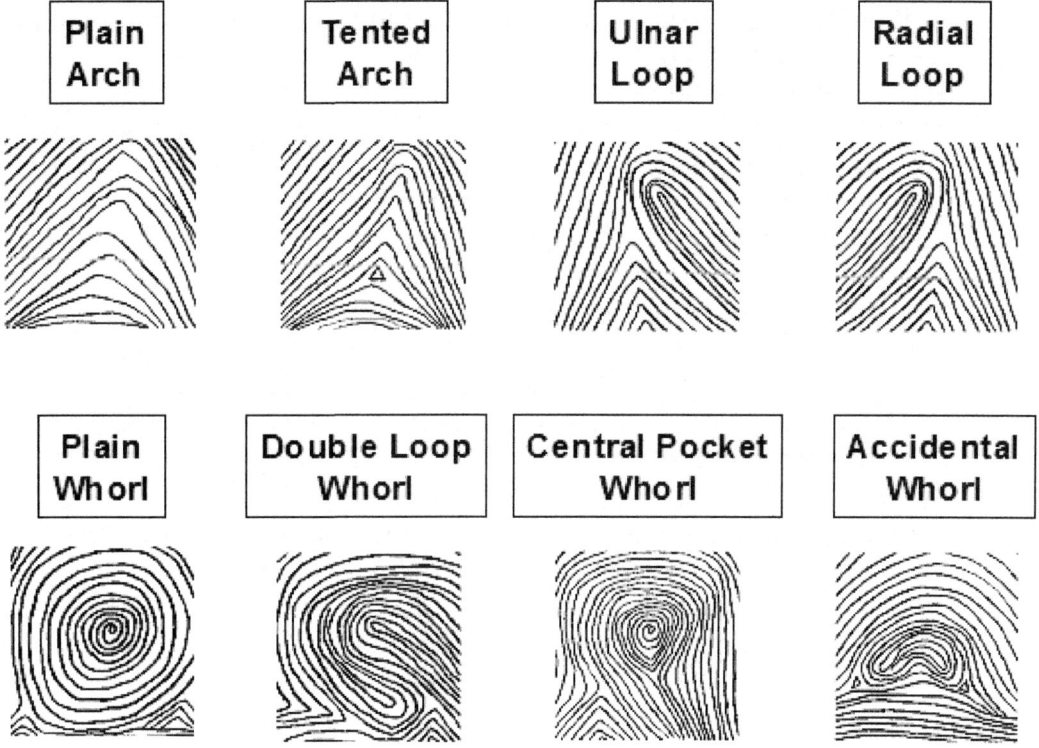

16.5 The FBI System for Archiving Fingerprints

Early in the history of fingerprinting, the FBI devised a system for classifying fingerprint cards. Fingerprint cards are paper cards upon which the fingerprints of both hands are collected for archival purposes. Cards need to be stored someplace where they are easily retrievable, normally a large number of file drawers sized appropriately for the fingerprint cards. Such large collections need to be organized in such a way that prints similar to an unknown print can be quickly retrieved. Consider for a moment how this could be done. Alphabetical indexing won't work because it is based on the name of a known person, not on the characteristics of the print, and if the name of the person who left the print is unknown it can not be retrieved.

The FBI system is based on scoring the fingerprint pattern of each finger. The system is arcane, but it served its purpose until computerized databases took over the chore. Here's how it works.

1. First score each finger as having a loop, arch, whorl.
2. Pair up the fingers into 5 pairs. Pair 1 is the right index finger and the right thumb, pair 2 is the right ring with the right middle finger, pair 3 is the left thumb with the right pinky, pair 4 is the left middle with the left index finger, and pair 5 is the left pinky with the left ring finger. Note that each pair has a first member (right index, right ring, left thumb, left middle, and left pinky) and a second member (right thumb, right middle, right pinky, left index, and left ring).
3. Next assign a value to each whorl you find. If the whorl is on one of the fingers in pair 1 assign it a value of 16. If it's in pair 2, assign it a value of 8. If it's in pair 3 assign it a value of 4. If it's in pair 4 assign it a value of 2, and if it's in pair five assign it a value of 1.
4. Total the scores for the five fingers that are the first member of their pair and add 1 to the total. Let's call this total the "row". Then, total the scores for the five fingers that are the second member of their pair and add 1. Let's call this total the "column". This gives two numbers each with a range of 1 to 32 and using these numbers we can define a table, organized as a grid with 32 rows (defined by the possible scores for the first members of the pairs) and 32 columns (defined by the possible scores for the second members of the pairs). There are 32 X 32 = 1024 cells in the table. Therefore, there are 1024 categories of prints.
5. Each person's fingerprints are assigned to a certain row and column of this table depending on their two numbers. The number of this cell, from 1 to 1024 is calculated and the fingerprint card is then filed in the appropriate drawer. The formula used to calculate the appropriate cell is [(row – 1) X 32] + column.

Here's an example. Let's say that we have a woman who has a whorl on her left thumb and on both middle fingers. Her left thumb is in pair 3 and receives score of 4. Her left middle finger is in pair 4 and receives a score of 2. Her right middle finger is in pair 2 and receives a score of 8. The left thumb and left middle finger are both the first member of their pair, so we get a total of 4 + 2 + 1 = 7 for the first members. Her right middle finger is the second member of its pair, so the score for the second members is

8 + 1 = 9. On the table with 32 rows and 32 columns, the woman's fingerprints would be classified into column 9 of row 7 of the table. This puts her in category [(7-1) X 32] + 9 = [6 X 32] + 9 = 192 + 9 = category 201 of the 1024 categories. Now, you would take her fingerprint card and file it in drawer 201 of your 1024 drawers.

I, for one, am glad that we live in the era of computerized databases and don't have to bother so often with these older archival methods. However, a substantial number of files of older fingerprint cards remain that are organized in this manner, so a fingerprint examiner needs to understand how the FBI system works.

16.6 Individual Characteristics of Fingerprints

Individual characteristics are small details of the ridge pattern, often called **minutia**. The locations of these details, and the relationships of these details to each other are used in matching. The types of minutia that can be looked for include ridge endings, ridge crossings, enclosures, bifurcations, dots (islands), and short ridges. Most fingerprint examiners focus on ridge endings, bifurcations, and dots, because these are the types of minutia that the computerized databases utilize. Forms of minutia are illustrated in figure 16.2.

16.7 Collection and Analysis of Fingerprint Evidence

The collection of fingerprint evidence is almost an art form. Prints can be found on many types of surfaces and in many forms. Each type of print may require a different method of processing and documentation.

Figure 16.2: Fingerprint Minutia (Individual Characteristics)

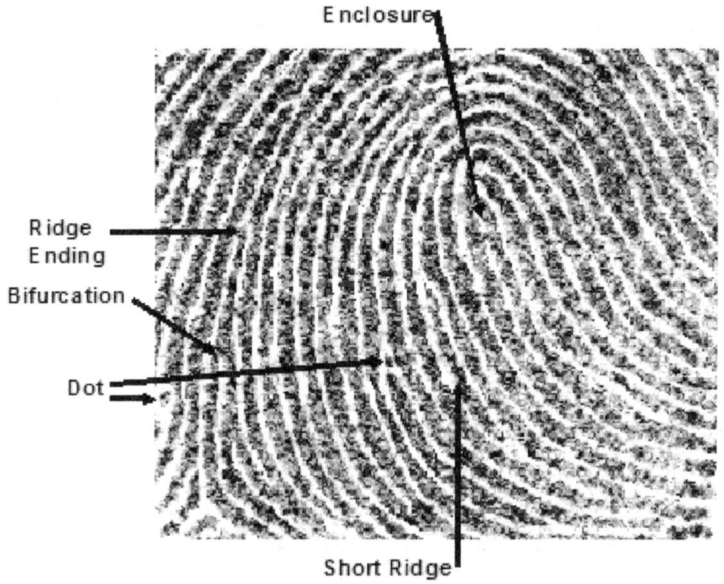

16.7.1 Types of Prints

There are three types of fingerprint evidence which demand separate methods of processing. These types of impressions are visible, plastic, and latent. **Visible fingerprints** are (as the name implies) visible to the naked eye. They are usually pressed onto some item by a finger that had some type of colored material on it such as blood, grease, or ink. **Plastic prints** are an impression made when a finger is pressed into a soft substance such as cheese, putty, or mud, and are also usually visible to the eye. **Latent prints** are invisible to the naked eye, or perhaps only slightly detectable. They are made up of natural body secretions of the hands and fingers, such as oil and sweat. Whenever a finger touches any surface, traces of these secretions are left behind as latent fingerprints. The same categories apply to palm prints and footprints.

Collecting visible prints is straightforward. First, the visible print is photographed for documentation. The preferred method for collecting visible prints is to cover the print with clear tape to preserve it, then package the entire item upon which the print is found and transport it to the crime lab. At the lab, the fingerprint examiners can work with the print under controlled conditions. If this procedure is not possible, then the investigators will usually attempt to lift the print with tape. The tape lifting procedure consists of placing clear tape over the print with the adhesive surface down, then peeling the tape away. The result is that the print is usually transferred to the tape, which can then be pressed onto an index card.

Plastic prints are collected in two ways, depending on the situation. In each case, the print is photographed first. In some cases, the item with the print in it is collected for evidence. In other cases a cast is made of the impression.

Latent prints are the most common. The visibility of latent prints depends on the physical condition of the person who left the print, on the surface of the object, and on the angle of reflection of the light by which they are viewed. The visibility of prints also depends on how much time that passed since they were placed, the amount of heat to which they have been exposed, and other factors. Over the time span during which latent prints remain on an object, they are affected by atmospheric conditions, air currents, and humidity; but even when the object has been exposed to adverse conditions for a period of time, it may still be possible to obtain prints.

16.7.2 Types of Surfaces

Not all surfaces will preserve latent prints. Surfaces that have coarseness that is larger in scale than the ridges on the fingertips usually won't preserve latent prints. Examples of these kinds of surfaces include egg cartons, bricks, and fabrics. However, there are cases in which prints on rough materials such as fabric have been rendered usable by computer enhancement.

The surfaces from which latent prints are collected can be divided into porous and non-porous. **Porous surfaces** such as paper, unfinished wood, cardboard, etc., are best for the preservation of prints because latent print residue can soak into the surface. **Non-porous surfaces**, such as plastic, glass, and metal, retain latent prints that are much more fragile because the latent print residue may simply be lying on the surface and not attached to it at all. Even the slightest handling can "wipe away" a latent print on a non-porous surface. The traditional method of dusting for latent prints with a brush may be disastrous when used on a non-porous surface, as it may result in wiping away the print. Glossy paper, such as is found in magazines is a non-porous surface and should not be handled like porous paper. It is sometimes even possible to find latent prints on human skin.

In modern times the standard procedure for both porous and non-porous items is to collect them for processing at the crime lab. In the lab the latent prints will be developed by exposing them to chemicals that bond to the latent print residues and make them visible. The items should be handled as little as possible because any latent prints on surfaces that are touched might be wiped off, even if gloves are worn. In some cases, it is possible to chemically treat the item in the field to make any latent prints visible; then the prints can be photographed before the item is collected.

16.7.3 Developing and Collection of Latent Prints

Rendering a latent print visible is called **developing** the print. The traditional method for developing and collecting latent fingerprints has been shown in innumerable police shows and movies. It consists of using a small soft brush to coat a surface with a dust, then blowing the dust away. If fingerprints are present some of the dust clings to the print, rendering it visible. Transparent tape is then pressed onto the print and lifted away. If done correctly, the tape will lift the fingerprint and can be pressed onto a paper card to collect the print. This method is still used in some situations. However, this method has mostly been superceded by visualizing the print chemically, usually by fuming with superglue and documenting it photographically.

In modern times, chemical methods of developing fingerprints have at least partly replaced dusting with powders. One of the most common methods for developing latent prints is **fuming with superglue**, a process that was introduced in 1982. The chemical in super glue, cyanoacrylate, interacts with the skin residues in the latent print (remember how quickly super glue will stick your fingers together). This interaction with the residues turns them white and visible, and may also help bond them to the surface they are on. When cyanoacrylate is heated it produces fumes that can be directed onto the latent print to cause this interaction to occur. Fuming wands, which can be used in the field, are now common. In the laboratory, a piece of evidence that might contain latent prints is placed in a cabinet into which cyanoacrylate fumes are introduced. After fuming with cyanoacrylate the prints can be further enhanced using fluorescent dyes such as rhodamine and RAM.

Iodine fumes can also be used to develop prints. Iodine reacts with the skin residues to turn them purple and make them visible. Prints can also be made visible by spraying them with ninhydrin, a dye that interacts with proteins in the skin residues to turn them black; or with silver nitrate, which reacts with salts in the print to produce a black or reddish-brown stain. Other chemicals are available, and experts in the analysis of latent prints have received training in which chemicals work best in each situation.

Alternative light sources, such as high intensity light, ultraviolet light, or lasers can be used to make the prints more visible under some circumstances. This can be particularly useful when preparing a print to be photographed.

16.8 Matching Fingerprints

Matching of fingerprints is something that takes an expert's knowledge and experience. I find it interesting that many TV shows and movies show fingerprints being matched by a computer. Computers can do many things, but accurate matching of fingerprints is not one of them. We will discuss the problems with computerized matching in the chapter on biometrics.

The process of matching fingerprints follows the methods which hopefully are now quite familiar to you. First the class characteristics, the fingerprint patterns, are compared to determine whether there is a quick exclusion or a consistent match. If the class characteristics match, then the fingerprint examiner will examine the individual characteristics, the minutia. The presence of minutia in certain locations and the relationships of the minutia to each other are used to judge whether there is a positive match or not. Fingerprint examiners find that the print of any area of skin with friction ridges from the hands or feet that is at least the size of a pencil eraser and at most the size of a dime, is sufficient to obtain enough minutia details to make a positive match.

Experts have debated since the beginning of this forensic science about how many points of similarity between two prints are necessary in order to say positively that they come from the same person. Numbers that range from 7 to 16 have been proposed, and some experts will give you a quick answer of 12 or 14. The International Association for Identification specifically states that no such criteria exist, which puts the burden on individual fingerprint experts to judge the reliability of a match. Occasionally, mistakes are made in matching fingerprints, and I believe that this is primarily due to not searching the entire available portion of the print for differences in addition to similarities.

16.9 Fingerprint Databases

The FBI oversees a system of computerized databases of prints, called the IAFIS/AFIS system. The prints stored in these databases were recovered as evidence or taken directly from criminals' fingers. This system is one of the primary tools of print examiners. The IAFIS/AFIS system will be discussed in more detail in the chapter on forensic databases.

16.10 Questions for Study and Review

1. In your own words, explain the principle of permanence as applied to fingerprints.
2. In your own words, explain the principle of individuality as applied to fingerprints.

3. List the 8 possible fingerprint patterns found in humans.
4. Say that a man has a whorl on his right index finger and his left ring finger. Into which file drawer would his fingerprint card be filed?
5. What types of minutia are commonly used in matching fingerprints?
6. in your own words, describe the difference between visible, plastic, and latent prints.
7. Describe how the nature of the surface upon which a print is located affects how it is handled.
8. Discuss the pros and cons of the various methods of developing and visualizing latent prints.
9. There is currently a debate in forensic science and in the courts over the number of similarities between two prints needed to make a positive match. One authority suggests that as few as seven similarities is enough. There exist, however, several cases of incorrect positive matches made on the basis of only seven similarities, so other authorities believe that more than seven similarities are required for confidence in matching. Why do you think the International Association for Identification has not taken a firm position on the number of similarities necessary for a positive match?

16.11 Sources

[Anonymous]. 2001. Ridges and Furrows [Internet]. [cited 2007 Oct 19]. Available from: http://www.ridgesandfurrows.homestead.com/.

[Anonymous]. 2005. Integrated Automated Fingerprint Identification System or IAFIS [Internet]. [cited 2007 Oct 19]. Available from: http://www.fbi.gov/hq/cjisd/iafis.htm.

Block EB. 1979. Science vs Crime. San Francisco: Cragmont Publications.

Brown EW. 1990. The Cyanoacrylate Fuming Method [Internet]. [cited 2007 Oct 23]. Available from: http://www.ccs.neu.edu/home/feneric/cyanoacrylate.html.

Budowle B, Buscaglia J, Perlman RS. 2006. Review of the Scientific Basis for Friction Ridge Comparisons as a Means of Identification: Committee Findings and Recommendations [Internet]. Forensic Science Communications 8(1). [cited 2007 Oct 21]. Available from: http://www.fbi.gov/hq/lab/fsc/backissu/jan2006/research/2006_01_research02.htm.

Futrell IR. 1996. Hidden Evidence: Latent Prints on Human Skin [Internet]. FBI Law Enforcement Bulletin 65(4). [cited 2007 Oct 19]. Available from: http://www.fbi.gov/publications/leb/1996/aprl965.txt.

German E. 1999. Latent Print Examination: Fingerprints, Palmprints and Footprints [Internet]. [cited 2007 Oct 19]. Available from: http://onin.com/fp/.

Jayasekara RW. 1998. Cyberguide to Basic Medical Genetics [Internet]. [cited 2007 Oct 19]. Available from: http://www.infolanka.com/org/genetics/cyberguide/bmg-home.htm.

Meloan CE, James RE, Saferstein R. 1998. Lab Manual for Criminalistics: An Introduction to Forensic Science. 6th Edition. Upper Saddle River (NJ): Prentice Hall.

Parisi KM. 1998. Getting The Most Of Your Fingerprint Powders [Internet]. [cited 2007 Oct 23]. Available from: http://www.geocities.com/cfpdlab/Sticky.htm.

Saferstein R. 1998. Criminalistics: An Introduction to Forensic Science. 6th Edition. Upper Saddle River (NJ): Prentice Hall.

Wargacki SP, Lewis LA, Dadmun MD. 2007. Understanding the Chemistry of the Development of Latent Fingerprints by Superglue Fuming. Journal of Forensic Sciences 52(5): 1057-1062.

Chapter 17
QUESTIONED DOCUMENTS EXAMINATION

In this chapter we will explore questioned document examination. This forensic science is very important in the modern world, where many legal questions revolve around the authenticity of documents. Document examiners are primarily involved in determining whether a written document is authentic or forged, and in determining who wrote the document. Document examiners are also experts on types of papers and inks. We think of them as only working with hand written documents, which are becoming more rare in our computerized world. Fortunately, however, they are also experts on computer generated documents, faxes, and photocopies. They are also able to develop latent handwriting, which is handwriting that is not visible, the most common example of which is an impression made in a surface by writing on a piece of paper laid on that surface.

The chapter begins with a list of the tasks of document examiners and the history of document examination. It continues with a discussion of the theory of handwriting and signatures. Next it will explore detection of altered documents and authenticating documents. Documents produced by typewriter, printers, and copiers will be given due consideration. We will look at the topic of document content before finishing the chapter with an investigation of methods for collecting standard samples of writing.

17.1 The History of Document Examination

Surely almost as soon as writing was invented, somebody got the idea to alter or forge a document, someone else suspected the alteration or forgery, thus disputing the document. To solve problems of this type, some people surely developed expertise in spotting the evidence of document alteration. Therefore, in a sense, questioned document examination is probably nearly as old as writing itself. There are innumerable accounts of questioned documents throughout history, and various methods, such as signature, seals, stamps, notarizing, etc. were developed as ways to assure the authenticity of documents.

So far as is known, the first person to scientifically approach the study of questioned documents was Hans Gross, in the late 1800's and early 1900's. The work of Albert S. Osborn in the early 1900's was discussed in chapter 1. In modern times, national and international repositories and collections of written, typewritten, and printed materials; inks; papers; and other materials of interest to document examiners have been built, which are of great help to document examiners.

17.2 Handwriting and Signatures

Handwriting and signatures remain one of the main types of evidence examined by questioned document examiners. Although more documents are produced using word processing software and printed by a printer, a surprising amount of handwritten material is still produced.

17.2.1 Theory: Handwriting Production and the Brain

Psychologists and physiologists have given us a pretty good idea of how handwriting is produced. The process of writing something by hand starts in the brain, where a mental template of what the resulting letters being written is formed. A **mental template** is an idea held in the mind, consciously or unconsciously, of how something being produced should look. The brain then sends signals to the hand, wrist, and arm to move in such a way as to produce the desired result.

The way we learn to write is fairly well known. In writing something, the brain sends signals via the nervous system to the muscles of the forearm and hand. The muscles interact with the writing instrument and the object being written on, and send feedback to the brain in the form of sensations that occur while writing. The eyes see the result and also give information to the brain. Eventually, the brain learns the correct movements in the correct sequence that will produce the desired result and stores it in memory. This pattern of commands that the brain learns is called a **motor program**. Once a motor program has been developed its use may become automatic. In other words we don't usually have to think about how to write each letter when we are writing. Once a skill like this is learned, it is not quickly forgotten. People say that once you learn to ride a bicycle, you never forget how. In reality, a person might forget how to ride a bicycle or how to write, but it takes a long time.

17.2.2 The Natural Variability of Handwriting

One of the unique challenges of analyzing handwriting is its **natural variability**. A person's handwriting changes in slight ways from moment to moment. Thus, a signature made at a certain time will be slightly different from a signature made at another time. This is a very different situation from, for example, fingerprints, which are permanent and do not change in substantial ways from one time to the next. Thus, a questioned documents examiner must take into account the natural variability in handwriting. If two samples of handwriting are exactly the same, then one (or both) of them was not produced by a human being. However, if two samples of handwriting are sufficiently different, then they probably were not made by the same person. The task of the document examiner, therefore, involves judging whether the variability in samples of handwriting is enough, but not too much, to say that the samples were produced by the same person.

Some of the variability in handwriting is physiological. Once a motor program has been learned, you can look at the body as a machine, composed of levers and manipulators that cause the task to be carried out under the direction of the brain. This means that any change in any of these systems is going to cause a change in the handwriting produced. Everything from the posture of the writer to whether they have a sore shoulder that day are going to effect the writing produced. Anything that affects the brain or its functioning, such as alcohol, drugs, accident, disease, or even fatigue, is also going to affect the handwriting produced.

In addition to factors of the brain and body, there are a variety of external variables that will affect any handwriting produced. The nature of the writing instrument is important. Different instruments have different amounts of friction with the writing surface. A pencil has quite a lot of friction, a ballpoint pen has less. Different instruments have differing rates of ink flow and point sizes. For example, a pencil's point is in a constant state of becoming larger as the pencil becomes dull, and when you sharpen it you are making the size of the point small again. The nature of the ink and paper affect the handwriting produced as well. Some inks absorb into the paper faster than others. Some inks dry faster than others. Some papers are smooth, and others are rougher. Some papers are more absorbent than others. All of these things add variability to the writing produced. The nature of the surface upon which the paper is placed for writing is important as well. Smooth surfaces are usually better than rough surfaces. Writing produced on a softer surface is different from writing produced on a harder surface.

Our personal writing continues to change over time as well as we gain more experience and therefore do it better. With practice, our writing may come to look more like our mental template of what it should look like. In addition, we can change our mental template of what our writing should look like, consciously or unconsciously, and our handwriting will change in response. As we age, however, our bodies become arthritic, and when combined with natural aging processes of the brain an elderly person's handwriting may be quite different from what it was in their youth.

This adds up to the fact that a person's handwriting differs in slight ways every time they write. As an experiment, try writing some short sentence, or even your signature, ten times in a row. Examine the writing and notice that (1) the writing is similar each time, and (2) there are slight differences each time.

17.2.3 Class and Individual Characteristics of Handwriting

There are relatively few class characteristics of handwriting, but many individual characteristics. One class characteristic is the type of script used. Some people prefer to print, others prefer cursive writing. Among both of these forms there are some variations. For example, I learned to print the lowercase letter 'a' like this: 'ɑ', but my stepdaughter learned to make it like this: 'a'. There are a variety of cursive styles as well, that vary through time and by region.

Individual characteristics result from the fact that we each have different brains, hands, nervous systems, muscles, etc. We also each have a different history of training and practice of handwriting. Therefore, everybody's handwriting is slightly different. We all differ in the way we form letters, the angles of the upright portions of certain letters, the ways in which we make loops and crosses, and a variety of other factors.

17.2.4 Comparing and Matching Handwriting

Given that so much variability exists in handwriting, from both internal and external factors, a document examiner must have considerable knowledge and experience in order to carry our his or her job correctly. As in other forms of criminalistics, the document examiner compares known samples of writing (called exemplars by document examiners) with questioned samples of writing. The amount of writing that the examiner is asked to compare varies greatly from just a signature, to an entire book.

Figure 17.1: Two Checks Purportedly Written by John Doe

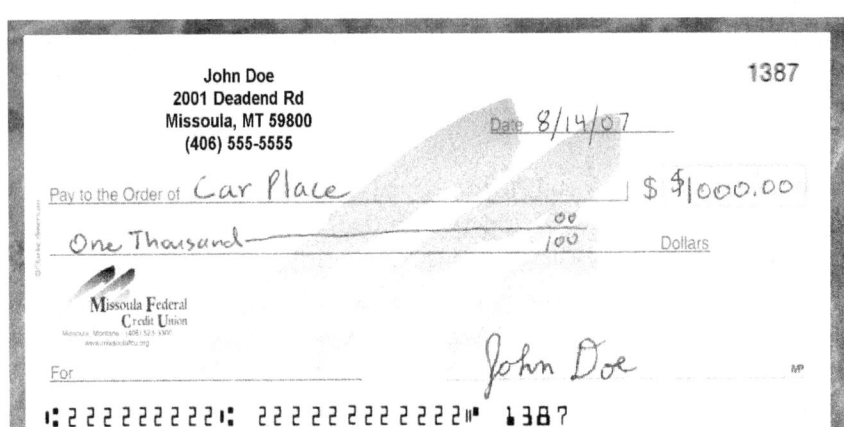

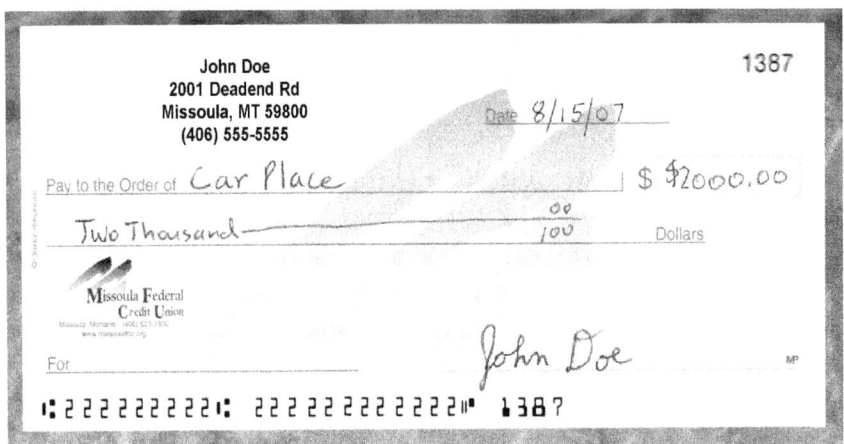

Most of the time the question asked will be of the form, "Did person X write this document?" To answer this question, the examiner has to compare the evidence sample with the standard sample of writing, taking into account all the things that can produce variability in a person's writing. The basic principle of matching remains the same as in other forensic sciences we have examined previously, that a positive match is obtained only if both the class and individual characteristics are the same. Otherwise there is an exclusion. Since there are more individual characteristics of handwriting than class characteristics, a result of consistent match is rarely obtained in document examination. However, the result of undetermined may be obtained for a fairly obvious variety of reasons, most commonly because the known samples of a person's writing are not large enough to allow identification of many individual characteristics.

In document examination there are no set rules for how many similarities must exist between a known and an unknown sample of writing to be able to say definitely that the two samples were produced by the same person or machine. As in the examination of fingerprints, this is left to the judgement of the examiner. The document examiner will weigh all the evidence and form an opinion that he or she will present in court if called upon to testify.

There are many types of questions that can be answered by handwriting comparison. Here are some of them:
- Was this document written by the person purported to have written it?
- Is the same person the author of several documents?
- Who wrote this anonymous letter or threat?
- Did the same person who signed the document also initial the changes to it?
- Was the entire document written by the same person, or by a group of people?

17.2.5 An Example: John Doe's Checks

Figure 17.1 illustrates a possibly typical questioned document case. John Doe (a fictional person, of course) claims he wrote the top check to his auto mechanic for repairs to his car. However, he claims that he did not write the bottom check the next day. He believes that he might be a victim of a type of fraud known as "check washing" which the National Check Fraud Center says costs Americans over 800 million dollars each year. A document examiner examines the two checks, paying particular attention to the signatures on the checks, and a sample of five of John's signatures found on other documents. These signatures are shown in Figure 17.2. The document examiner concludes that John is indeed a victim of fraud, but not check washing. How can he or she tell? First, check washing consists of altering the information written on the original check. In this case however, there are two checks. The critical piece of information is the two signatures on the checks, which are identical and do not show any of the natural variation we would expect in two signatures and which we see in the five signature of Figure 17.2. This means that one of the two signatures on the checks must have been produced through some form of copying, by a machine, and conclusively demonstrates that John did not write both checks. Most likely, someone scanned John's original check, made some adjustments using graphics software, and printed the second check using a good quality printer, onto paper similar to the paper used in checks.

Figure 17.2: John Doe's Known and Purported Signatures

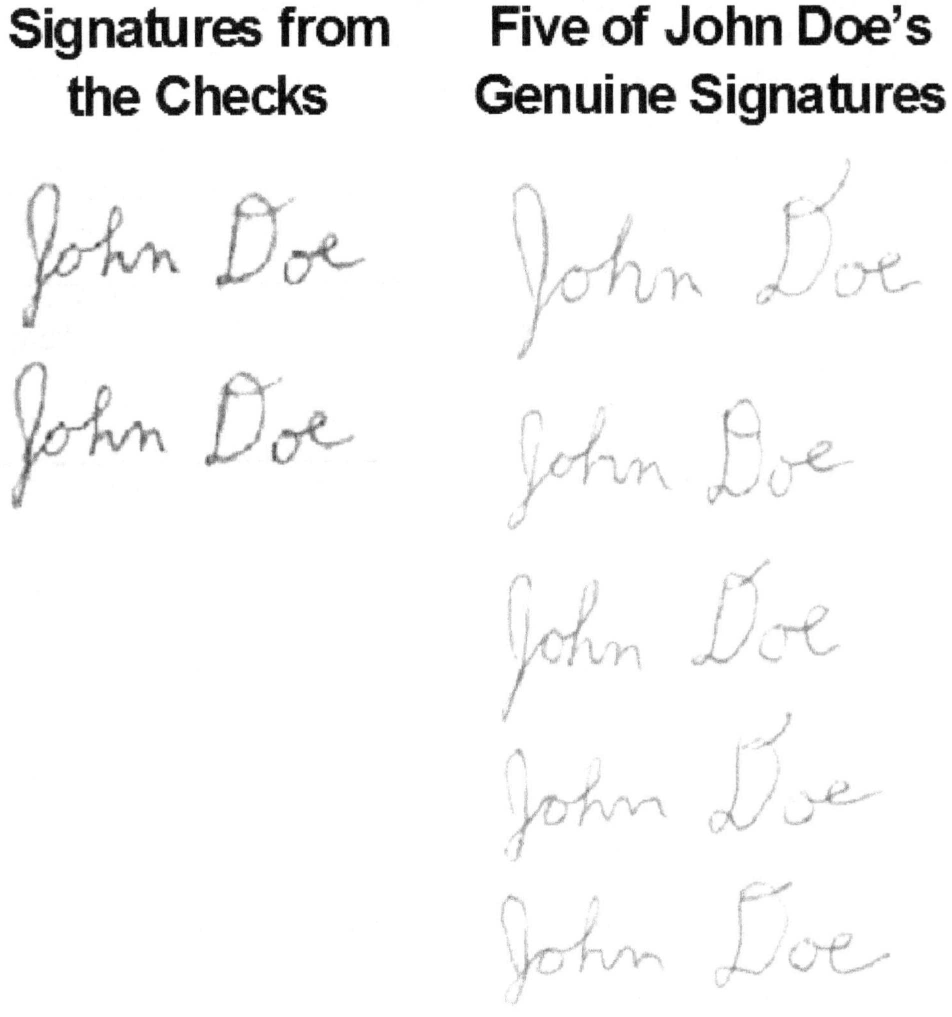

Signatures from the Checks

John Doe

John Doe

Five of John Doe's Genuine Signatures

John Doe

John Doe

John Doe

John Doe

John Doe

17.2.6 What Can and Can't Be Determined From Handwriting

Due to the connection between mental state and handwriting, a document examiner can sometimes make a determination about the mental state of the person who wrote a document. Therefore, a document examiner may be able to determine whether a document was written by someone who was ill, or mentally ill, or under the influence of drugs or alcohol. They might also be able to tell whether the document was written under **duress**, which means that someone was threatening to do the writer harm unless they wrote or signed the document.

There are, however, many things that many people assume a document examiner would be able to tell about the person doing the handwriting that are actually not possible to determine reliably. We all think that we can tell whether a person is right handed or left handed by looking at how their writing slants. However, this is not always true, and in many cases the handedness of a writer can't be determined. At best, a document examiner can determine that a person was likely to have been right or left handed. Likewise, contrary to popular opinion, it is not usually possible to determine the age or sex of the writer.

17.3 Detecting Forgery and Document Alterations

Common methods of forgery, such as tracing or copying a signature, can usually be spotted by a document examiner. Usually, what they spot is small corrections, or starts and stops, that the forger makes as they are trying to do the tracing or copying. This is one type of analysis in which a document examiner may use a microscope. Most analyses, however are done without magnification.

Another common task of a document examiner is to detect alterations in documents. There are several ways in which alterations are made, including erasures, write-overs, and substituting pages in a multi-page document.

Erasures can be detected by the fact that it is impossible to erase any mark completely without damaging the paper. All forms of ink, and even the graphite of pencil lead, soak or penetrate into the paper to at least a small degree. This means that the mark can't be removed without also removing the layers of paper that the ink has penetrated. Therefore, an erasure always leaves a thin spot in the paper which can be easily detected by shining a light through the paper from behind. The erasure will show up as an area of the paper that lets more light through than the rest of the paper.

Write-overs occur when someone simply writes over the top of existing writing in a way that makes it look like the added writing was the original. There are several ways to detect write-overs. First, the layer of ink will be thicker where the new writing coincides with areas of the old writing. This can be detected in a variety of ways, perhaps most simply by the old trick of shining a light through from behind. Also, in making a write-over the ink may not be matched perfectly and this difference might be detectable directly or by using an alternative light source. Use of an alternative light source, usually ultraviolet light, is effective because different inks that may look similar under visible light will look very different under ultraviolet. Third, write-overs never look perfect, unless you are turning a number 3 into a number 8. Even in this case, however, the number 8 made by closing the loops of a number 3 is often not the same as a number 8 produced normally by that person.

One similar task is to determine what was originally written underneath an erased, over-written, or crossed-out portion of a document. The difficulty of this varies considerably, but, document examiners are usually successful at it. The basic technique usually involved examining the altered area using an alternative light source, usually ultraviolet, and magnification. Remaining portions of the original writing can usually be detected and are often enough to reconstruct what was originally written there.

Some criminals may try to alter a document by inserting entire pages into it. This can only successfully fool someone who is inattentive or who doesn't know what to look for. The number of things that have to be matched in order to successfully insert a page are overwhelming and include the printing style (font or handwriting), the type of paper, the type of ink, the page numbering, and the intensity (darkness or lightness) of the original. Even if these things are matched, it is impossible to match such factors as the age-related changes that occur in the paper and ink over time. Therefore, this type of alteration of a document is usually fairly easy to spot.

17.4 Dating and Authentication of Documents

Another type of question a document examiner may address is, "When was this document written?" This question has a lot of applications from judging the authenticity of wills, to authenticating ancient documents and maps. **Authentication**, as used here, refers to determining that the document is consistent with having been produced at the time and place, and in the manner suggested by some informant.

Dating a document relies primarily on analysis of the inks, papers, and writing instruments. Through history, inks have changed considerably in composition, as have papers and pens. For example, nobody would believe that something claimed to be an ancient religious text was genuine if it was written on inkjet paper using a ballpoint pen. However, if it was written on papyrus paper or parchment with a quill pen and an ancient variety of ink, then there is a possibility that it is genuine.

The basic task in authentication or dating of a document is to determine whether the ink, paper, and writing instrument used are consistent with those used at the time the document is claimed to have been made. If these are all consistent, then you can be fairly certain that the document dates to that time period. Databases of inks, papers, and writing instruments exist, and can be accessed by a document examiner to get information needed to do dating or authentication of documents.

17.5 Typed Documents

Typewriters have been around for more than 100 years, and methods to identify a particular document as having been typed on a particular typewriter are well established. Details of the font are class characteristics and can help narrow down the brand and model of typewriter used. In addition, each individual typewriter has a unique history and a unique set of mechanical defects that can be identified in the typed material. For example, I used to have a typewriter that would never completely print the letter 'r' no matter how hard I punched the 'r' key. One day I looked and saw that the corner of the 'r' was broken off of the hammer that struck the ribbon and paper when the 'r' key was pressed. Most typewriters have similar small defects that allow them to be individualized.

In the case of typewriters, it may also be possible to determine who it was who did the typing. A human interfaces with a typewriter directly, and different individuals press the keys on a typewriter differently, causing slight differences in the output from the same typewriter when used by different typists. This can sometimes allow the identity of the author of a document to be determined, if the pool of potential authors is small.

17.6 Documents Produced Using Printers, Faxes, Copiers, and Other Modern Machines

Computer printers, printing presses, FAX machines, and copiers all vary a surprising amount in the output they produce. They each have class characteristics and individual characteristics that are produced by the particular history of a device as it ages. Since these are machines, they are always in a continuous process of wearing out, and like we discussed for firearm manufacturing devices, every item output from them is slightly different.

Many people believe that in the modern age of computers and printers documents produced on these sorts of machines are anonymous. Nothing could be further from the truth, and document examiners can often not only narrow down the device that produced a document to a certain brand and model of printer, but can often identify exactly which printer was used.

The analysis of the output of these types of machines differs from handwriting analysis and typewriting analysis, however, in that it is usually not possible to determine from the printed output who actually typed the document into the word processor.

Photocopiers present unique challenges for document examiners. There are two primary types of questions asked about these types of machines. First, "Is this document an original or a copy?", and second, "What photocopier was used to produce this copied document?" Copied documents are usually easy to identify. It used to be true that anybody could tell an original document from a photocopy, and this is still true in many cases. However, new high-end photocopiers can produce copies that are so similar to the original that they are difficult to recognize as a copy by most people. Document examiners are not ordinary people, however, and can recognize a copy from an original fairly easily. Photocopier toners are usually distinctively different from the inks and laser toners used to produce original documents. There is also an unavoidable loss of the sharpness of the output when it is copied that can be identified by a person who knows how.

Since photocopiers are machines, the same processes of constant change apply to them as to all machines. Therefore, a document examiner can often determine which exact photocopier produced a given copied document.

Databases of printers, faxes, photocopiers, and similar instruments are maintained and can be accessed by document examiners who are working with the output of one of these sorts of machines.

17.7 Document Content

The content of a document can often be nearly as informative as the form of the writing. Each of us has distinctive content to our writing that consists of the words we use and the way we use them. This is what makes Shakespear's writing different from that of J. K. Rowling. An expert can often recognize a certain style of writing, in the same way that a literature professor might be able to read a small portion of an unfamiliar book and be able to guess its author.

Vocabulary is one aspect of content. Everyone uses a slightly different vocabulary depending on where they grew up, their level of education, their occupation, how much reading they have done and who they have read, and a variety of other factors. If a person uses a particular word or phrase in an unusual way, it may be possible to link them to a certain document in which that word or phrase is used in that way. Spelling, punctuation, and grammar may also be distinctive. Some people intentionally or unintentionally make these types of errors consistently in a way that can help link them to their writing.

There is also a less tangible aspect of writing called "style". The writing can be straightforward, flowery, or convoluted. It can have many short sentences, or fewer long sentences. It can use active voice or passive voice, or some combination of both. It can be written in first person perspective where the writer is writing as if they themself were the actor, or it can be in third person perspective where the writer is writing as if they were observing the action rather than participating in it. All of these are aspects of style and can help link a writer to their writing.

Finally, there is an aspect of content called special knowledge. If the writer reveals knowledge of things or events that are not commonly known, then it may help in linking the writer to a certain crime. The writing may also reveal that the writer has certain special knowledge such as how to build a bomb or how to make methamphetamine, which might be important.

17.8 Collecting Standards

The single most important factor in the success of document examination is having an adequate number of good standard items, **exemplars**, of the known origin. This is especially important for comparing handwriting, but also important for comparing typewritten or computer printed documents.

Collection procedures for exemplars are important. The main challenge with exemplars of handwriting is that a suspect may try to alter or disguise their natural handwriting. Therefore, there are a few commonsense rules to follow that make it likely that you will obtain good exemplars. Many people may not think of these offhand. First, the exemplars need to be collected in such a way that there is no doubt as to their authenticity. You don't want a defense lawyer to be able to say "that isn't my client's writing". Second, the exemplars need to be collected in such a way that they will be accepted as evidence in court, which means using accepted evidence collection methods including obtaining a search warrant if necessary.

The more exemplars the better. If the document examiner has many exemplars of a person's writing, they can get a good idea of the range of variation in that person's writing. There are no real standards for how many exemplars are normally collected, other than as many as reasonably possible.

There are two types of exemplars: collected and requested. Collected exemplars are those that are already in existence and which an investigator collects from a suspect's office, home, or other location. Requested exemplars are those that the suspect is requested to give as part of the investigation. Collected exemplars are much more reliable, because a suspect will almost always intentionally disguise or alter their handwriting on a requested exemplar.

Collected exemplars that are produced close to the time of the event being investigated are best. Since a person's writing changes over time, and the output of typewriters and printers also changes, exemplars that were produced long before the relevant time may not give good results. This is especially important if there has been a change in the health or mental condition of the person whose writing is being collected.

17.9 Questions for Study and Review

1. When I was a child, my grandmother once told me that the process of learning how to write consisted of training the hand to form letters. Explain why this explanation of learning how to write agrees or disagrees with what we know today about how people learn to write.
2. A person's fingerprints don't change (theory of permanence), but handwriting exhibits natural variability. Discuss how this difference makes handwriting examination different from fingerprint examination.
3. List some important internal variables (i.e. effects on or conditions of a person's body) that lead to variability in handwriting.
4. List some important external variables (i.e. things that are not effects on a person's body) that have an impact on the appearance of handwriting.
5. What does it mean for a document to be written "under duress"? Why would this be an issue in some types of cases?
6. I have a scanned graphic of my signature that I import into certain types of documents that I intend to send by email. Given a collection of my emailed documents, how could a document examiner determine that I did not individually sign them all?
7. Discuss those characteristics about a person that can and can not be determined by analysis of their handwriting.
8. Graphology is the study of how a person's character, and possibly their future, are revealed by their handwriting. It is a popular form of entertainment at carnivals and similar events, where a graphologist will have a booth in which he or she charges a few dollars to examine someone's handwriting. Sometimes, this examination is done by a machine. Discuss why graphology does not provide evidence that can be used in court.
9. Make a list of ways in which a criminal may try to alter a document and how that type of alteration can be detected by a document examiner.
10. Why is it often possible to determine who typed a document using a typewriter, but almost never possible to determine who typed a document using a word processor and printer?
11. Discuss how document content can help narrow the list of suspects that might have produced a certain document.

12. What is an "exemplar"? Is "examplar" just another term for a standard item of evidence?
13. What precautions would you take if you were asked to collect a requested exemplar of a suspect's handwriting?

17.10 Sources

Block EB. 1979. Science vs Crime. San Francisco: Cragmont Publications.
Brunelle RL, Reed RW. 1984. Forensic Examination of Ink and Paper. Springfield (IL): Charles C. Thomas.
Held DE. 2001. Handwriting, Typewriting, Shoeprints, and Tire Treads: FBI Laboratory's Questioned Documents Unit [Internet]. Forensic Science Communications 3(2). [cited 2007 Oct 21]. Available from: http://www.fbi.gov/hq/lab/fsc/backissu/april2001/held.htm.
Hilton O. 1993. Scientific Examination of Questioned Documents, Revised Edition. Boca Raton (FL): CRC Press.
Kam M, Fielding G, Conn R. 1997. Writer Identification by Professional Document Examiners. Journal of Forensic Sciences 42 (5): 778-786.
Kam M, Lin EW. 2003. Writer Identification Using Hand-printed and Non-hand-printed Questioned Documents. Journal of Forensic Sciences 48 (6): 1391-1395.
LaPorte G. 2004. The Use of an Electrostatic Detection Device to Identify Individual and Class Characteristics on Documents Produced by Printers and Copiers - a Preliminary Study. Journal of Forensic Sciences 49 (3): 610-620.
National Check Fraud Center. 2006. Check Washing: What is It? [Internet]. [cited 2007 Nov 8]. Available from: http://www.ckfraud.org/washing.html.
O'Connor T. 2006. Questioned Document Examination [Internet]. [cited 2007 Oct 19]. Available from: http://faculty.ncwc.edu/TOCONNOR/425/425lect05.htm.
Saferstein R. 1998. Criminalistics: An Introduction to Forensic Science. 6th Edition. Upper Saddle River (NJ): Prentice Hall.
Sita J, Found B, Rogers DK. 2002. Forensic Handwriting Examiners' Expertise for Signature Comparison. Journal of Forensic Sciences 47 (5): 1117-1124.
Wikipedia contributors. 2007. Questioned document examination [Internet]. Wikipedia, The Free Encyclopedia; 2007 Oct 2, 17:55 UTC [cited 2007 Oct 19]. Available from: http://en.wikipedia.org/w/index.php?title=Questioned_document_examination&oldid= 161834305.
Will EJ. 2007. Theory: Handwriting and Signatures - Some Basic Facts and Theory [Internet]. [cited 2007 Oct 19]. Available from: http://qdewill.com/modules.php?name=Content&pa=showpage&pid=1.
Will EJ. 2007. Some Typical Document Examination Applications [Internet]. [cited 2007 Oct 19]. Available from: http://qdewill.com/modules.php?name=Content&pa=showpage&pid=2.
Will EJ. 2007. Famous Cases [Internet]. [cited 2007 Oct 19]. Available from: http://qdewill.com/modules.php?name=Content&pa=showpage&pid=3.

Chapter 18
Basic Science
ANALYTICAL METHODS AND INSTRUMENTS

Up to this point in our study of the forensic sciences, the main analytical instrument we have discussed is the microscope in its various forms. As we move beyond the criminalistic forensic sciences, we will find other instruments in use and the main goal of this chapter is to provide a brief explanation of how they work. We will cover analytical methods and instruments for determining what a suspect substance is, and for determining how much of the substance is in a mixture. These instruments are used by forensic chemists, toxicologists, and serologists, and will be referred to often in the chapters on those forensic sciences. Before I delve into these methods and instruments, however, I first need to explain the metric system of measurement and the scientific system for expressing very large and very small numbers.

18.1 The Metric System of Measurement

In general, Americans are hostile to the **metric system** of measurement, for reasons that I completely fail to understand. We insist on our pints, pounds, and feet, even though most countries in the world use a much simpler and more logical system – the metric system of measurement. Even in the U.S. however, all scientific work is done using the metric system, and those students interested in the forensic sciences should learn it.

Many people in the U.S. get upset when this system is mentioned as a replacement for the English Imperial system currently in common use. In fact, the U.S. government passed a law in the 1970s to convert the U.S. to the metric system, but later gave up on our citizens as "too stupid" to understand it and repealed the act. The lack of metric system use in the U.S. is really an aberration; the rest of the world uses the metric system. America is more English than England in sticking to the obsolete English measures. In fact, since the U.S. is the only country still using this system, we should call it the American system. Over time, a few items with metric measurements are creeping into general use in the U.S. For example, I often buy my soft drinks by the liter, which is similar enough in size to a quart that most people wouldn't notice the difference. However, It would be foolish to present the results of forensic analyses measured using the metric system in a courtroom, because the judge, attorneys and jury probably won't understand it. Therefore, forensic scientists very often have to convert from one system to another.

There are two important reasons to use the metric system. First, the metric system is universal. All scientists worldwide agree on what constitutes the measures taken within this system, and use this exact same system. This means that scientists can compare results they obtain with results obtained by any other scientist without conversions or errors. Second, the metric system is simple. The metric system has only one basic unit in each category of measures. This basic unit is converted to smaller or larger units by using powers-of-ten multipliers.

These are the **basic units of the metric system**:
- Distance: the meter. A meter is about 39 inches, just a little more than an American yard. One of the reasons the U.S. gave up on the metric system is that the National Football League protested that they would have had to expand football fields to 100 meters instead of 100 yards, which would have added about 300 inches (25 feet) to the length of a football field and many football fields wouldn't be able to fit within their stadiums.
- Volume: the liter. A liter is close enough to a quart that it doesn't make any practical difference.
- Weight: the gram. A gram is a very small weight. I've heard it said that a standard metal paperclip weighs about one gram.
- Temperature: degrees Celsius. The Fahrenheit temperature scale is completely illogical and arbitrary. The Celsius or centigrade scale, on the other hand, is based logically on the freezing and boiling points of water. Water freezes at 0 degrees and boils at 100 degrees. There are 100 degrees between these two points. Human body temperature is 37 degrees Celsius.
- Other units exist for measures of force, heat, torque, etc.

There are a set of **modifying prefixes** that work with all basic units. These prefixes are the same, regardless of whether you are talking about length, volume, weight, or any other type of measurement. The prefixes are based on multiples of ten or tenths. Therefore, instead of having to remember that a yard contains three feet, each of which contains 12 inches, we get to remember that a meter has 10 decimeters, a decimeter contains 10 centimeters, and a centimeter contains 10 millimeters.

These are the basic prefixes, from smallest to largest. Others exist as well for even smaller and larger quantities.
- nano = billionth
- micro = millionth
- milli = thousandth
- centi = hundredth
- deci = tenth
- deka = ten
- hecto = hundred
- kilo = thousand
- mega = million
- giga = billion

Once you know what the basic unit of measurement is and learn the prefixes, you can measure anything easily and precisely.

18.2 Scientific Notation for Numbers

For many cases the familiar decimal system of expressing numbers works just fine. The decimal system is what we all learned in grade school. You have some number followed by a decimal point, followed by some more numbers. To interpret a decimal number you start at the decimal point. The first position to the left of the decimal point is the one's position, and each additional position to the left is 10 times the one to the right of it. Similarly the first position to the right of the decimal point is 1/10th of the position to its left, and this progresses as you go farther right from the decimal point. For most of us, we are so familiar with this system that we would find it difficult to describe it as I just have (I found it difficult).

For an example of how the decimal system works, let's take the number 1234.5678. We start at the decimal point. The number to the left of the decimal point is the number of ones (there are 4 ones), to the left of it is the number of 10's (there are 3 tens), to the left of the tens is the number of hundreds (there are 2 hundreds), and to the left of the hundreds is the number of thousands (1 thousand). This can be extended indefinitely. The first number to the right of the decimal point is the number of tenths (there are 5 tenths), the next number to the right is the number of hundredths (there are 6 hundredths), next is the number of thousandths (7 of them), and finally the number of ten-thousandths (8 of them). So, we can read this number from left to right as 1 thousand, two hundred, thirty-four, 5 tenths, 6 hundredths, 7 thousandths, and 8 ten-thousandths.

Scientists often deal with numbers that are so very large or very small that the decimal system becomes difficult to use. For example, Avogadro's number is a very large number that can be written as 602200000000000000000000. Putting in commas helps, so we could write this number as 602,200,000,000,000,000,000,000. This is 602 sextillion, 200 quintillion. Numbers this large are clearly difficult to handle using the decimal system. Similar problems are encountered when dealing with very small numbers.

Scientific notation relies on the concept of powers of ten, just as the decimal system does. However, scientific notation expresses the powers of ten as the number 10 raised to some power. In this system 10^0 is 1, 10^1 is 10, 10^2 is 100, 10^3 is 1000, and so on. Note that an number raised to the 0 power is 1, any number raised to the 1 power is that number (10), $10^2 = 100$, and $10^3 = 1000$. Numbers that would be written to the right of the decimal point in decimal notation are expressed as negative powers of 10. So, 10^{-1} is tenths, 10^{-2} is hundredths, 10^{-3} is thousanths, and so on. Note that any number raised to a negative power is exactly one divided by that number, so $10^{-1} = 1/10^1 = 1/10$. Similarly $10^{-2} = 1/10^2 = 1/100$ and $10^{-3} = 1/10^3 = 1/1000$.

In cases of very large or very small numbers scientists denote them using a number multiplied by the power of 10. What they do is take the leftmost (most significant) number and add a new decimal point to the right of it. Follow that decimal point by any other digits remaining until you reach a string of all zeros, then multiply it by 10 raised to a power that corresponds to the number of digits to the left of the original decimal point minus one. I know that this is difficult to comprehend from the written description, but remember that the decimal system is also difficult to describe textually. Here is an example that should make things clearer. Avogadro's number, written in decimal above, would be 6 followed by a decimal point, followed by 022, times 10^{23} (23 because there were originally 24 digits in Avogadro's number). So, Avogadro's number is 6.022×10^{23}, which is a much more compact way to write it and has the advantage of directing attention to the part of the number which is most significant, the 6.022.

Similar notation is used for very small numbers which in decimal form consist of a decimal point followed by a long string of digits. I won't burden you with the method of converting these to scientific notation, but it's simple and essentially the same method as for large numbers.

18.3 Testing Substances

Now we will look at the question of how forensic scientists determine what chemical substances are present in a sample of some material submitted for analysis. Generally, if given an unknown sample, the forensic scientist wants to do the following things:
1. Separate the components of a mixture to determine what substances are present in the sample;
2. identify each substance that is a component of the sample;
3. quantify the amount of each substance in the sample to determine how much of it is present.

A variety of techniques and instruments exist for testing unknown sample. These include various forms of chromatography (gas, liquid, thin layer), electrophoresis, spectrophotometry, and mass spectrometry. Chromatography and electrophoresis are methods for separating the substances in a sample. Spectrophotometry and mass spectrometry are methods for detecting the presence of a substance and determining what substance it is. Additionally, spectrophotometry can be used to quantify the amount of the substance present.

18.3.1 Chromatography

Chromatography is a basic and standard method for separating substances in a mixture, based on their weights and their solubilities. Its ability to separate components of a mixture is based on the fact that each substance has a different combination of weight and solubility in a given solvent.

The weight of a molecule of some substance depends on the atoms that make it up. The more atoms and the heavier those atoms are, the heavier the molecule will be. The weight of an atom is generally referred to as its **atomic weight**. For example, the atomic weight of hydrogen is 1 and the atomic weight of oxygen is 16. The sum of the atomic weights of the atoms in a molecule is called the **molecular weight** of that molecule. Therefore, the molecular weight of water (H_2O) is $(2 \times 1) + 16 = 18$. What this tells you is that if you take a certain large number of water molecules, the resulting amount of water will weigh 18 grams. The number of molecules you need is

Avogadro's number (6.022×10^{23}). So, if you gather up 6 sextillion, 200 quintillion water molecules in a cup (and they would all fit in a cup) you would have a quantity weighing 18 grams. Chemists refer to Avogadro's number of molecules of a molecule as a **mole** of that molecule.

Solubility refers to how easily a substance dissolves in a given solvent. Let's take table salt (sodium chloride or NaCl) for an example. Table salt dissolves easily in water. You put some in, stir it around, and voila! you have salty water. However, table salt doesn't dissolve well in alcohol. If you buy isopropyl rubbing alcohol at the store, salt will dissolve in it because what you buy at the store is only about 70% alcohol and the rest is water in which the salt will dissolve. However, if you took pure isopropyl alcohol or pure ethyl alcohol (the alcohol in alcoholic beverages) and added table salt to it, the salt would just fall to the bottom of the container and sit there because salt isn't very soluble in alcohol. Similarly, vegetable oil doesn't dissolve well in water. If you mix oil and water, the oil floats to the top. However, vegetable oil dissolves fairly well in alcohol.

Chromatography is a way to separate the different substances in a mixture by taking advantage of differences in weight and solubility. By choosing the correct solvent or combination of solvents, you can separate almost any mixture into its components. There are many forms of chromatography. My intention here is not to give a complete list, but just to cover those most commonly used in a crime lab.

18.3.1.1 Paper Chromatography

The simplest form of chromatography is **paper chromatography**, which is a staple of chemistry lab classes but for the most part too crude for use in a crime lab. However, it can provide us with an example of how all forms of chromatography work. An example of an apparatus for paper chromatography is shown in figure 18.1. To separate a mixture by paper chromatography place a drop of some mixture of colored substances on a piece of industrial style paper towel (the soft and absorbent types of paper towels normally used in a kitchen don't work as well). Now, suspend the paper towel from something so that one end is immersed in a container of solvent and the other end up in the air. The solvent will begin to wick its way up the paper towel, moving from the container slowly up the towel. As the solvent wicks up the towel past the drop of mixture you placed on it, the various substances in the mixture will separate. The substances that dissolve more easily in that solvent would be carried farther upward by the moving solvent, and the substances that dissolved less easily would not move as far. Heavier molecules don't move as far as lighter ones, even if both are equally soluble. Once the solvent has reached the top of the paper towel, or as high as it will get after a few hours, the towel can be taken out of the container of solvent and placed somewhere to dry. If the substances in the mixture were colored, separate dots, lines or smudges of color on the towel would reveal where each component of the mixture ended up. You can try this at home using a mixture of food colorings as the test mixture and water as the solvent.

If it is found that not all of the components of the mixture have separated, a further separation can be performed by using a different solvent and letting it wick across the paper towel, in a direction at right angles to the direction the solvent wicked the first time. This amounts to placing what was originally one of the sides of the towel in a container of solvent and proceeding as above. This process is called two-dimensional paper chromatography.

Figure 18.1 Paper Chromatography

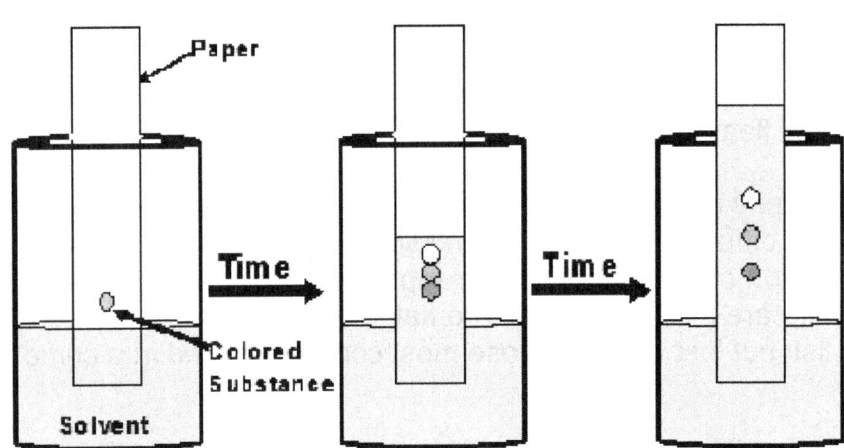

Now, assuming that all the components of the mixture have been separated, they can be recovered from their spot on the paper towel and subjected to further tests to determine what each component is, and how much of it there is.

All chromatography is based on these principles. The various types of chromatography differ in what the form of the solvent is and in what takes the place of the paper towel. In general, the solvent, or whatever substances moves and carries the components of the mixture with it at various speeds is called the **mobile phase**. Whatever takes the place of the paper towel, serving as a substrate for the mobile phase to move across is called the **stationary phase**. Several different kinds of things can be used as mobile and stationary phases, but whatever they are the result is still a form of chromatography. The different forms of chromatography differ in what constitutes the mobile phase and what constitutes the stationary phase.

18.3.1.2 Gas Chromatography

Let's look at gas chromatography, which, unlike paper chromatography is extremely precise and is commonly used in crime labs to separate components of mixtures. **Gas chromatography** (GC) separates substances based on their solubility in a liquid vs their solubility in a gas. In gas chromatography there is a tube partly filled with a liquid, which serves as the stationary phase. The mobile phase is a gas, which is is blown through the tube, past the liquid. A substance is placed in the liquid at one end of the tube. Substances that are more soluble in the gas and less soluble in the liquid move through the tube faster, and those that are less soluble in the gas and more soluble in the liquid move through the tube more slowly. At the end of the tube a detector reads when substances are coming out of the tube.

The time it takes for a component of a mixture to get to the far end of the tube of a gas chromatograph and be detected is called its retention time. The detector is usually hooked up to a recorder, which makes a permanent record of what is detected at each moment of time by causing a pen to move across a moving sheet of paper. Figure 18.2 illustrates a gas chromatograph.

Figure 18.2: Diagrammatic Representation of a Gas Chromatograph

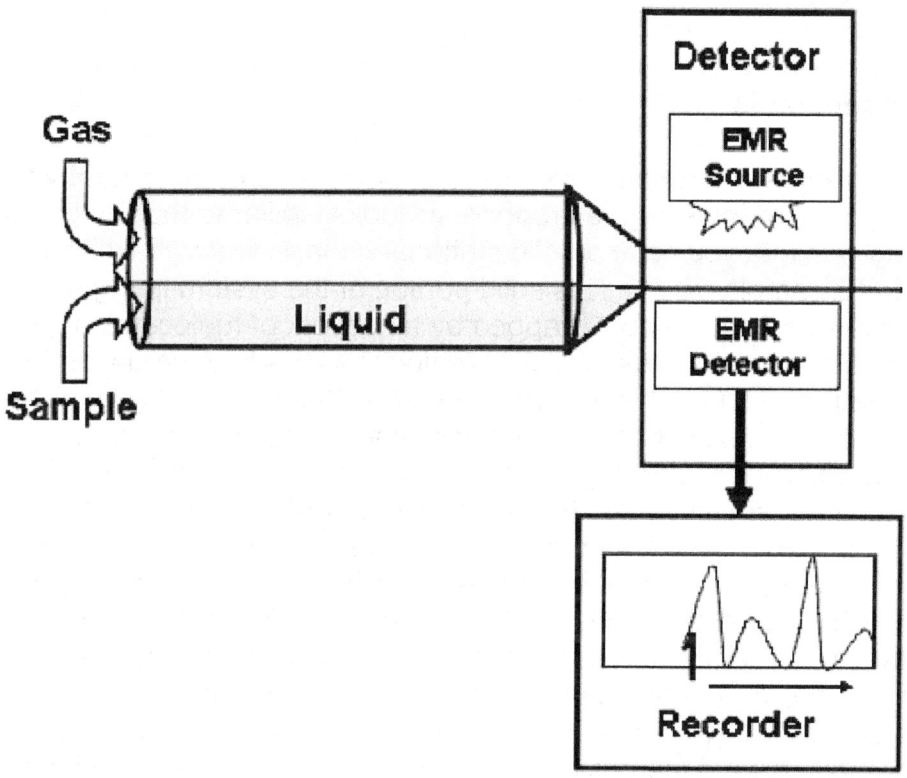

Gas chromatography is widely used in crime labs because it is sensitive, meaning it can detect small quantities of a substance. Also, it can identify substances based on each substance's characteristic retention time, and it is quantitiative, meaning that the detector will tell you how much of the substance you have, which is a function of the height and width of the corresponding peak on the recording paper. There are other forms of gas chromatography, including one in which the stationary phase is a solid instead of a liquid.

18.3.1.3 HPLC & TLC

High Performance Liquid Chromatography (HPLC) is a form of chromatography in which a liquid is pumped through a column filled with beads of a solid material. The HPLC apparatus also includes a detector, usually one that records on a moving paper.

Thin layer chromatography (TLC) is similar to paper chromatography. A plate is covered with a thin layer of a solid medium, such as silica gel or alumina gel. One end of the plate is then placed in a solvent, which wicks up the plate. If a drop of sample is place on the plate, the moving solvent will separate it into its components. TLC typically doesn't utilize a detector. Instead a stain is placed on the plates. The stain interacts with the separated components to make them visible.

18.3.2 Electrophoresis

Electrophoresis extends the separating power of chromatography by adding an electric field. The apparatus for electrophoresis looks similar to that for paper chromatography in that you have a solid stationary phase, through which the components of a sample move. This solid portion of the system is a **gel**, which consists of a considerable amount of liquid, trapped by a network of molecules of plastic, starch, or some other substance. Think of "Jello", which is a gel of flavored water trapped by a network of gelatin molecules. The liquid trapped by the gel is always a water-based solution and is called a **buffer**. It serves to carry an electrical current through the gel.

Normally, both ends of the gel are immersed in buffer. In contrast to a paper chromatography system, however, the buffer is not moving through the gel, but is instead simply a way to deliver an electric charge to the gel. An electric power source is added to the apparatus in such a way that the positive terminal is at one end of the apparatus and the negative terminal is at the other end. Thus, an electrical current flows from the power supply, through the buffer, and through the gel. This allows the substances in a mixture to be separated based on their **electrical charge** in addition to their weight and solubility. Electrical opposites attract, so a more negatively charged molecule will move toward the positive terminal and a more positively charge molecule will move toward the negative terminal. Figure 18.3 illustrates an electrophoresis apparatus.

Electrophoresis is especially useful with biological mixtures, such as blood, blood plasma, or DNA. This is because many biological molecules have net electrical charges. For proteins, the charges come from their constituent amino acids. Three of the 20 amino acids have a negative charge and three have a positive charge, the remaining 14 amino acids have no charge. So, if the protein is formed using more of the negative amino acids than the positive ones, it will have an overall negative charge, and vice versa. Electrophoresis capitalizes on these excess electrical charges to cause mixtures to separate. Because electrophoresis is the method of choice for working with biological mixtures, it is one of the most common types of apparatus used by serologists and DNA analysts.

Figure 18.3: A Electrophoresis Apparatus

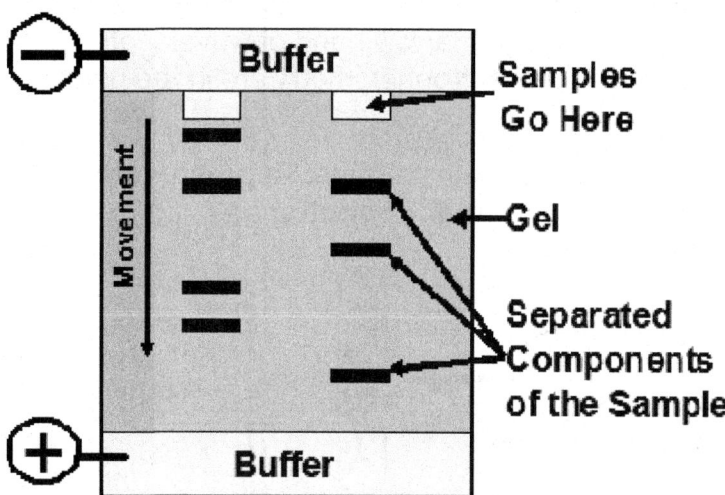

18.3.3 Spectrophotometry

Spectrophotometry is a method for detecting the presence of substances, based on the fact that all substance absorb EMR, such as visible light or ultraviolet light. Further, each substance absorbs EMR of certain wavelengths but not others. Spectrophotometers determine not only that a substance absorbs EMR at a certain wavelength, but also how much of the EMR the substance absorbs at that wavelength. With proper calibration this can allow a spectrophotometer to both identify and quantify substances.

A spectrophotometer has an EMR source that can be set at a variety of wavelengths, and a detector that measures the amount of EMR that strikes it. A small container of the substance to be analyzed is placed in a position between the EMR source and the detector. The detector then measures how much of the original EMR passes through the sample of the substance to reach it. An illustration of a simple spectrophotometer is shown in Figure 18.4.

We can easily see how this works using a light bulb as the EMR source and our eyes as the detector. Cola drinks absorb visible light well, but water does not. If you were to hold a plastic bottle of cola between your eye and the light, you would have no problem seeing that there is something in the bottle. Water in the same bottle is less easy to see, but you could still tell that you don't have an empty bottle. Now, let's say that you added some water to the cola in the bottle and held it between the light and your eyes. It would be possible to see that the cola has been diluted because it now allows more light to pass through it and appears lighter in color. In this experiment you are acting as a simple and crude spectrophotometer. The human eye isn't sensitive enough to serve as a very good detector, though with practice you would become able to distinguish various amounts of adulteration of your cola with water (or adulteration of your water with cola, if you prefer). Also, a light bulb emits EMR over a wide portion of the infrared and visible spectrum, but in a good spectrophotometer the EMR that is emitted has to be limited to a very narrow band of wavelengths. Most of the wavelengths typically used in spectrophotometry are in the ultraviolet portion of the spectrum.

Figure 18.4: Diagrammatic Representation of a Spectrophotometer

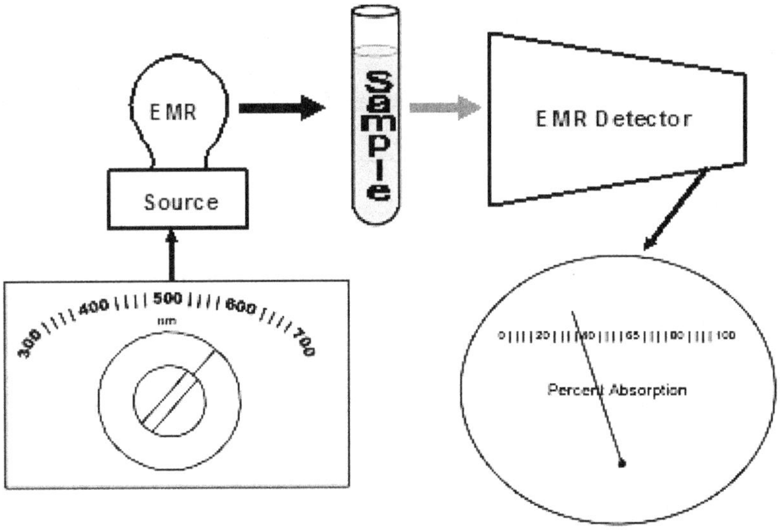

Often, the **detector** for a GC or HPLC apparatus is a spectrophotometer. In the case of detectors used in GC and HPLC apparatus, the light source is on one side of a clear tube and the light detector is on the other side. As substances are separated they flow down this tube and the spectrophotometer measures their presence as they pass by. As discussed previously, the presence of these substances is usually recorded by a **recorder** consisting of a pen on a moving strip of paper. The result of a substance moving through the tube past the detector usually looks like a mountain peak. As the substance first starts moving into the area "seen: by the detector there is very little of it so the recorder's pen only raises a little above the baseline. As more of the substance moves into the detector's "view" the recorder's pen moves higher (farther from the baseline), and as the main body of the substance moves past the detector on its way out of the tube the amount detected decreases so the pen moved back to the baseline. The more of a certain substance, the higher and broader the peak on the recording will be.

Spectrophotometry can also be used to identify a certain substance. Each molecule has chemical bonds between its constituent atoms and a certain shape that causes it to absorb EMR at certain wavelengths, but not others. The pattern of which wavelengths are absorbed by a molecule and which are passed through, and in what amounts, is unique to each different molecule. This pattern of absorption and transmission of EMR of various wavelengths is called the **absorption spectrum** (plural spectra) of a substance. Advanced spectrophotometers are designed to measure and record the absorption spectrum of substances. Databases of the absorption spectra of most of the known substances in the world exist, allowing many substances to be easily identified.

18.3.4 Mass Spectrometry

A **mass spectrometer** (MS) provides a different way of identifying substances. It uses the divide and conquer method, whereby the molecules of a substance are broken apart into fragments by bombarding them with a beam of high energy electrons. The fragments are then separated by their size and electrical charge in a magnetic or electric field. The fragments are detected and the amount of each fragment is measured (often by spectrophotometry). This produces a characteristic fingerprint of fragments for each substance, called its mass spectrum. Databases of the mass spectra of most of the known substances in the world have been compiled, allowing most substances to be easily identified. The mass spectrum of ethanol is shown in figure 18.5.

Figure 18.5: Mass Spectrum of Ethanol

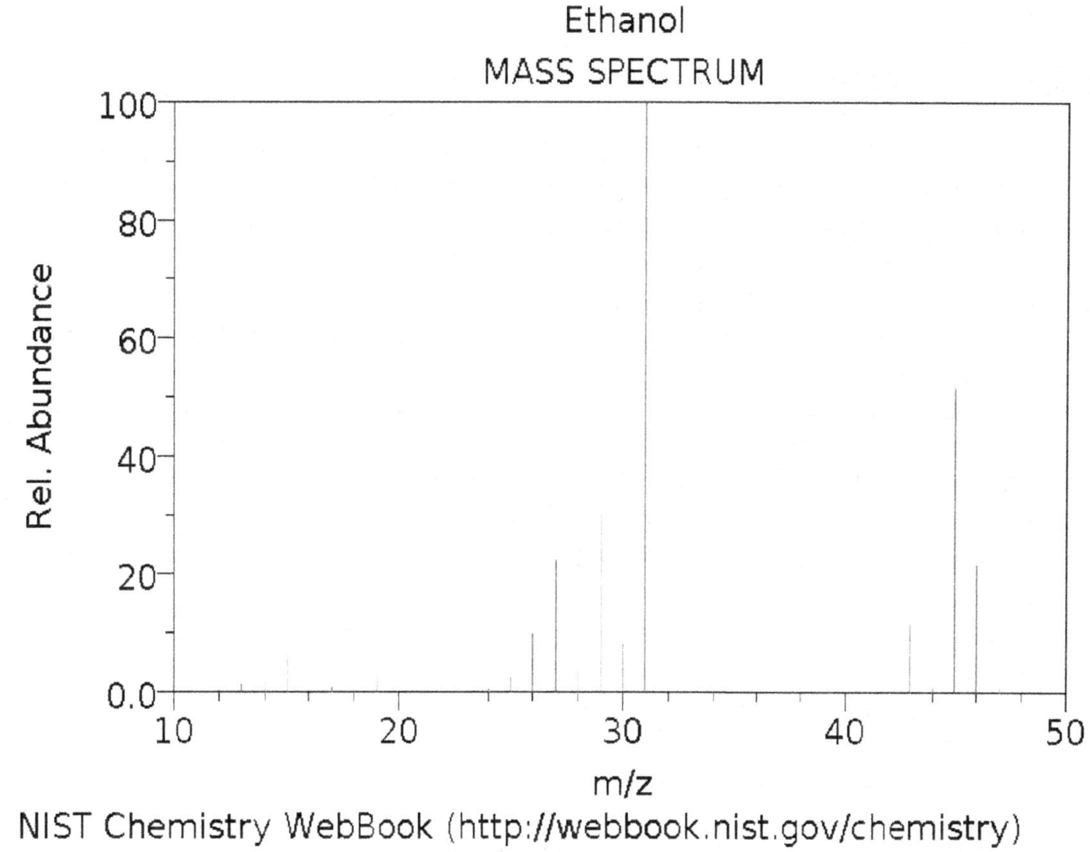
Ethanol
MASS SPECTRUM
100
80
60
40
20
0.0
Rel. Abundance
10
20
30
40
50
m/z
NIST Chemistry WebBook (http://webbook.nist.gov/chemistry)

18.3.5 GCMS Systems

For those types of chromatography or electrophoresis that do not lead to identification of the components of the mixture, the components can be recovered and identified using spectrophotometry or mass spectrometry. Some instruments used in the crime lab combine a method of separation based on some form of chromatography with a form of detection and indentification based on spectrophotometry or mass spectrometry to provide an accurate and convenient way to rapidly identify the components of a mixture.

One of the most common of such pairings is gas chromatography with mass spectrometry (**GCMS**). This gives a rapid and accurate way to determine the composition of an unknown sample. Using a GCMS, a mixture can be separated, and the amount and identity of each component of the mixture can be determined. For example, if a forensic chemist is given a suspicious white powder, he or she could use a GCMS to determine that the power was actually 62% heroin and 37% sugar with traces of other chemicals (which they would also be able to identify and quantify).

GCMS devices are common and reliable. They form the backbone of a forensic chemist's or toxicologist's arsenal of equipment for determining the nature of materials submitted to them.

18.4 Questions for Study and Review

1. If you live in the United States, imagine a typical day in your life and discuss what things you might encounter that would be different if the U.S. used the metric system instead of the English Imperial system of measurement.
2. The speed of light (in a vacuum) is 299,792,458 meters per second. Express this number in scientific notation.
3. List the analytical methods and instruments that are available for separating the components of a mixture.
4. List the analytical methods and instruments that are available for identifying each component in a mixture.
5. List the analytical methods and instruments that are available for quantifying the amount of each component in a mixture.
6. What physical or chemical properties of substances allow them to be separated using chromatography?
7. List at least four types of chromatography and describe the stationary and mobile phase for each of them.
8. What physical or chemical properties of substances allow them to be separated by electrophoresis?
9. In your own words, describe what a spectrophotometer does and how it does it.
10. In your own words, describe mass spectrometry.
11. In your own words, describe a GCMS system.

18.5 Sources

Chasteen T. n.d. Coupling Gas Chromatography to Mass Spectrometry [Internet]. [cited 2007 Oct 23]. Available from: http://www.shsu.edu/~chemistry/primers/gcms.html.

Douglas F. n.d. GC/MS Analysis [Internet]. Scientific Testimony: An Online Journal. [cited 2007 Oct 23]. Available from: http://www.scientific.org/tutorials/articles/gcms.html.

Lodder RA. n.d. High Performance Liquid Chromatography (HPLC): A Users Guide [Internet]. [cited 2007 Oct 23]. Available from: http://www.pharm.uky.edu/ASRG/HPLC/hplcmytry.html.

Meloan CE, James RE, and Saferstein R. 1998. Lab Manual for Criminalistics: An Introduction to Forensic Science. 6th Edition. Upper Saddle River (NJ): Prentice Hall.

Peter Piper Publishing Inc. 1998. Paper Chromatography [Internet]. [cited 2007 Oct 23]. Available from: http://www.yesmag.bc.ca/projects/paper_chroma.html.

Reusch W. 1999. Mass Spectrometry [Internet]. [cited 2007 Oct 23]. Available from: http://www.cem.msu.edu/~reusch/VirtualText/Spectrpy/MassSpec/masspec1.htm

Saferstein R. 1998. Criminalistics: An Introduction to Forensic Science. 6th Edition. Upper Saddle River (NJ): Prentice Hall.

Tissue BM. 1996. Thin-Layer Chromatography (TLC) [Internet]. [cited 2007 Oct 23]. Available from: http://elchem.kaist.ac.kr/vt/chem-ed/sep/tlc/tlc.htm.

University of Utah, Genetic Science Learning Center. 2007. Gel Electrophoresis [Internet]. [cited 2007 Oct 23]. Available from: http://learn.genetics.utah.edu/units/biotech/gel/.

Wikipedia contributors. 2007. Chromatography [Internet]. Wikipedia, The Free Encyclopedia; 2007 Oct 22, 01:24 UTC [cited 2007 Oct 23]. Available from: http://en.wikipedia.org/w/index.php?title=Chromatography&oldid=166182866.

Wikipedia contributors. 2007. Electrophoresis [Internet]. Wikipedia, The Free Encyclopedia; 2007 Oct 19, 20:40 UTC [cited 2007 Oct 23]. Available from: http://en.wikipedia.org/w/index.php?title=Electrophoresis&oldid=165704767.

Wikipedia contributors. 2007. Gas-liquid chromatography [Internet]. Wikipedia, The Free Encyclopedia; 2007 Oct 23, 07:56 UTC [cited 2007 Oct 23]. Available from: http://en.wikipedia.org/w/index.php?title=Gas-liquid_chromatography&oldid=166467717.

Wikipedia contributors. 2007. Spectrophotometry [Internet]. Wikipedia, The Free Encyclopedia; 2007 Sep 19, 15:51 UTC [cited 2007 Oct 23]. Available from: http://en.wikipedia.org/w/index.php?title=Spectrophotometry&oldid=158969449.

18.6 Acknowledgements

The figure of the mass spectrum of ethanol is courtesy of the National Institute of Standards and Technology, NIST Chemistry Webbook, http://webbook.nist.gov/chemistry. NIST is a federal agency and this graphic is therefore in the public domain.

Chapter 19
SEROLOGY OF BLOOD

With this chapter we will begin a series of three chapters on serology. In this chapter, the focus will be on the serology of blood. In the next chapter the focus will be on a variety of issues in serology, and the final chapter in the series will focus on DNA analysis. As discussed in chapter 2, some people consider serologists to be criminalists, and other don't. It is my impression that many serologists do not consider themselves to be criminalists, yet their methods and goals are the same as those of criminalists – linking a suspect to a crime scene, in this case by some form of biological evidence such as blood or semen.

This chapter will begin with a discussion of what serology is, and what serologists do. After a brief history of serology we will take a look at the composition and nature of blood. Next, the focus will shift to the practice of serology, after the discussion of which we conclude with an exploration of blood groups and blood typing.

19.1 What is Serology

One of the most concise definitions of **serology** is that it is the study of genetic markers in tissues, blood, and other body fluids (serums). In this usage, DNA analysis is a form of serology. Other authorities consider serology to be restricted to "traditional serology" which is those subjects covered in this chapter and the next one, and consider DNA analysis to be something different. I observe that in most cases all of these things are done by the same people, who do distinguish between the pursuits of serology and DNA analysis, even though they do it all.

There are two or three primary sources of genetic markers that serologists work with, depending on whether you are talking to someone who considers DNA analysis part of serology, or someone who considers it a separate forensic discipline. One type of genetic marker that serologists use is blood groups. Blood groups are a feature of the immune system, and are investigated using immunological methods. The second source of genetic markers is proteins found primarily in blood plasma or blood cells. Many blood proteins exhibit genetic variation that is inherited. Most of this variation is neutral, but in some cases one variant will work better or worse than another variant. The third source of genetic markers is the DNA itself. Each of us has a unique DNA sequence.

The primary work of serologists is to identify genetic markers of these types in blood or other body fluids left at a crime scene and attempt to match them to the genetic markers of a suspect.

One interesting problem encountered by serologists is that genetic markers are always class characters. There are no individual characters in serology or DNA analysis. Therefore, in order to make a match, serologists have to do a statistical process of estimating the probability that two individuals have exactly the same genetic markers. If several markers are used, then the probability of two people having the same set of markers is small, and the court will accept the evidence as linking the suspect to the crime scene. This procedure will be discussed in the next chapter.

19.2 The History of Serology

The deduction that if you find blood there must have been a violent incident is probably older than humanity itself. The use of blood evidence to determine where a crime occurred is also very ancient. For example, the story of Cain killing Able in the Bible tells us that God said that Able's blood cried out to him from the ground. It was not until the twentieth century, however, that the methodology for detecting genetic markers became reliable enough for serology to be taken seriously as a forensic science.

By the middle 1800's, scientists such as Paul Jeserich of Germany had worked out methods for detecting blood stains. In 1901, Karl Landsteiner announced one of the most significant discoveries of this century – the typing of blood, a finding that led to his earning the Nobel prize 29 years later. Working on the problem of blood transfusions, Landsteiner discovered the ABO blood group, which is still used for matching transfusions today. By the late 1930's other blood groups were becoming known, and today we know of many of them, only a few of which are used by serologists. During the 1980's, advances in DNA research led to techniques such as DNA fingerprinting, which could be used to match suspects to their blood with great certainty.

19.3 Blood and Its Components

If you take blood and spin it in a centrifuge it will separate into 3 fractions based on weight: plasma, white cells, and red cells. A centrifuge is a device that is loaded with test tubes, which it then spins in a circle. The various components of the substance placed in the test tubes will separate based on how heavy they are. Heavier things will be forced to the bottom of the test tubes, and lighter things will stack up on top of the heavier things in order of weight. The components of blood are shown in Figure 19.1.

Red blood cells (erythrocytes) are the heaviest component of blood. They make up about 40% of the volume of blood. There are several features of red blood cells that can be considered genetic markers. First, red cells have substances attached to the external surface of their cell membranes which interact with the immune system. These substances are called **antigens** and are the basis for blood groups. Second, red cells contain several proteins, which exhibit genetic variation. Interestingly, red cells have no DNA. Red cells are produced by an unusual process that leaves them without a cell nucleus and therefore without DNA. If a person's red cells have nuclei, then that person probably has the disease mononucleosis.

White blood cells (leukocytes) are a fairly rare component of blood, comprising about 1% of the volume of blood. White blood cells are intimately involved with the immune system. There are several types of white cells, some that eat foreign cells, such as bacteria, some that help in triggering an immune system response to some foreign organism or tissue, and some with other similar functions. They have antigens attached to their surfaces just like red cells do.

Plasma is a translucent yellowish fluid that contains all the dissolved substances in blood, including enzymes, nutrients, clotting factors, waste products, antibodies, and many other things. Plasma is important to serologists because it carries several proteins that exhibit genetic variation. **Antibodies** are a type of protein designed to seek out antigens. Also known as the globulin group of proteins, antibodies circulate

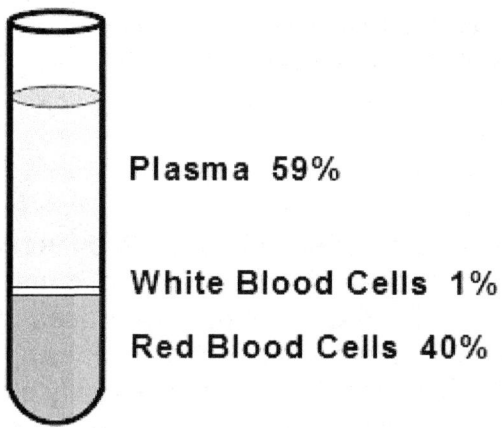

Figure 19.1: The Components of Blood When Centrifuged

Plasma 59%

White Blood Cells 1%

Red Blood Cells 40%

through the blood stream carried by the plasma.

One of the proteins in plasma is fibrinogen, which causes clots to form when it reacts to the air. If you remove the clotting factors from plasma, the fluid you are left with is called **serum**. Serum (plural sera) is the root of the word serology. Often the term serum is broadened to include other body fluids, such as saliva and seminal fluid.

19.4 The Practice of Serology

Blood and blood stains are produced during a wide variety of crimes. Not only are they produced when a person is wounded by another, but are also produced if the perpetrator receives a scratch or cut while involved in their criminal activities. When a forensic scientist processes a crime scene there are four bloody questions that have to be asked: (1) Is there any blood here? (2) Is it really blood? (3) Is it human blood? and (4) Whose blood is it?

19.4.1 Locating Blood Stains

The first question is whether blood is present or not. Often blood will be visible, but sometimes the quantity of blood is so small that it isn't visible to the naked eye. In other cases the criminal might have cleaned the area and removed all visible blood. Even with the most thorough cleaning, however, minute traces of blood likely remain because blood is difficult and sometimes impossible to remove from all surfaces. Blood may also have soaked into the cracks between floorboards, or have been absorbed by carpeting, wood, concrete, or other porous material.

When blood is suspected to exist, but isn't readily visible, and even sometimes when it is visible, the person processing the crime scene will often use a chemical substance called **luminol** to reveal any blood that may be present. Luminol is a liquid that is sprayed onto an area suspected to contain blood. Luminol reacts with blood to make it fluorescent under ultraviolet light. The luminol is sprayed, then the lights are dimmed, and an ultraviolet light source is used to illuminate the area. Any patches of blood present will glow under the ultraviolet light. The luminol method is extremely sensitive and is capable of detecting blood that has been diluted up to 10,000 times, so it can detect even the most minute traces of blood.

19.4.2 Is it Really Blood?

The question of whether the substance in question is really blood or something else is not trivial. Several substances, such as ketchup, stunt blood as used in movies, and red paint can look like blood under some circumstances. Fortunately, a variety of techniques exist to determine whether a suspect substance is really blood. The luminol test is a good test for whether or not the substance is really blood, as luminol's fluorescent effect is very specific to blood. Luminol won't react with ketchup or red paint. The luminol test does have a severe problem, however. It is known to destroy many of the genetic markers in blood, rendering it unusable for matching to a suspect or victim. Therefore, it is most commonly used for detecting traces of blood, which wouldn't be useful for matching anyway, and less commonly used for demonstrating that some red sticky substance is really blood.

There are other tests that can be used to determine whether a suspect substance is blood. Some of these are color tests, that work by mixing the suspect substance with a chemical agent that changes color in the presence of blood. One standard color test is called the **Kastle Meyer test**. When performing this test the serologist mixes blood with a combination of two substances, a chemical called phenolphthalein and hydrogen peroxide. This test is based on the fact that the hemoglobin of blood (which is the protein in red cells that transports oxygen) can mimic the action of a class of enzymes called peroxidases. Peroxidases speed up an oxidation reaction of certain organic substances by utilizing the extra oxygen present in hydrogen peroxide. Hydrogen peroxide is H_2O_2, which can easily decompose to release free oxygen and water. The free oxygen oxidizes its target molecules in such a way that the solution becomes more acidic. That's where the phenolphthalein comes in. Phenolphthalein is a marker for increased acidity and turns pink in an acidic environment. Therefore, if a mixture of the suspect substance, phenolphthalein, and hydrogen peroxide turns pink, the suspect substance is blood. The Kastle Meyer test does have some drawbacks, however. The most severe problem is that other common substances can also have peroxidase-like activity. Some of these substances are potatoes and horseradish. However, it is unlikely that someone will confuse potatoes or horseradish with something likely to be blood (unless someone put ketchup on the potatoes).

Another type of test used to determine whether a substance is blood or not is based on the fact that many proteins form distinctive crystals in the presence of certain chemicals. Hemoglobin is one such protein that forms a distinctive crystal. Tests that seek to confirm the presence of a substance by finding characteristics crystals are called **crystallization tests**. There are two common tests of this sort for blood, the Takayama test and the Teichmann test. They are similar but use different chemicals and procedures to crystallize the hemoglobin.

Which is better, color tests or crystallization tests? Both have advantages and disadvantages. Crystalline tests are more accurate, but they require a larger quantity of blood. Color tests can give misleading results if there is stray horseradish present, but require less blood. So, the serologist will use his or her judgement to choose the correct test.

19.4.3 Is it Human Blood?

Once the serologist has determined that a certain suspect substance really is blood, she or he has to answer the question of whether it is human blood. A criminal might claim that the blood came from cutting up a chicken for dinner, or that their dog had a bloody nose, or some other story suggesting that the blood is not human.

The standard tests for determining whether blood is human or not is the group of tests known as **precipitin tests**. Precipitin tests are based on the fact that when the blood of one species is injected into another species, the immune system of the animal that received the blood will perceive it as foreign and make antibodies against it. Later in this chapter we will look at this process in more detail. For our immediate purpose however, we can just assume that this happens. This principle is utilized by injecting a non-human animal, usually a rabbit, with human blood. After the rabbit's immune system has produced antibodies to human blood, the rabbit is killed, its blood is removed, and the serum of the rabbit's blood, which contains the antibodies, is separated and bottled. Because of the antibodies against human blood, this serum has the ability to interact with human blood to cause a type of reaction called agglutination, wherein the human blood cells are clumped together by the antibodies to form dark particles, called **precipitin**. If this reaction takes place in a test tube the precipitin will fall to the bottom of the test tube – "precipitate" in the language of chemistry, hence its name. There are different types of precipitin test, but all of them detect the formation of precipitin in different ways. Figure 19.2 shows a simple type of precipitin test, in which a line of precipitin forms in between a drop of human blood and a drop of anti-human serum due to diffusion of antigens and antibodies across a gel.

Serum can also be produced that reacts with the blood of other common animals, such as dogs, cats, and deer.

19.4.4 Whose Blood is it?

Now, the serologist has shown that there is blood present, and that the blood is human. What remains is to determine whose blood it is. This question is answered by attempting to match genetic markers present in the blood with those present in the blood of a victim or suspect.

Figure 19.2: A Simple Precipitin Test for Human Blood

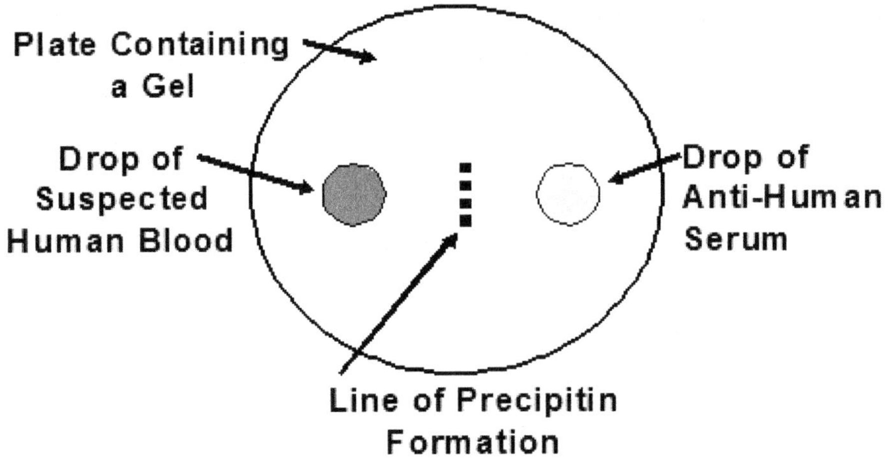

- Plate Containing a Gel
- Drop of Suspected Human Blood
- Drop of Anti-Human Serum
- Line of Precipitin Formation

19.5 Blood Groups

One category of genetic markers is blood groups. Both red cells and white cells have blood group markers. The blood groups exist because there are substances called antigens attached to the surfaces of these cells, which probably function as a means for various cells to recognize and communicate with each other. Exactly which substances are attached to the surfaces of cells is determined by a person's genes. Different alleles code for different substances. Hence blood groups are genetic markers. Most of the blood groups examined by serologists are red blood cell groups.

19.5.1 Antigens and Antibodies

The substances attached to the surfaces of cells, which determine blood groups, are called antigens. Antigens are the targets of antibodies. Antibodies are created by the immune system to attack antigens, by grabbing onto them and holding tight. One law of immunology is that a person can't (or at least shouldn't) make antibodies that attack antigens present in their own bodies, but can (and perhaps should) make antibodies against antigens that are foreign to their body.

For example, if a disease-causing microorganism enters a person's body, it will hopefully have antigens on its surface that are different from those of the cells in the person's body and the immune system will recognize this and build antibodies that attack the microorganism by grabbing on to its antigens. This is how the anti-human serum produced by a rabbit forms precipitin. Precipitin is human blood cells that have been latched onto by antibodies and clumped together.

Antibodies react with cells bearing foreign antigens in two ways, both of which rely on the fact that antibodies have a structure similar to the letter 'Y', where both of the branches of the 'Y' have an area that can bind to an antigen. Thus, the antibody can either bind twice to one cell's antigens, or it can bind to antigens on two different cells. What probably determines whether an antibody will bind to the same cell twice or to two different cells is the distance between the antigens on the cell's surface. If the antigens are far apart, then the antibody can't bind to two of them on the same cell and has to bind to two of them on different cells. If the antibody binds to two antigens on the same cell, then it will use this double hold to tear the cell apart. This is an immunological reaction called **lysis**, which in general refers to bursting cells. If the antibody binds to two different cells, large numbers of cells get stuck together to form a clump of cells (precipitin). This is an immunological reaction called **agglutination**. The agglutination of cells with antigens on their surfaces by antibodies is illustrated in Figure 19.3.

Since antibodies are simply proteins floating in the plasma, the plasma, or more often serum, can be obtained from blood using a centrifuge and used to test for the presence of certain antigens present on cells. A specially prepared serum of this type is called an **antiserum**. The serum from a rabbit injected with human blood is a type of antiserum.

Figure 19.3: A Agglutination Reaction Between Antibodies and the Antigens on Cell Surfaces

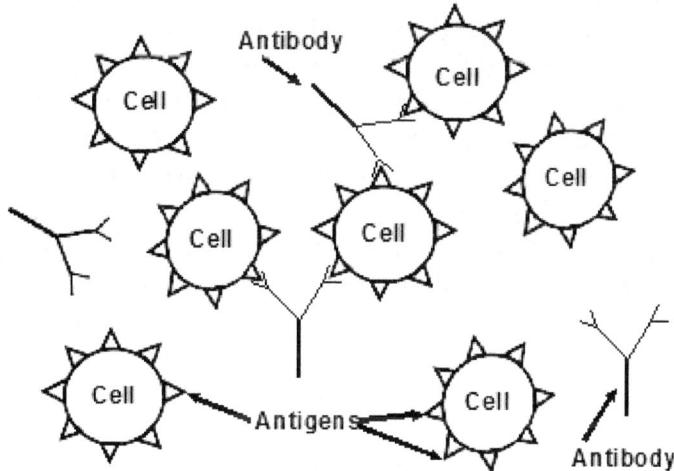

19.5.2 An Example: The ABO Blood Group

We encountered the transfusion blood group, or ABO blood group in chapter 15. The ABO blood group is a red cell blood group, meaning that the four different phenotypes A, B, AB, and O refer to different antigens that are present on the surfaces of red cells. These four phenotypes are called "**blood types**" and are determined by a person's genotype as discussed in chapter 15. A system of genetically controlled cell antigens of this sort is called a "**blood group**". Hence A, B, AB, and O are the blood types possible for the ABO blood group.

The antigens for the ABO system are a set of molecules containing both protein and sugar components, called A antigen, B antigen, and H antigen. Almost everybody has some H antigen, with the exception being people with the Bombay blood type, who always test as type O regardless of their alleles for the ABO locus. Otherwise, a person who has only antigen H will test as blood type O. For people with an A or a B allele, some of the antigen H has an additional sugar molecule attached to it. The allele for blood type A codes for the attachment of a sugar called N-acetylgalactosamine to the antigen H. The allele for blood type B codes for the attachment of a sugar called galactose to the antigen H. It is these additional sugars that constitute antigen A and antigen B. People with two O alleles have no antigen A or B, so all of their antigens H is unmodified. People with the AB genotype have some antigen H with antigen A attached, and some antigen H with antigen B attached, and therefore present both antigens to the immune system.

A person with blood type O makes antibodies against both the A antigen and the B antigen. People with blood type A make antibodies against the B antigen, but not the A antigen (because it's present in their own bodies). People with blood type B make antibodies against the A antigen, and people with blood type AB do not make antibodies against any of these antigens.

Outside of forensic science the main importance of the ABO system is in blood matching for transfusions. If the donor and recipient are not matched for their ABO blood types then the antibodies in the recipient's plasma cause the red cells from the donor to agglutinate, causing rapid death of the recipient. Note, however, that people with blood type AB have no red cell antibodies, so they can receive blood from anybody. For this reason they are called universal recipients. People with blood type O have only the H antigen, and no A or B antigen. This means that nobody's antibodies can agglutinate their cells. For this reason type O people are called universal donors.

There are several blood groups, of which less than a dozen are commonly used by serologists. All of them have antigens of different sorts, coded for by their alleles. Therefore, all blood groups are determined in a manner similar to the ABO blood group.

19.5.3 Blood Testing

In order to determine the blood type of a sample of blood, the proper set of antisera are obtained. These antisera contain antibodies that target the antigens involved in the blood group being tested for. A drop of each antiserum is applied to a small quantity of the blood. If agglutination or lysis occurs, then the blood cells must have the antigen that the antibodies in the antiserum react with, and this tells the serologist the blood type of that sample of blood.

In our ABO blood group example, there are two necessary antisera, anti-A, which has antibodies that target antigen A, and anti-B, which has antibodies that target antigen B. A drop of anti-A is added to a drop of blood, and a drop of anti-B is added to another drop of blood. The possible results are shown in Table 19.1, and reveal the blood type of the blood sample.

Table 19.1: Blood Typing for the Transfusion (ABO) Blood Group

Agglutination By Anti-A	Agglutination By Anti-B	Blood Type Result
Yes	No	Type A
No	Yes	Type B
Yes	Yes	Type AB
No	No	Type O

19.6 The Absorption/Elution Method for Typing Dried Blood

Dried blood poses a problem for blood typing. When blood dries out, the cells burst and may even break apart into fragments. Thus, there are no cells remaining to be agglutinated or lysed, and even if you mixed the dried stain with water in an attempt to rehydrate it you will cause the cells to break up even further. Unfortunately, many blood stains that are discovered have already dried by the time a forensic scientist arrives. It would be great if blood were always fresh, but this just isn't the case.

When cells burst apart, however, their antigens remain intact. So, the antigens are there and can potentially react with an antiserum. The problem is how to detect the reaction since there are no intact cells to agglutinate or lyse. The blood type can not be determined directly in the manner described above for blood typing, but an alternative method exists.

Forensic scientists get around this limitation using an ingenious technique called the **absorption-elution method** to type the blood. It works like this. First, you add antiserum to the dried blood. If the dried blood contains antigens of the type targeted by the antiserum, then the antibodies in the antiserum will attach to the antigens. This uses up the antibodies and makes the antiserum incapable of reacting with normal blood cells that it was designed to react with. The antiserum is drawn off (eluted) using a pipette or similar appropriate instrument and added to cells of the type that the antiserum would normally react with. If these cells are agglutinated or lysed, then there were no antigens of the type that the antiserum targets present in the dried blood. However, if these red cells fail to agglutinate or lyse, then antigens of the target type were present in the dried blood and have reacted with all the antibodies and used them up.

Let me illustrate this with our ABO example. A drop of anti-A antiserum is mixed with a bit of the dried blood, and a drop of the anti-B antiserum is mixed with another bit of the dried blood. The two antisera are then drawn off, leaving as much of the dried blood as possible behind. The anti-A antiserum that was drawn off is now added to red cells that are known to be type A. If the cells agglutinate, then antigen A was not present in the dried blood. If the cells do not agglutinate, however, then antigen A was present in the dried blood. The same procedure is followed with the anti-B antiserum that was drawn off after reacting with the dried blood. The results can be used to determine the blood type of the dried blood as shown in Table 19.2.

Table 19.2: Blood Typing for the ABO Blood Group Using the Absorption/Elution Method

Known Type A Blood Cells Are Agglutinated By Anti-A That Has Been Allowed To React With the Dried Blood	Known Type B Blood Cells Are Agglutinated By Anti-B That Has Been Allowed To React With the Dried Blood	Blood Type Result For The Dried Blood
No	Yes	Type A
Yes	No	Type B
No	No	Type AB
Yes	Yes	Type O

Observant readers will note that Table 19.2 is exactly the opposite from Table 19.1. That is, every "Yes" entry in Table 19.1 is a "No" entry in Table 19.2, and vice versa.

Only the ABO system is routinely typed by the absorption/elution method. Although the absorption-elution method should work for any blood group, it has not been demonstrated to be reliable for any except the ABO system.

19.7 Questions for Study and Review

1. In your own words, define "genetic marker" and list the three categories of genetic markers commonly used in forensic science.
2. What are the three main components of blood, and what types of genetic markers does each of them provide?
3. What tasks can luminol be used for?
4. List the methods that can be used to confirm that a substance is really blood, and describe the basics of how each method works.
5. What tasks can a precipitin test be used for, and how do precipitin tests work?
6. In your own words, describe antigens and antibodies.
7. In your own words, describe the two types of antigen-antibody reactions.
8. Describe the antigens present on the cells of people with each of the four ABO blood types, and discuss how these antigens allow the blood types to be identified.

9. In your own words, explain how the absorption-elution method of blood typing works.
10. In what situations is the absorption-elution method used instead of the conventional method for typing blood?

19.8 Sources

Alberts B, Johnson A, Lewis J, Raff M, Roberts K, Walter P. 2002. Molecular Biology of the Cell. 4th Edition. New York: Garland Science.

[Anonymous]. 2006. Antibody and Antigen [Internet]. [cited 2007 Oct 23]. Available from: http://www.bookrags.com/research/antibody-and-antigen-wog/.

[Anonymous]. n.d. Absorption-elution technique [Internet]. [cited 2007 Oct 23]. Available from: http://meaganpickles.tripod.com/forensicbiology101/id5.html.

Barni F, Lewis SW, Berti A, Miskelly GM, Lago G. 2007. Forensic Application of the Luminol Reaction as a Presumptive Test for Latent Blood Detection. Talanta 72(3): 896(18).

Block EE, 1979. Science vs Crime. San Francisco: Cragmont Publications.

Boaz NT, Almquist AJ. 1999. Essentials of Biological Anthropology. Upper Saddle River (NJ): Prentice-Hall.

Harris T. n.d. How Luminol Works [Internet]. [cited 2007 Oct 23]. Available from: http://science.howstuffworks.com/luminol.htm.

O'Connor T. 2006. Forensic Serology [Internet]. [cited 2007 Oct 23]. Available from: http://faculty.ncwc.edu/TOCONNOR/425/425lect13.htm.

Saferstein R. 1998. Criminalistics: An Introduction to Forensic Science. 6th Edition. Upper Saddle River (NJ): Prentice Hall.

Tagliasacchi D, Carboni G. 1997. Let's Observe the Blood Cells [Internet]. [cited 2007 Oct 23]. Available from: http://www.funsci.com/fun3_en/blood/blood.htm.

Villa A, Drago F, Misto R, Morelati F, Poli F, Sirchia G. 1996. Abo Genotyping in Italian Blood Donors. Haematologica 81: 492-496.

Wikipedia contributors. 2007. ABO blood group system [Internet]. Wikipedia, The Free Encyclopedia; 2007 Oct 22, 20:13 UTC [cited 2007 Oct 23]. Available from: http://en.wikipedia.org/w/index.php?title=ABO_blood_group_system&oldid=166356729.

Wikipedia contributors. 2007. Blood type [Internet]. Wikipedia, The Free Encyclopedia; 2007 Oct 10, 10:43 UTC [cited 2007 Oct 23]. Available from: http://en.wikipedia.org/w/index.php?title=Blood_type&oldid=163538405.

Chapter 20
OTHER ISSUES IN SEROLOGY

In this chapter we will explore several additional issues related to serology. The chapter will begin with an examination of blood typing of body tissues and fluids other than blood, with a detailed look at the serological questions involving semen. The chapter will continue with a brief overview of blood proteins as genetic markers, then tackle the issue of how combinations of genetic markers are used to narrow down the probability that anyone other than a certain suspect left the evidence. We will look at how genetic evidence is used in paternity testing, then the chapter will conclude with a brief introduction to blood spatter evidence.

20.1 Tissue and White Cell Antigens

Antigens are not restricted to red blood cells. White blood cells and the cells of most tissues of the body have antigens as well. The most important of these antigen systems is called the **HLA** (human leukocyte antigen) or **MHC** (major histocompatibility complex) set of antigens. These two names refer to exactly the same system of antigens. Two names exist because the antigen system was discovered independently by scientists working with white blood cells and scientists working with organ transplants. The HLA/MHC system includes antigens found on almost every cell of the body, including white blood cells, but not including red blood cells. The main utility of the HLA/MHC system is in matching organs for transplants. Rarely is this system used forensically, though there are enough alleles that it provides a large system of useful genetic markers.

20.2 ABO Blood Typing of Other Body Fluids and Tissues

ABO blood typing is not restricted to just blood. Most people, about 80% of the population, have at least one copy of a dominant allele for the **secreter locus**, also known as the Lewis locus. (If you have forgotten what alleles and loci are, a review of chapter 15 is recommended.) People with this dominant allele secrete their ABO blood group antigens in all their body fluids, including saliva, semen, vaginal fluid, the gastric juices produced by the stomach, and even sweat. In fact, saliva and semen can have a higher concentration of antigens A and B than blood. This is very useful, because it means that saliva left on a cigarette butt, semen left on the body or clothing of a victim, sexual stains on bedding, and vomit can all be blood typed. In addition, skin, muscle, and other solid tissues have blood in them, and can be blood typed. For all of these fluids and tissues the antigens are present, but red blood cells are not present, so the absorption-elution method is used.

20.3 Semen

Semen testing is quite common because many cases that are serious enough for evidence to be collected and sent to a crime lab involve some form of sexual assault. When testing for semen there are a similar set of questions for the serologist to address as there are when testing blood.

20.3.1 Detecting & Identifying Semen

The first question that has to be asked is whether the stain or fluid in question is semen. Semen is a sticky looking cloudy to white liquid when fresh, that dries to form a crusty stain. However, many other substances can have a similar appearance, especially when dry, such as egg white, milk, or mayonnaise.

Fortunately, there are a variety of tests that can be used to determine whether a stain is or isn't semen. The first of these is simply the fact that dried semen fluoresces under ultraviolet light. Fluorescing stains, especially in certain obvious locations are likely to be semen. However, other things besides semen can fluoresce, and a clever defense attorney may dispute a simple fluorescence test.

One of the most straightforward tests is to find sperm in the semen. A normal male releases 2.5 to 6 milliliters of seminal fluid during an ejaculation. Each milliliter of ejaculate contains 100 million or more sperm. A serologist can examine a fresh sample of evidence directly under the microscope. If he or she is working with a dried stain, scrapings or a piece of the material containing the stain can be put in water and swished around to free the sperm. The water is then examined under the microscope. You would think that it would be easy to find sperm in a sample of semen, but it's actually not. Sperm are fragile and easily destroyed, especially as the semen dries out. Also, many men have had vasectomies, wherein the duct that carries sperm from the testicles has been cut or blocked. Some men have a condition known as oligospermia, which means "low sperm count". The semen of these men will have few or no sperm even when fresh.

Because of the problem of spermless semen there are a set of chemical methods that can be used to determine whether a sample or stain is really semen, which don't rely on detecting the presence of sperm. The most popular of these is the **acid phosphatase color test**. Acid phosphatase is an enzyme that is produced in the prostate gland and is a major component of semen. Although it is found in many other body fluids and tissues, its concentration in semen is about 400 times greater than anywhere else in the body. Acid phosphatase reacts with sodium alphanaphthylphosphate and Fast Blue B dye to produce a color change (to blue). Such a color change produced by this test is acceptable in court as evidence that the substance in question is really semen.

There are some crystallization tests for detect the presence of those proteins that are found in high concentration in semen, such as choline and spermine. These are similar to the crystallization test for hemoglobin in blood. However, these tests are not as specific to semen and the acid phosphatase color test is considered better.

20.3.2 Determining Whose Semen It Is

Once the fluid or stain has been determined to be semen, the process of determining whose semen it is may begin. As explained above, most men secrete their ABO blood type antigens in their semen. However, in modern times most determination of semen source is accomplished using DNA analysis, which will be presented in the next chapter.

20.4 Blood Proteins

There are many proteins carried by the blood plasma (serum), and are called **serum proteins**. Yet other proteins are present inside red blood cells, and are called **red cell proteins**. For example, the globulins proteins (antibodies) are serum proteins, while hemoglobin is a red cell protein. Many of these proteins exhibit variation that is the result of inheritance of alleles at the loci that control their production. Thus, they are genetic markers.

20.4.1 An Example: PGM, a Serum Protein

A thorough exploration of the full list of serum and red cell proteins is beyond the scope of this testbook, so let's focus on one of them as an example. One system that illustrates the general concepts is **PGM** (phosphoglucomutase), which is an enzyme found in blood serum. There are four alleles for PGM, which code for slightly different forms of the PGM protein that can be distinguished by how far they migrate during electrophoresis. The four alleles are: PGM 1+, PGM 1–, PGM2+, and PGM 2–. This gives us 10 possible genotypes: 1+1+, 1+1–, 1–1–, 1+2+, 1+2–, 1–2+, 1–2–, 2+2+, 2+2–, and 2–2–. The protein coded for by PGM 1+ migrates fastest and farthest during electrophoresis, followed by the proteins coded for by PGM 1–, PGM 2+ and PGM 2–, in order. This translates the 10 genotypes into 10 phenotypes that can be detected as one or two bands in specific locations when the serum is subjected to electrophoresis and stained.

For those people with PGM genotype 1+1+, only one protein, PGM 1+, is produced. This show up as one dark band on the electrophoresis gel at the location to which PGM 1+ normally migrates. These people with the PGM genotype 1+1– exhibit two bands of lesser intensity at the normal location of PGM 1+ and PGM 1– on an electrophoresis gel. Similar patterns of one or two bands on the electrophoresis gel are produced by the other eight genotypes.

There are at least a dozen additional blood proteins that exhibit variability that can be detected and characterized in this way. Many of these systems are used forensically.

20.5 Narrowing Down Probabilities of Combinations of Genetic Markers

The fact is that genetic markers are all class characters, whether blood types, blood protein types, or DNA itself. There are no genetic markers that are unique to an individual. What is unique about individuals is the combinations of genetic markers that they possess. This presents a problem when matching blood to a certain blood source, because individual characteristics, that arise out of some combination of the inherent characteristics of something (the genetics in this case) plus some environmental effect, are required for positive matching.

Since genetic markers are all class characteristics, serologists can never achieve a positive match between evidence and standard biological samples. In order to deal with this problem, they have devised a way to utilize statistical calculations based on the known frequencies of genetic markers in order to give a realistic estimate of how certain a match is. The resulting level of certainty is express as how likely it is for the exact combination of genetic markers seen to occur together by chance.

20.5.1 The Concept of Frequency in a Population

In order to show you how serologists do statistical calculations of match certainty, I need to explain the concept of frequency in a population. **Frequency** refers to how common a particular allele, genotype, or phenotype is in the population. Frequency is expressed as a number between 0 and 1, or alternatively between 0% and 100%, and is the number of people who have the characteristic of interest divided by the total number of people in the population. For example, if I say that the frequency of the PGM 1+1+ phenotype is .36, or 36%, then I am saying the 36 out of each hundred people have the banding pattern for PGM 1+1+.

Let's continue this example by looking at the frequencies of all the PGM electrophoresis band patterns (phenotypes). Their approximate frequencies are shown in table 20.1.

Table 20.1: Frequencies of PGM Electrophoresis Band Patterns

Phenotype	Frequency
PGM 1+1+:	.36
PGM 1+1-:	.18
PGM 1-1-:	.02
PGM 1+2+:	.24
PGM 1+2-:	.06
PGM 1-2+:	.06
PGM 1-2-:	.02
PGM 2+2+:	.04
PGM 2+2-:	.02
PGM 2-2-:	.0025
	1.0025

Notice that the frequencies add up to around 1.0 (100%) because everybody (100% of the population) has one of these banding patterns.

20.5.2 Probabilities of Combinations of Markers

The second thing I need to explain for this method of narrowing down probabilities is basic probability theory, especially as it relates to the word "and". If we were taking about a person with a certain combination of genetic markers we would say that the person has marker A "and" marker B "and" marker C, etc. Whenever one encounters the word "and" in a probability statement it means that you need to multiply. In this case we would multiply the frequencies of the markers involved. For example the frequency of blood type O in the American population is about 43%. We already know that the frequency of PGM1+1+ is about 36%. In order to calculate the probability of finding a person who was both blood type O and PGM 1+1+ one simply multiplies the two frequencies together.

$$.43 \times .36 = .1548$$

Now we know that the probability of finding a person who has both these genetic markers is about 15%. Notice that the probability of finding these two markers in combination is less than the probability of finding either one alone. This is because all frequencies are less than 1.0, and so when you multiply them together, you always get a result that is smaller than any of the original values.

Let's continue this line of reasoning. Let's assume that a certain person has the most common variant for 11 genetic marker systems commonly used by serologists. What is the probability of finding someone with this exact combination of markers, even though each marker is the most common of its respective marker system? The 11 genetic marker systems and their most common variant are shown in table 20.2.

Table 20.2: The Most Common Variant for 11 Genetic Marker Systems

Genetic Marker System	Type	Frequency (approx.)
ABO (transfusion blood group	O	43%
ADA (Adenosine DeAminase)	1	90%
AK (Adenylate Kinase)	1	91%
EAP (Erythrocyte Acid Phosphatase)	BA	43%
EsD (Esterase D)	1	79%
GLO I (GLycOxylase I)	2-1	52%
Gc (Group specific component)	1	50%
Hp (Haptoglobin)	2-1	49%
PGM (PhosphoGlucoMutase)	1+1+	36%
6PGD (6 Phosphogluconate Dehydrogenase)	A	96%
Tf (Transferrin)	CC	98%

To find the probability that a single person would exhibit all of these most frequent markers is obtained by multiplying the frequencies of each of the markers together.

.43 X .90 X .91 X .43 X .79 X .52 X .50 X .49 X .36 X .96 X .98 = .00516

This result means that 516 out of every 100,000 people (about 1 out of every 194 people) will have this combination of genetic markers.

20.5.3 Interpreting the Probability

Narrowing down the probability of a person having a certain combination of genetic markers is certainly useful. If the combination in the example above were found, then only 516 out of every 100,000 would have it. The City of Missoula, MT, has a population of approximately 100,000 people (in the daytime, fewer people actually live in town). This means that if a crime occurred that resulted in blood with this combination of markers as evidence, only a few more than 500 people in town that day would be suspects.

But, how does an investigator choose between those 516 suspects? He or she does it by using general methods of investigation to determine who had access to the crime scene. Not all crime scenes are publicly accessible, and in the case of a public crime scene, many people will have alibis demonstrating that they were not there at the time of the crime. This allows a reasonable assessment of who the suspect might be. If, for example, 10 people had access to the crime scene around the time of the crime, and one of them had the combination of markers in our example, then the investigator can be fairly sure that this is the correct suspect. Use of probabilities always relies on investigation to narrow down the number of people who could have had access to the crime scene. With a careful combination of serology and investigative work a crime can sometimes be easily solved and a good court case built against a suspect.

20.6 Paternity Testing

Paternity testing is performed to determine who the father of a child is. Fatherhood isn't a crime, but failure to fulfill the obligations of being a father is a violation of civil law in most states. Paternity testing gives us another example of the use of genetic markers in forensic science and also allows us to take a look at a procedure more typically used in civil court than criminal court. Maternity testing, determining who was the mother of a child, is occasionally done, but not as often because it is less often in dispute.

Most cases of disputed paternity arise when a woman seeks public assistance for herself and her child. Agencies that offer public assistance will usually require her to name the father of her child, or at least a list of possibilities, so that the agency can recover from the father some of the costs of supporting the child. In this situation a paternity dispute arises when the woman gives the requested information, but when the authorities contact the man in question, he denies being the father of the child.

Because a child's genetic markers are inherited from both its mother and its father, serology can often show that a given man has a certain probability of being the father of the child. In more complex cases DNA testing will be used.

One of the most useful applications of serology in this context is quick exclusion of paternity. Exclusion of paternity is a demonstration that the man in question could not have been the father of the child. As you remember from previous chapters, this is one way in which class characters in general can be used in an investigation. It is usually as important to know who didn't do the deed as to know who did do it. If the child has a genetic marker that is not also present in its mother or in the man in question, then the man in question could not have been the father of the child.

The genetic marker system most commonly used for quick paternity tests is the HLA/MHC system. The probabilities of the HLA markers are such that if a man can not be excluded as the father on the basis of the HLA system, then there is a 90% probability that he is the father of the child. This probability can be raised to more than 95% by also testing for the man's ABO type.

Again, since genetic markers are all class characteristics, serology can not show that a particular man is unquestionably the father of a given child. However, using combinations of genetic markers it can be shown that it is very unlikely for any other man to have been the father. Therefore,
paternity testing, like the use of serology in a criminal investigation, relies on a combination of probabilities and investigative work that combine to build a case that a certain man is the father of the child. If the case is tight enough, then a civil court will rule that the man is the father and he may be required to pay support for his child.

20.7 Blood Spatter Evidence

Investigators have always considered the location and pattern of blood stains to be clues in reconstructing the event of a crime. Since the late 1900's, however, intensive scientific experimentation has allowed the analysis of the nature and location of blood stains to become a routine, useful, and exact science. This field of investigation is usually referred to as **blood spatter analysis**. Because of their familiarity with blood, many serologists have developed expertise with blood spatter analysis, but other types of forensic scientists also practice it, and it will probably develop into its own forensic specialty in time.

The nature of the surface upon which blood lands influences the appearance of the resulting stain. Surfaces can be hard or soft, non-porous or absorbent, smooth or rough, etc. The nature of the surface has to be taken into account when interpreting blood spatter, and much of the recent blood spatter research has focused on developing methods of accounting for the nature of the surface.

The shape of the stain caused by a flying blood drop reveals its direction of travel. As a blood drop lands on a surface, the blood that touches the surface sticks to it, but the blood on the other side of the drop is still moving. This causes the production of a "**tail**" on the stain produced by the blood drop. Blood spatter analysts use the term "tail" or "trail" for this projection from the main blood stain, but it would be better to think of it as a "nose", because it points in the direction that the blood drop was traveling. In general, this is the opposite direction from the body of the person who was the source of the blood.

If the direction of travel of several blood drops can be determined, the investigator can determine whether the origins of the drops of blood converge on a certain location. Therefore, the investigator will sketch the location of the blood drops and plot a line that leads from each drop in the direction of the source of the blood (opposite the direction of the tail). If these lines converge at a single point or area, then this is the location of the wounded body part that produced the blood.

It is possible to determine the angle of impact of a blood drop on a surface by looking at the distortion of the stain. A blood drop that lands at right angles to a surface will produce a circular stain. As the angle at which the blood drop impacts the surface departs from 90 degrees, the stain will become elongated. Using a principle of convergence similar to that for direction of travel of a blood drop, the investigator can determine how far from the wall the person was when they started losing blood.

Combining these two forms of information, location of the wound and distance from the surface upon which the blood lands, the investigator can get a good idea of where the victim of a bloody crime was located when they began bleeding. They might also be able to get an idea of the severity of the injury, and to what part of the body it occurred.

As a bleeding person moves around after being wounded, their blood will be shed at different locations. This can allow a victim's movements to be tracked which may reveal where they were first attacked, whether the attack continued after the victim was wounded, whether the victim tried to help themselves, and other items of important information about how the crime occurred.

20.8 Questions for Study and Review

1. Discuss how the blood type of a suspect is determined from their saliva, semen, and other bodily fluids other than blood.
2. List the tests that can be performed to determine whether a suspicious substance is semen.
3. What would you conclude if a substance was shown to be semen, but lacked sperm?
4. List the possible genotypes for PGM and the phenotypes that result from each.
5. In your own words, describe how the range of possible suspects is narrowed using serology.
6. A certain suspect has the following genetic markers: PGM 1+2–, Blood type A, and ADA type 1. The frequencies of these markers are 0.06, 0.19, and 0.90, respectively. What is the probability of randomly encountering someone with this exact combination of markers?
7. What types of tests are commonly used in paternity testing?
8. Describe how the location of a bleeding person can be determined by analysis of blood spatter.

20.9 Sources

Akin LL. 2005. Blood Spatter Interpretation at Crime and Accident Scenes: A Basic Approach [Internet]. FBI Law Enforcement Bulletin 74(2). [cited 2007 Oct 19]. Available from: http://www.fbi.gov/publications/leb/2005/feb2005/feb2005.htm#page21.

Akin LL. 2005. Blood Spatter: Interpretation at Crime Scenes. The Forensic Examiner 14(2): 6(5).

[Anonymous]. n.d. History of Paternity Testing [Internet]. [cited 2007 Oct 23]. Available from: http://www.paternity-answers.com/history-paternity-test.html.

Block EB. 1979. Science vs Crime. San Francisco: Cragmont Publications.

Boaz NT, Almquist AJ. 1999. Essentials of Biological Anthropology. Upper Saddle River (NJ): Prentice-Hall.

Genetic Identity LLC. 2003. Semen/Sperm Identification [Internet]. [cited 2007 Oct 23]. Available from: http://www.genetic-identity.com/Semen_Sperm_ID/semen_sperm_id.html.

Saferstein R. 1998. Criminalistics: An Introduction to Forensic Science. 6th Edition. Upper Saddle River (NJ): Prentice Hall.

Schiro G. n.d. Bloodstain Photography [Internet]. [cited 2007 Oct 23]. Available from: http://www.crime-scene-investigator.net/phoblood.html.

Wikipedia contributors. 2007. Human leukocyte antigen [Internet]. Wikipedia, The Free Encyclopedia; 2007 Oct 9, 17:29 UTC [cited 2007 Oct 23]. Available from: http://en.wikipedia.org/w/index.php?title=Human_leukocyte_antigen&oldid=163368464.

Chapter 21
DNA ANALYSIS

In this chapter we will investigate DNA analysis. Since the development of forensic DNA analysis in the late 1980's it has led to the solution of innumerable crimes. We will begin the chapter with a consideration of whether traditional serology and genetic markers are still useful in the era of DNA analysis. We will continue with an examination of the methods of DNA analysis and matching. I will briefly present a section on collection of blood evidence to round out this series of chapters on serology. The chapter will finish with an advanced section on genetic profiling.

21.1 Traditional Serology vs DNA Analysis

The trend is clear. Traditional serology is on its way out. DNA analysis has proven so effective and reliable that many crime labs have abandoned the use of traditional genetic markers such as blood groups and serum proteins. The Montana State Laboratory of Criminal Investigation is one such. Is traditional serology still useful in the era of DNA analysis? Actually there are several reasons for continuing to use these older methods.

First, although DNA analysis is effective and reliable, it is neither quick, easy, nor cheap. In contrast, many of the traditional methods, especially blood typing, are quick, cheap, and easy. What could be quicker, cheaper, and easier than adding drops of antisera to drops of blood and looking for an agglutination reaction. Second, the older methods continue to give useful, accurate information. They can be especially useful if used for quick exclusions.

Third, a judge and jury are likely to be more impressed with a case if it is based on a variety of different kinds of evidence, rather than a single type of evidence. Since blood types, blood proteins, and DNA are all separate and independent lines of evidence, a case in which all of these methods have been shown to give the same identification is stronger than a case based on only one of these methods alone. It is a shame, therefore, that many crime labs have abandoned the use of traditional genetic markers.

The hope of most DNA analysts is that improved technology will make DNA analysis ever quicker, cheaper, and easier. This is a realistic expectation, because such has been the trend though the 1990's and 2000's. Will DNA analysis ever be as easy as mixing a drop of antiserum with a drop of blood. Possibly so. As any of you who are familiar the several Star Trek TV series have no doubt noted, in the future, DNA analysis may be as easy as simply pointing a tricorder at a blood stain. Who knows what the future may bring, but DNA analysis will certainly continue to improve.

21.2 The History of DNA Analysis

DNA testing is one of the newer technologies in forensic science. Unlike so many of the other forensic sciences we have covered or will cover, DNA analysis does not go back to ancient times, or even to the 1800's. The first practical method for matching DNA evidence from a crime scene to DNA from a suspect was developed in the mid 1980's. This method is technically known as RFLP (for restriction fragment length polymorphism), but more popularly called DNA fingerprinting. DNA fingerprinting was developed by a team of geneticists in England, headed by Alec Jeffreys.

DNA typing was quickly picked up by forensic scientists. By the early 1990's, DNA comparisons had become acceptable evidence in most criminal courts in the country. Many crime labs began setting up DNA laboratories, and private companies emerged with laboratories equipped to do DNA typing for law enforcement agencies. The Montana State Crime Lab was a little slow in setting up a DNA lab, and didn't get it fully operational until the late 1990's. Now, they are completely swamped with more work than they can possible do.

21.3 DNA as a Set of Genetic Markers

In reality, DNA simply provides a large set of genetic markers. These markers are no different in theory from blood groups and serum proteins. There are two major advantages to DNA however. First, DNA is present in almost every fluid, tissue, and cell of biological origin. In general, the same techniques can be used whether you have blood, kidney, skin, semen, or another type of fluid or tissue. Also, certain structures composed of keratin, such as hairs, can trap cells as they grow and thus can be used as a source of DNA. This gives investigators a wide range of biological materials that can be tested. Second, DNA provides a large set of genetic markers. The number of genetic markers provided by a DNA test is usually large enough to narrow down the probability of having the combination of markers present in the sample to the point that it nearly excludes every other human being on the planet. The probability of any two people having the same exact DNA type is sometimes 1 in several billions. When you get to that level of probability, there is usually no difficulty in convincing a court that there has been a successful DNA match to a suspect.

Despite its utility, DNA typing is not the magical bullet that it is sometimes believed to be. We must still keep in mind that these are genetic markers, therefore class characteristics, and although they provide a strong form of consistent match with high certainty, these probabilities are based on statistical arguments and do not constitute a positive match. Caution must still be maintained and general investigation carried out to demonstrate that the suspect did have access to the crime scene and that a large number of other people did not.

21.4 The Technology of DNA Analysis

There are three methods of DNA typing that are use to some extent or another today. RFLP, STR, and mitochondrial DNA typing. Of these, the STR method is most common, and the forensic DNA databases in use today focus on this method. While it is tempting to plunge straight into STR methodology, taking the time to explore RFLP will give me the chance to introduce some concepts that are important for all forms of DNA analysis. Mitochondrial DNA is a specialized technology that is only used in certain circumstances, which I will discuss below.

21.4.1 Restriction Enzymes and DNA Fingerprinting Using RFLP

Let's look first at Restriction Fragment Length Polymorphism (**RFLP**) analysis, or DNA fingerprinting. These four words that are abbreviated as RFLP no doubt sound to you like four random terms strung together by a madman, but they actually describe fairly accurately the basis of the method. The meanings and relevance of these words will shortly become clear.

The RFLP method It is based on the fact that there are certain enzymes, called **restriction enzymes**, that function to cut the DNA strand each time they encounter a certain sequence of DNA bases. Here is the word "restriction", which describes the enzymes used in the method. For example, E. coli restriction enzyme 1 (ECOR1) recognizes the sequence:

$$-G-A-A-T-T-C-$$
$$:\ :\ :\ :\ :\ :$$
$$-C-T-T-A-A-G-$$

and breaks the bonds between G and A on both strands to yield two fragments, the ends of which have these sequences.

$$-G- \qquad\qquad -A-A-T-T-C-$$
$$:\qquad\qquad\qquad\qquad\ :$$
$$-C-T-T-A-A- \qquad\qquad -G-$$

A sequence of DNA base pairs that a restriction enzyme recognizes as a place to cut the DNA is called a **restriction sequence** or restriction site. The two fragments produced by cutting the DNA strand are called **restriction fragments** because they were produced by restriction enzymes. Restriction fragments differ in length. Variation in length is referred to technically as **length polymorphism**. Now you should be able to grasp the idea that the RFLP is based on differing lengths of restriction fragments produced by restriction enzymes that cut the DNA strand. Restriction fragments are identified using electrophoresis, which, as you surely remember from chapter 18, separates molecules based partly on weight. Longer restriction fragments weigh more, and do not travel as far on the electrophoresis gel.

Different individuals have different DNA fingerprints because they have different DNA sequences. Where person A has a restriction sequence recognized by a restriction enzyme as a place to cut the DNA strands, person B may have a slightly different sequence, which is not recognized as a restriction sequence by a restriction enzyme. Also, person B may have a restriction sequence at a DNA location where it is not present in person A. Thus, treatment of both persons' DNA with restriction enzymes will produce restriction fragments of different lengths. Each restriction fragment's length constitutes a genetic marker.

A typical analysis performed with a few different restriction enzymes will produce a large number of restriction fragments. The pattern of bands formed on the electrophoresis gel by the restriction fragments is called a **DNA fingerprint**, although it is really not as unique as a fingerprint. The DNA fingerprints of evidence DNA from a crime scene can be placed next to a suspect's DNA on an electrophoresis gel to give a side by side comparison. Such a comparison of evidence DNA and suspect DNA is illustrated in Figure 20.1.

Ultimately, these differences between individuals were produced by mutations in their DNA or in ancestors' DNA which were passed down through the generations. A **mutation** is a change in the DNA base sequence. If a mutation occurs that causes a change in any of the 6 base pairs in the ECOR1 restriction sequence, then the DNA will not be cut at that location. Similarly, if a mutation occurs that changes a similar sequence into the sequence recognized by ECOR1, then a new restriction site will have been created and ECOR1 will cut the DNA strands at that point.

Figure 21.1: Diagrammatic Representation of a DNA Comparison Between Evidence DNA and the DNA of Two Suspects. The Evidence Matches Suspect 2.

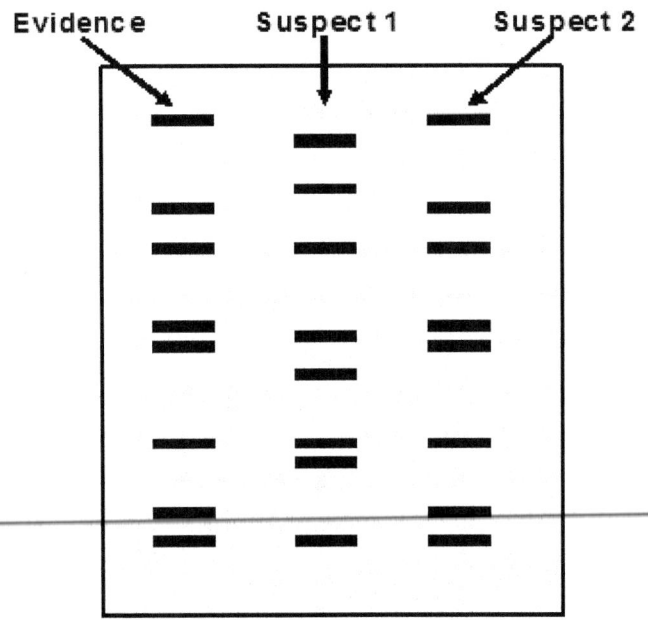

Since the goal is to find a lot of differences between individuals, so their DNA fingerprints are as distinct as possible, DNA analysts focus on those regions of the DNA that have accumulated a lot of mutations. These highly mutated regions are not those that are part of a locus which holds instructions for some metabolically important protein, because most mutation at these regions result in a protein that does not work as well, and the forces of evolution work to keep these types of mutations from being passed on. However, not all of human DNA consists of loci that have alleles which code for something. Large stretches of it are nonfunctional in this sense, and these are the regions that accumulate large numbers of mutations. Some of the nonfunctional regions are called repetitive DNA regions, and are sections of the DNA strands that have the same sequence of base pairs over and over again. These are the regions that DNA analysts focus on when using the RFLP method.

RFLP gives a crude but quick DNA fingerprint that can be used to compare samples of biological evidence. However, it has some limitations and is not used as much as STR analysis.

21.4.2 Amplifying DNA Using PCR

One of the disadvantages of the RFLP method is that it requires a relatively large amount of DNA in order to give nice clear banding patterns. Sometimes it is hard to acquire enough DNA to use. Even a substantial sample of blood may be too small, because red blood cells have no nuclei, and hence no DNA. Plasma lacks DNA as well, so the source of DNA in blood is the white blood cells, which comprise only about 1% of the volume of blood.

In the 1990's a method of copying DNA was developed called PCR. **PCR** stands for polymerase chain reaction. Polymerase is an enzyme that triggers the process of DNA replication. To increase the amount of DNA available to work with, a small amount of polymerase is added to the DNA, along with an abundant supply of free bases. A reaction is then set in motion that causes the DNA to be copied. This reaction is described as a chain reaction in that the product that results from the reaction, a new DNA strand, itself undergoes the reaction. Thus, starting with one DNA strand, the PCR reaction produces two strands, then each of them is copied to produce four strand, the four become eight, and so on until the free bases are used up or the DNA analyst stops the reaction.

Although most people would describe this process as "copying" DNA, its inventors used the term "**amplifying**" the DNA. Perhaps they had in mind turning up the volume on a stereo system. For this reason, DNA analysts use the term amplification of DNA when speaking of the PCR process.

The PCR technique is also useful when working with old or degraded samples. DNA is fragile and breaks down easily. If most of the DNA in a sample has broken down, then RFLP will probably not be useful. However, if even a little DNA is still intact, it can often be amplified using PCR.

21.4.3 STR: Short Tandom Repeats

One form of repetitive DNA has been found particularly useful for forensic DNA analysis, and forms the basis of the method most commonly used in crime labs today. This form of repetitive DNA is found in STR regions of the chromosomes. **STR** stands for short tandem repeats. The "repeats" refers to the fact that some DNA base pair sequence is repeated a certain number of times. The word "tandem" refers to the fact that the repeated sequences follow one after another like boxcars in a train. The word "short" means that the repeated sequences are relatively short, usually consisting of three to seven base pairs rather than sequences hundreds of base pairs long.

Remember from chapter 15 that a haplotype is a named or notated DNA sequence at some region of a single chromosome, and that it is the correct term to use when the region is not a locus with one or more alleles that code for some protein. Since STR regions do not code for any protein product the variant sequences are haplotypes. The difference between STR haplotypes is how many times the short base pair sequence is repeated. For example, one STR haplotype may consist of 6 repeats of the sequence and another may consist of 9 repeats. We each inherit two of each STR region from our parents (one from Mom, one from Dad).

The target STR region is extracted from the DNA and amplified using PCR. The variation at that region is then characterized by electrophoresis. When combined with information from other STR regions, the forensic scientist can narrow down the probabilities of having a certain combination of STR types in the same way as for traditional genetic markers.

Here's an example. One STR region commonly used in forensic DNA analysis is the HUMTH01 region. This DNA segment contains between 5 and 11 repetitions of the sequence A-A-T-G. Therefore, there are 7 different HUMTH01 haplotypes that a person could inherit, corresponding to 5, 6, 7, 8, 9, 10, or 11 repetitions of A-A-T-G. Everybody has two HUMTH01 regions, one on the chromosome they inherited from their mother and the other on the chromosome they inherited from their father. This give 49 possible combinations (7 for each chromosome, and 7 X 7 = 49). The frequencies of each of these combinations is known. For example, the frequency of having 6 repeats on one chromosome and 8 on the other is about .035 (3.5%).

The number of repeats in a person's STR regions is easily identifiable using electrophoresis, because the more repeats, the longer and heavier the STR region is, and thus STR regions with different numbers of repeats will separate on an electrophoresis gel. A person with a combination of 6 and 8 HUMTH01 repeats will exhibit two bands on the electrophoresis gel, one corresponding to the haplotype with 6 repeats, which will move farther, and the other corresponding to the haplotype with 8 repeats.

STR is the newest method for doing DNA typing and has several advantages over RFLP. First it produces a more complex DNA type, which means that the probabilities can be narrowed down more. Second, it requires a smaller amount of sample than RFLP. Third, it is more useful on degraded samples such as decayed corpses or dried blood stains than RFLP is.

The materials required for STR analysis are nowadays available as kits, which include all the enzymes and other components necessary for analyzing several STR regions at the same time. Specialized analytical equipment is also available to streamline the process of doing the analysis.

21.4.4 Mitochondrial DNA Analysis

The mitochondria are small structures within a cell that function to supply and regulate energy for the cell. The ultimate origin of mitochondria is thought to be parasitic or symbiotic bacteria that invaded the cells of the ancestors of plants, animals, and fungi a couple of billion years ago, then became recruited to serve their current function. One of the reasons that the origin of mitochondria is though to be from an invading microorganism is that each mitochondrion has its own DNA, separate from that of its host.

The DNA in the mitochondria, called **mitochondrial DNA**, can also be used for forensic DNA matching using methods similar to those discussed previously. Mitochondrial DNA is often abbreviated as MtDNA. It contains highly repetitive regions, in which mutations accumulate rapidly. MtDNA has some advantages over conventional DNA, which is called **nuclear DNA** because it is the DNA of chromosomes which reside in the nucleus of the cell.

The most important advantage of mitochondrial DNA is that while each cell has at most one nucleus and hence one set of chromosomes, most cells have thousands of mitochondria. Therefore, there is thousands of times more MtDNA than nuclear DNA in a cell. Even red blood cells, which lack a nucleus and nuclear DNA have mitochondria. This fact makes MtDNA more useful in situations where nuclear DNA is difficult to find. For example, old bones have very little residual nuclear DNA, because bones don't have as many cells as most tissues and after time the nuclear DNA present degrades (i.e. breaks apart). There have been some recent successes in recovering and analyzing STR's from the nuclear DNA of old bones, but in most cases MtDNA is preferable because there is more of it to recover. Also, the DNA strands of MtDNA are shorter and less likely to fragment than the strands of nuclear DNA.

Another characteristic of MtDNA, is that is is inherited only from one's mother. Fathers do pass on mitochondria in their sperm, but recent research shows that these paternal mitochondria are flagged with an antigen that instructs the ovum to destroy them. Thus, the mitochondria in our bodies comes only from the mitochondria in our mother's ovum. Maternal inheritance is a simplified pattern of inheritance, which is an advantage in some cases, and a disadvantage in others. For example, MtDNA can not be used for paternity testing.

21.5 Collecting Blood Evidence

Blood is good evidence. Its spatter characteristics and the genetic markers within it are very useful in a criminal investigation. However, blood is quite difficult to collect, transport, and preserve. It breaks down easily, and in the process loses some of its utility. Further, blood is hazardous, because it may carry infectious agents such as the viruses that cause AIDS or hepatitis. For these reasons it's imperative that we take a brief look at methods for collecting blood evidence.

As we know already, all evidence should be documented before being moved or otherwise disturbed. With blood, the general excitement and need for haste sometimes influences the person processing the crime scene to skip documentation of the blood before it is collected. This is not a good thing. Many courts will insist on seeing a photograph or sketch of the blood before it was collected. Therefore, the crime scene investigator needs to photograph the blood, especially close up, sketch the location where the blood was collected, and record notes about the location and collection of blood evidence. If it appears that the blood spatter pattern might be informative, then the locations and nature of the stains have to be carefully and thoroughly photographed and sketched, preferably by an expert in blood spatter interpretation.

Perhaps the most important consideration in collection blood evidence is to avoid contamination of the blood by external products. Blood is subject to contamination from a variety of sources. Saliva, snot, sweat or dead skin cells from an investigator may fall into the blood, creating confusion later. Also, blood can be cross contaminated, meaning that blood samples from two different person have come into contact, compromising them both. At most crime scenes there is the possibility that some of the blood is from the victim and some is from the suspect. The location where the victim's blood was located versus the location where the suspect's blood was located, and places where the two mingle is critical information. Therefore, samples obtained from different areas of the crime scene must be handled and packaged in such a way that they don't come into contact with each other.

Another form of contamination that must be avoided is contamination of the person collecting the evidence. It is important to keep in mind at all times that blood might be infectious, and to always treat it as if it is a biological hazard. Therefore, the investigator needs to wear latex gloves and in some particularly bloody crime scenes even wear a hazmat suit. If contact does occur, the blood needs to be washed off and the affected area cleaned with soap and water as quickly as possible. Further, the incident should be discussed with a physician, such as the crime lab's forensic pathologist. I can not stress this enough. When I was in graduate school, one of my fellow graduate students became ill with a rare and deadly form of hepatitis from exposure to monkey blood. The poor fellow nearly died, and even after recovery his mental capacities were not what they once were. At that time we were operating under the mistaken assumption that monkey diseases could not be passed to humans. Nowadays we know that several terrible maladies, including aids, had their origin in our primate relatives.

Blood evidence may be present on a variety of items, including the victim's clothing, the suspect's clothing, bedding, and many other predictable places. However, blood may also be found in unusual places. For example, if a suspect treated a cut over a sink, there might be blood mixed with water remaining in the p-trap of the sink drain that can be collected and characterized. The criminal may have wiped their hands on a towel, handkerchief, or rag, so these items need to be collected and inspected for blood evidence. Cracks or crevices in the floor or other surfaces may have collected blood that the criminal was not able to remove.

Wet blood is considered easier to work with than dried blood. Therefore, it needs to be collected if present at a crime scene. It can be collected with a medicine dropper and stored in a glass container. However, bacteria invade blood immediately when it is exposed to the atmosphere, and enzymes within the blood itself cause it to decompose quickly. Therefore, it needs to be refrigerated while waiting to be taken to the crime lab. I have been told by crime lab serologists that blood should never be kept for more than two hours in the field without refrigeration before being taken to the lab. If the quantity of wet blood is too small to collect using a medicine dropper it can be wiped up with cotton pads or swabs and packaged. However, you need to avoid packaging this material in airtight containers, because the buildup of moisture causes the bacteria degrading the blood to work even faster. Paper bags are preferred to plastic for this reason. It is also acceptable to dry these types of samples before packaging them. This reduces their utility somewhat, but not as much as letting it rot.

The best way to collect dried blood evidence is to take the whole bloodstained item to the crime lab. There, it can be handled under more controlled conditions than is possible at the crime scene. If the entire item can't be collected. For example, if the blood occurs on the wall of a room, then it can be collected by rubbing with a cotton pad or swab that has been moistened with water. The pad or swab can then be allowed to dry, packaged in a well-ventilated container and taken to the crime lab.

By following the guidelines that the crime lab provides, you will avoid becoming sick from someone else's blood, and you will avoid having the serologists yell at you.

21.6 DNA Databases

Databases of the loci of the human genome and the distribution of nuclear, mitochondrial, and Y-chromosome DNA haplotypes are maintained by scientific organizations. The NDIS/CODIS system of databases of DNA profiles from evidence and criminals is maintained by the FBI. These databases will be described in more detail in the chapter on forensic databases.

21.7 Advanced Topics: Genetic Profiling

Genetic profiling is a newer technique in which the STR and MtDNA haplotypes of a person are compared to known frequencies of these haplotypes in various races and geographic regions. The best statistic match is made between the person's haplotypes and these frequencies to assign the person to a certain race or geographic origin. In cases where the suspect is unknown, genetic profiling is sometimes used to characterize the race of the person, as a means of narrowing down the list of possible suspects for peace officers to check out.

Although it is indisputable that genetic profiling has helped solve some crime,s I have to say that I am extremely skeptical of this technique. Everything in my training as a physical anthropologist (those people who are supposed to know the most about race) and as a statistician tells me that this approach is misguided. Allow me to share with you the reasons for my skepticism.

First, speaking as a physical anthropologist, race is not the set of biologically based categories that most Americans think it is. Since the middle 1900's, it has been clearly shown that race is a socio-cultural construct, which takes its legitimacy from our culture's belief in it, and not from scientific fact. Humans do vary in a remarkable number of ways, including skin color and genetics, but there is nothing in the patterns of these characteristics that offers any support for races as American conceive of them. A thorough treatment of this subject is beyond the scope of this book, but for anyone interested in this subject I recommend a very short book written in 1994 by Eugenia Shanklin, titled *Anthropology and Race*, published by Wadsworth (ISBN-13: 9780534192181). Unfortunately, it costs about $30, but it marshals the evidence to convincingly destroy the common American idea of race as biological categories.

Second, forensic anthropologists, such as myself, have long used similar methods, based on a statistical procedure called discriminant functions analysis, to try to determine the sociocultural race affiliation of individuals from measurements of their skulls and other bones. Even though forensic anthropologists usually don't believe in race, they often try their best to determine the race of a person from their skeletal remains in order to help law enforcement agencies narrow down their missing persons list. Although this technique is surprisingly accurate in some cases, it is remarkably inaccurate in others. I am convinced that this is because, again, race is a sociocultural construct and not a system of biological categories. Therefore, it is rare to find an individual who shows all of the characteristics of their race and none of the characteristics of people of other races. People who study human variation have long known of a phenomenon called the "continuity effect" which refers to the fact that genes are spread around the world due to the common human phenomenon of people choosing mates from the next village over, who choose mates from the next village over from them, and so on around the world.

Third, as a statistician I do not believe that the statistical methods in use today are powerful enough to do this task with a great deal of precision, whether it be based on skull measurements or on DNA haplotypes. I note that many statisticians, though not all, agree with me on this. There is a tendency for people not highly trained in mathematics to view statistics as magic, and I think this is the situation with genetic profiling.

Fourth, genetic profiling is usually offered for commercial gain, by those companies that market STR kits. They offer genetic profiling as a service for those laboratories that buy their kits. While there is nothing wrong with this in principle, we all know that companies tend to inflate claims of the efficiency of their products. What does disturb me is that the exact methods used are in most cases considered proprietary and are not shared with the overall scientific community. There have been some in-house studies of the validity of genetic profiling by the companies themselves, but for something as critical as forensic science such techniques should be replicated many times by several independent scientists, as is true with other scientific methods.

Summing up, although genetic profiling is certainly exciting and useful if it works, I would certainly hate to see someone convicted of a crime based on genetic profiling evidence alone. There is some merit to the method, and perhaps someday independent scientists will have the chance to evaluate how well it actually works.

21.8 Questions for Study and Review

1. In your own words discuss why some crime labs no longer do conventional serological tests, such as blood typing, while other crime labs continue to perform these tests.
2. In your own words, describe RFLP analysis.
3. The result of a DNA analysis is often referred to as a "DNA fingerprint". Explain why a positive match between DNA fingerprints is not possible, given that it is possible with actual fingerprints.
4. Discuss where differences in the DNA fingerprints comes from and how these differences are produced.
5. Discuss how and why PCR is used in a DNA analysis.
6. In your own words describe what an STR is.
7. What is the characteristic of an STR that can vary among individuals?
8. Discuss why most modern DNA analyses utilize STR's.
9. The LPL STR locus has between 7 and 14 repeats of TTTA. If these haplotypes are designated as 7, 8, 9, A, B, C, D, and E, respectively, what possible genotypes could be found in an individual? How many genotypes does this add up to?
10. Discuss the advantages and disadvantages of MtDNA analysis, and some situations in which this form of DNA analysis would be preferred.
11. Discuss why you would or would not use MtDNA analysis in paternity testing.
12. Discuss why working with blood is hazardous.
13. Discuss the ways in which a blood sample can become contaminated.
14. Discuss the advantages and disadvantages of working with wet blood versus dried blood.
15. In your own words, describe genetic profiling?
16. Discuss some of the reasons why genetic profiling may give an inaccurate result.

21.9 Sources

[Anonymous]. 1998. DNA Profiling Advancement [Internet]. FBI Law Enforcement Bulletin 67(2). [cited 2007 Oct 19]. Available from: http://www.fbi.gov/publications/leb/1998/febleb.pdf.

Boaz NT, Almquist AJ. 1999. Essentials of Biological Anthropology. Upper Saddle River (NJ): Prentice-Hall.

Budowle B, Chakraborty R, Carmody G, Monson KL. 2000. Source Attribution of a Forensic DNA Profile [Internet]. Forensic Science Communications 2(3). [cited 2007 Oct 21]. Available from: http://www.fbi.gov/hq/lab/fsc/backissu/july2000/source.htm.

Butler JM. 2005. Forensic DNA Typing : Biology, Technology, and Genetics of Str Markers. Burlington (MA): Elsevier Science & Technology Books.

Byrd M. 1999. DNA, The Next Generation Technology is Here! [Internet]. [cited 2007 Oct 23]. Available from: http://www.crime-scene-investigator.net/dna.html.

DNA Advisory Board. 2000. Statistical and Population Genetics Issues Affecting the Evaluation of the Frequency of Occurrence of DNA Profiles Calculated From Pertinent Population Database(s) [Internet]. Forensic Science Communications 2(3). [cited 2007 Oct 21]. Available from: http://www.fbi.gov/hq/lab/fsc/backissu/july2000/dnastat.htm.

Isenberg AR, Moore JM. 1999. Mitochondrial DNA Analysis at the FBI Laboratory [Internet]. Forensic Science Communications 1(2). [cited 2007 Oct 21]. Available from: http://www.fbi.gov/hq/lab/fsc/backissu/july1999/dnalist.htm.

Isenberg AR. 2002. Forensic Mitochondrial DNA Analysis: A Different Crime-Solving Tool [Internet]. FBI Law Enforcement Bulletin 71(8). [cited 2007 Oct 19]. Available from: http://www.fbi.gov/publications/leb/2002/august2002/august2002leb.htm#page_17.

Riley DE. 2005. DNA Testing: An Introduction For Non-Scientists: An Illustrated Explanation [Internet]. Scientific Testimony: An Online Journal. [cited 2007 Oct 23]. Available from: http://www.scientific.org/tutorials/articles/riley/riley.html.

Rudin N, Inman K. 2001. Introduction to Forensic DNA Analysis. Boca Raton (FL): CRC Press.

Saferstein R. 1998. Criminalistics: An Introduction to Forensic Science. 6th Edition. Upper Saddle River (NJ): Prentice Hall.

Wilson C. 2002. Ready for Your Close-up? Working out What Someone Looks like from Only a Dna Sample Is No Longer Science Fiction. You'd Be Surprised What Forensics Experts Can Already Do. New Scientist 175(2352): 34(4).

Chapter 22
FORENSIC CHEMISTRY AND TOXICOLOGY

In this chapter we will explore three closely related topics: forensic chemistry, forensic toxicology, and forensic pharmacology. Almost every time I teach a class on forensic science I will have a student try to convince me that forensic chemistry and forensic toxicology are exactly the same thing. I agree that they are similar and that the methods they use are similar, but they are considered different things by most crime labs, which put their chemists and their toxicologists in different sections. For example, the Montana State Crime Lab puts their chemists in the Drug Chemistry section and their toxicologists in the Toxicology section. Similarly, the distinction between forensic toxicology and forensic pharmacology is often not clear. At first glance forensic toxicologists are interested in poisons and drugs of the recreational sort, where forensic pharmacologists are interested in drugs of the medicinal sort. There is more to the distinction than this, which I will leave for the discussions that follow.

Among the many things these two fields share is the fact that one of their most common tasks is the identification of drugs within a sample. In most cases the instruments used are the same as well, with GCMS systems being a mainstay of both disciplines. The main way in which they differ is in the nature of the samples they analyze. Forensic toxicologists work with biological samples, fluids and tissues, but forensic chemists analyze powders, solids, liquids, and gasses that are not parts of a body. So, if an investigator finds a bag of white powder and wants to know whether it is heroin, methamphetamine, sugar, baking soda, fertilizer, or exactly what, he or she will submit it to a forensic chemist. On the other hand, if the question involves the concentration of alcohol in someone's blood, whether a dead person had ingested a poison, whether a suspect is under the influence of drugs, or whether a dead person had forgotten to take their required medication, they will submit their samples to a toxicologist.

22.1 Forensic Chemistry

As discussed in chapter 11, many forensic scientists consider chemistry to be the heart of their discipline. I agree that unless one understands at least the terminology and basic subject matter of chemistry they will find most branches of forensic science to be difficult to practice. Fingerprint examiners need to be able to understand how the developing agents they use work, document examiners need to understand the composition of inks and papers, and trace evidence examiners need to understand the chemistry of many things, such as paints and fibers.

Forensic chemistry can be defined simply as the application of chemistry to develop evidence used in investigations. In particular it is the application of analytical chemistry, which is the branch of chemistry that focuses on determining the composition of an unknown sample.

22.1.1 The History of Forensic Chemistry

Chemistry as we know it today emerged from the alchemy of the Middle Ages, which was primarily a pursuit of magic. Between the 1500's and 1700's a few people emerged with a true scientific approach to the subject, including replicability of experiments. From the efforts of these people the science of chemistry began to emerge. During the 1700's and 1800's the theories and methods of chemistry as we know it today were developed. Of course, new ones have emerged since, but these were the basics of the field.

The analytical devices that forensic chemists rely on today were mostly developed during the late 1800's and 1900's. As the pace of invention of new technology accelerated during the 1900's, a variety of new instruments emerged. Also during the late 1800's and 1900's chemistry began to be used to analyze materials of interest in investigations. Today, much of forensic chemistry involves the detection of illegal drugs. In fact, the Montana State Laboratory of Criminal Investigation calls their forensic chemistry section simply Drug Chemistry.

22.1.2 Analysis for Drugs

Several tests exist for analyzing a suspicious substance to determine whether it consists of or contains drugs. Usually, the drugs of interest are the illegal recreational drugs rather than medications.

Several tests exist for determining the presence of drugs in a sample. One of the most popular and simple is the many types of **color tests** that are packaged as test stick or strips. All of the chemicals required for the test are placed on an area of the stick or strip, and the stick or strip is simply placed in contact with the substance or dipped into a portion of the substance that has been dissolved in water. A color change or similar result indicates that the substance being tested for is present. The sticks and strips don't, however, give an acceptable quantification of the amount of the target substance in the sample, so if the presence of drugs is detected using one of these simple tests, it will usually be analyzed with a GCMS system. GCMS systems are capable of determining the exact amount of all substances in the sample, as described in chapter 18.

Drug testing technology will be explored in more detail in the chapter on drugs and drug testing.

22.2 Forensic Toxicology

Forensic toxicology is one of the busiest branches of the forensic sciences, especially if we consider those people who deal with breath or blood alcohol to be toxicologists. It would be difficult to carry out forensic science as we know it today without forensic toxicologists.

The root of the name toxicology is "toxic", which refers to poisons. At first, toxicology was simply the study of poisons and their effects on the body. However, it has become broadened over time to include not only poisons, but substances that have an intoxicating effect, such as alcohol and drugs. Its application to drugs has also been broadened to include legal medicinal drugs in addition to illegal recreational drugs. So, maybe the best definition of toxicology at this time is that it is the detection of any unusual substance in the human body and the study of the effect of unusual substances on the human body.

Forensic toxicology is defined as the application of toxicology to developing evidence for investigations. This seems like a logical and straightforward definition, which can include both its application to criminal law and its application to civil law.

22.2.1 The History of Forensic Toxicology

The use of poisons goes far back into prehistory. Socrates, the great Greek philosopher was required to drink a cup of hemlock. Cleopatra VII, Queen of Egypt is thought to have committed suicide by allowing a poisonous snake to bite her. The first scientific paper on the detection of poisons was published In 1814 by Matthieu Orfila. In 1836, James Marsh of England developed a method for the detection of arsenic poisoning.

During the late 1800's and 1900's, the history of forensic toxicology has primarily been defined by the invention of new instruments and tests for detecting and measuring the quantity of substances in the body. Francis Aston developed the first mass spectrometer in the 1900's and received the Nobel Prize for it in 1922. Advanced chromatography techniques were developed in the middle and late 1900's.

Forensic toxicology has also received attention for its ability to determine whether certain celebrities died from drug overdoses. Among these are Marilyn Monroe (she did) and Elvis Presley (he didn't - he died from a heart attack).

One of the greatest spurs to the development of forensic toxicology was President Reagan's "War on Drugs", which was launched in early 1980's. The early 1980's was also the time of the beginnings of MADD (Mothers Against Drunk Driving) and their campaign against driving under the influence of alcohol.

22.2.2 Types of Forensic Toxicologists

There are three branches to forensic toxicology: post-mortem forensic toxicology, human performance forensic toxicology, and forensic drug testing. **Postmortem forensic toxicologists** primarily work with forensic pathologists in death investigations. They analyze samples of body fluids and tissues that were collected during an autopsy to determine whether drugs (medicinal or recreational), alcohol, poisons, or other substances were present in the person's body at the time of their death. Detecting the presence of these sorts of substances in the body of a dead person can have a great impact on how the death is interpreted. Depending on the amount and nature of the toxic substance, it may have killed the person through poisoning or overdose.

Alternatively, it may have impaired the person's judgement or their ability to drive and therefore contributed to their death. Or, it may give information about what the person was doing at the time of their death or had done earlier. It may also give clues about the person's activities that may lead investigators in a certain direction.

Human-performance forensic toxicology also focuses on detecting evidence of alcohol or drugs in the human body, but in this case a living human. This type of evidence is used primarily to determine whether the person committed a crime such as illegal use of drugs or driving under the influence of alcohol or drugs. It may be a factor in whether a homicide is determined to be deliberate or negligent. If a person is under the effects of alcohol or certain drugs they might kill someone without intending to, which is negligent homicide rather than deliberate homicide. Also, what might be ruled an accidental death, such as a death in a traffic accident, if the person is not under the influence may be ruled negligent homicide if the person is under the influence of something. Human performance forensic toxicologists often examine blood, breath, or urine specimens to determine drug or alcohol use.

Toxicologists who work in what is called **forensic drug testing** are to some extent considered problematical by other toxicologists. These scientists work in private commercial laboratories that do drug tests for a fee on samples submitted to them. For example, if a trucking company routinely tests its drivers for illegal drug use, chances are that samples of the drivers' urine are sent to a commercial drug testing laboratory for this purpose. The criticism leveled at forensic drug testing toxicologists by more mainstream forensic toxicologists is that they often use only the quick, cheap, and simple drug tests, rather than the more reliable tests using GCMS and similar systems as is done in a crime lab. Whether this scorn is warranted or not is unclear, since these laboratories do provide a valuable service and any toxicology done for the purposes of civil court must be done by a commercial laboratory rather than a crime lab. However, as will be discussed in the chapter on drugs and drug testing, the use of simple tests often results in a considerable number of false positive results, which suggest that a person has been using drugs when in fact they have not. Urine, blood, perspiration, and hair can all be analyzed for the presence of drugs and their metabolites, to demonstrate prior use or abuse.

22.2.3 Alcohol Testing Technology

Alcohol testing is one of the most common forensic toxicological tests because of its application to enforcing drunk driving laws. Alcohol (C_2H_5OH) is a drug that affects the nervous system and even in small amounts (less than the legal limit) can reduce judgement and reduce driving skills. On average, about one third of all driving accidents are connected with alcohol. Many crime labs have a whole section devoted exclusively to alcohol testing, or as in the case of the Montana State Laboratory of Criminal Investigation, to providing maintenance of alcohol testing units and training in their use.

Alcohol testing is primarily done via testing a person's breath or their blood. The definitions of illegal amounts of alcohol in the body while driving are based on alcohol concentration in the blood. Breath alcohol testing relies on a physiological principle known as Henry's Law, which states that the amount of alcohol in a person's breath is dependant on the concentration of alcohol in their bloodstream. Henry's Law, however, does not provide a precise way to determine the amount of alcohol in the blood from the alcohol in the breath for all people under all conditions. Also, breath testing devices work on principles that are less accurate than determining the amount of alcohol in the blood using chromatography or a GCMS device. Therefore, breath tests are not nearly as reliable as blood tests, which use a gas chromatograph or a GCMS unit to quantify the exact amount of alcohol in a person's blood.

It is impractical, however, for peace officers to carry around a gas chromatography unit in their patrol car. For one thing, these are large instruments, and in addition the analyses need to be performed under controlled conditions by a scientist trained in the use of the instrument. Therefore, alcohol testing in the field is usually done by breath tests or by a field sobriety test. Field sobriety tests, still common in some jurisdictions, consist of a set of tasks that are difficult to do in an impaired state. Many people believe that field sobriety tests are overly difficult, and indeed, I don't think I could pass a field sobriety test stone cold sober. Since these tests are not as reliable as blood tests, a person who fails them is usually taken to a facility where their blood is drawn for blood testing.

The traditional form of breath alcohol testing device is the "tube and bag" type breathalyser. This device is not very accurate, but can give a peace officer a reason to take a suspected drunk driver in for a blood test. Tube and bag breathalysers are not commonly used anymore, having been superceded by the more accurate fuel cell type of breath alcohol testing device. The **tube and bag breathalyser** consists of a tube that feeds into a plastic bag. The tube contains crystals of potassium dichromate ($K_2Cr_2O_7$) which is normally yellowish orange in color but which changes color to green upon contact with alcohol. Drivers suspected of exceeding the legal amount of blood alcohol are required to blow into the tube, and the amount of potassium dichromate that changes color reflects the amount of alcohol in their breath. The tube containing the potassium dichromate is marked with approximate blood alcohol concentrations which can be read by referring to how far along the tube the color change extends. If the color change extends beyond the mark that designates the legal limit of blood alcohol concentration, then the driver is taken to have their blood drawn for confirmation using gas chromatography.

The most commonly used type of breathalyser today is based on a **fuel cell** that can convert chemical energy combined with oxygen straight into electrical energy. The fuel required by the cell is provided by the alcohol present in the driver's breath, and the voltage produced corresponds directly to the amount of alcohol in the breath. This voltage is measured and the device shows whether the driver is above or below the legal limit by flashing one or the other of a set of two lights that are labeled "pass" and "fail". Other lights on the device signal when the driver is blowing hard enough and when the test has been completed.

The fuel cell based breathalyser offers some advantages. First, it can be used repeatedly, thousands of times by simply changing the used mouthpiece for a new, sterile one after each test. This is much more practical than the tube and bag breathalyzer which has to be recharged with fresh potassium dichloride after each use. Second, the simple pass/fail result system avoids the temptation to be overly precise about the amount of alcohol in the breath or blood. A result of fail is simply used as cause to take the driver to have their blood drawn, and not to state that the driver's blood alcohol concentration was a certain value based on the breathalyser reading.

22.2.4 Other Roles of Forensic Toxicologists

Forensic toxicologists have some additional roles that they carry out during the course of their job. Also, the field is evolving in some new and interesting directions, such as environmental toxicology.

One less often recognized role of forensic toxicologists is as guardians of public health. If there is an increase in deaths or other problems associated with drug use, such as a toxic batch of heroin on the street or even a bad prescription medication, it is a forensic toxicologist who is likely to recognize the problem first. The proper authorities can then be contacted to attempt to solve the problem.

A new frontier in toxicology is environmental toxicology. This branch of the discipline studies the effects of environmental pollutants on the body. As the amount of human-generated pollution, toxins, and irritants increases in the environment, they cause problems for people. An environmental toxicologist can often determine what substance is causing the problem, which may lead to steps being taken by the government to increase regulation of the sources of that substance. Environmental toxicology can also help people make good choices about foods to eat. For example, some fish are so contaminated with mercury that toxicologists recommend not eating them more than about once per week. Environmental toxicologists provide their expertise in civil suits more often than in criminal trials.

22.2.5 Trends in Forensic Toxicology

Some forensic toxicologists have expressed alarm recently at an apparent lack of growth in the number of professionals in the field. They note that while the number of non-forensic toxicologists is growing, the number of forensic toxicologists is not. For example, the major U.S. organization of general toxicologists, the Society of Toxicology, numbered over 1200 members in 2005 and continues growing rapidly, while a corresponding group of forensic toxicologists, the Toxicology Section of the American Academy of Forensic Sciences, had fewer than 300 members that year, a number that has changed little in recent years.

Part of this phenomenon may relate to the fact that environmental toxicologists, who form the most rapid growing segment of forensic toxicology, do not generally seek membership in the AAFS, since they are primarily oriented toward regulatory and civil law enforcement rather than criminal law enforcement. It may also be that many budding forensic toxicologists are lured into the commercial forensic drug testing industry by a combination of higher salaries and less stressful workload. These forensic toxicologists no doubt follow the advances in the field that are published in forensic science journals, but are less likely to join the AAFS than are those forensic toxicologists who work at a crime lab.

Putting this together, it is not clear whether there is cause for alarm over a static or dwindling number of forensic toxicologists, but it possibly serves as a notice to those governments that fund crime labs that they need to keep salaries and working conditions similar to those in the private sector or they may have trouble filling toxicologist positions in the future.

22.3 Forensic Pharmacology

Forensic pharmacology is the application of pharmaceutical and medication sciences to investigations. Forensic pharmacists mostly become involved with civil or regulatory cases, but occasionally contribute to the criminal justice system.

A few pharmacists are full-time, and work for a government agency such as the federal Food and Drug Administration, the federal Drug Enforcement Administration, or a state regulatory agency. Most forensic pharmacists, however, are part-time consultants who work full-time in a pharmacy or hospital.

Although forensic pharmacists share an interest in drugs with toxicologists and drug chemists, the work of a forensic pharmacist usually lies outside that of toxicologists and drug chemists. Therefore, it qualifies as a forensic technology in its own right. While, forensic chemists and toxicologists are mostly concerned with illegal, recreational drugs, forensic pharmacologists are mostly concerned with issues surrounding the use of medicinal prescription drugs. Issues of interest to forensic pharmacists include prescription forgery, use of drugs in suicide or assisted death (euthanasia), medication errors, adverse drug reactions and drug interactions, drug impaired driving and drug induced violence, poisoning, the effect of psychoactive medications on competency to stand trial, assessing the magnitude of pain in an individual by analyzing dosages of analgesic medications, appropriate use of chemical restraints and tranquilizers given to patients in hospitals, and psychopharmacological effects as a mitigating factor in criminal activity.

22.4 Questions for Study and Review

1. Compare and contrast the methods, samples, and goals of forensic chemistry, forensic toxicology, and forensic pharmacology.
2. Describe the two main approaches to determining whether a sample of a substance contains illegal drugs. Under what conditions would both of these types of tests be performed?
3. How has the range of substances of interest to toxicologists become broader over time?
4. In your own words, describe the three types of forensic toxicologists, and how each type is distinctive.
5. Discuss the advantages and disadvantages of breath testing versus blood testing for determining blood alcohol content.
6. Discuss how a modern fuel cell breathalyser differs from a traditional tube and bag breathalyser.

7. Discuss some of the new directions and modern trends in forensic toxicology.
8. List some of the topics handled by forensic pharmacologists that are not handled by forensic toxicologists.

22.5 Sources

Anderson PD. n.d. What is a forensic Pharmacist [Internet]. [cited 2007 Oct 23]. Available from: http://hometown.aol.com/PAnder7291/forensic-pharmacist.index.html.

California Department of Justice Bureau of Forensic Services, n.d.,Physical Evidence Bulletin, Toxicological Analysis [Internet]. [cited 2007 Oct 23]. Available from: http://ag.ca.gov/bfs/evidence.php.

Gombos J. n.d. Drug Testing FAQ [Internet]. [cited 2007 Oct 23]. Available from: http://www.urban75.com/Drugs/testing2.html.

Cravey RH, Baselt RC. n.d. An Introduction to Forensic Toxicology [Internet]. [cited 2007 Oct 23]. Available from: http://www.soft-tox.org/default.aspx?pn=Introduction.

Proctor, Alissa, Dale, Mike, and Willimans, Joel, 1998. Evidence: The True Witness. Forensic Toxicology [Internet]. [cited 2007 Oct 23]. Available from: http://library.advanced.org/17049/gather/cgi-bin/document_get.cgi?path=/Toxicology/00Intro.

Saferstein, R. 1998. Criminalistics: An Introduction to Forensic Science. 6th Edition. Upper Saddle River (NJ): Prentice Hall.

Wikipedia contributors. 2007. Forensic chemistry [Internet]. Wikipedia, The Free Encyclopedia; 2007 Oct 21, 10:02 UTC [cited 2007 Oct 23]. Available from: http://en.wikipedia.org/w/index.php?title=Forensic_chemistry&oldid=166023814.

Wikipedia contributors. 2007. Forensic toxicology [Internet]. Wikipedia, The Free Encyclopedia; 2007 Oct 21, 14:21 UTC [cited 2007 Oct 23]. Available from: http://en.wikipedia.org/w/index.php?title=Forensic_toxicology&oldid=166053720.

Chapter 23
Advanced Topics
DRUGS AND DRUG TESTING

In this chapter we will explore drugs and drug testing. Drug related crimes and their evidence are usually the second largest source of the caseload at a crime lab (the number one source is usually alcohol related evidence). Many states, as well as the federal government, have entire law enforcement agencies that are dedicated to the pursuit of drug related crime. These agencies and other law enforcement agencies are highly successful, and our prisons are now full with drug offenders who have been sent there by mandatory sentences required by "get tough on drugs" laws.

23.1 A Rant on The Problems With Drug Laws

I don't use recreational drugs, and I wish that nobody did. However, as an anthropologist, I know that it is the nature of human beings to seek out mind altering substances. There is even some evidence that non-humans, such as chimpanzees, seek out natural intoxicants in their environments. Given that it is human nature to seek intoxication, it is unfortunate that drug use is illegal; and especially unfortunate that the penalties for drug use or possession are extreme. There is a growing sentiment within the justice system, which I have heard or read expressed by attorneys, judges, forensic scientists of many sorts, and even (though less often) by peace officers, that the present system of drug law enforcement is too harsh and extreme. The practical problem with the present system become clear when we see news reports which tell us that prisons are so full of drug offenders that there is often no room in them for the violent offenders who truly need to be removed from the streets. Cases abound of people in prison for a long sentence due to possession of very small amounts of drugs. Further, the system is heavily weighted against non-White people and less wealthy people. When you add the observation that there is no evidence whatsoever that these strict drug laws are a significant deterrent to drug use, one wonders why the American public and its lawmakers allow the current practice to continue.

There is a lesson to be learned from American history. During the era of prohibition, 1920 to 1933, the sale, manufacture and transportation of alcohol were constitutionally banned throughout the United States. Prohibition began in 1920, after the passage of the 18th Amendment to the U.S. Constitution in 1919. Rather than significantly curbing the use of alcohol, the result of prohibition was the rise of massive organized crime networks to supply people's desire for liquor. The gangsterism of the 1920's and 1930's, with all of its violence, was a direct result of prohibition. Finally, the American public had enough and Congress passed the 21st Amendment to the U.S. Constitution, which repealed the 18th Amendment, in 1933. We are in the same situation today with gangs and other organized crime operations fighting for control of the sales and supply of drugs. The solution is clear – decriminalization and regulation of the safer recreational drugs, thereby knocking the legs out from under organized crime, generating huge tax revenues, making our society safer, freeing up prison cells for violent offenders, and leading to a better system of justice.

One factor that keeps the war on drugs going is that it provides a significant source of income for law enforcement agencies. In many cases, if a person is convicted of a serious drug related offense, their property and possessions are confiscated and sold, with the money going into the operating budget of the law enforcement agency. This is our fault, fellow citizens, for insisting on keeping taxes low, resulting in slashed budgets for law enforcement. Law enforcement has to fund itself some way, and if they don't get the necessary funding from their governments, they become revenue driven in their enforcement of the law. Drug related crimes are the prime example of this, but anyone who has ever received a nuisance traffic ticket, such as for not coming to a full and complete stop at a stop sign even if there was no other traffic on the road, has been a victim of revenue driven law enforcement. Many peace officers feel humiliated by the requirement to generate the money that pays their own salaries and pays for the equipment they use. Don't we owe it to our law enforcement agencies and to ourselves to properly fund law enforcement so that our peace officers can concentrate on the important parts of their jobs?

23.2 Types of Drugs

There is a huge variety of illegal drugs. Not only are there the traditional heroin, marijuana, cocaine, and LSD of my youth, but many others have appeared in more recent times. A variety of prescription drugs are also abused. I have chosen here to approach this overview of drugs by dividing them into commonly recognized categories: narcotics, hallucinogens, depressants, stimulant, and other. Forensic pharmacists would recognize these categories, as would toxicologists. The advantage of using this category system is that I can group drugs together based on their effects, which allows me to say things about all the drugs in that category.

23.2.1 Narcotics

The term "narcotic" is derived from a Greek word meaning sluggishness or sleepiness. Pharmacologists define **narcotics** as those drugs that bring relief from pain and promote sleep. Unfortunately, narcotic has come to be popularly associated with any drug that is socially unacceptable. As a consequence, many drugs are called narcotics that aren't. In this chapter I will stick with the scientific pharmacological definitions and classifications of drugs - not the popular or legal ones.

All narcotics have the potential to cause physical drug dependence. Since they depress the central nervous system, it is possible to overdose and die from their use.

23.2.1.1 Opiates

The largest group of narcotics is the **opiates**, a category that includes those drugs derived from opium. Opium is a gummy juice exuded through a cut made in the unripe pod of an opium poppy. Opium poppies grow throughout much of Asia and provide substantial income for local farmers in many regions of that continent. Opium and all its derivatives are addictive, leading to drug dependence. Morphine is the active substance in opium, which contains between about 4% and 21% morphine. The morphine can be purified and is a powerful prescription pain reliever. Codeine is synthesized from morphine, and is another prescription pain killer. Codeine is only one sixth as strong as morphine.

Heroin is another drug synthesized from morphine. Heroin isn't naturally stronger than morphine, but it is water soluble, so it can be dissolved and injected under the skin. Typically a heroin user will dissolve the heroin in a small quantity of water in a spoon. The process can be speeded up by heating the spoon with a candle, match, or other heat source. This solution is then drawn into a syringe for injection into a vein or beneath the skin. As bought on the street, heroin is usually diluted with some other substance. Typical substances used to dilute or cut heroin include quinine, starch, lactose, Novocain, and mannitol. Since the 1980's the typical street bag of heroin has actually contained more heroin in it, around 35%, than was true during the 60's and 70's when a typical bag contained only 15% to 20% heroin.

23.2.1.2 Other Narcotics

Other drugs, not derived from the poppy, but having similar actions, are called **opioids**. They are all synthetic, in that they are totally the product of a laboratory – not derived from a plant. There are many opiates, the two most widely known of which are methadone, which is used in some heroin treatment programs, and Darvon (propoxyphene).

Other narcotic drugs tested for by drug chemists include : Etorphine, Naloxone, Fentanyl, Oxycodone, Meperidine, Oxymorphone, Hydrocodone, Pentazocine, Hydromorphone, Ethylmorphine, and Tramadol. Other narcotics exist and more are being developed for medicinal purposes. As time goes on these inevitably end up being used recreationally.

23.2.2 Hallucinogens

The term **hallucinogen** means something that can generate hallucinations. Hallucinogens are drugs that cause alterations in normal thought processes, perceptions, and moods, leading in some cases to hallucinations and other experiences that are outside of normal reality. As a group, hallucinogens are noted for being the safest of the recreational drugs. Unsafe hallucinogens certainly exist, but the majority neither cause drug dependence nor pose a high risk of overdose and death.

23.2.2.1 Marijuana

The most popular hallucinogen, probably for the past several thousand years in places where it grows, is marijuana. Marijuana is first mentioned in a pharmacy book written about 2737 BC in China, so it has been used for nearly 5000 years, at least. It is the most widely used illegal drug in the U.S., and has been for decades. According to a recent survey, 43 million Americans have tried marijuana, including former U.S. President Bill Clinton (who claims never to have inhaled). As many as 20 million Americans are probably regular users of marijuana.

Marijuana bought on the street is the leaves, flowers, stems, and/or seeds of the marijuana plant, Cannabis sativa. The active ingredients are forms of cannabinol, primarily tetrahydrocannabinol (THC). There are many varieties of marijuana, that differ in potency due to the genetics of the plant, the soil and other environmental conditions where it was grown, and the cultivation practices used. The average potency is about 1.5% THC. Cannabis has male plants and female plants, with female plants being more potent. One of the most potent forms of marijuana is sinsemilla, which is the unfertilized flowering tops of female cannabis plants. Sinsemilla has a potency of about 3.5 to 4% THC. Marijuana is normally smoked, rolled up in a cigarette paper to form a "joint", but other ways of smoking it are known and it can also be eaten.

Hashish, often known simply as "hash", is a derivative of marijuana, specifically the resin from the external surface of the marijuana plant, which is high in THC, about 3.5%. There are at least two ways to prepare it. The traditional way to prepare hashish is to have children run naked through the marijuana fields. The resin sticks to their bodies and is scraped off with a stick or rubbed of by hand. The other way to make hashish is to extract the resin from ground-up marijuana using solvents. This yields hashish oil, which can be between 20% and 65% THC. Hashish oil is so potent that the usual way to use it is to place a small drop on a tobacco cigarette, which is then smoked.

The effects of marijuana usually include a feeling of well-being, hilarity, a dreamy state of relaxation, alterations of sensory perceptions, hunger and a craving for sweets, and some changes in speech. At normal dosages, these effects are barely noticeable by an observer. There is some evidence that marijuana use causes physical or psychological drug dependence, but the withdrawal symptoms, if any, are poorly documented. One thing is certain, that nobody has ever died from an overdose of smoking marijuana "joints" – ever. This is not to say, however, that marijuana has not contributed to fatalities. Under the influence of marijuana some people are tempted to do incredibly stupid things, such as try to drive their car. Some of these acts can lead to fatal accidents.

Marijuana also has some medical uses. In particular it reduces excessive internal eye pressure in people with glaucoma. It also helps the nausea caused by some powerful anticancer drugs. Recent research has detected the presence of cannabinol receptors in one of the more common pain pathways, and it is likely that cannabinol is effective in relieving certain types of chronic pain. Some states have decriminalized the use of marijuana for medicinal purposes.

23.2.2.2 LSD

LSD is an old and still popular hallucinogen. It can cause profound sensory changes and hallucinations. LSD is derived from Lysergic acid, which is found in ergot, a fungus that grows on wheat and other grains. It is very potent, with tiny amounts causing vivid hallucinations that may last for 12 hours or more. The drug also causes changes in mood, with the user laughing or crying at the slightest provocation. Feelings of fear and tension almost always accompany the use of LSD.

Despite its profound effects, LSD is remarkably safe. It is almost impossible to overdose to a point leading to death, though a person taking a large dose may be on a trip for days. It does not cause physical dependence. Reports of "flash backs" occurring months or years after use are probably myths in most cases. There is, however, some medical evidence suggesting that LSD causes chromosome damage in the children of users.

LSD is taken by eating it. Typically it is available on blotters, small patches of filter paper, which are chewed or sucked on to dissolve and ingest the LSD. This is another thing that has changed since the days of my youth. The typical dosage on a blotter today is only about 1/10,000 the amount on a blotter in the 1960's and 1970's.

23.2.2.3 Other Hallucinogens

PCP (phenycyclidine), or angel dust, is a dangerous hallucinogen. PCP is eaten, smoked, or sniffed. In contrast to LSD, you can overdose on PCP and die. Even moderate use can cause psychological changes, such as depression and paranoia, that last a long time and can recur days after use. PCP is synthesized in a lab.

Ecstacy, also known as MDMA, is one of the newer drugs on the market. First developed as a diet pill, it was found to bring about feelings of happiness and relaxation. It enhances self-awareness and decreases inhibitions, but also increases heart rate and blood pressure, produces nausea, and can cause psychological difficulties such as confusion, anxiety, and paranoia. Some of these side effects can be fatal.

Other hallucinogens include mescaline (from the peyote cactus), psilocybin (from a type of mushroom), and several synthetic drugs. Peyote and psilocybin have long been associated with certain religious and magical beliefs. Chewing peyote is considered a sacrament of the Native American Church and is allowed in that context. Peyote chewing is relatively safe, because the user vomits up the peyote almost immediately, thereby avoiding a lethal overdose.

23.2.3 Depressants

Depressants, or "downers" are drugs that cause a pleasant sleepiness, but do not provide the pain relief associated with narcotics. Most, if not all, can cause physical dependence, and can be fatal in overdoses.

23.2.3.1 Alcohol

Alcohol, also known as ethyl alcohol, grain alcohol, or ethanol, is the most widely used depressant. Although effects vary, most users experience expansion of personality and increased confidence at the cost of impairment of judgement, memory, concentration, and motor skills. Long term use can lead to alcoholism, and overdoses can kill. One of the toxicologists who works at the Montana Laboratory of Criminal Investigation gives public talks to anyone who will listen (including about once a year my forensic science class) about the dangers of alcohol. He sees more deaths due to alcohol use than to any other form of drug use. Do not believe that alcohol is safe simply because it is available legally.

23.2.3.2 Barbiturates

Barbiturates are another class of depressant. There are about 25 different barbiturates, which are all derivatives of barbituric acid, first synthesized over 100 years ago. They produce relaxation, a feeling of well-being, and sleepiness. These drugs are normally taken by mouth. All of them can cause physical dependence with terrible withdrawal symptoms if usage is then stopped. There are five prescription barbiturates: amobarbital, secobarbital, phenobarbital, pentobarbital, and butabarbital. Recreational "barbs" are usually named for the color of the capsules, for example: yellow jackets, blue devils, or reds.

23.2.3.3 Tranquilizers

Tranquilizers are depressants that produce relaxation without significantly impairing the higher thinking facilities or causing sleepiness. They are divided into "major" tranquilizers that are used only by mentally ill people (reserpine, chlorpromazine), and "minor" tranquilizers that are used by the rest of us. Some well-known minor tranquilizers are Miltown, Librium, Valium, Alprazolam, Fluphenazine, Midazolam, Thioridazine, Chlordiazepoxide, Flurazepam, Nitrazepam, Trazodone, Chlorpromazine, Flunitrazepam, Nordiazepam, Trifluoperazine, Clonazepam, Hydroxyzine, Oxazepam, Triazolam, Desalkylflurazepam, Lorazepam, Promazine, Zolpidem, Diazepam, Mesoridazine, and Temazepam. These are prescribed by physicians as well as being available on the street. All tranquilizers can cause physical dependence over time and overdoses can be fatal.

23.2.3.4 Inhalants and Other Depressants

The inhaling of volatile solvents and other components of glues and other substances has grown in popularity since I was a child in the 1960's, and is commonly known as "sniffing" or "snuffing". Substances that are inhaled in order to achieve and intoxication effect are known as **inhalants**. Nowadays, most of the attention is on inhaling aerosol gasses, such as freon. There seem to be two effects involved, the effect of depriving the brain of oxygen and the effect caused by the substance inhaled. Depriving the brain of oxygen may feel good temporarily, but it can quickly cause death and causes irreversible brain damage if done regularly. Most of the substances inhaled are depressants. Some of the substances popular to snuff to achieve a cheap high are toluene, naphtha, methyl ethyl ketone, gasoline, and trichloroethylene.

The immediate effects of snuffing are a feeling of exhilaration and well being, combined with slurred speech, double vision, and impaired judgement. Long term use can cause organ damage, especially liver damage, and even lead to death.

Other depressants at least occasionally tested for include: Amobarbital, Carisoprodol, Chloral Hydrate, Methaqualone, Phenytoin, Ethylchlorvynol, Metharbital, Butalbital, Glutethimide, Methyprylon, Carbamazepine, and Mephobarbital. As with narcotics, more are being developed and eventually finding their way onto the street.

23.2.4 Stimulants

Stimulants are chemicals that stimulate the central nervous system, causing the heart rate and blood pressure to increase. They cause a feeling of well being, increased alertness, and a feeling of power. America's favorite legal drug, caffeine, is a mild stimulant. As a category, stimulants are the most dangerous of any drugs. Their use can easily lead to a fatal overdose, and many are also physically or psychologically addictive.

23.2.4.1 Amphetamines

Amphetamines are a group of synthetic stimulants commonly referred to on the street as uppers or speed. When taken in low dosages they cause decreased fatigue and decreased appetite, in addition to general stimulant effects. Therefore, most of them got their start as diet pills. The most common form is methamphetamine, also known as "meth" or "crank". Methamphetamine can be produce easily, and meth labs, wherein methamphetamine was produced used to be very common. One of the necessary ingredients for making methamphetamine is found in over-the-counter cold and flu medications, which have now been placed under higher security in many states, so that in order to purchase them one has to ask a store employee. Records are also kept of who buys these medications. Amphetamines can be injected, smoked, or inhaled. The smokable form is the newest and is often called "ice".

Amphetamines are among the most dangerous types of drugs of all. They can cause death if used in high dosages, and the long term effects on the body are grim. The cardiovascular system in particular is compromised by long term amphetamine use, and severe dental effects are a widely known consequence as well. After the effects of a large dose wear off, the user will often sleep continuously for one or two days, followed by several days of depression. In order to combat this sleepiness and depression the user will often take more amphetamine. Repeated use leads to a psychological dependence, but probably not a physical dependence.

23.2.4.2 Cocaine

Cocaine is a widely used drug in the U.S. Although use of most drugs is decreasing, the use of cocaine continues to rise. The reason for the popularity of cocaine is that it directly stimulates a pleasure center in the brain that is connected to areas responsible for emotions. Therefore, cocaine produces the most intense sensations of feeling good of any drug.

Cocaine is derived from the coca plant, native to South America. Several influential people in the 1800's praised it, including Mark Twain and Sigmund Freud. Cocaine used to be a component of Coca Cola, and some of the flavoring agents in Coca Cola still come from the coca plant.

Traditional cocaine is a white powder that is sniffed or snorted into the nose, where it is absorbed through the mucus membranes. A newer form, called "crack", freebase, or "rock" is a smokable form. Crack cocaine is described as the ultimate experience of well-being because of how quickly it stimulate the pleasure center in the brain.

Cocaine has the reputation of being harmless, but this is not true. There is a definite psychological dependence, though possibly not a physical dependence. After the rush is over, users often feel very bad, with restlessness, irritability, and anxiety. As with amphetamines, one common practice is to take more cocaine in order to combat these symptoms, which can lead to almost daily use. At high doses, cocaine can cause cardiac arrest or seizures that can lead to respiratory arrest and death.

23.2.4.3 Other Stimulants

Other stimulants include: Benzphetamine, Diethylpropion, Methamphetamine, Phenmetrazine, Caffeine (legal and unregulated), Methcathinone, Strychnine, Cathinone, Methylphenidate, Theophylline, Chlorphentermine, MDA, and Nicotinamide. As is true for narcotics and depressants, new stimulants are constantly being discovered and over time make their way into recreational usage.

23.2.5 Other Drugs

Several other types of drugs are abusable. Some of these drugs are placed in the category of "other" because they have mixed effects. Many of these drugs are illegal, or at least regulated.

The drug I abuse every day is nicotine, the drug in tobacco products. Nicotine is a stimulant, but has other properties, such as a potent antianxiety effect, which causes it to be placed in the "other" category. Nicotine can be smoked, chewed, or snorted up the nose for absorption by the mucus membranes. I prefer my nicotine chewed.

Antidepressant drugs, such as Amitriptyline, Prozac, Haloperidol, Lexapro, and many others have mood enhancing properties and also fall into this category.

Anabolic steroids cause an increase in muscle mass and other masculine characteristics. Therefore, they are occasionally abused by athletes and other sports enthusiasts. Many sporting organizations now have regular steroid testing programs for their athletes that are designed to reduce the use of these drugs.

I have placed designer drugs in the "other" category for lack of a better place for them. Designer drugs are new drugs, designed by clever chemists to mimic the properties of the better known drugs. New ones are being concocted all the time, and some of the drugs discussed above were designed in this way. In the past, designer drugs were often completely legal for a while, because the drug laws were designed as lists of prohibited substances, and if something didn't appear on the list, it was legal. Recently, however, the laws have been changed so that in order for a smoked, ingested, or inhaled substance to be legal to use it has to appear on a list of legal substances.

23.3 Drug Testing

For a variety of reasons, it may be necessary to establish whether or not a certain person has taken illegal drugs recently. If a crime is thought to be involved, then the drug testing is usually performed by a forensic toxicologist at a crime lab. For civil and regulatory purposes, commercial labs that employ forensic drug testing toxicologists will perform the test for a fee. Drugs and their metabolic byproducts, called metabolites, can be detected in urine, blood, hair, feces, and even perspiration. I will approach this overview of the technologies involved categorized by what body product is being tested.

23.3.1 Urine Testing

The most common type of drug test is the urine test. The primary target of urine testing is the popular recreational drugs, such as marijuana, cocaine, and methamphetamine, though other drugs can be detected as well.

Urine tests are usually done using an **immunoassay**. This type of test, which can be done in several ways, relies on having antibodies to drugs in question. These antibodies effectively treat the drug as an antigen and bind to it as described in chapter 19. Assuming that this occurs, you need some way to detect the reaction. The different immunoassay tests use various ways to detect a reaction. One technology is RIA (radioimmunoassay), which uses radioactive iodine that binds to the antibody to detect a positive reaction to a drug. The EMIT test (Enzyme Multiple Immunoassay Technique) uses an enzyme to do the same. The FPI (Fluorescence Polarization ImmunoAssay) method uses a fluorescing dye.

All of these tests are fairly inaccurate, and give a lot of **false positive results**. A false positive result occurs when the test says the person has used drugs when they have not. Studies show that there are up to 25 to 30% false positives. The number of false negative results is not known because negative results are not followed up on.

Many people claim to have "beaten" a urine test, by using one or more tricks that I will not discuss here.

Urine can also be tested using various types of chromatography. Gas chromatography, GCMS, HPLC, and TLC are all useful for testing urine. These tests are more accurate, with vastly fewer false positives and presumably fewer false negatives as well, but are more expensive.

23.3.2 Hair Testing

When drug metabolites are in the blood, they go through the blood vessels in the head, and are a part of the blood that nourishes the follicles that produce the hair. Drug metabolites, especially THC metabolites from marijuana, become incorporated into the hairs produced by these follicles and remain in the hair as a long term record of the drug use.

Hair testing is more expensive than urine testing and is, therefore, less commonly done. Hair tests are widely used in the casino industry, however. A typical procedure is to cut 50 strands of hair from the scalp, and send it in to the testing lab. In the lab a hair sample is dissolved in a series of solvents which extract the drug metabolites, which are analyzed via GCMS.

On the average, most peoples' hair grows about 1/4 inch per month. Typically hair tests will use only the hair one and a half inches from the scalp, which is equivalent to testing for drug use during the past 6 months. Some labs will use enough hair to test for up to a 3 year interval.

Hair testing has some problems. For one thing, what would they do with someone like me who is partially bald and keeps the rest of his hair cut very short. A similar problem will be encountered with someone who shaves their head. I don't even want to think about where they would obtain hair from for a hair test on someone like me. Some reports suggest that hair analysis is more accurate than urine analysis for detecting drug use. It is likely that hair analysis also generates a large number of false positive results.

23.3.3 Perspiration Testing

Another approach is to test for drugs or metabolites in a person's perspiration. Such products only exist in perspiration for a short time, so this test in designed to identify drug use over the period of time defined by the perspiration collection device. The typical device used to collect perspiration is a band-aid type patch, which is worn for a week or more. If illicit drugs are used during the time the sweat patch is worn, the patch will test positive when the lab analyzes it. The accuracy of perspiration testing has not been published, but there is no reason to expect it to be any better than urine testing.

23.3.4 Testing for External Residues

Yet another approach to detecting illicit drug use is to test for residual traces of drugs on the body, clothing, or possessions of a person. This is not commonly done forensically, due to the prevalence of trace amounts of drugs in the environment. For example, it is widely quoted that up to 90% of all paper currency contains traces of cocaine. Further, the technology upon which these tests are based is proprietary and therefore nothing in known about its accuracy other that what the company claims.

23.4 Questions for Study and Review

1. The U.S. Bureau for International Narcotics and Law Enforcement Affairs (BINLEA) advises various federal officials and agencies on matters related to the international drug trade and international crime. One internationally traded drug of interest to BINLEA is cocaine. Discuss whether the use of the term "narcotic" in the name of BINLEA is correct.
2. Rank the common opiate drugs from most powerful to least powerful in terms of achieving a narcotic effect.
3. How does an opioid differ from an opiate?
4. How does a hallucinogen differ from a narcotic?
5. List the forms and ways in which a drug user might get THC into their bloodstream.
6. Make a list of hallucinogens that are commonly used recreationally.
7. How does a depressant differ from a narcotic?
8. Into which class of drug does alcohol (ethanol) fall?
9. In your own words, define "tranquilizer."
10. In your own words, define "inhalant" and list some of the substances commonly used as inhalants.
11. How does a stimulant differ from a hallucinogen?
12. Discuss why amphetamines are often described as the most dangerous drugs of all.
13. Why do many cocaine and amphetamine users feel the need to use the drug often, even though the use of these drugs is unlikely to lead to physical dependence?
14. What are the effects of nicotine?
15. What are the effects of anabolic steroids?
16. In your own words, explain the concept of "designer drugs."
17. List the body products or byproducts that can be tested for evidence of drug use, and rank them in terms of reliability.
18. In your own words, define "immunoassay."
19. Discuss the problem of false positive results. How common are false positive results in commercial drug testing?

23.5 Sources

[Anonymous]. n.d. Top 20 Drugs and Their Street Names [Internet]. [cited 2007 Oct 23]. Available from:
http://www.casapalmera.com/articles/top-20-drugs-and-their-street-names/.

Batchelder T. 2001. Drug Addictions, Hallucinogens and Shamanism: the View from Anthropology. Townsend Letter for Doctors and Patients p74(4).

Cohn JE. 1989. Narcotics: a primer. Current Health 2, a Weekly Reader publication 15(7): 11(3).

Cole SO. 2005. An Update on the Effects of Marijuana & its Potential Medical Use: Forensic Focus. The Forensic Examiner 14(3): 14(10).

Gombos J. n.d. Drug Testing FAQ [Internet]. [cited 2007 Oct 23]. Available from:
http://www.urban75.com/Drugs/testing2.html.

Goldstein A, Kalant H. 1990. Drug policy: striking the right balance.Avram Goldstein and Harold Kalant. Science 249(4976): 1513(9).

Hanwell D. 2007. Police Practice: Cooperative Investigations of Methamphetamine Laboratories [Internet]. FBI Law Enforcement Bulletin 76(8). [cited 2007 Oct 19]. Available from:
http://www.fbi.gov/publications/leb/2007/august07/august07leb.htm#page18.

McElhatton PR. 2000. Fetal Effects of Substances of Abuse. Journal of Toxicology: Clinical Toxicology 38(2): 194(2).

Musshoff F, Driever F, Lachenmeier K, Lachenmeier DW, Banger M, Madea B. 2006. Results of Hair Analyses for Drugs of Abuse and Comparison with Self-reports and Urine Tests. Forensic Science International 156(2-3): 118(6)

Proctor, Alissa, Dale, Mike, and Willimans, Joel, 1998. Evidence: The True Witness. Forensic Toxicology [Internet]. [cited 2007 Oct 23]. Available from:
http://library.advanced.org/17049/gather/cgi-bin/document_get.cgi?path=/Toxicology/00Intro.

Scientific Working Group for the Analysis of Seized Drugs. 2005. Methods of Analysis/Drug Identification [Internet]. Forensic Science Communications 7(1). [cited 2007 Oct 21]. Available from:
http://www.fbi.gov/hq/lab/fsc/backissu/jan2005/standards/2005standards11.htm.

Chapter 24
DEATH INVESTIGATION

This chapter begins a series of four chapters on the investigation of deaths. In this chapter we will explore the process of death investigation and the people involved in it, especially coroners and medical examiners. We will pay particular attention to those questions that must be answered on a death certificate: cause, manner, circumstances, and time of death. Chapter 25 will continue this theme with an examination of forensic pathology and autopsies. Chapter 27 presents forensic anthropology, including the recovery of buried bodies and the identification of deceased people from their skeletal remains. In chapter 28 we will take a look at forensic odontology and how deceased people are identified from their dental remains. Further, we examined postmortem forensic toxicology in chapter 22, and part of chapter 29 will be devoted to forensic entomology as a method for estimating time since death.

The chapter will begin with a discussion of the participants in a death investigation, particularly coroners and medical examiners. We will explore the important questions in a death investigation, then move on to a more detailed treatment of coroners and medical examiners. The chapter will conclude with a look at a "mixed system" state, in which both coroners and medical examiners are found.

24.1 Participants in a Death Investigation

Many people, both peace officers and forensic scientists, participate in a death investigation. However, the two principle characters that I want to focus on in this chapter are the coroner and the medical examiner. The main distinction between these two investigators is that a coroner is usually a senior detective with vast experience in investigation. The medical examiner, on the other hand, is a medical doctor with vast medical knowledge of how people die.

Some states have coroners, some have medical examiners, and some have both. In 2005, 22 states and the District of Columbia had only medical examiners, 10 states had only coroners, and 18 states had both. Many people consider the coroner system to be outdated and favor abolishing it in favor of a nationwide system of medical examiners. The reasoning behind this is sound as far as it goes. Cause, manner, and timing of death are medical matters and therefore the person ruling on them should be a medical doctor, preferably a pathologist. Other people favor keeping the coroner on in some capacity, such as being the detective in charge of death investigations. The trend, nationally, has been to replace coroners with medical examiners

Another observation of note is that coroners are usually county officials, whereas medical examiners are usually state level officials. Therefore, in states where coroners are replaced by medical examiners, there is a shift from county centered death investigation based on general police investigative methods to a state led death investigations based more on medical and scientific principles. There are advantages and disadvantages to making this transition, and it is difficult to say which system is better. I live in Montana which has retained the older system of county level investigations led by a coroner, but added a state medical examiner who interacts with the coroners in a variety of complex ways in a death investigation. As I will discuss further below, this may be the best system.. Figure 24.1 shows which states have coroner only systems, which have medical examiner only systems, and which have mixed systems of death investigation.

Figure 24.1: States with Medical Examiner Only, Coroner Only, and Mixed Systems of Death Investigation

■ Medical Examiners

□ Coroners

□ Both (Mixed System)

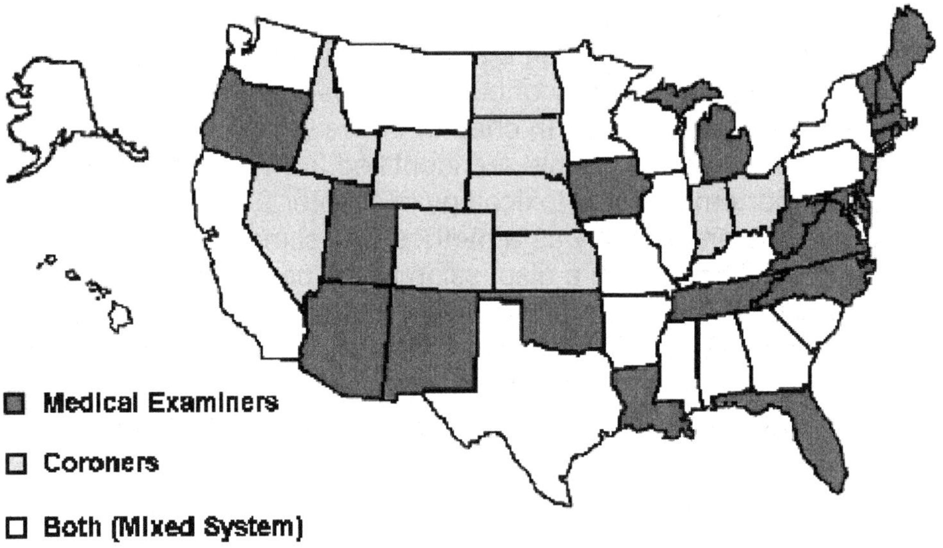

24.2 Important Questions in a Death Investigation

The important questions in a death investigation are those that relate to how the death occurred and the legal status of the death. These questions must be answered on a death certificate for the deceased person. Almost universally these questions relate to the cause of death, the manner of death, the circumstances of the death, and the time of death.

These questions have to be answered in order to complete a death certificate. The death certificate is a legal document that lists the time, date, place, cause, and manner of death, along with identifying information such as the name, age, and sex of the deceased person. If any injury was involved with the death, its details are recorded. If any surgery or medical treatment occurred around the time of death, it will also be recorded. If an autopsy was performed, that fact will be recorded as well. If the death occurs in a hospital or other institution with medical doctors on staff, the death certificate will most often be signed by a physician. In all other situations it is most often signed by the coroner or medical examiner. A death certificate is required by a funeral home prior to burying or cremating a body, and is also filed with county and state vital records departments.

24.2.1 Cause of Death

There are many causes of death. Heart attack, cancer, electrocution, drowning, gunshot injury, crushing injury, and poisoning, among thousands of others are all causes of death. The **cause of death** is what caused the heart to stop beating and the brain's electrical activity to cease.

Sometimes, the cause of death relates to an event that occurred in the recent or even distant past. For example, if a person was injured in a traffic accident, which caused them to be a quadriplegic, and after sitting in a wheelchair for many years the person developed bed sores that became infected and killed the person, the original traffic accident is as much a part of the cause of death as the infection. The cause of death may be listed on the death certificate as sepsis of decubitus lesions due to quadriplegia due to a traffic accident in 1980.

Cause of death is usually determined from a variety of types of evidence. Testimonial evidence is important, and can consist of witnesses statements, a suicide note, testimony by acquaintances that the person was a drug courier, and similar information. Examination of the body is also important in establishing cause of death and this often means performing an autopsy. Autopsies will be discussed in more detail in the next chapter.

24.2.2 Manner of Death

In contrast to the nearly infinite number of ways to die, the official status of the death must fall into one of five or six categories called the **manner of death**. Five of these manners of death are universal and are: homicide, suicide, accident, natural, and undetermined. Homicide is the manner of death when a person's death was caused by another human being, intentionally or through negligence. Suicide is the manner of death in which a person takes their own life. Accident is the manner of death when the death was due to an act of God or nature (being stuck by lightning, being mauled by a bear, being crushed by a falling tree, etc.), or when the death was due to unforseen and unpreventable circumstances. Natural death is death due to a disease or other pathological condition, including simple old age. In general, a death can not be considered natural if there is any wounding or other trauma involved with it. Some jurisdictions add therapeutic complication as a manner of death for those situations in which a person dies as a result of a medical procedure designed to save or improve their life. If the manner of death can not be unambiguously assigned to one of these categories, then it must be listed as undetermined. Having to list the manner of death as undetermined is something that coroners and medical examiners strive to avoid, but sometimes it is unavoidable.

The accidental category of manner of death is perhaps the most problematical in some ways. Causes of death such as falling down stairs may have to be investigated closely because the person might have been pushed or might even have committed suicide. When the actions of another human are involved, then the nature of these actions and their motive have to be taken into account. For example, if someone was shot by a hunting buddy, was the shooting unintentional, was it due to the shooter's negligence, was there a defect in the gun that caused it to fire unexpectedly, or was the shooting intentional. Some of these factors lead to a ruling of accidental and others lead to a ruling of homicide (either negligent or deliberate). Deaths resulting from traffic accidents also warrant serious consideration. If only a single vehicle was involved, and it hit an icy patch, then the manner of death is accidental. Even in this case, however, homicide by sabotaging the car and suicide have to be ruled out. If two cars are involved then the condition of the other driver and the events leading up to the accident need to be examined. Were one or both of the drivers intoxicated? Were one or both of them negligent? If the other driver was intoxicated or negligent the manner of death may have to be listed as homicide.

It is the manner of death that determines whether further investigation will be done. If the death is natural or accidental there is nothing further to accomplish by investigation. Similarly, therapeutic complication provides no grounds for a criminal investigation, though a civil suit may be filed against a doctor or hospital. Suicide is the most sensitive manner of death, because many life insurance policies will not pay in the case of suicide. However, suicide is another manner of death that does not call for further investigation, because the perpetrator is dead. Homicide is the manner of death that leaves the door open for further investigation. If the death was caused by another person, then efforts should be made to identify and punish that person.

24.2.3 Circumstances of Death

The **circumstances of death** are the miscellaneous details of the situation in which the death occurred, such as whether the person was alone when they died, what attempts were made to save the person, who discovered the body, what weapons were present, whether drugs or drug paraphernalia were present, and similar details.

These circumstances are often important in deciding on the manner of death. For example, if a person is found dead of a gunshot wound to the head, there was a 38 caliber revolver present, no other people were in the room at the time, and there was a note nearby written by the deceased in which they say goodbye to all their friends, then the likely manner of death is suicide. However, if we have the same circumstances with the exception of removing the note and replacing it with another person in the room, then homicide has to be considered carefully as a possible manner of death.

24.2.4 Time of Death

Time of death seems self-explanatory. It is the time that the person died, determined as accurately as possible. Time of death estimations are based on a combination of testimonial evidence and changes in the appearance and characteristics of the body after death. The most recent time the person was seen alive is an important bit of testimonial evidence. There are a few physical changes that a body undergoes after death which provide evidence that can be used to estimate the time of death. They include the following.

- **Livor mortis** is discoloration of the body due to settling of blood in the lower portions of the body. This occurs within a few hours of death. The word "livor" is amusing to me in that it can be pronounced either "liver" as in the organ, or "live or" as in "alive or not". Whichever way you pronounce it in the presence of another forensic scientist they will say, "Do you mean livor mortis?", giving the other pronunciation. I have repeated this experiment 22 times as I write this book, including alternating my pronunciation in the presence of the same forensic scientist, who will invariably respond with the alternative pronunciation, whatever my initial pronunciation was. People are funny sometimes, and forensic scientists have an especially interesting culture of their own.
- **Rigor mortis** is stiffening of the body after death. Rigor mortis occurs as proteins that normally keep the muscles from stiffening get used up. The timing of the onset of rigor mortis depends on several factors, but on the average begins about 3 or 4 hours after death and the body becomes completely stiff about 12 to 48 hours after death. Over the course of several days, depending on conditions, rigor mortis will disappear as the proteins in the muscles decompose.
- **Algor mortis** refers to change in body temperature after death. Upon death the body begins to cool from its normal approximately 37 degrees Celsius to room temperature. Traditionally, the rate of cooling has been said to be about 1.5 degrees Celsius per hour. However, the size of the body, the external temperature, how well the body is insulated, and a variety of other factors have to be taken into account.
- Other useful indicators of time since death include changes in the chemical composition of various body liquids and other decompositional changes.

In general, the longer the time since death, the less accurate the time of death determination. A good rule of thumb is that the methods and indicators described above work best in the first approximately 72 hours after death. If the time since death is longer than 72 hours, it is likely that a forensic entomologist can generate a more accurate time since death based on insect infestation of the body. After the body becomes completely skeletonized, time since death determination relies on the expertise of forensic anthropologists, usually resulting in extremely crude estimates.

24.3 Coroners

In the traditional American system of justice the person who handles death investigations is a **coroner**. The nature of coroners varies between jurisdictions, often within the same state for those states that have them. My home state of Montana is typical in that the coroner is a county level elected official. In some counties, the coroner is the same person as the sheriff, while other counties have separate offices for the two. One thing that confused me when I first began working on forensic anthropology cases is that the person who acts as the county coroner, and often is called the county coroner, is not actually the Coroner. This arises from the fact that the

Coroner or Sheriff/Coroner is a very busy person, who doesn't have the time to become personally involved in every death investigation. Therefore, he or she will appoint deputy coroners from among the detectives, who are actually the people who will do the investigations. What I found is that the person who was presented to me as the "county coroner" actually had a title something like "**chief deputy coroner**". Although confusing, this is not actually misleading, because the chief deputy coroner is, for all intents and purposes, acting as the county coroner.

24.3.1 The History of Coroners

The office of Coroner is over 800 years old. It was first created in England in the year 1194 by a body of law titled the Articles of Eyre. These first coroners were elected and each county had three knights and one clerk who served as coroners. Coroners today are still elected officials, as they were in the beginning, though there tends to be only one of them per county, rather than four. The original title of the coroner was "Custos Placitorum Coronae" which translates as keepers of the pleas of the crown. Their main purpose seems to have been to guard the King's financial interests from corrupt local officials, including the sheriff. Therefore, the modern tendency for the sheriff and coroner to be the same elected official is a difference from the original office. The original duties of coroners included not only death investigations but a variety of other things in which the King stood to gain financially. At this period in the history of England the ruling class was the Normans and the underclass was the Anglo-Saxons. One of the primary tasks of coroners was to determine whether a murder victim was Norman or Anglo-Saxon. If Norman, not only was the murderer subject to stiffer penalties, but the community in which the murder occurred was required to pay a fine, called a "murdrum".

Between the 1200's and the 1500's the powers and duties of the coroner were transferred to other public officials, with the exception of death investigation, which remained the task of the coroner as it still remains today in those American states that have them.

24.3.2 Requirements and Duties of the Coroner

The various states have somewhat differing requirements for the coroner. In most states, there are no specific requirements of a coroner, since they are elected officials. Upon election, however, they are required to take coroner training. In reality, the "real" coroners, or chief deputy coroners are chosen for their investigative skill. Louisiana is unusual in that they retain the office of the coroner, but the coroner has to be a medical doctor. In my opinion, these "coroners" are actually medical examiners, and therefore I classify Louisiana as a medical examiner only state instead of a coroner only state, which is its official designation.

Similarly the various states define the duties of the coroner differently, with the single common duty being to perform inquests. **Inquest** is the term used for an official death investigation. In some jurisdictions an inquest is conducted like any other investigation. In other jurisdictions, however, an inquest takes on a more court-like atmosphere, with formal hearings presided over by the coroner, subpoenaed witnesses, and expert testimony. One thing that is widely misunderstood about an inquest is that it is not a trial. The focus of an inquest is not to show the guilt or innocence of a suspect, or really even to identify a suspect at all. Rather, the goal is to answer the fundamental questions about the death itself, which must be entered in some form or another on a death certificate. These questions are the cause, manner, circumstances, and time of death as discussed above. The determination of innocence or guilt of a suspect is made during a trial in a court.

24.4 Medical Examiners

In contrast to a coroner, who is usually a detective, a **medical examiner** is a medical doctor, usually a forensic pathologist. I will speak more about forensic pathology in the next chapter. Being a medical doctor, the medical examiner is often more qualified than a coroner at determining the medical facts surrounding the death. Surprisingly, one state, Minnesota, does not require their medical examiners to be medical doctors. In my opinion this makes them coroners with a strange name. Minnesota is listed as a state with both coroners and medical examiners, and I have counted it as such even though it may be questionable whether the Minnesota "medical examiners" are actual medical examiners in the conventional sense.

In contrast to coroners, who are elected, medical examiners are usually hired or appointed by a senior state official on the basis of their qualifications. For example, Montana's State Medical Examiner is appointed by the State Attorney General.

24.4.1 The History of Forensic Medicine

It is likely that the origins of forensic medicine are ancient. Anthropologists observe that in traditional hunting/gathering societies, such as those the ancestors of all of us lived in until the invention of domestic plants and animals about 10,000 years ago, that the shaman (often referred to as medicine man or witch doctor) was often both the group's healer and the group's keeper of tradition and law (or at least social rules). Archaeological research has shown that as people entered into civilization the priesthood was usually charged both with keeping records of the law and with healing. Over time, healers became more secular, but still probably contributed to some death investigations.

The code of Hammurabi of Babylon, dating to about 2200 BCE, is regarded as the oldest written body of law. It defines the the rights and duties of medical practitioners as regards the enforcement of law. The ancient civilizations of India and China also recognized what we would regard as forensic medical practitioners today. Of the ancient civilizations, Greece appears to have been an exception in the acceptance of legal medicine. Because dead bodies were regarded as sacred, autopsies were forbidden, but over a five year period ending in 35 BCE. the "father of medicine", Hippocrates, spoke extensively on the legal responsibility for wounds and their relation to death. This attitude toward the "sacred dead" carried over into Roman times as well. However, in the assassination of Julius Caesar, a post mortem examination was performed on his body. The physician who conducted the examination concluded that only one of Caesar's twenty-three wounds was fatal.

Ironically, it was the "barbarian" Germanic and Slavic peoples which overthrew the Roman Empire in western Europe in the 400's C.E. who were the first in the Western tradition to require expert medical opinion in cases of deaths. It was during this era that the importance of medicine in civil proceedings became established as well, because an individual found guilty of injuring another was obliged to pay a fine, the amount of which depended on the nature and severity of the wound, as testified to by a medical expert.

In the late 1500's a French surgeon, Ambroise Pare, wrote a book about the contribution that medical experts could make to cases involving death or injury. Paulus Zacehias (Italian) and Samuel Parr (English) also wrote books relating to this subject, Zacehias in the middle 1600's and Parr in the 1700's. Despite their enlightened views, during much of the history of medieval Europe the medical contribution to death investigation consisted primarily of superstitious and magical practices. For example, King James the VI of Scotland (who also commissioned the King James translation of the Bible) approved of a method of testing murder suspects in which they were required to touch the corpse of the victim. It was believed that if that person was the murderer, then blood would flow from the victim's wounds during this touch.

As we have seen for other forensic sciences discussed previously, forensic medicine as we know it today is primarily a product of the late 1800's and the 1900's. Thoughout the 1800's and 1900's, and continuing on into the present century, forensic medicine and forensic pathology have developed the scientific aspects of investigating death and injury. Advances in medicine in general, new types of diagnostic equipment, and the emergence of forensic toxicology have all had an impact on forensic medicine and forensic pathology. Many people and many advancements in medicine have brought us to the medical examiner of today, as well as to those forensic medical experts who primarily render their opinions in civil court.

Images of forensic medical doctors are common in popular media, ranging from Sherlock Holmes' sidekick Dr. Watson, through the Quincy TV program of the 1970's, to agent Scully of the X Files. They are also common in more recent movies and TV shows.

24.5 An Example of a Mixed System State: Montana

Earlier in this chapter I asserted that the mixed system of death investigation, which includes both coroners and medical examiners, was perhaps the best system of all. I say this after nearly 20 years' experience with Montana's mixed death investigation system, working both with county coroners and with the State Medical Examiner (and more recently the Associate State Medical Examiner). Montana seems typical of a mixed system state, and I will use it as an example.

24.5.1 Powers of the Coroner

In Montana the coroner is an elected official charged with several responsibilities, some or all of which may be delegated to deputy coroners. I find it interesting that in Montana state law, coroners have "powers" while medical examiners have "duties". Some of the powers of coroners as set forth in state law are as follows.
- Pronounce the fact of death of a human being;
- Certify and amend death certificates;
- Issue subpoenas for inquest proceedings;
- Order autopsies;
- Conduct examinations and tests necessary to determine the cause, manner, and circumstances of death and identification of a dead human body;
- Order a dead human body to be disinterred or removed from its place of disposition;
- Conduct inquests;
- Order cessation of any activity by any person or agency, other than the law enforcement agency having jurisdiction, that may obstruct or hinder the orderly conduct of an inquiry or the collection of information or evidence needed for an inquiry;
- Seize and preserve evidence, ask for a search warrant, subpoena witnesses and documents, and direct experts to perform tests.

24.5.2 Duties of the Medical Examiner

In Montana we have a State Medical Examiner and an Associate State Medical Examiner. The duties of the State Medical Examiner are specified in state law as follows.
- Provide assistance and consultation to associate medical examiners, coroners, and law enforcement officers;
- Provide court testimony when necessary;
- Stimulate and direct research in the field of forensic pathology;
- Maintain an ongoing educational and training program for associate medical examiners, coroners, and law enforcement officers;
- Appoint associate medical examiners;
- Perform autopsies as requested.

Other sections of state law give the State Medical Examiner jurisdiction over a case in which the Coroner has not stepped forward to fulfil his or her duty, and the power to collect evidence and specimens for educational purposes.

24.5.3 How the Mixed System Works

I have found that the coroners of the 56 counties in Montana and the State Medical Examiner work together well in most cases. It remains the county coroner's responsibility to investigate and rule on cause, timing, and manner of death, but the State Medical Examiner provides critical information upon which those determinations are made, primarily through the results of autopsies. In many ways this is an ideal system. The coroner is a detective and is familiar with the process of conducting an investigation in which the various forms of evidence collected are used to build a case that is presented in court. The medical examiner is not normally an experienced investigator of this sort, but is, of course, an expert on the scientific methods of establishing cause of death. The mixed system works well in that the coroner uses the information provided by the medical examiner as one line of evidence in the overall case. Each person gets to do the work they are best qualified for, and this can not help but lead to better investigative results.

24.6 Questions for Study and Review

1. Compare and contrast coroners with medical examiners.
2. Describe the nature and function of a coroner in most jurisdictions of the U.S.
3. Describe the nature and function of a medical examiner in those jurisdictions in which there is no coroner.
4. Describe the nature and function of a medical examiner in those jurisdictions that have both a coroner and a medical examiner.
5. List the important questions in a death investigations (i.e. those for which something must be written on a death certificate).
6. What is the most common method used by a forensic scientist to determine cause of death?
7. List the five manners of death and give a brief definition for each that distinguishes it from any other manner of death.
8. Cases in which a person takes too large a dose of a drug and dies from an overdose are sometimes considered difficult to classify as to manner of death. Cases of this sort have been listed on death certificates as suicide, "involuntary suicide", and accidental. The modern consensus is that the correct manner of death is accidental. Discuss why suicide and "involuntary suicide" are not acceptable manners of death in most cases of this sort. Why is "involuntary suicide" a contradiction in terms?
9. Describe algor mortis, livor mortis, and rigor mortis; and discuss how these changes are used to estimate time since death.
10. Compare and contrast coroners and medical examiners in terms of the qualifications for the office and how one becomes a coroner or medical examiner.
11. In your own words describe an inquest and how it differs from a trial.
12. Describe your state's system of death investigation. Are there coroners, medical examiners, or both?

24.7 Sources

[Anonymous]. n.d. The Coroner - A Historical Sketch [Internet]. [cited 2007 Oct 23]. Available from: http://www.columbianacounty.org/Coroner/History.htm#The_Coroner_-_A_Histori cal_Sketch_.

[Anonymous]. 1993. The first postmortem recorded in the country. The Journal of the American Medical Association 270(16): 1891(1).

Cohen JI. n.d. General Information for Individuals [Internet]. [cited 2007 Oct 23]. Available from: http://www.forensiconline.com/generallink.htm.

DiMaio VJM, Hanzlick R. 1997. Medical examiners, forensic pathologists, and coroners. The Journal of the American Medical Association 277(7): 531(2).

Edwards JB. Homicide InvestigativeStrategies [Internet]. FBI Law Enforcement Bulletin 74(1). [cited 2007 Oct 19]. Available from: http://www.fbi.gov/publications/leb/2005/jan2005/jan2005.htm#page11.

Hanzlick R, Combs D. 1998. Medical examiner and coroner systems: history and trends. The Journal of the American Medical Association 279(11): 870(5).

Hanzlick R. 2006. Medical examiners, coroners, and public health: a review and update. Archives of Pathology & Laboratory Medicine 130(9): 1274(9).

Harris County Medical Examiner's Office. n.d. So You Want to be a Medical Examiner [Internet]. [cited 2007 Oct 23]. Available from: http://www.co.harris.tx.us/me/Medical.aspx#SectionViewer1_section143.

Kesselring K. 2006. Detecting 'Death Disguised'. History Today 56(4): 20(7).

National Association of Medical Examiners. 2007. General FAQs [Internet]. [cited 2007 Oct 23]. Available from: http://thename.org/index.php?option=com_content&task=category&sectionid=3&id=7&Itemid=42

Plunkett J, Thomas LC. 1998. Coroner and Medical Examiner Systems. The Journal of the American Medical Association 280(4): 325(1).

Ubelaker DH, Buchholz BA. 2006. Complexities in the Use of Bomb-Curve Radiocarbon to Determine Time Since Death of Human Skeletal Remains [Internet]. Forensic Science Communications 8(1). [cited 2007 Oct 21]. Available from: http://www.fbi.gov/hq/lab/fsc/backissu/jan2006/research/2006_01_research01.htm.

24.8 Acknowledgements

The map clipart upon which Figure 24.1 is based is courtesy of Graphic Maps.com, http://graphicmaps.com/clipart.htm.

Chapter 25
FORENSIC PATHOLOGY

This chapter will continue the theme of death investigation with a consideration of forensic pathology. First we will situate forensic pathologists within the framework of the medical field and discuss the training of these most highly educated forensic scientists. We will examine how forensic pathologists use autopsies to determine cause of death, and how autopsies are performed. The chapter will conclude with an introduction to the subject of identifying unknown dead people, which will be continued in the following two chapters.

25.1 Pathology and Forensic Pathology

Forensic pathology is considered a part of forensic medicine. Forensic medicine is a larger field that also includes forensic podiatry, forensic chiropractic, forensic orthopedics, and a variety of other specialties. The role of those who practice forensic medicine is to give expert opinions in court. As discussed previously, most of these experts typically testify in civil court, where they give opinions about the nature of the injury or illness that is the issue in a law suit. Forensic pathologists and forensic podiatrists are the only members of this group more often found in criminal court.

Let's approach the subject of what a forensic pathologist is and does by first looking at what a general pathologist is and does. Then we will see how this area of expertise is applied to criminal investigations. The history of forensic pathology is part of the history of forensic medicine as presented in chapter 24.

A **pathologist** is a type of medical doctor who specializes in why people become sick and why they die from their illnesses. You may be thinking the question, "Don't all doctors do that?" Actually, most physicians focus on the management of symptoms – not on determining the underlying causes of illnesses. When you or I go to a doctor we want a treatment for our symptoms, with considerations like the actual cause of the condition being secondary. Sometimes, the doctor won't actually know what is causing the symptoms, say a bad cough, but does know that certain treatments are effective for those symptoms, whatever their cause. Most of us encounter a pathologist in their role as diagnostician, in cases of a serious condition that has not responded to ordinary symptom management strategies. In this case, special tests may be conducted and a pathologist studies the case to pinpoint the underlying cause so that it can be treated appropriately.

The non-forensic types of pathologists most often work in a hospital, but a substantial number work for private laboratories and for a government or private research institution.

There are two main branches of pathology - anatomic and clinical. Let's examine these briefly.

25.1.1 Anatomic Pathology

Anatomic pathology encompasses surgical pathology, autopsy pathology, and diagnostic cytology. Surgical pathologists examine organs and tissues, usually recovered by a biopsy or during surgery, for the purpose of making a diagnosis of the patient's illness. Autopsy pathologists perform autopsies on individuals who have died from their illness in order to determine the cause of death and to answer questions raised by their physicians and family members. This autopsy is usually performed in the hospital, and the manner of death is most often clearly natural causes or therapeutic complication. In cases where the manner of death is not clearly one of these two categories, a forensic pathologist may be asked to perform the autopsy. Cytologists examine the cells from body tissues under the microscope, generally in order to diagnose them as cancerous (malignant) or non-cancerous (benign).

25.1.2 Clinical Pathology

Clinical pathology includes specialties usually performed within a hospital's laboratory. These specialties include hematology, transfusion medicine, microbiology, immunology, clinical chemistry, and toxicology. Most clinical pathologists specialize in only a few of these areas, or even just one, because there is simply too much to know for one person to be an expert at everything. The director of the hospital laboratory is commonly a broadly trained and experienced clinical pathologist as well, which qualifies her or him to oversee the analyses of the other pathologists.

Hematologists specialize in blood and diseases of the blood. Those who specialize in transfusion medicine are also interested in blood, but focus on issues of blood transfusions rather than diseases of the blood. Microbiologists are experts in the identification of disease causing microorganisms, such as bacteria and viruses. Immunologists specialize in the workings of the immune system and its disorders. Clinical chemistry is the branch of clinical pathology that specializes in performing and interpreting the blood chemistry tests and other similar tests that a doctor may request for their patient. Generally, most of the routine chemistry tests are performed by a technician, but the clinical chemist is there to oversee the analysis and perhaps to conduct some of the more demanding tests. Toxicologists of the pathologist type are similar to the toxicologists discussed in chapter 22. Their specialty is poisons and other foreign substances that might be present in the body.

25.1.3 Forensic Pathology

Some authorities consider forensic pathology to be a branch of anatomic pathology, and this makes sense given that this branch includes autopsy pathology, and forensic pathologists perform a lot of autopsies. However, most forensic pathologists also receive training in the clinical branch of pathology. Forensic pathology is probably best thought of not as a branch of pathology, but as the application of the field of pathology to the system of justice, primarily to answer the question of what caused a person to die. The forensic pathologist's main tool is the autopsy, which is performed to determine the cause, and which might also yield an obvious manner of death.

25.2 The Education of Forensic Pathologists

In general, forensic pathologists are the most highly educated of forensic scientists, though some of the university oriented forensic scientists may have as much of more education. The education of a forensic pathologist begins with a four year bachelor's degree with a focus on the pre-medical curriculum. The student's major at this point is not terribly important, though I observe that most pre-med students major in biology. However, most of the other scientific fields, including chemistry, physics, psychology, and even anthropology, attract some students who intend to go on to medical school.

The next step in the education of a forensic pathologist is medical school, which is typically a four or five year program of study. After graduation from medical school, the person has an M.D. degree and is considered a doctor. However, all physicians who wish to practice medicine must specialize in one of the many branches of medicine by doing a residency in that specialty. A residency normally lasts three to five years, and can be regarded as a form of apprenticeship in which the young doctor works with established doctors to learn those practical aspects of their branch of medicine that weren't taught to them in medical school. In most cases, the aspiring forensic pathologist will do a residency in both anatomic and clinical pathology, which means that this phase of their education is more likely to take five years than three. After completing the residency, a forensic pathologist undertakes a specialized training program in forensic pathology that lasts at least a year, in which he or she will work under the direction of an established forensic pathologist, usually in a large city.

Let's total this up. Four years for the bachelor's degree, four years for medical school (minimum), five years for the combined anatomic and clinical pathology residence, and another year for the specialized training in forensic pathology. According to my calculator, this totals up to 14 years of schooling, which beats my 10 years of formal education to receive a Ph.D. in physical anthropology by a considerable margin. My hat is off to the forensic pathologists, and they deserve every penny of the relatively large salaries they receive.

But wait, there's more! In order to be seriously considered for a job, the forensic pathologist has to become "board certified". This means that they present their credentials to the American Board of Pathology and take a set of examinations. If they pass, they are certified as knowing forensic pathology.

25.3 The Practice of Forensic Pathology

Forensic pathologists have a variety of types of work they perform. One of the main distinctions is between forensic pathologists who work at a crime lab and forensic pathologists who work for independent labs.

Forensic pathologists who work for an independent lab most often function as consultants who give second opinions about a death that has already been investigated. If a family is convinced that the cause and manner of death have not been determined accurately, they can hire an independent forensic pathology to review the evidence, make additional tests, or whatever needs to be done to arrive at the best possible characterization of the cause and manner of death.

Crime labs do process a variety of cases that don't involve deaths, and therefore don't involve the forensic pathologist. When it comes to death investigation, however, I observe that the forensic pathologist takes on a leadership role. Not only is he or she doing their own examinations, but they are also coordinating various related analyses being done by other forensic scientists working at the crime lab and elsewhere, especially toxicologists, serologists, anthropologists, and odontologists. These scientists supply information to the forensic pathologist to help him or her make a scientific decision about the death.

25.4 Autopsies

The primary way that a medical examiner determines cause of death is by conducting an autopsy. An **autopsy**, also known as a post-mortem examination, is an examination of the body after death. It usually consists of an external and an internal examination of the body. The purpose is to identify and document any natural disease processes and/or injuries which may have contributed to the death of the individual.

In most jurisdictions an autopsy is required, at the discretion of the County Coroner or the Medical Examiner when the death is "unattended". An death is **unattended** when it occurs outside of a hospital, nursing home, or other facility overseen by medical doctors, and the person has not been examined by a medical doctor in some other context recently enough that the cause of death is clearly some natural disease process. When autopsies are performed, the cause of death is often the one suspected, but in a surprisingly large number of cases, the true cause of death turns out to be something nobody expected.

An autopsy takes between 30 minutes and several hours, with an average of around two to three hours. They are performed at a variety of places including the crime lab's morgue, a police morgue, the coroner's office, a hospital, or even a funeral home.

25.4.1 External and Internal Examinations

An autopsy usually consists of both an external and internal examination of the body, although there are cases where the external examination is sufficient. The external examination documents identifying features such as scars, tattoos or other markings which may assist in the identification of the body. Any external wounds or other evidence of cause of death that are visible on the outside of the body are also recorded and described. If there is a question about the identity of the body, fingerprints and samples for DNA analysis are usually taken.

During the internal examination the internal organs are examined, sampled, and weighed to document any natural disease processes and/or injury. The internal examination is performed in a standard manner, regardless of who is performing it or where it is performed. A large Y-shaped incision is made from each shoulder to just below the navel. Then the organs are removed from the chest and abdomen, examined, and samples taken for further evaluation. Organs not used for additional study are returned to the body for burial. If the brain is to be examined, an incision is made along the back of the head, the skull is opened using a specialized saw, and the brain removed. After the entire procedure is completed, the body is stitched closed. Although this procedure sounds extremely invasive, it doesn't disfigure the face and does not interfere with embalming the body. The damage to the body is such that with proper preparation it may still be displayed in an open-casket funeral.

Samples of organs and other specimens collected at the time of autopsy may be examined further, either by the forensic pathologist or by another forensic scientist. Collection of samples of body fluids and tissues for toxicological analysis is routine.

25.4.2 Other Benefits of Autopsies

Historically, human tissue samples removed during autopsy have helped advance medical knowledge. For example, pathologic observations helped confirm the link between cigarette smoking and lung cancer, and the nature of diseases like heart disease and viral hepatitis. Autopsy studies also contributed to an early medical understanding of Legionnaire's disease and toxic shock syndrome. Researchers also have learned a great deal from autopsies about the effects of toxic chemicals and industrial hazards on the body.

Autopsies contribute to developing accurate statistics about causes of death, prevalence of illnesses, and to an understanding of trends in the increase or decrease of these things. Accurate death statistics are important because they are used in governmental decision making and influence the amount of money allocated for research and health care. They also are used in the formulation of safety standards and rules, and in the design of medical equipment.

Further, autopsies contribute to the education of medical doctors. Even with the best quality visual aids and imaging equipment there is simply no substitute for seeing the effects of an illness in the tissues of a cadaver.

25.4.3 Trends in Autopsies

The trend through time is for fewer autopsies to be done. In the 1950's about half of all people who died were autopsied, whereas in the 2000's the percentage has fallen to around 10%. Due to the decline in the percentage of autopsies performed, many hospitals, especially small or rural facilities, don't even provide facilities for them anymore. Autopsy rates at medical teaching and research institutions tend to run higher. For example, the Mayo Clinic performs autopsies on about a third of the people who die at its hospitals.

The declining autopsy rate is taken as good news by some people and regarded with dismay by others. There are multiple causes of this trend, but the most important is likely to be the fact that more than 80 percent of Americans die in hospitals, nursing homes, or other health care institutions. An autopsy is usually not required under these conditions because diagnostic tests done during the life of the patient establish the cause of death so well that an autopsy is unnecessary. Modern diagnostic technology includes several sophisticated types of imaging devices that can be used to peer into the body to detect the effects of illnesses or injuries. If these have been used before death, the need for an autopsy is less clear.

There is also a sociocultural aspect to this trend. Modern America is experiencing a resurgence of unscientific, pseudoscientific, and religious beliefs and attitudes that are expressed in everything from the health food movement to belief in abduction by extraterrestrials. One aspect of this is a re-emergence of the idea of the "sacred dead", whose bodies should not be mutilated to serve justice, or even examined closely. As an experiment, do a Google search for "sacred dead" (with the quotes) to see how many times this exact term shows up on the internet and the vast number of different contexts in which it appears. A search through the known religions of all the regions of the world shows that only a tiny number of actual religions have ever held dead bodies to be sacred, though a substantial number consider dead bodies to be dangerous (I'm willing to take the risk). Most authentic religions agree with scientific observations that a dead body is merely an empty shell, devoid of a soul or whatever important part that once gave it life.

Personally, I believe that the fault for this lies more with the media that anywhere else. On my satellite TV system I get several channels that purport to be scientific, yet the content of most of their programming is anything but. When people watch a TV program about the lost city of Atlantis, which portrays it as an actual reality, it is no wonder that people don't know the difference between what is grounded in genuine knowledge and what is based on the personal beliefs of some marginal group.

Whatever the cause, the widespread sentiment that the dead are sacred and taboo results in the fact that, in many cases, families are reluctant to ask for or permit an autopsy. In other cases, the families consider the procedure degrading or feel that the deceased has "suffered enough." Others may worry that an autopsy will delay or otherwise adversely affect funeral plans. Worries about the cost of an autopsy also may have contributed to the decline. If a death occurs at a hospital, especially at a teaching facility, there is usually no charge for an autopsy. If a family wants to learn the cause of death of someone who has died in another setting (perhaps at home) the cost could be as high as $3000 or more. The family does not have to pay for an autopsy if it is requested or required by the legal system -- it is done at public expense in this case.

Some health officials find the decline in the rate of autopsies to be cause for some alarm. It has been found historically that autopsies reveal important undetected and unsuspected causes of death in about 20% of the patients who are autopsied. Therefore, there is considerable concern that the real cause of death is being missed in a substantial number of deaths. However, it is unlikely that the situation will change, simply because with our current large population the sheer number of deaths is overwhelming. Morgues around the country are swamped with cases as it is, and little improvement to this situation seems likely. Therefore, it is likely that the autopsy rate will remain where it is, or even continue to drop.

People on both sides of this controversy are interested in the new technology of "virtual autopsies". Virtual autopsies are detailed scans of the body using computerized tomography (CT) scanning. CT scanning uses x-ray technology to take a multitude of images of a body (alive or dead) that can be recombined by computer software to give a detailed three-dimensional set of images of the body. Initial studies of virtual autopsies are positive, and this may be the wave of the future.

25.5 Identifying Deceased Persons

Before a person's body may be buried, cremated, or otherwise laid to rest, either the identity of the deceased has to be determined, or it must be determined that their identity can not be determined. In most cases identification is easy. Most people carry some form of identification with them. If somebody is missing, friends or family will usually report it to the police. These people can identify the deceased person by looking at a photograph of their face or by looking at the body itself. In most cases, peace officers will ask relatives or acquaintances of the deceased to identify the person from photographs of the face.

Even if a body is recognized, however, the identity is usually confirmed by at least some investigation. Also, there is usually at least some attempt to make sure that the external appearance of the body is consistent with the medical and dental history of the person it is supposed to be. There have simply been too many cases of mistaken identity to rely too heavily on testimonial evidence in identifying a deceased person.

If the facial features have been damaged or decayed past recognition, the process of identification becomes more challenging. Except in cases of completely skeletonized remains, it may be possible to take fingerprints or DNA samples. In the case of completely skeletonized remains it may be possible to recover Mitochondrial DNA, or even nuclear DNA, from the bones to use for identification. However in most of these cases secure identifications can only be made by matching features observable on the skeleton with identical features recorded in medical or dental records while the person was alive. Forensic anthropologists and forensic odontologists can help with this process as will be discussed in coming chapters on these subjects. In cases where nothing else leads to the person's identity, the pathologist may request a facial reconstruction to reconstruct the person's facial features for circulation to law enforcement and the media as discussed in chapter 7.

25.6 Questions for Study and Review

1. Describe the field of pathology and its branches.
2. Describe the education of a forensic pathologist.
3. Under what conditions is it likely that an autopsy will be performed on a deceased person? Under what conditions is it unlikely that an autopsy will be performed?

4. Describe the process of doing an autopsy.
5. What benefits do autopsies provide to society other than in determining the cause of death?
6. Are more people being autopsied at death today than in the past? Discuss some pros and cons that arise from this trend.
7. List the methods that exist for identifying an unknown deceased person. Under what conditions would each method be used, and what type of forensic scientist would performs any required examinations or procedures for each method?

25.7 Sources

[Anonymous]. 1993. The first postmortem recorded in the country. The Journal of the American Medical Association 270(16): 1891(1).

Block EB. 1979. Science vs Crime. San Francisco: Cragmont Publications.

Byard RW. 2005. Who's Killing the Autopsy? A New Tool for Assessing the Causes of Falling Autopsy Rates. The Medical Journal of Australia 183(11-12): p654-655.

Cohen JI. n.d. General Information for Individuals [Internet]. [cited 2007 Oct 23]. Available from: http://www.forensiconline.com/generallink.htm.

DiMaio VJM, Hanzlick R. 1997. Medical examiners, forensic pathologists, and coroners. The Journal of the American Medical Association 277(7): 531(2).

Hanzlick R, Hutchins GM. 1999. History Repeats Itself (Sometimes). Archives of Internal Medicine 159(16): 1837.

Hooper JE, Geller SA. 2007. Relevance of the autopsy as a medical tool: a large database of physician attitudes. Archives of Pathology & Laboratory Medicine 131(2): 268.

National Association of Medical Examiners. 2007. General FAQs [Internet]. [cited 2007 Oct 23]. Available from: http://thename.org/index.php?option=com_content&task=category&sectionid=3&id=7&Itemid=42

Park K. 1994. The criminal and the saintly body: autopsy and dissection in Renaissance Italy. Renaissance Quarterly 47(1): 1(33).

Ravakhah K. 2006. Death Certificates Are Not Reliable: Revivification of the Autopsy. Southern Medical Journal 99(7): 728(6).

Sachs JS. 2004. Why Give a Dead Man a Body Scan? Forensic Scientists in Switzerland Are Pioneering a Whole New Way to Do Autopsies. No Scalpel Required. Popular Science 265(4): p50.

Smith SS. 2007. Autopsy patterns and trends. Public Health Reports 122(4): 565(1).

Wilson ML. 2006. Infectious diseases and the autopsy. Clinical Infectious Diseases 43(5): 602(2).

Yamazaki K, Shiotani S, Ohashi N, Doi M, Kikuchi K, Nagata C, Honda K. 2006. Comparison Between Computed Tomography (Ct) and Autopsy Findings in Cases of Abdominal Injury and Disease. Forensic Science International 162(1-3): 163(4).

Chapter 26
Basic Science
SKELETAL AND DENTAL ANATOMY

The topic of this chapter is skeletal and dental anatomy. My goal is to present the fundamentals of the science of skeletal anatomy, especially as it is used by forensic anthropologists and forensic odontologists. I will attempt to cover the basic concepts without becoming distracted by the details. The chapter will begin with a consideration of general anatomical terms for regions of the body, direction and location, and other preliminaries. The focus will then move to the anatomy of the human skeleton. The chapter will finish with a look at the terminology and anatomy of teeth.

26.1 Terms of Direction and Location

Anatomists have devised a system of directions that work well for describing the locations of parts of the body and their relationships to each other. All of these terms are defined relative to standard **anatomical position**, in which the person is standing erect with the eyes forward and the palms of the hands facing to the front. There are three standard planes that divide the body (in anatomical position). The **sagittal plane** is a vertical plane through the body at its midline that divides the body into symmetrical right and left halves. The **coronal plane** is another vertical plane, at right angles to the sagittal plane and directed side to side, that divides the body into more or less equal front and back portions. The **transverse plane** is a horizontal plane, at right angles to both the sagittal and coronal planes, that divides the body more or less equally into upper and lower portions.

The terms for direction on the body reference anatomical position and the sagittal, coronal, and transverse planes. Perhaps the best way to present the terms of direction is as a set of contrasting directions. These are superior vs inferior, anterior vs posterior, and medial vs lateral, proximal vs distal, and palmar/plantar vs dorsal.

Superior means in the direction of the top of the head (upward), and **inferior** is the opposite direction, toward the sole of the foot (downward). **Anterior** means toward the front of the body, where the front of the body is what you would see if a person were facing you. **Posterior** is the opposite of anterior and means toward the back of the body. Medial and lateral are defined in terms of the sagittal plane, where **medial** is toward the sagittal plane (toward the midline of the body) and **lateral** is away from the sagittal plane (toward the sides of the body). Proximal and distal are directions that apply to the limbs, where the **proximal** direction is toward where the limb joins the body (i.e. toward the shoulder or pelvis) and **distal** is the direction away from where the limb joins the body (i.e. toward the tips of the fingers or toes). Palmar vs dorsal are directions that apply only on the hands, where **palmar** is toward the palm of the hand and **dorsal** is toward the back of the hand. Similarly, plantar and dorsal apply only to the foot, where **plantar** is toward the sole of the foot and dorsal is toward the top of the foot.

26.2 Features of Bones

Bones have complex surfaces, and every little bump and groove has a name. Such detail is beyond the scope of this class, but we can explore some regularities in bone features.

A **process** is a projection of bone from its surrounding surface. Processes occur in a variety of shapes and sizes. A process consisting of a ridge of bone that is thinner than it is long is often (but not always) described as a crest. A sharp, thin process is often described as a spine. A low, rounded process (like a hill arising from a field) might be described as a tubercle, or if it is large enough as a tuberosity. A condyle is a process shaped more or less like a doorknob. A head is the rounded end of a limb bone. These terms are not designed to be precise in application, but rather to serve as general guidelines to the shapes of processes.

A depression in the surface of a bone is generally called a **fossa**. If it is long and thin, it is called a groove. If it is an actual hole into or through the bone it is called a foramen. If it is an air space in a bone, it is called a sinus. Again, these categories are not sharply defined, and are more like guidelines for terminology than a formal taxonomy.

26.3 Gross Types of Bone Tissue

There are three gross types of bone tissue. Here the word "**gross**" is not used as in slang to mean crude or disgusting, but instead is used in its traditional meaning of something that can be seen with the naked eye rather than requiring a microscopic to be seen. The three types of bone tissue that can be distinguished without resorting to the use of a microscope are compact bone, cancellous bone, and subchondral bone.

Compact bone, also called cortical bone, is commonly found on the external surface of bones, where it forms a hard layer. **Cancellous bone**, also called spongy bone, is commonly found on the interior of most bones, where it is usually formed by plates of bone that intersect with each other to form roughly square cells. **Subchondral bone** is found where the bone surface is covered with cartilage. It usually looks darker and less dense that compact bone.

26.4 Shape Categories of Bones

There are several shape categories of bones. Some bones are placed into different categories by different authorities, and it is occasionally difficult to decide exactly which category certain bones belong to. Don't worry about these inconsistencies – nobody really cares that much.

Long bones are bones that are, well ..., long. They are more-or-less tubular and relatively thin. Most of the limb bones are long bones. Short bones are – you guessed it – short and roundish or squarish. The bones of the wrist and ankle are short bones. Flat bones are flat. Get the picture? Most of the bones that form the braincase of the skull are flat bones. Irregular bones are bones that are not long, short, or flat, but have irregular shapes. The vertebrae of the spinal column are classic irregular bones. Sesamoid bones are usually small bones that develop within a muscle tendon. Their shape is often similar to that of a sesame seed, though they are larger than a seed. The largest and best known sesamoid bone is the patella or knee cap.

26.5 Region and Segment Divisions of the Body

Anatomists have developed systems for classifying and naming specific portions of the human body. Some of the systems used differ in purpose from others. We will focus in this chapter on those commonly used to describe the skeleton and dentition.

One very commonly used system divides the skeleton into cranial and postcranial portions. The **cranial skeleton** is the head, including the mandible (jaw bone) and hyoid (a small bone of the throat). The **postcranial skeleton** is everything else, from the neck vertebrae to the toes. Alternatively, the skeleton can be divided into axial vs appendicular portions. The **axial skeleton** forms the axis or core of the skeleton and consists of the skull, vertebral column, ribs, and sternum (breast bone). The **appendicular skeleton** is the limbs, including the shoulder and pelvic bones.

In addition to these approaches, it is conventional and convenient to divide the body into regions, and further divide some of the regions into segments. The major regions of the body are the head, neck, thorax (chest), abdomen, perineum (the area where your "privates" are located), upper limb, and lower limb.

The upper limb contains three major joints: the shoulder, elbow, and wrist. Segments of the upper limb are defined in terms of these three joints. The shoulder girdle is that part of the upper limb that lies between the thorax and the shoulder joint. The arm lies between the shoulder and the elbow, a terminology that contrasts with the common usage of the word "arm" in English to refer to the entire upper limb. The forearm lies between the elbow and the wrist joint. The wrist segment occurs at the wrist joint. The hand is that portion of the upper limb that lies beyond the wrist joint, including the body of the hand and the fingers.

The lower limb similarly has three major joints: the hip, knee, and ankle. Segments of the lower limb are defined relative to these three joints. The pelvic girdle is the area between the abdomen and the hip. The thigh is the segment that lies between the hip and the knee. The leg is the segment that lies between the knee and the ankle, another term that contrasts with the common English use of the word "leg" to refer to the entire lower limb. The anatomical term "leg" corresponds to the common English term "shin". The ankle segment is located at the ankle joint. The foot segment lies beyond the ankle joint and includes the body of the foot and the toes.

26.6 Portions of the Skull

The skull is a complex structure that has many functions. The skull holds and protects the brain. It is also the seat of four of the five senses: vision, hearing, smell, and taste. Further, it has to be able to point the eyes in a variety of directions if vision is to be useful. Also, it contains the apparatus for breathing and eating. The eating apparatus is particularly complex, consisting of teeth, jaws, muscles to move the jaws, and the tongue. Because of the complexity of the skull, anatomists divide it into a set of regions or portions, in a manner similar to how the limbs are divided into segments. Here for the first time we find disagreement in terminology between different branches of anatomy. I use the term "medical anatomy" to describe the set of anatomical terms learned and used by medical doctors, dentists, and others who are only interested in the anatomy of humans. I use the term "comparative anatomy" to refer to the set of anatomical terms learned and used by those anatomists that study a wider range of species, including anthropologists and zoologists. For the purposes of this book only I will use the terms current in medical anatomy, because these are the terms and usages that are necessary to know when interacting with forensic pathologists and forensic odontologists. Forensic anthropologists might use a slightly different set of terms, but chances are they too will be familiar with the usage of terms by forensic pathologists and odontologists.

The term "skull" refers to the entire set of fused bones of the head, plus the mandible (lower jaw). The cranium, also known as the neurocranium or the vault is the portion of the skull that encloses the brain. The base is the bottom part of the skull. The face, or splanchnocranium is the anterior part of the cranium including the eye sockets and everything below them.

Figure 26.1: Lateral View of the Skull

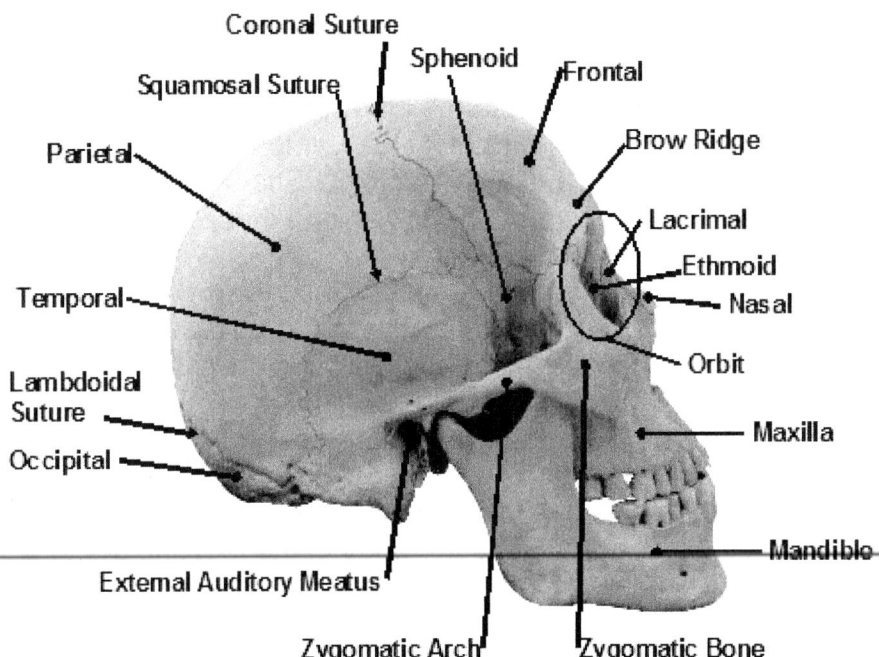

26.7 Bones of the Skull

Let's start our exploration of the bones of the body with the bones of the skull – the cranial skeleton. Figure 26.1 is a lateral view of the skull, so named because the skull is being viewed from the side or lateral direction.

The bones of the skull visible in lateral view are the frontal, a single bone best viewed from the front; one of the two parietal bones, which occur as a right and left pair; one of the two temporal bones, which also occur as a right/left pair; a small portion of the occipital bone, a single bone that makes up most of the back and some of the base of the skull; a small portion of the sphenoid bone, which makes up part of the base of the skull; one of the two zygomatic bones, which occur in a right/left pair; one of the maxilla bones, which occur in a right/left pair, one of the nasal bones, which occur in a right/left pair, and the mandible, a single bone. One of the orbits, or eye sockets, is visible in this view. Focusing on what is visible of the interior of the one visible orbit you can see part of the ethmoid bone, which sits dead center above and behind the nose; and one of the two lacrimals, which are paired small bones that occur on the anterior medial wall of the eye socket.

Some of the cranial sutures, or joints between the skull bones, can be seen in lateral view. The coronal suture divides the frontal from the parietals (it runs in a plane parallel to the coronal plane). The lambdoidal suture divides the occipital from the parietals. The squamosal sutures (one on each side of the skull) divide the temporals from the parietals.

Figure 26.2: Frontal View of the Skull

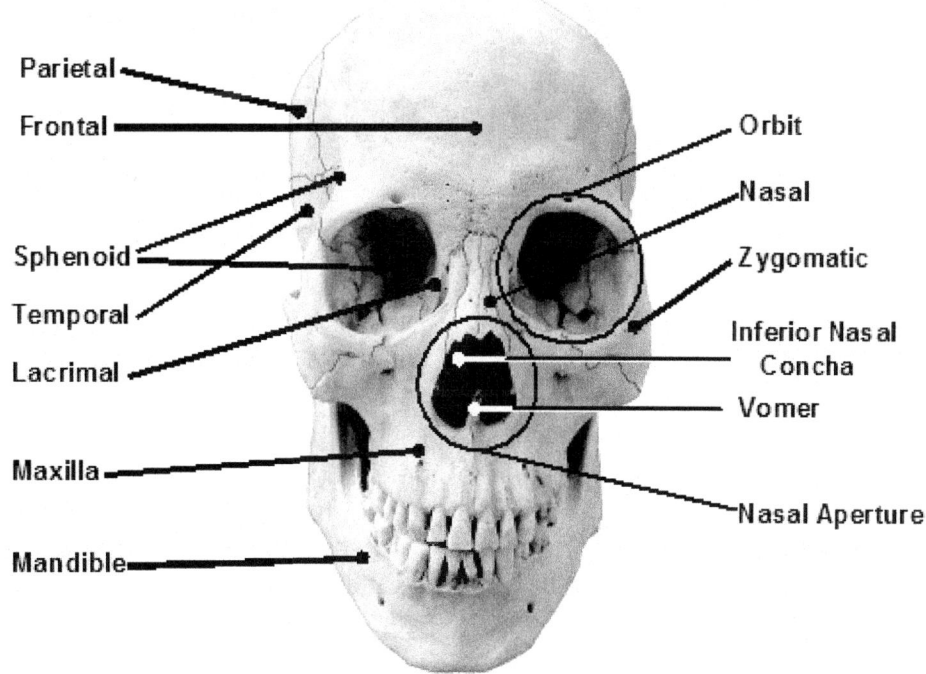

287

Also visible in lateral view is one of the two zygomatic arches, formed by processes of the temporal bone and the zygomatic bone; one external auditory meatus, or ear hole; one of the two mastoid processes, located on the temporal bone behind the ear; and part of the brow ridges, which occur just over the orbits.

Figure 26.2 presents a frontal view of the skull. From this view the two orbits and the nasal aperture can be easily seen. The only bones visible from this direction that were not visible in lateral view are the two inferior nasal concha bones and the vomer. These are smaller bones that are contained within the nasal aperture.

Figure 26.3 presents an inferior view of the skull, which is the view of the skull seen from below. The only bones visible here that were not seen in previous views are the palatine bones, which occur just posterior to the maxilla bones and form the posterior portion of the palate. In this view the foramen magnum (literally "big hole") can be seen, through which the spinal cord runs to join to the brain. The occipital and sphenoid bones can be seen better in this view.

Figure 26.3: Inferior View of the Skull

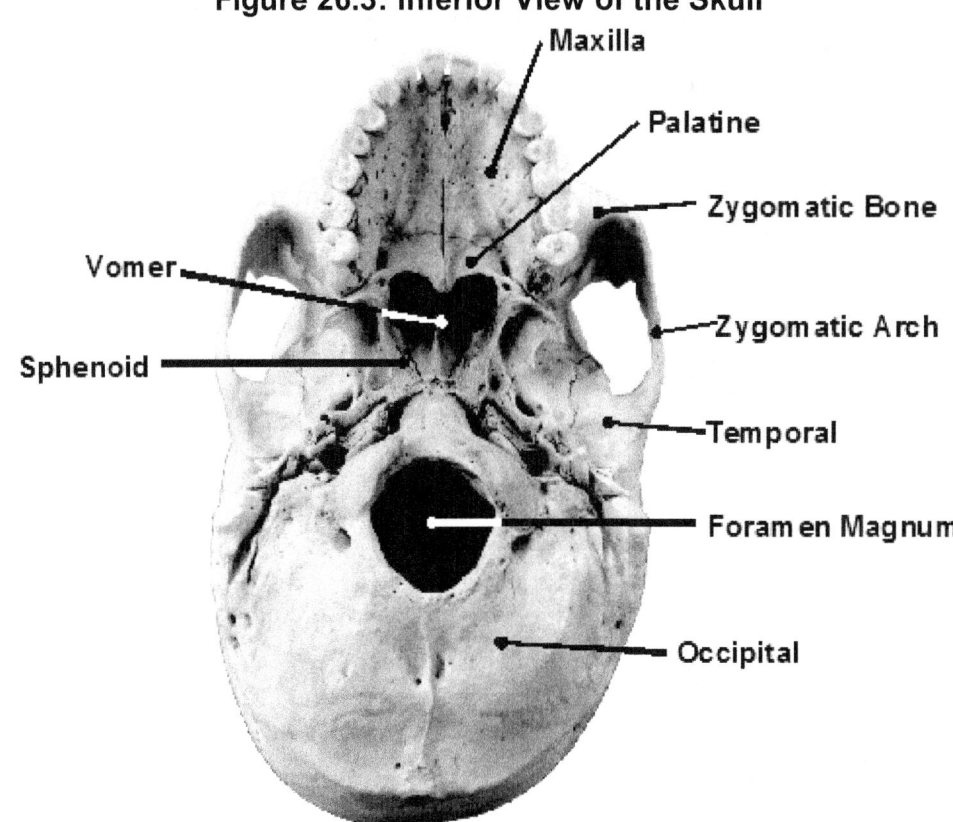

Figure 26.4 presents a superior view of the skull, which is the view of the skull seen from above. No new bones are visible here, but two of the largest sutures of the skull, the sagittal suture and the coronal suture, are seen clearly. Because the sagittal suture lies at the top of the skull, exactly in the sagittal plane, this is the only view in which it can be clearly seen.

Figure 26.4: Superior View of the Skull

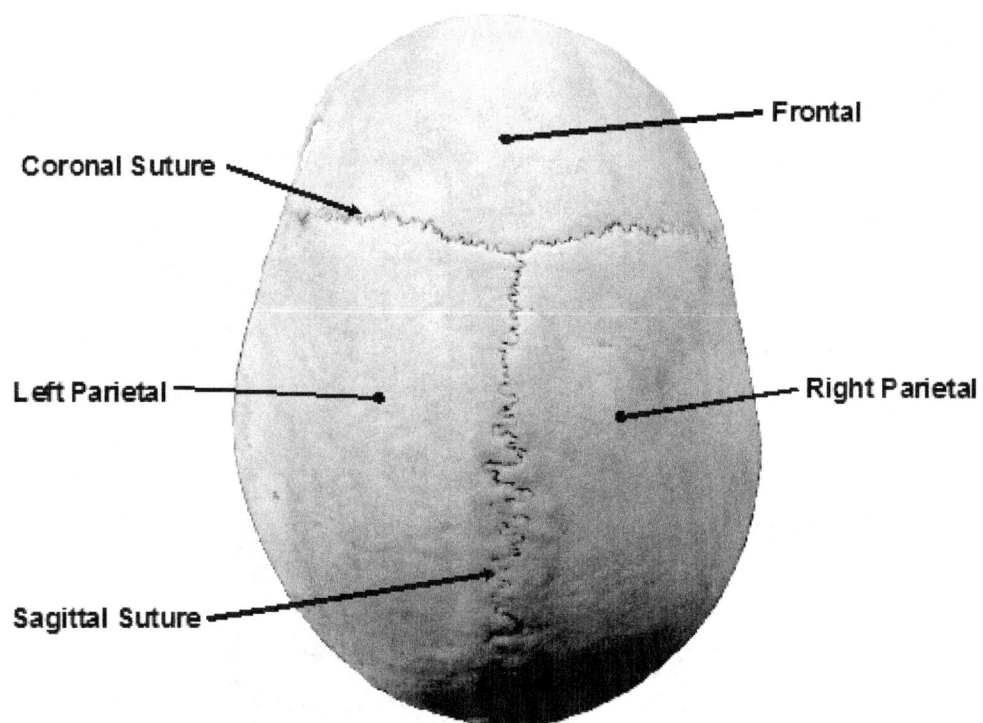

26.8 Bones of the Postcranial Skeleton

The main framework of the axial portion of the postcranial skeleton is provided by the vertebral column, which consists of a stack of 32 vertebrae (singular vertebra). Figure 26.5 is a posterior view of the upper portion of the skeleton that illustrates the vertebral column. The vertebral column is divided into five sections: the 7 cervical vertebrae of the neck region, the 12 thoracic vertebrae of the thorax region, and the 5 lumbar vertebrae of the abdominal region are normally separate vertebrae. The sacrum is a bone consisting of 5 sacral vertebrae that are fused together. The sacrum, along with the two coxal bones forms the pelvis. The coccyx, known informally as the tail bone, consists of 3 coccygeal vertebrae fused together to form a single structure.

In the thorax region, ribs arise in right/left pairs from the thoracic vertebrae, and wrap around anteriorly to form the structure of the chest. The origin of the ribs from the thoracic vertebrae may be seen in Figure 26.5. The attachment of the ribs to the sternum may be seen in Figure 26.6. The first 10 pairs of ribs attach, either directly or via a strip of cartilage, to the sternum. The sternum is known informally as the breast bone – a flat bone located in the midline of the anterior chest. The 11th and 12th ribs are floating ribs and do not attach to the sternum. The ribs and sternum are also part of the axial skeleton.

Figure 26.5: Posterior View of the Upper Skeleton

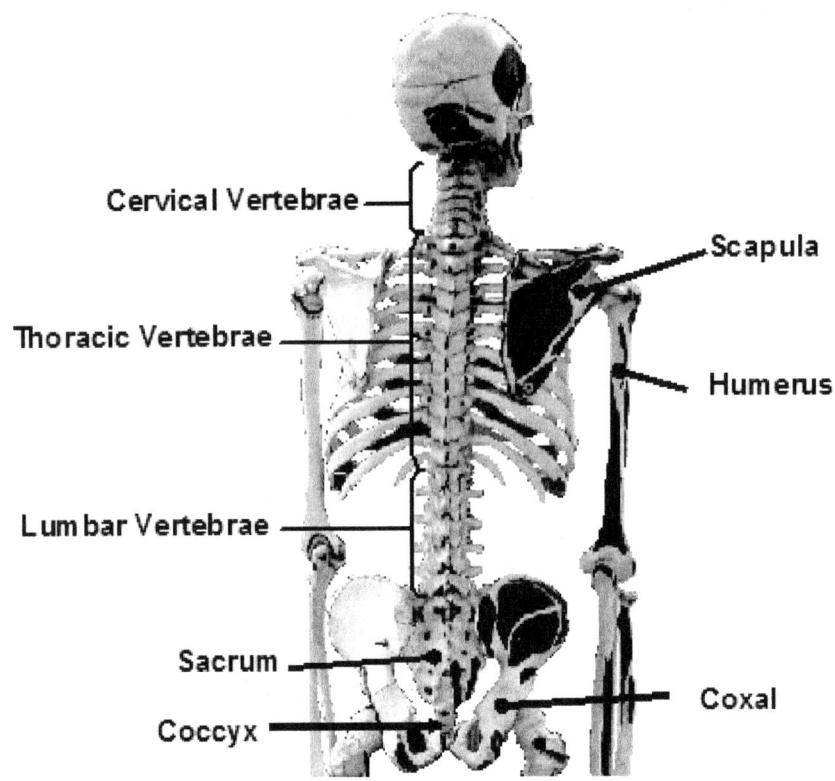

The appendicular skeleton consists of the bones of the upper and lower limbs. All bones of the appendicular skeleton occur as right/left pairs. We will start with the upper limb. The bones of the shoulder girdle are the clavicle, known informally as the collar bone; and the scapula, known informally as the shoulder blade. The single bone of the arm segment is the humerus. There are two bones of the forearm, the radius and the ulna. When the hand is held in anatomical position the radius is the bone on the thumb side of the forearm and the ulna is the bone on the pinkie finger side. The wrist segment is composed of eight small bones known as carpals, that articulate in complex ways to form the wrist joint. Each of the carpal bones has a name but we do not need this detail for our purposes. The body of the hand is formed of 5 metacarpal bones, each of which runs from the wrist to the base of each finger (including the thumb). The thumb is composed of two bones, a proximal phalanx (plural phalanges) and a distal phalanx. The other fingers are composed of three bones, proximal, middle, and distal phalanges. A joint exists between each of the phalanges of a given finger.

Moving on to the lower limb, the bones of the pelvic girdle are the coxal bones, one on each side. The sacrum, discussed above as part of the vertebral column, is located between the two coxals, which articulate with it to form the pelvis. The single bone of the thigh is the femur. The leg segment has two bones on each side, the tibia and the fibula. The tibia is the large bone, which is informally called the shin bone. Situated lateral to the tibia is the more slender fibula. The ankle is formed by 7 bones, called the tarsals; the largest of which is the calcaneus, known informally as the heel bone. The distal end of the tibia sits on the talus, known informally as the ankle bone, and this joint between the tibia and the talus is where the primary motions of the ankle occur. The other five short bones are functionally part of the foot and we do not need to delve into the details of their names. The body of the foot is formed by 5 metatarsal

Figure 26.6: Frontal and Lateral Views of the Skeleton

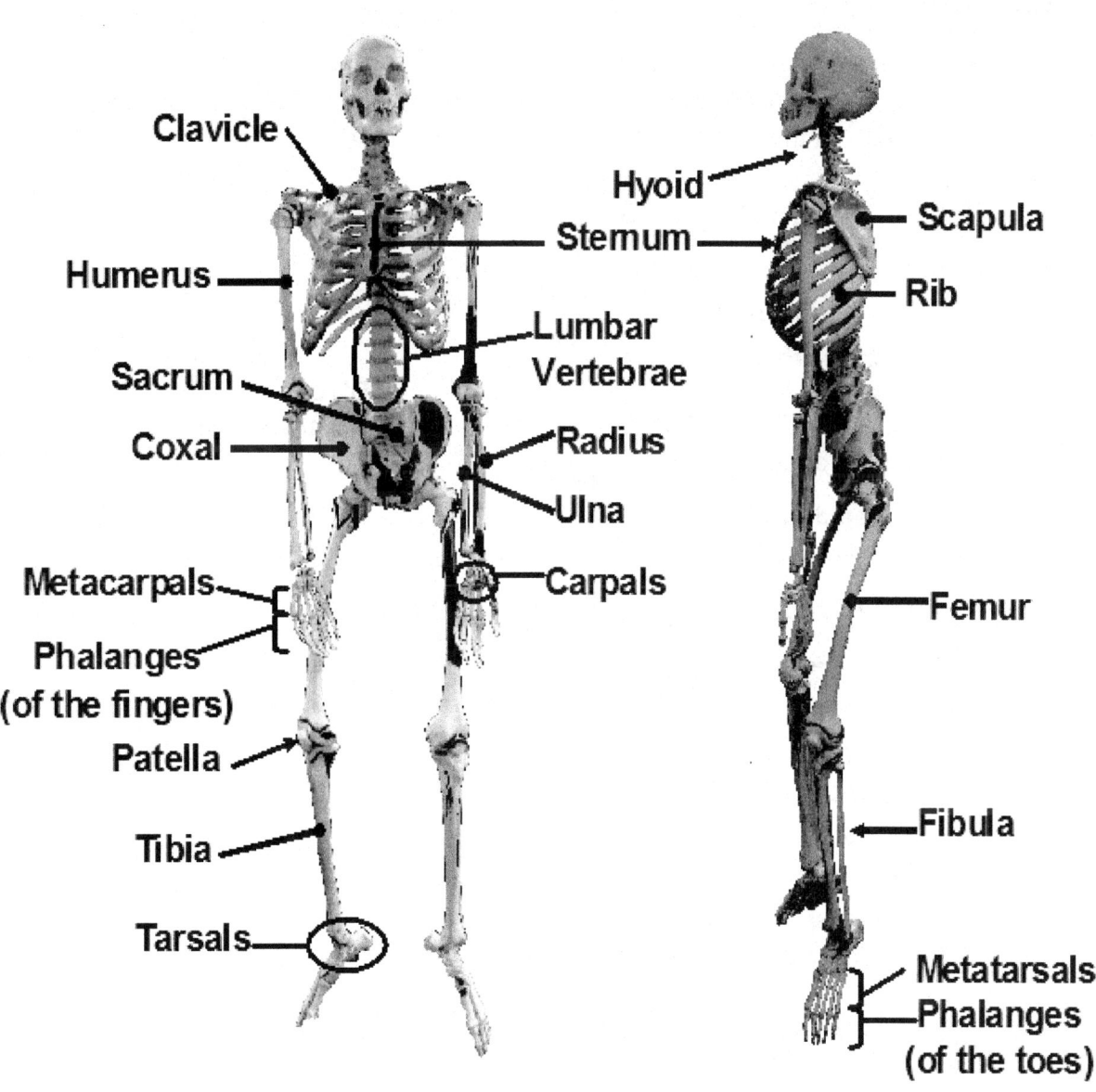

bones, running from the ankle to the base of each toe. The toes have the same configuration of phalanges as the fingers, two phalanges for the big toe (hallux) and three for each of the other toes.

The hyoid bone sits in the throat or neck region, about at the crease where the neck skin joins the skin that runs under the jaw. It is not the Adam's apple, which is actually a cartilage capsule that protects the thyroid gland. The hyoid does not attach to any other bone, but is important in that many small muscles involved with speech are anchored to it. Because of its location, and its small, fragile nature, a broken hyoid bone discovered at autopsy is usually taken to be an indicator that the person was strangled. The hyoid is formally a cranial bone, but it is best seen in the lateral view of the entire skeleton in Figure 26.6.

26.9 Teeth

The teeth are a subject of interest in their own right. Though we are often barely aware of them, due to their being hidden inside our mouths, they are complex and fascinating structures. In this chapter, however, I will continue to keep details to a minimum.

Figure 26.7: The Human Dentition: Tooth Names and Directions

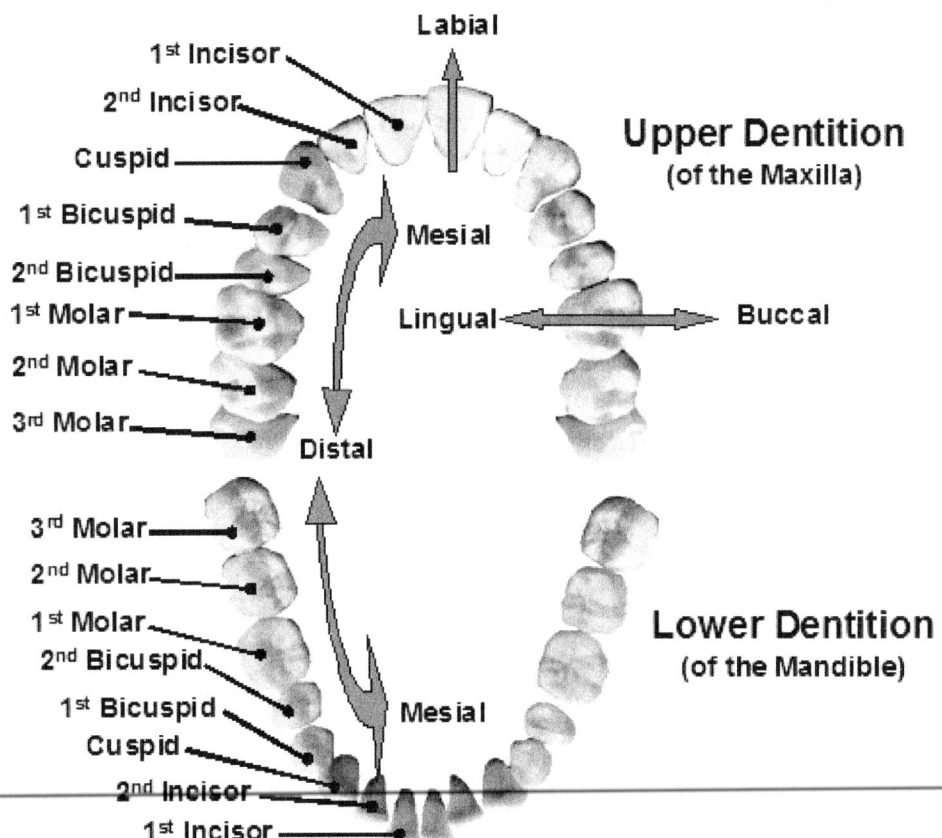

Because of the way the mouth is laid out, there are special terms of direction and location for the teeth. Figure 26.7 illustrates these directions. Note that some of the teeth run in a line nearly parallel to the coronal plane, across the front of the mouth, while others run in a line more parallel to the sagittal plane, on the right and left sides of the mouth. **Mesial** is the direction in the mouth that moves toward the exact midline of the front of the tooth rows (the sagittal plane). Mesial is a direction that does not form a straight line, rather it curves, following the teeth as they lie in the jaws. Be careful not to confuse mesial with medial or middle. The opposite of mesial is **distal**. Yes, this is the same as the term of direction that means away from where a limb joins the body. I don't know the reason for this "recycling" of terms, but the difference in context (teeth vs limbs) usually allows the two usages of distal to be unambigious. **Buccal** is the direction that lies at right angles to mesial for a given tooth and which is the direction toward the cheek. Toward the front of the mouth, where the cheeks leave off and the lips begin the equivalent direction is **labial**, meaning toward the lips. The opposite direction from buccal or labial is toward the tongue and is called **lingual**.

Teeth are composed of two parts. The **crown** is the part of the tooth that is covered with hard, white enamel. The **root** is the portion of the tooth that fits into the tooth socket and which is not covered with enamel. The junction between the crown and the root is known as the **neck** of the tooth. The surfaces of the teeth that meet when the jaw is closed and which are the surfaces that chew food are called the **occlusal surfaces**. For the teeth in the very front of the jaw, the occlusal surface consists of a more or less sharp edge, called the incisal edge. The occlusal surfaces of

Figure 26.8: Cross Section of a Tooth

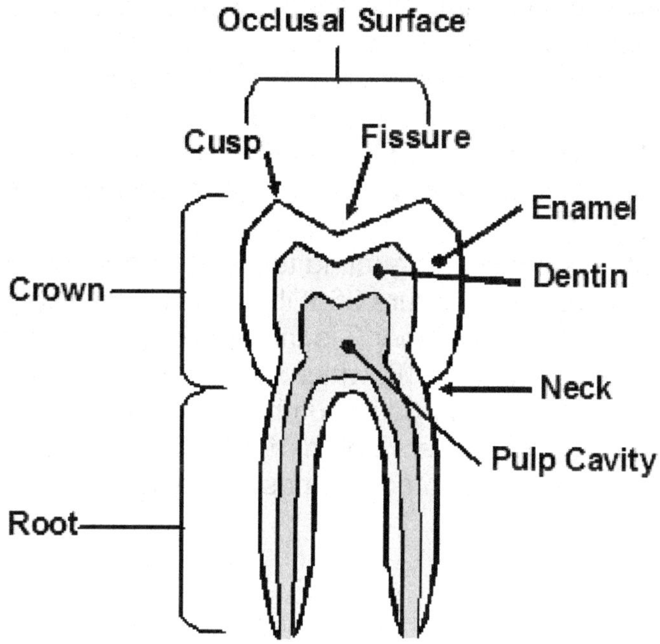

most teeth have considerable detail, especially the molars. Projections from the occlusal surface are called **cusps**, and the grooves between the cusps are called **fissures**. Figure 26.8 illustrates these features.

If one were to cut a cross section through a tooth, say in a plane parallel to the sagittal plane to divide the tooth into right and left halves, the layered nature of a tooth's construction would be clear. Figure 26.8 illustrates this structure. The crown is composed of three layers: an outer layer of enamel, a middle layer of dentine, and an inner pulp cavity where blood vessels and nerves are found. Enamel is the hardest substance in the human body and dentin is the second hardest.

Like all mammals, humans are both diphodont and heterodont. **Diphodont** means that we have two sets of teeth over our lifetimes, a set of deciduous teeth (also called milk teeth or baby teeth), which fall out during childhood; and the adult teeth (or permanent teeth) which appear in our mouths during middle to late childhood and which are kept for the rest of one's life.

The term **heterodont** means that we have four differently shaped types of teeth in our mouths. Adults whose teeth have all erupted (and none have been pulled) have 32 teeth in their mouths. The types of teeth are symmetrical, both side to side and up to down. This means that the teeth can be divided into four quadrants, upper right, upper left, lower right, and lower left, and that the number and types of teeth are the same in each quadrant. In each quadrant most people have two incisors, one cuspid, two bicuspids, and three molars, for a total of eight teeth. Since each quadrant is one fourth of the entire mouth, each person has 8 incisors, 4 cuspids, 8 bicuspids, and 12 molars in their mouth.

Within a quadrant, the teeth of each type are distinguished by number. Therefore, in each quadrant we have, from mesial to distal, a first incisor, a second incisor, a cuspid, a first bicuspid, a second bicuspid, a first molar, a second molar, and a third molar. Informally, the first molars are often called the 6-year molars, because that is the age at which they appear in the mouth. Similarly, the second molars are informally called 12-year molars, and the third molars are informally called wisdom teeth. The deciduous dentition includes in each quadrant two incisors (first and second), a cuspid, and two molars (first and second). The deciduous molars occur in the locations of the adult bicuspids and are replaced by the adult bicuspids. The adult molars erupt distal to the deciduous molars as the jaws elongate through the years of childhood growth. Figure 26.7 shows the names of the teeth.

The most current system of designating teeth does not name them at all, but instead gives each one a number. Figure 26.9 illustrates this system. Tooth number one is the upper right third molar, number 2 is the upper right second molar, and so on around the upper dentition to number 16, which is the upper left third molar. Tooth number 17 is the lower left third molar, and numbering continues around the lower dentition to tooth 32, which is the lower right third molar. This system was developed by the U.S. Army, and has been adopted my most dentists, doctors, and other people who work with human anatomy.

Figure 26.9: The Army System of Tooth Designation

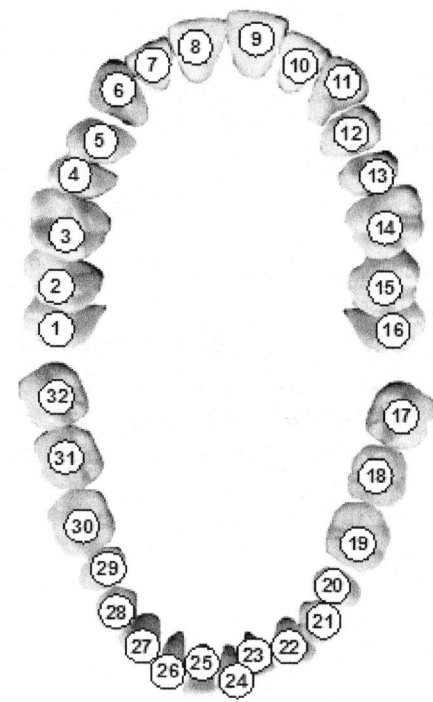

26.10 Questions for Study and Review

1. In your own words, describe "anatomical position."
2. Describe the three major planes of the body and the directions that reference them.
3. Describe the various types of bone processes and fossas.
4. List the three types of bone tissue and where each is most commonly found.
5. In your own words, describe the distinction between the cranial and the postcranial portions of the skeleton.
6. In your own words, describe the distinction between the axial and the appendicular portions of the skeleton.
7. List the seven major regions of the body (skeleton).
8. Describe the joints and segments of the upper limb.
9. Describe the joints and segments of the lower limb.
10. Describe the portions of the skull.
11. If given an unlabeled graphic similar to those in this chapter, be able to label it with the names of the main bones visible in it.
12. Describe the directions used when referring to teeth.
13. Describe the parts of a tooth and the features of the crown of a tooth.
14. If given an unlabeled graphic similar to those in this chapter, be able to label it with the names of the teeth that are visible.

15. Define the terms heterodont and diphodont.

26.11 Sources

Bass WM. 1987. Human Osteology: A Laboratory and Field Manual. 3rd Edition. Columbia (MO): Missouri Archaeological Society.
Scott GR, Turner CG. 1997. Human Teeth: Dental Morphology and Its Variation in Recent Human Populations. Cambridge (United Kingdom): Cambridge Univ. Press.
Shipman P, Walker A, Bichell D. 1985. The Human Skeleton. Cambridge (MA): Harvard University Press.
Steele DG, Bramblett CA. 1988. The Anatomy and Biology of the Human Skeleton. College Station (TX): Texas A&M Press.
White TD, Folkens PA. 1991. Human Osteology. San Diego (CA): Academic Press.
Woodburne RT, Burkel WE. 1994. Essentials of Human Anatomy. 9th Edition. New York: Oxford University Press.

Chapter 27
FORENSIC ANTHROPOLOGY

In this chapter I will present my specialty, forensic anthropology, which is concerned with the discovery, recovery, and identification of skeletal remains. First we will look at some terminological issues concerning the branches of forensic anthropology. Next we will delve into forensic archaeology, and examine methods for locating and recovering buried evidence. The chapter will then conclude with a consideration of forensic osteology and methods for generating clues about the age, sex, and other characteristics of a person from examination of their skeletal remains.

27.1 The Branches of Forensic Anthropology

Forensic anthropology is currently considered to have two branches, forensic archaeology and forensic osteology. **Forensic archaeology** includes the activities involved with searching for and recovering skeletal remains or other buried evidence. **Forensic osteology** includes the laboratory activities geared toward estimating sex, age, and other class and individual characteristics of deceased persons from their skeletal remains.

Before the 1980's, forensic anthropology was synonymous with what we now call forensic osteology. A typical case scenario would begin with the discovery of a skeleton, which was then recovered by peace officers and sent to an osteologist's lab for analysis. During the 1980's however, a group of people associated with Florida State University, including the late Dan Morris, published a book titled *The Handbook of Forensic Archaeology*, which illustrated the value of archaeologists in searching for and collecting buried evidence. The development of American forensic archaeology owes a debt to the forensic archaeologists of the United Kingdom, who have been tireless in their promotion of archaeologists as the most qualified people to locate and recover buried evidence.

Forensic anthropologists who were trained before the middle 1980's, such as myself, tend to be much better osteologists than archaeologists. People trained more recently are well trained both in osteology and in archaeology.

27.2 The History of Forensic Anthropology

Forensic anthropology presents an interesting history because there were times during the 1900's in which forensic anthropology was not considered a legitimate or reliable forensic science. Even today, forensic anthropology is sometimes regarded as the "lite beer" of forensic science. The main reason for this is that physical anthropology, of which forensic anthropology is a branch, was plagued by problems during the 1800's. Their biggest goal at that time was to show the superiority of White peoples, a subject riddled with patches of both ethical and scientific quicksand.

Further, this was a sterile pursuit because no scientific evidence of superiority or inferiority of any race has ever been replicable.

Therefore, when forensic anthropology's first major cases occurred, it is not surprising that in some ways the forensic anthropologists did not live up to our modern expectations of them. The first forensic anthropology case on record, occurred on the campus of Harvard University, when one faculty member killed another, apparently to avoid paying off a large debt. Parts of the deceased professor's body were discovered in his office and privy, and were examined by yet another member of the Harvard faculty. Interestingly, the best evidence leading to the identification of the body parts was a set of dentures, which were identified by the victim's dentist. This underscores the fact that forensic anthropologists and forensic odontologists have often teamed up since the very beginnings of their specialties. In some ways this case was poorly handled, as was a case in the late 1890's involving a Chicago sausage maker who murdered his wife and disposed of the body by boiling it with caustic potash in a sausage vat. The major problem in both of these cases was that a higher level of identification (matching) was claimed for the bodies on the basis of the anthropological evidence than what the evidence actually warranted. In the Chicago case, for example, the sausage maker's wife was positively identified on the basis of one finger bone and a ring. The ring may or may not have been unique, but the finger bone was not. Did the sausage maker really kill his wife, or had she run away out of disgust for the smell of sausages? The sausage maker was convicted of murder, and many of us are comforted by the report that he later confessed to it. In another similar case, however, the wife might be enjoying the good life in the Florida Keys while her husband is executed for her murder.

As we have seen for many other forensic sciences, forensic anthropology began to acquire its modern scientific basis in the late 1800's. From the late 1800's on, however, a variety of anthropologists worked to gather the basic information upon which our current methods of estimating age, sex, race, and stature are based. Forensic anthropology began to gain respect in the 1930's when prominent anthropologists consulted on a variety of cases for the FBI that arose from the gangster violence of that era. Dr. Wilton Marion Krogman did a considerable amount of this consulting, and in 1939 published "A Guide to the Identification of Human Skeletal Material" in the FBI's *Law Enforcement Bulletin*. The re-emergence of forensic anthropology began with this publication. In the 1960's, Dr. J. Lawrence Angel was hired to work at the Smithsonian Institution, which happens to be across the street from FBI headquarters in Washington, DC, making it convenient for the FBI to ask the Smithsonian scientists for help with skeletal remains. During his career at the Smithsonian, Angel consulted on a total of 565 forensic cases.

During the 1970's forensic anthropology can be said to have entered its modern period. Forensic Anthropology was recognized as a forensic science by the American Academy of Forensic Sciences in 1972. By 1978, the American Board of Forensic Anthropology had been formed to certify the skills of forensic anthropologists. During the 1980's forensic anthropologists began to become involved in the process of documenting human rights abuses by national governments. This was pioneered by Dr. Clyde Snow, who worked in Argentina, and is an integral part of the field of forensic anthropology today. It was also during the 1980's that a national databank of data from forensic cases around the country was founded by the University of Tennessee, Knoxville.

In the 1990's forensic anthropology received a major setback with the passage in 1991 of the Native American Graves and Repatriation Act (NAGPRA). This federal legislation requires repatriation of Native American skeletal remains. However one looks at that issue philosophically and ethically, it was a blow to science in that it has led to the loss of a substantial amount of research potential. I see it as yet another aspect of the modern American attitudes about the "sacred dead", about which I ranted in chapter 25. In the 21st century, forensic anthropology is alive and well. Like my fellow forensic scientists at the Montana State Laboratory of Criminal Investigation, my caseload is larger than ever, and I only see it increasing into the future.

27.3 Forensic Archaeology

Let us start our investigation of forensic anthropology with the field aspect of the discipline, forensic archaeology. This is a natural place to start, because a skeleton can not make it to a laboratory until it is first located and collected.

27.3.1 Searching for Buried Evidence

One of the tasks for which forensic archaeologists are best suited is searching for buried evidence. This buried evidence is usually a buried body, but other kinds of buried evidence can be located by the same methods. The best way to search for buried evidence depends on who you ask. If you ask a forensic archaeologist, they will usually say that they best way to search is to systematically walk around an area suspected to contain a grave, looking for evidence of disturbance of the ground. When disturbance of the ground is identified, the forensic archaeologist will evaluate it to try to determine its cause. If there doesn't appear to be a logical reason for the ground to be disturbed at a certain location, then it is regarded as suspicious.

Figure 27.1: The Process of Creating a Grave

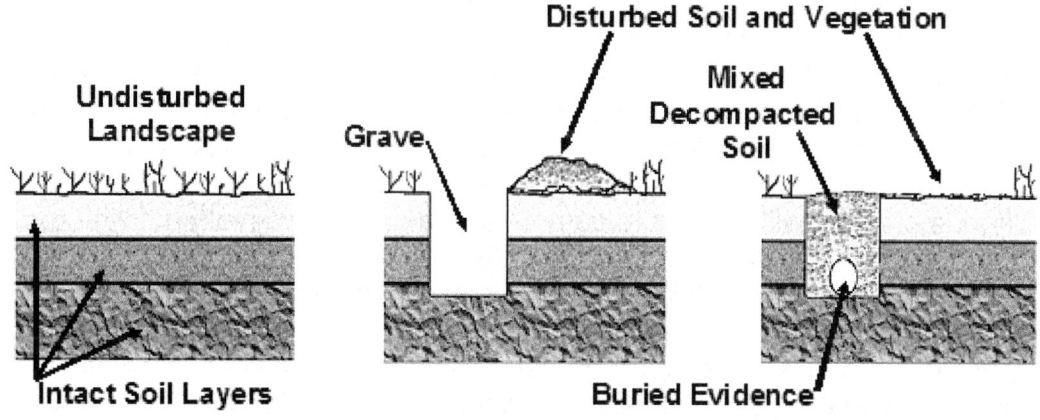

Figure 27.1 illustrates how **disturbance of the ground and vegetation** occurs as a grave is dug, a body is placed in it, and the hole is filled in. First, digging the hole disturbs both the soil and the vegetation that was growing on the ground in that area. The hole itself only accounts for part of the disturbance, however, because inevitably the dirt removed from the hole has to be piled somewhere, which also disturbs the soil and the vegetation.

What makes the type of soil disturbance produced by digging a grave distinctive, is that as the grave gets deeper and the soil pile gets taller, the soil from the deepest part of the grave ends up on top of the pile. If this could be done precisely, then perhaps the soil could be returned to the grave in its original order. However, this is never possible, and the result is that soil from beneath the surface becomes scattered on the surface of the ground. Subsurface soil usually has a different color or texture, or both, from the soil found at the surface. No natural process can produce this type of evidence, except the activities of burrowing animals. Because of the difference in size between a grave and a gopher mound, however, few people confuse the two.

What makes a grave stand out even more, is that removing the soil from the hole **decompacts** it. Gardeners know that spading their garden loosens the soil, and this happens as a grave is dug as well. Therefore, the soil removed from the grave is greater in volume that it was when in its natural state in the ground. Placing a body inside the grave further uses up space, and therefore when it comes time to fill in the grave there is too much soil to fit in the hole, no matter how forcefully the criminal tries to stomp it in. Now the criminal has a choice of whether to mound the soil up over the grave, a dead giveaway that something is buried there, or to level the surface of the grave off with the rest of the ground. If the second option is chosen, then there is extra soil that must be disposed of, and in the vast majority of cases it is simply scattered around, making the area of disturbance even larger and more visible to the archaeologist's trained eye.

Once the grave has been filled in, the natural process of soil compaction resumes due to the force of gravity, and is helped along by any rainfall that occurs. This will cause a depression to form at the area of the grave. The depression is often about eight feet long and six feet wide for an adult (smaller, of course, for a child). Further, a **secondary depression** may occur over the area of the abdomen of the deceased person. One of the first stages of decay of a body involves bloating of the abdomen. As this occurs in the buried person, the soil over the abdomen is pushed upward against the overlying soil, which results in compacting it fairly tightly. As decay progresses and the bloating ceases, this compact soil falls into the abdominal cavity leaving a void above it. Over time, soil compaction processes will cause the soil to fill in this void, causing it to be transferred to the surface, where it is visible as a small depression within the larger depression of the grave.

Often, a criminal will attempt to camouflage a grave by covering it with rocks or brush. In most cases this simply makes it stand out more. The archaeologist sees a pile of brush in the center of a field and wonders why it is there, which prompts her or him to check that area first.

Peace officers rarely search for graves by looking for soil and vegetation disturbance. Instead they prefer a variety of alternative techniques. One such is the use of metal rods to probe suspect areas in an attempt to "feel" the buried bones. This method is not recommended because it can cause damage to the bones. Some peace officer strategies are more high tech, such as using metal detectors, ground penetrating radar, or devices that detect the methane produced during decay of a body. It is also popular to use cadaver dogs, which will be discussed further in the chapter on pseudoscience. A thorough treatment of these methods is beyond the scope of this book, but a former graduate student of mine did his thesis on how effective archaeologists, peace officers, and cadaver dogs were at locating graves (buried pigs in this case). Not surprisingly, the archaeologist was more successful than the peace officer, but surprisingly (to me at least) the archaeologist was also more successful than the cadaver dogs.

27.3.2 Crime Scene Issues Applied to Forensic Archaeology

The place where a skeleton is discovered is a crime scene and should be treated as such. This is true regardless of whether the skeleton is within a grave, or scattered over the surface. The normal procedures of crime scene processing need to be followed. An especial problem with a surface bone scatter is the possibility of trampling the evidence.

Once a grave or surface bone scatter is discovered, the area needs to be secured and protected by cordoning it off and leaving a single entrance/exit area. As usual in crime scenes, the goal is to preserve as much of the evidence present as possible. A common mistake is to only think of the bones as evidence that might be found. Other forms of evidence are commonly found, however, including footwear and tire impressions, drag marks, ammunition components, cigarette butts, etc. This evidence can be obliterated by someone walking on it.

If there is a grave, it should be regarded as the primary area of the crime scene, and the area surrounding it for a distance of at least 15 meters (about 50 feet) should be considered the secondary area. If the skeleton is scattered on the surface, the area bounded by the most widely separated bones to be the primary area and an area 15 meters in every direction beyond the bones should be considered the secondary area. The secondary area should be searched first, with all evidence documented before being collected. Once the evidence in the area surrounding the grave or bone scatter is safe, work can begin on documenting and collecting the bones and the evidence found with them.

27.3.3 Excavation: Recovering Buried Evidence

The excavation of a buried body or a skeleton should proceed according to good archaeological excavation methods. There are, however, some differences from standard archaeological practice.

The grave itself is the primary area of the crime scene, as discussed above. The first main departure from standard archaeological practice is that the recovery should not be conducted by excavating square units defined by a grid system. Instead, the goal is to remove only the material from within the grave and to expose the original sides and bottom of the grave. The grave itself is evidence, since it was produced during the crime of hiding a dead body, and it may preserve evidence that is only recoverable if its original surfaces are revealed during the excavation. For example, the sides of the grave may preserve toolmarks from the shovel used to dig it. The bottom of the grave may preserve footwear impressions. Therefore, the goal should be to focus on the grave itself, working carefully to reveal its original contours and surfaces.

One requirement that is more stringent than standard archaeological excavation technique is that forensic archaeologists try to collect trace evidence such as hairs and fibers. Therefore, all the dirt from the excavation should be passed through a small mesh screen.

When excavating a skeleton, the archaeologist will **pedestal** it. This means that the dirt is removed from around the bones, leaving the bones themselves in place until they have all been revealed. When this is done correctly, it is possible to see the original arrangement of the body in the grave, which may be important evidence. For example, are the hands tied behind the back, or are they folded across the chest? These differences can be interpreted in terms of the relationship between the deceased person and the person who buried him or her. If the person doing the burying did not care about the deceased person, they might just dump them into the grave. However, if the person doing the burying had feelings for the deceased person, they may take time to arrange the body in a conventional or peaceful position. Other items encountered during excavation, such as firearm components, should also be pedestaled if possible so that there is a chance to see any relationships between the bones and other items in the grave.

27.4 Forensic Osteology

What forensic osteologists do is interpret the skeleton. Given the appropriate bones, a forensic osteologist can usually determine several things about the individual who is represented by the skeletal remains. Forensic osteology is done in a laboratory. It is important to avoid snap judgements based on observations in the field. The Montana State Medical Examiner is fond of saying, "I don't do field autopsies," and we forensic anthropologists should likewise avoid doing field osteological examination, despite the pressure any peace officers present will put on us to make an immediate pronouncement.

It is also important that the bones arrive at the laboratory packaged intelligently and with any information about the grave and its contents that was discovered during the recovery of the skeleton. I am always puzzled by the fact that peace officers seem to forget all their training in crime scene processing when recovering skeletal remains. In the vast majority of cases that I handle, the bones come to me shoved into paper bags or cardboard boxes, with no rhyme or reason to their packaging and no information about the circumstances of their recovery. First of all, this wastes my time, because I have to go through and do a considerable amount of sorting of the bones, placing the right hand bones together, separating the right ribs from the left ribs, etc. Since it is likely that the bones were discovered in the proper anatomical relationships in the grave, I do not understand why peace officers are so careless with the evidence as to scramble the bones during collection and packaging. Second, my analysis can only be better if I am supplied with the photographs and other documentation of the recovery.

27.4.1 Class and Individual Characteristics of a Human Being

Interestingly, a human being, or at least their remains that are in evidence, are like many other forms of evidence in that they have both class and individual characteristics. The class characteristics of a human being that can be estimated by a forensic osteologists include their sex, their age when they died, their race (though see my rant on race in the advanced topics section on genetic profiling in chapter 21), and their stature (standing height).

We have seen that the individual characteristics of other types of evidence come from uniqueness of manufacture and from use or abuse of the item. The individual characteristics of human beings are similar in that they come from uniqueness of birth or arise from events that happen to them over their lifetimes, such as diseases and injuries.

The identification of an unknown deceased person is the same as in any form of matching in forensic science. Here, the evidence items are the class and individual characteristics identified by the osteologist. The standard items are records of the person's characteristics that were made while he or she was still alive. Since most of the characteristics forensic osteologists can determine are class characteristics, most of the identifications they make are consistent identifications rather than positive identifications.

In order to make a positive identification of a person from their skeletal remains, some of the person's individual characteristics must be detected, and this must be matched to information about the person that was collected while they were still alive. The main sources of information about a person's individual characteristics in life are medical records and dental records. Therefore, positive identification of a person from their skeletal remains usually involves collaboration between the forensic osteologist and a forensic pathologist or forensic odontologist.

Therefore, the process of identification often goes like this. First, the osteologist determines the class characteristics, say that this is a White male about 50 years old and around 5' 11" in height. The law enforcement agency with jurisdiction in the case will then query missing person's lists to find one or more individuals who fit this general description. Once these candidates are identified, their medical and dental records will be obtained, and the osteologist will work with a forensic pathologist to match details of bone fractures, maxillary sinus patterns, or other disease or injury data that might be preserved in the medical records. The teeth will likely be sent to a forensic odontologist, who will try to match them to the data recorded in the dental records.

27.4.2 Age, Sex, Race, and Height From the Skeleton

A person's age is primarily recorded as the results of body growth and deterioration processes. In the skeleton, this generally means that you are looking at progressive **ossification** (deposition of bone) through time. We all start life as a boneless embryo, and if we live long enough will become an arthritic elderly person by the end of our lives. The rate of ossification varies – it is much more rapid in the embryo than in the elderly person – but it is a constantly occurring phenomenon. During childhood there are many features of growth that can be examined to estimate age. One of these features is bone fusion. For example, we are born with our frontal bone in two parts, which fuse together to form the single frontal around the age of two years. Another feature that can be examined is the eruption of the teeth. For example, as discussed in chapter 25, the first molars erupt at about age 6 years and are informally known as the six-year molars. After full skeletal growth is complete, usually when the person is between 22 and 25 years old, the processes of bone fusion continue. For example, the cranial sutures eventually begin fusing, and in some individuals can become completely fused. The extent of cranial suture fusion can be used to generate a rough estimate of age. Other processes record degeneration or wearing out of the skeleton. For example, the occlusal surfaces of teeth accumulate wear over a person's lifetime, as they are used for chewing. Unlike other tissues of the body, the enamel of teeth can not regrow or repair itself, so this wear gets more and more pronounced with increasing age. Within a certain population, which eats a certain type of diet and has a typical rate of tooth wear, this process can be used to give an estimate of a person's age. Figure 27.2 illustrates dental attrition. In this figure the enamel has been worn through at a pinpoint-sized location on one cusp of this tooth from a middle-aged individual exposing the underlying dentin, which is darker in color.

Figure 27.2: Dental Attrition

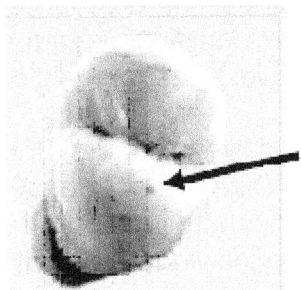

Sex is primarily recorded as some aspect of **sexual dimorphism**, which is defined as those ways in which the bodies of females and males differ. Genitals, and other sexually specific body parts are not part of the skeleton, and despite an often repeated bit of nonsense, both males and females have the same number of ribs. In humans, three factors of sexual dimorphism are useful for estimating sex from the skeleton. The first is differences in body size, with males on the average being larger than females. The second is differences in the shape of the pelvis due to the fact that human infants have big heads that have to fit through the birth canal of their mother's

Figure 27.3: Female and Male Pelves

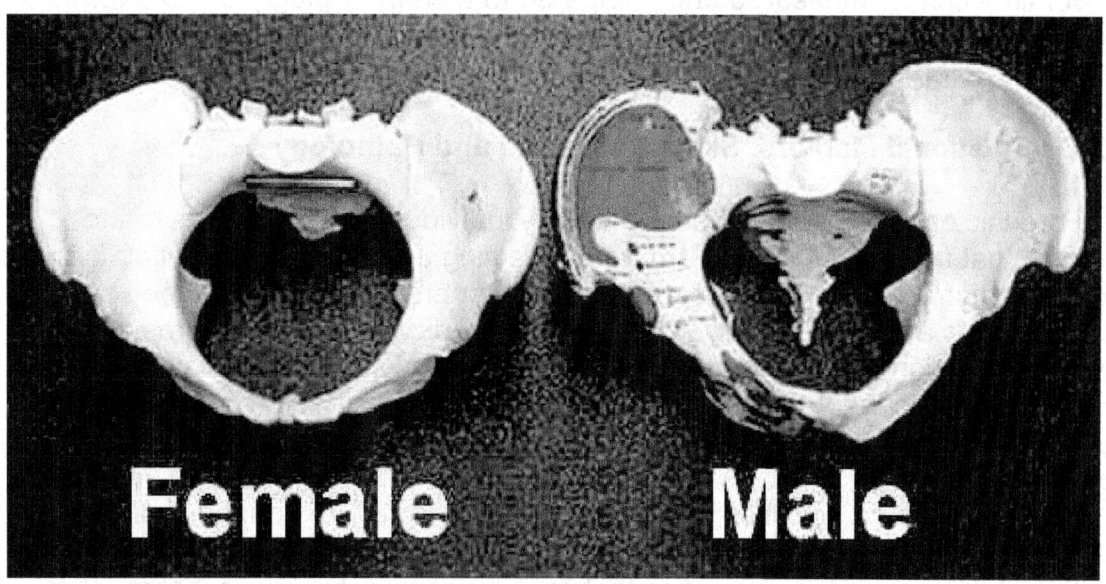

pelvis. Females therefore have a pelvis with plenty of room in the birth canal, which is a shape difference that translates to sex specific differences in the shapes of the coxal bones and the sacrum. Males do not give birth, and the shapes of their coxals and sacrum differ from those of females. Figure 27.3 illustrates the difference between female and males pelves. Third, due to hormonal systems, males are more muscular than females and this translates to larger muscle attachment sites on the bones.

Race is very difficult to estimate from the skeleton. I estimate that of all the cases that come to my lab, I can estimate race unambiguously in only about 25% of them. In about half the cases, I can exclude one race, to say for example that this person was not Caucasian, but that is all. In the final 25% of cases I can not estimate a race for the person at all – they appear completely generic. One of the biggest problems is that anthropologists tend to believe the scientific evidence that race is a sociocultural phenomenon and not a biological phenomenon. Regardless, the features in which people of different races differ are pretty much only skin deep and are not reflected in the bones. Primarily it is the region in the center of the face that contains a few subtle characteristics that are partially diagnostic for race. One of the best ways to think about the differences in the faces of people with different ancestries is to imagine the generic human face as a flat surface. People of different races have faces that differ from the generic human face in that some region of it has been pulled forward. The peoples of Europe and their descendants look as if someone has grabbed the generic human face by the nose and pulled that region of the face forward. The peoples of Africa and their descendants, look as if someone has grabbed them by the mouth and pulled that region forward. The indigenous peoples of Asia, the Pacific Islands, and the Americas look as if someone has grabbed them by the sides of the face and pulled the zygomatic bones forward.

A person's stature is estimated by measuring the lengths of limb bones. This is possible because of genetic relationships between the length of a person's limb bones and their overall height. This is most clearly seen in the fact that the length of the femur and tibia contribute directly to a person's height. The length of the bones of the upper and lower limb can be measured and compared to a chart or plugged into a formula that provides estimates of the person's stature from these measurements.

27.4.3 Occupational Markers, Skeletal Trauma and Pathology

Forensic osteologists can usually identify individual characteristics of the skeleton. What usually prevents them from obtaining a positive identification is the lack of documentation of these characteristics while the person was alive. When such documentation is available, usually in the form of medical or dental records, then a positive identification may be attempted in collaboration with a forensic pathologist or odontologist as described above. Individual characteristics that forensic osteologists recognize fall into three broad categories: occupational markers, skeletal trauma, and pathology.

Certain repeated activities leave characteristic signatures on the skeleton. These are often referred to as **occupational markers**, though we certainly recognize that not all of them arise from activities that are part of a person's occupation. For example, squatting facets, which are expanded areas of the joint surfaces of the ankle and knee, are often found in the skeletons of people whose activities caused them to spend a lot of time squatting. Certain habits, such as clenching the stem of a tobacco pipe between the teeth while smoking can leave a record in the form of characteristic tooth wear. People who ride horses often have enlarged muscle attachment sites for the adductor muscles of the thigh, which are used when squeezing the horse's body with the thighs and knees while sitting in a saddle. Similar features are produced by a variety of activities.

Trauma is the result of an injury, and some types of trauma are visible on the bones, such as bone fractures and bullet or knife wounds. The role of forensic osteologists in identifying trauma is complex. On one hand, it has been demonstrated repeatedly that osteologists are better able to distinguish bone abnormalities from normal bone features than any medical professionals except perhaps orthopedic surgeons and some forensic pathologists. On the other hand, however, osteologists lack the medical training to be able to correctly diagnose or identify the cause of the abnormality. Therefore, this is another area in which collaboration between the forensic osteologist and their friendly neighborhood forensic pathologist is necessary. Over the years the State Medical Examiner and I have developed a relationship of mutual trust and respect as he has sought my expertise on many occasions and I have likewise sought his. The result, of course, is a better characterization of the nature and cause of the skeletal abnormality.

One thing that forensic osteologists are recognized as having the expertise for is determining the timing of a trauma relative to the time of death. Traumas, can be described as **premortem** (or antemortem), meaning that they occurred significantly before the time of death; **postmortem**, meaning that they occurred significantly after the time of death; and **perimortem**, meaning that they occurred around the time of death. Perimortem trauma might be related to the cause of death. Postmortem trauma can not be related to the cause of death, though premortem trauma might be in some cases, as discussed in chapter 24. Even if trauma is judged to be perimortem, this does not mean that it must have been involved with the cause of death, because the features used to distinguish the three forms of trauma are not sensitive enough to distinguish small differences in time, and the term perimortem is probably best though of as including that time period between about two weeks before death and about two months after death.

Premortem trauma is distinguished by the fact that some process of healing is visible. This may be in the form of rounding off the edges of a break as natural bone destroying cells called osteoclasts clean up the damaged areas, to actual bone repair. Postmortem trauma is normally diagnosed by a difference in color or texture of the broken surface. After death, bones begin to "weather", taking on colors and textures primarily determined by the environment in which the bone is located at the time. If a bone is broken after enough time has elapsed for these processes to begin, then the broken surface will not show the changes in color or texture to the same degree as the normal surface of the bone. If the trauma occurs to a bone long enough after death that the bone has dried out, then the break will be sharp and clean, like breaking a dry twig, and this pattern of breakage is called **dry stick fracture**. In contrast, premortem and perimortem trauma to a bone will result in some bending of the bone, similar to that seen when one attempts to break a freshly plucked twig. This pattern of breakage that shows a bending component is called **green stick fracture**.

If a trauma does not show the characteristics of either premortem trauma or postmortem trauma, then it occurred soon enough before death that no healing processes are visible or at the time of death, or soon enough after death that changes in the color or texture of the bone surfaces had not progressed to a detectable degree. Figure 27.4 illustrates postmortem and premortem trauma. In the upper panel of this figure rodent gnawing of a tibia from an individual who died several years previously has removed the stained outer layer to reveal inner bone that is a different color. The bone in the bottom panel of Figure 27.4 shows a thickening in the middle of the tibia due to a healed fracture that occurred while the person was alive.

Figure 27.5: Closeup View of Porotic Hyperostosis, A Pathological Condition that May Result from Anemia

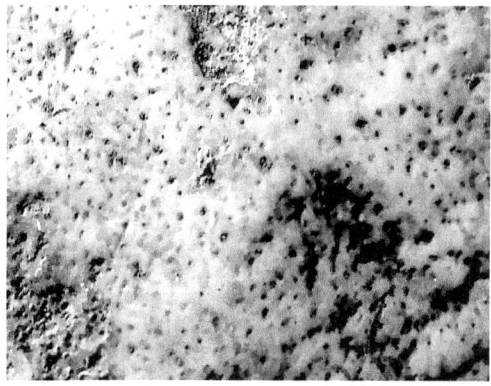

Figure 27.4: Postmortem and Premortem Trauma

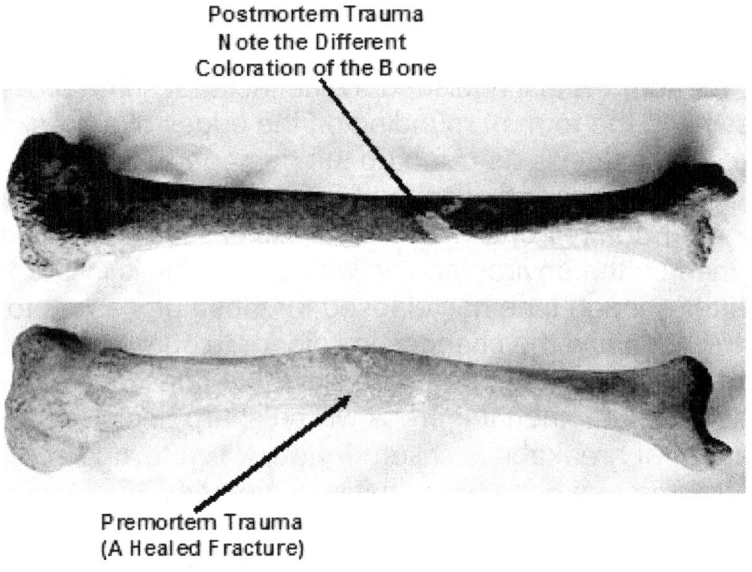

Postmortem Trauma
Note the Different
Coloration of the Bone

Premortem Trauma
(A Healed Fracture)

The term **pathology** refers to a disease process, rather than to the result of an injury or the result of some commonly repeated activity. Some diseases leave traces on the skeleton, and certain diseases occur in such a way that they leave characteristic patterns of lesions on the bones. The list of diseases that can be diagnosed from the skeleton includes tuberculosis, syphilis, leprosy, various infections that affect the bones, some vitamin deficiencies, arthritis, and osteoporosis, among many others. Figure 27.5 is a closeup of the pincushion appearance of the parietal bone near the sagittal suture. This is a pathological condition called porotic hyperostosis, which can result from anemia. As is the case for trauma, osteologists often find themselves in a position where they recognize the abnormality of the bone, but do not have the medical knowledge to actually diagnose the disease that produced it. There are a few books now published that describe the characteristic lesion patters of a wide range of disease, and these are helpful. However, it is usually a better idea to seek the opinion of a forensic pathologist in making a diagnosis.

27.5 Questions for Study and Review

1. In your own words, describe the difference between forensic archaeology and forensic osteology.
2. Discuss why the pre-scientific pursuit of race differences by physical anthropologists had lasting consequences on the respectability of forensic anthropology.
3. In your own words, describe how the process of burying evidence creates soil and vegetation disturbances that can easily be detected.
4. In your own words, describe how a secondary depression forms in some graves.

5. List some of the methods other than simply searching for soil and vegetation disturbance that are used in locating buried evidence.
6. List and discuss the ways in which the excavation of a grave differs from standard archaeological excavation methods.
7. Discuss why it is important to determine the relationships between items in a grave, in addition to simply recovering the items.
8. If you were excavating a grave, how would you package the bones and other items in a way that will save time when the forensic osteologist analyzes them?
9. What are the class characteristics of a human being that can sometimes be estimated from examination of their skeleton?
10. What are the individual characteristics of a human being that can sometimes be estimated from examination of their skeleton?
11. In your own words, define ossification and sexual dimorphism.
12. Discuss why a person's race is the most difficult thing to determine from examination of their skeleton.
13. What is the difference between a pathology and a trauma?
14. Discuss how a forensic anthropologist distinguishes between premortem, postmortem, and perimortem trauma.

27.6 Sources

Baldwin HB. 1998. The Recovery of Human Remains: A Crime Scene Perspective [Internet]. [cited 2007 Oct 23]. Available from: http://www.feinc.net/cs-recover.htm.

Batchelor J. n.d. Solving The Osteological Conundrum Of Buried Human Remains: What can your local archaeologist offer? [Internet]. [cited 2007 Oct 23]. Available from: http://www.soton.ac.uk/~jb3/forearc2.html.

Bass WM. 1987. Human Osteology: A Laboratory and Field Manual. 3rd Edition. Columbia (MO): Missouri Archaeological Society.

Boyd, RM. 1979. Buried body cases; ESI Law Enforcement Bulletin February 1979,M-7.

Briggs C. 2007. The Changing Role of Forensic Anthropology in the 21st Century. Homo – Journal of Comparative Human Biology 58 (3): 238-239.

Brothwell DR. 1981. Digging up Bones. 3rd Edition. Ithaca (NY): Cornell University Press.

Buikstra JE. 2007. Forensic Anthropology and Medicine: Complementary Sciences from Recovery to Cause of Death. International Journal of Osteoarchaeology 17 (4): 434-436.

Byrd M. 2000. The Corpse as a Scene [Internet]. [cited 2007 Oct 23]. Available from: http://www.crime-scene-investigator.net/corpse.html.

Cattaneo C. 2007. Forensic Anthropology: Developments of a Classical Discipline in the New Millennium. Forensic Science International 165 (2-3): 185-193.

Hackens T. 1985. How science unlocks the secrets of the past. UNESCO Courier, July 1985 p12(7).

Morse D, Duncan J, Stoutamire J. 1983. Handbook of Forensic Archaeology and Anthropology. Published by the editors. Available from Jack W. Duncan, 1830 Big Pond Way, Tampa, FL 33647.

Murdo MD. 2001. A Comparison of 3 Methods for Locating Buried Remains. Thesis submitted in partial fulfillment of the requirements for the Master of Arts degree in Anthropology at The University of Montana – Missoula.

Newman M. 2007. We found bodies with hands tied. Times Higher Education Supplement 1810: 8(8).

Shipman P, Walker A, Bichell D. 1985. The Human Skeleton (MA): Harvard University Press.

Steele DG, Bramblett CA. 1988. The Anatomy and Biology of the Human Skeleton. College Station (TX): Texas A&M Press.

Strongman B. n.d. GPR [Internet]. [cited 2007 Oct 23]. Available from: http://www.soton.ac.uk/~jb3/gpr/gpr.html.

Ubelaker DH. 1978. Human Skeletal Remains: Excavation, Analysis, Interpretation. Revised Edition. Washington (DC): Taraxacum Press.

Ubelaker DH. 2000. A History of Smithsonian–FBI Collaboration in Forensic Anthropology, Especially in Regard to Facial Imagery [Internet]. Forensic Science Communications 2(4). [cited 2007 Oct 21]. Available from: http://www.fbi.gov/hq/lab/fsc/backissu/oct2000/ubelaker.htm.

Wright R. 1995. Investigating War Crimes - The Archaeological Evidence. The Sydney Papers 7(3): 39-45.

Chapter 28
FORENSIC ODONTOLOGY

The topic of this chapter is forensic odontology, also known as forensic dentistry. Forensic odontology includes two applications to forensic science. The first is identifying the remains of deceased people from their teeth. The second is matching bite marks. These two specialties, though both within the expertise of dentistry, are quite different in application. The identification of deceased persons is part of the realm of death investigation, while bite mark matching is a criminalistic enterprise. There are also some forensic odontologists who specialize in injury analysis and dental malpractice issues in civil court.

After discussing forensic odontology, this chapter will continue with an advanced topics section that presents the "golden team" of death investigators (forensic pathologists, anthropologists, odontologists, and sometimes including DNA analysts) and their work in mass disasters and similar disaster-like scenarios such as wars and human rights violations.

28.1 The History of Forensic Odontology

Cases in which the identification of dead persons have been made by means of teeth are known from ancient Egypt, dating to around 2.500 BCE; and famous examples have been documented from Rome, dating to around 45 – 70 CE. Two large fires in the late 1800's, one in Vienna (1881) and the other in Paris (1897), resulted in large numbers of people who were burned beyond recognition and needed to be identified dentally. Most of the bodies from the fire in Paris were identified by Oscar Amoedo, a Cuban. Almoedo wrote a book about his experiences, which is regarded as the first text in the specialty of forensic odontology. Through the 1900's, forensic odontologists became established as the scientists to turn to for identification of deceased persons. More initially unknown deceased persons have been identified by forensic odontology than by any other method.

Bite mark matching emerged during the late 1800's and 1900's, as technologies appeared for documenting bite marks and for collecting the bite mark patterns of suspects.

28.2 Identification of the Deceased from Dental Records

Teeth and dental work are often very important for identifying dead bodies. This is so because the recognizable soft tissues of the body, such as the face, often decay after only a couple of weeks. Teeth, however, are quite durable and are usually the last part of the body to disappear.

Forensic odontologists believe that no two individuals have identical teeth. First, everyone's teeth are slightly different in size and shape. Second, the size, shape, nature, and location of dental pathologies and their treatments, such as cavities and fillings, are individual characteristics because they arise from processes that are random. However, unless information is collected on these features of a person's teeth while they were alive, it will not be possible to identify them on the basis of these characteristics when they are dead. Fortunately, dentists do routinely record the location, and often the size and shape of fillings. If crowns or bridges are used to restore damaged or missing teeth, their nature and location are recorded. Further, dental x-rays record information on sizes and shapes of teeth. The task of the death investigator now becomes finding the appropriate dental records to which to compare the teeth of a deceased person. This is almost impossible unless there is some clue as to the person's identity. There is no national database of forensic odontological information to turn to. Often, a forensic anthropologist can generate information that allows a small number of individuals to be chosen from a missing persons list, which simplifies the task of locating dental records.

28.2.1 Dental Records

Dental records occur in two forms, charts and radiographs. **Dental charts** are paper diagrams on which the nature and location of fillings and other dental restorations are made. Although most dentists try their best to keep accurate dental charts, it is impossible to record the exact three dimensional shapes of fillings in this manner. However, the presence or absence of fillings on specific teeth, their locations, and their approximate shapes can be matched.

Radiographs are dental x-ray pictures of the teeth. Most dentists will take a full panel of radiographs for their patients every five to ten years, and at other times where the status of a particular tooth is in question. Radiographs give the best information about the shapes and locations of fillings and other dental work. In addition, they give information about the details of tooth sizes and shapes, tooth sockets, tooth roots, and the jaw bone, which also may provide individual characteristics for matching.

There are three major limitations on the availability of dental records. First, as discussed above, investigators need to know whose dental records to ask for. Second, not all people seek regular dental care and dental records for a certain individual may not exist. In fact, due to advances in preventive dentistry, such as fluoridation of water, many younger people have never had a cavity and may never even have had a comprehensive dental examination that includes radiographs. Third, dentists do not keep patient records forever. Most states require dentists to keep patient records for at least several years, but older ones may be discarded. Dentists do not have unlimited storage space and must periodically clean out their records, discarding those of patients who they have not seen for years.

28.3 Bite Mark Analysis

The second major focus of forensic odontology is bite mark matching. Since each of us has teeth that are unique, each of us produces a unique mark when we bite something, called a **bite mark**. The unique features of the bite mark include the sizes and shapes of the biting surfaces of the teeth, the spacing between the teeth, and the rotations of the teeth. Bite marks are relatively often found on or in food, or food-related items such as styrofoam cups. Also, biting is an unexpectedly common form of attack and also an unexpectedly common form of defense. I guess we are not so far removed from our primitive ancestors after all. Because of this, bite marks can often be found on human bodies, both dead and alive. The body of the victim may bear an attacker's bite marks and the body of the attacker may bear the victim's bite marks.

The material being bitten influences what the bite mark looks like. For this reason, bite mark evidence is most often used to exclude than to make a positive match to a suspect. However, there are many cases in which the bite marks are clear enough that a positive match can be made.

28.3.1 Is it a Bite Mark, and is it Human?

When a forensic odontologist is presented with something that might be a bite mark, there are a set of questions that need to be answered. First, is the feature in question really a bite mark? Bite marks will usually consist of punctures, tears, bruises, and reddened areas, but rarely cuts.

If the mark is determined to be a bite mark, the next question is whether it is a human bite mark or not. Many animals, ranging from dogs to mountain lions, will bite humans. The bite of a carnivore can usually be distinguished from the bite of a human by comparing the sizes of the incisor and cuspid teeth. Cuspids in non-humans are called **canine teeth**. Carnivores, such as dogs, bears, and mountain lions have small incisors and large canines. Therefore, their bites usually consist of two deep punctures with six small indentations in between, which correspond to the three incisors per quadrant of most carnivores. In humans, however, the cuspids are relatively small and the incisors are relatively large, especially the first incisors. So, a human bite will usually consist of two broad tears or punctures in the center of the mark, corresponding to the two first incisors; with two smaller indentations on each side corresponding to the second incisors and the cuspids.

28.3.2 Collecting Bitemark Evidence

Once a mark has been determined to have been produced by a human, it has to be collected as evidence. Bite marks on a living person will disappear with time, and bite marks will also sometimes disappear from the body of a dead person over time due to decay processes. The first step in collecting bite mark evidence is photography. It is especially important to use a scale in the photo, so that the bite marks can be measured. If the person has recently been bitten, the area of the bite mark should be swabbed to collect saliva that can be used for serology or DNA analysis. If the bite mark has depth to it, a cast of it can be made which will often give a very accurate reproduction of the teeth that made the bite.

Next, a sample of a suspect's bite is taken (or the victim's bite if it was the suspect who was bitten). The investigator could offer their arm to the suspect to bite, but more often the suspect is asked to bite a flat plate of hard wax. Their teeth make an impression in the wax that records their bite. Other methods for collecting this information also exist, such as making a cast of the suspect's teeth.

28.3.3 Matching Bite Marks

Finally, the known bite (a standard item) can be compared to the evidence bite mark. The most convincing method to do this involves photographically transferring the details of the standard bite to a sheet of clear plastic material. Then the forensic odontologist attempts to superimpose the details of the standard bite mark over a photograph of the evidence bite mark. The sizes of these two items will have to be adjusted, but this is easy to do using digital imaging methods. If the standard and the evidence bite marks can be superimposed in such a way that the details of both line up perfectly, the match is considered positive. If, however, the standard bite mark can't be plausibly matched to the evidence bite mark evidence, then there is and exclusion. Occasionally, the details of the evidence bite mark are not clear enough to allow a positive match, and a result of indeterminate or even undetermined is reported in this situation.

28.4 Advanced Topics: Forensic Scientists in Mass Disasters

Unfortunately, we live in a world in which mass disasters happen more than occasionally. Nature remains unpredictable and uncontrollable, resulting in floods, earthquakes, volcanos, hurricanes, tsunamis, and other disasters. Disasters caused by humans, or which occur due to malfunctions in technology, such as fires, bombings, and airplane crashes, are even more common and seemingly becoming more common all the time. Disasters of this sort often leave unidentified bodies and body parts strewn around, which forensic scientists must attempt to identify.

28.4.1 The "Golden Team"

A mass disaster is clearly a situation that calls for teamwork among law enforcement and several types of forensic scientists. This is, however, no place for untrained volunteers. There is a role for almost any type of forensic scientist to play in the aftermath of a disaster, yet one group of them has been found to be so useful in identifying the victims that they have been informally designated the "golden team" of victim identification. The "golden team" has traditionally consisted of forensic anthropologists, forensic odontologists, and forensic pathologists. DNA analysts have joined the team as valuable members in recent times.

28.4.2 Disasters that Could Happen: Spokane, WA as an Example

Mass catastrophes with a large number of simultaneously dead or missing persons are increasingly common around the world and the potential for them happening is enormous. As part of their emergency preparedness planning, the State of Washington, in 1996, made a survey of what sorts of mass disasters could occur in the area of Spokane. Although it is not in the exact center of the country, Spokane serves as a good example of a typical city because it is neither too large nor too small, neither too near the coast nor too far away from it, neither too high in elevation nor too low, etc. It also has a nearby river, as do most major cities in the US. What this amounts to is that those disasters that could happen in Spokane could happen almost anywhere. Table 28.1 presents a partial list of the disasters that the State of Washington identified. Different types of disasters may be more or less likely in other cities, depending on their location. For example, the probability of having a tornado is quite low on the list for Spokane, and would be higher in, for example, Oklahoma City.

**Table 28.1: Disasters that Could Happen in Spokane, WA
In Order from Most Likely to Least Likely**

Winter Storm
Wildfire
Hazardous Materials Accident (Chlorine Spill, for example)
Urban Fire
Transportation Incident (Airplane Crash, for example)
Flood
Hazardous Materials Facility Accident
Civil Disorder
Radiation Incident
Dam Failure
Earthquake
Volcano
Landslide
Tornado
Avalanche
Hurricane
Tsunami

28.4.3 Who Gets Involved and What Do They Do?

During a mass disaster, forensic anthropologists usually take responsibility for working with emergency services personnel to impose order upon the chaos of the scene. Forensic archaeologists are especially well prepared to do this because the profession of archaeology often involves imposing order upon an archaeological site, sometimes a large one. The forensic osteologists, due to their familiarity with the human body in all stages of decomposition, often have a very low "yuck factor", meaning that they less often become ill or emotionally unable to continue when

presented with the reality of death on a large scale. Some of the things forensic anthropologists might do include the following.
- Imposing a grid system upon the location in order to provide a system for organization of who does what where, and for recording where remains are found.
- Organizing the recovery of remains (including excavation if necessary) and recording where they are found.
- Distinguishing human from non-human remains.
- Determining which body parts belong together, if appropriate.
- Doing preliminary, consistent identifications based on sex, age, race, etc. In large mass disasters this may involve simple sorting people and body parts into age/sex categories.

The roles of forensic odontologists, forensic pathologists, and forensic DNA analysts is normally to take primary responsibility for identification of victims. Forensic anthropologists often provide a quick analysis of the class characteristics of a disaster victim, allowing the others to make appropriate choices of medical or dental records; or in the case of DNA analysis, which relatives to ask for DNA samples.

28.4.4 DMORT Teams

DMORT stands for Disaster Mortuary Operational Response Teams. DMORT's are a set of local and regional teams that consist of forensic scientists and other professionals who have volunteered to help in the case of a disaster. DMORT's are voluntary associations, though when a forensic scientist responds to a disaster as a part of a DMORT, their expenses are paid. DMORT's are only mobilized when requested by a government agency and they only do assistance with identification of remains in situations where the number of fatalities exceeds the ability of local resources to handle them. As of 2005, there are about 1,200 people in the US who were members of a DMORT.

The list of things DMORT's can help with is large, and include an array of expert services provided by the forensic scientists, as well as many more general services. A partial list is shown in table 28.2. It is worth your time to survey this list because it gives a brief overview of the vast number of complex tasks that have to be performed in response to a disaster.

Table 28.2: DMORT Services
Mobile Morgue Operations, embalming, casketing
DNA Acquisition
Remains identification
Search and recovery
Scene documentation
Medical/psychology support
Family Assistance center
Antemortem & postmortem data collection
Records data entry

Database administration
Personal effects processing
Coordination of release of remains
Provide liaison to government agencies
Provide communications equipment

28.4.5 Other Disaster-Like Scenarios

There exist other situations that produce a large number of bodies and body parts. One of these is war, which certainly qualifies as a disaster for any country or community in which one occurs. Another disaster-like scenario is human rights violations, such as genocides or government repressions. As in mass disasters, those who handle the identification of victims in the aftermath of these situations are likely to be the "golden team" of forensic anthropologists, odontologists, pathologists, and DNA analysts.

In examining Amnesty International's 2007 report, I found references to internal or external armed conflict within the borders of about 40 of the world's nations during 2006. Most authorities count 194 nations in the world, including the 192 members of the United Nations, the Vatican, and the Republic of Taiwan. During the past century, the United States has been involved in several military actions, none of which were on U.S. soil. The U.S. government promises to each and every member of the armed services that if they die overseas, every effort will be made to recover and return their remains. As near as I can tell, the government has kept this promise. Forensic anthropologists have been involved with war dead since World War II (1941 – 1945), and other members of the "golden team" have been involved since at least the Vietnam War (1965 – 1975). The U.S. Army's Joint POW/MIA Accounting Command (**JPAC**) takes primary responsibility for identifying U.S. military casualties, and employed the necessary forensic specialists to get the job done.

From information in Amnesty International's 2007 report, I estimate that there were government perpetrated human rights violations that led to the unauthorized death of civilians in about 50 of the world's countries. By unauthorized death I mean deaths that were not part of a war and not the result of a sentence received during a government sanctioned trial. In cases where citizens have been killed by their governments, the "golden team" can often recover skeletal remains that document the trauma a person experienced, and sometimes establish the identity of the victim. When a person's identity and the form of abuse they suffered can be documented, a case can sometimes be developed against the perpetrators of the human rights violations.

The major documents that establish basic human rights for all the people living in the world were developed by the United Nations. The United Nations issued the Universal Declaration of Human Rights in 1948, which contains 30 articles that list minimal human rights. For example, article 3 says, "Everyone has the right to life, liberty and security of person." This document is fairly short, and worth reading. It is online at http://www.un.org/Overview/rights.html. Another important UN document is the "Principles on the Effective Prevention and Investigation of Extra-legal, Arbitrary and Summary Executions". This document is also short and worth the read, and also online at the UN website. A third UN document establishes the relevance of forensic science to investigating human rights violations. This is the Commission on Human Rights Resolution on Human Rights and Forensic Science (2000), which is also available online.

The types of violations that forensic scientists have helped in documenting include massacres, mass murders, genocide, summary execution, political "disappearances", the existence of "death camps" and the existence of slave labor camps. The group most active throughout the world in documenting human rights atrocities is the Equipe Argentino de Antropologia Forense (**EAAF**), which translates to English as the Argentine Forensic Anthropology Team. They have been involved with this work in several Latin American countries, as well as countries in the Pacific, the Middle East, Africa, and Europe. Other teams also carry out these types of investigations, including teams of Americans and teams of Australians. Such teams have also documented human rights abuses in a variety of countries. This effort continues to be of interest to forensic scientists, who will no doubt continue their efforts into the future.

28.5 Questions for Study and Review

1. List and discuss situations in which it would be impossible to identify a person from their dental remains.
2. What class and individual characteristics exist for teeth?
3. Discuss why dental radiographs provide better evidence than dental charts.
4. How would you decide whether a bite mark was caused by a dog or by a human?
5. Discuss how bite marks are matched.
6. Which forensic scientists are part of the "golden team" of mass disaster responders?
7. Make a list of mass disasters that could occur in the city in which you live. Try to list them in order from most likely to least likely.
8. Describe what a DMORT is.
9. Describe how forensic scientists can help in the aftermath of war or human rights violations.
10. What agency has defined human rights on a worldwide basis?
11. What is the EAAF? In which country is it based? What does it do?

28.6 Sources

Amnesty International, 2007. Amnesty International Report 2007 [Internet]. [cited 2007 Oct 3]. Available from: http://report2007.amnesty.org/eng/Homepage.

[Anonymous]. n.d. How it All Started [Internet]. [cited 2007 Oct 24]. Available from: http://www.dmort.org/DNPages/DMORTHistory.htm.

Argentine Forensic Anthropology Team. 2007. Argentine Forensic Anthropology Team [Internet]. [cited 2007 Oct 24]. Available from: http://www.eaaf.org.

Bowers CM, Johansen RJ. 2001. Digital Rectification and Resizing Correction of Photographic Bite Mark Evidence [Internet]. Forensic Science Communications 3(3). [cited 2007 Oct 21]. Available from: http://www.fbi.gov/hq/lab/fsc/backissu/july2001/bowers.htm.

Burns KR. 1999. Forensic Anthropology Training Manual. Upper Saddle River (NJ): Prentice Hall.

Byrd M. 2000. Disaster Management: Lost Innocents [Internet]. [cited 2007 Oct 23]. Available from: http://www.crime-scene-investigator.net/disaster.html.

Department of Defense, Defense Prisoner of War/Missing Personnel Office. n.d. Vietnam War, Personnel Missing -- Southeast Asia (PMSEA Database) [Internet]. [cited 2007 Oct 24]. Available from: http://131.84.1.34/dpmo/pmsea/files_full.htm#name.

Glass RT. 2002. Body Identification by Forensic Dental Means. General Dentistry 50(1): 34-38.

Gunby P. 1994. Medical team seeks to identify human remains from mass graves of war in former Yugoslavia. The Journal of the American Medical Association 272(23): 1804(2).

Hackens T. 1985. How science unlocks the secrets of the past. UNESCO Courier, July 1985 p12(7).

Hannibal K. 1987. Forensic experts aid Philippine search for disappeared. Science 235: 535(2).

Heussner T, Holland T. 1999. Worldwide CILHI Mission To Bring Home Missing Heroes. Quartermaster Professional Bulletin Summer 1999 [Internet]. [cited 2007 Oct 24]. Available from: http://www.qmmuseum.lee.army.mil/mortuary/worldwide_cilhi_mission.htm.

Joyce C, Stover E. 1991. Witnesses From the Grave. Boston: Little, Brown, and Co.

Kieser J. 2006. Tsunami: Tragic Task: Dental Identification after the Thai. The Forensic Examiner 15(1): 46(4).

Maples WR, Browning M. 1994. Dead Men Do Tell Tales. New York: Doubleday.

Menez LL. 2005. The Place of a Forensic Archaeologist at a Crime Scene Involving a Buried Body. Forensic Science International 152(2-3): 311(5).

Needell BA, Kruse-Feldstein K. 2006. Hurricane Katrina: Morgue Operations: Two Forensic Professionals' Perspectives. The Forensic Examiner 15(3): 16(8).

Neiburger EJ, Patterson BD. 2002. A Forensic Determination of Serial Killings by Three African Lions. General Dentistry 50(1): 40-42.

Petju M, Suteerayongprasert A, Thongpud R, Hassiri K. 2007. Importance of Dental Records for Victim Identification Following the Indian Ocean Tsunami Disaster in Thailand. Public Health 121(4): 251(7).

Joint POW/MIA Accounting Command. n.d. Joint POW/MIA Accounting Command [Internet]. [cited 2007 Oct 24]. Available from: http://www.jpac.pacom.mil/index.htm.

Poole RM. 2006. Lost over Laos: Scientists and Soldiers Combine Forensics and Archaeology to Search for Pilot Bat Masterson, One of 88,000 Americans Missing in Action from Recent Wars. Smithsonian 37(5):38(10).

Skolnick A. 1992. Game's afoot in many lands for forensic scientists investigating most-extreme human rights abuses. The Journal of the American Medical Association 268(5): 579(3).

Snow CC, Stover E, Hannibal K. 1989. Scientists as detectives: investigating human rights. Technology Review 92(2): 42(9).

Stover E. 1997. The Grave at Vukovar: a War Crimes Tribunal Sent Forensic Scientists to Investigate Mass Graves like this One in the Former Yugoslavia. What Happened Here?. Smithsonian 27(12): 40(12).

Sweet D. 2002. The Wide Range of Forensic Dentistry. General Dentistry 50(1): 8-10.

Tidball Binz M. 1991. Argentina: Forensic Investigation of past Human-rights Violations. The Lancet 337(8757): 1593(2).

Tuller H, Duric M. 2006. Keeping the Pieces Together: Comparison of Mass Grave Excavation Methodology. Forensic Science International 156(2-3): 192(9).

United Nations General Assembly. 1948. Universal Declaration of Human Rights [Internet]. [cited 2007 Oct 24]. Available from: http://www.un.org/Overview/rights.html.

United Nations Economic and Social Council, The Commission on Human Rights. 2000. Human Rights and Forensic Science [Internet]. [cited 2007 Oct 24]. Available from: http://www.unhchr.ch/Huridocda/Huridoca.nsf/TestFrame/3f49328de45e94fd802568d400300bd8?Opendocument.

United Nations High Commissioner for Human Rights. 1989. Principles on the Effective Prevention and Investigation of Extra-legal, Arbitrary and Summary Executions. Recommended by Economic and Social Council resolution 1989/65 of 24 May 1989 [Internet]. [cited 2007 Oct 24]. Available from: http://www.unhchr.ch/html/menu3/b/54.htm.

Wright R. 1995. Investigating War Crimes – the Archaeological Evidence. The Sydney Papers 7(3): 39.

Chapter 29
FORENSIC ENTOMOLOGY, BOTANY, AND GEOLOGY

In this chapter we will explore three forensic sciences that are practiced by scientists who usually work at a university or college: forensic botany, forensic entomology, and forensic geology. Forensic botanists apply their knowledge of plants to examining evidence that may lead to the solution of a case. Forensic geologists are experts in dusts, soils, rocks, and petrochemicals, which are items of evidence in some cases. Forensic entomologists specialize in insects and insect evidence. Forensic botanists and geologists can be viewed as high powered trace evidence examiners in a certain sense. Trace evidence examiners occasionally work with pollens, plant fragments, dusts, and soils, and have some expertise in working with them. For very complex or serious cases, however, the additional expertise of a forensic botanist or forensic geologist may be exactly what is needed to develop crucial evidence. The largest, and most visible task of forensic entomologists is in estimating time since death from insect evidence associated with a corpse.

The chapter will start with an investigation of forensic entomology and how it contributes to death investigations and other types of cases. Next we will look at forensic botany and its applications. The chapter will conclude with an overview of the things that forensic geologists can contribute to investigations.

29.1 Forensic Entomology

Entomology is the study of insects. Forensic entomology applies knowledge about insects to develop evidence for use in criminal and civil cases. The use of forensic entomology goes back at least 700 years, but only since the 1980's has it been developed as a modern forensic science. Forensic entomology is commonly divided into three branches: urban entomology, stored products entomology, and medicolegal entomology. **Urban entomology** refers to cases in which insects and related animals are important in cases related to homes, restaurants, and other buildings. **Stored products entomology** refers to cases involving insects that have infested stored commodities such as grains. **Medicolegal entomology** is the branch that seeks to make a determination of the time since death.

29.1.1 Estimating Time Since Death

By far the most common usage of forensic entomology in the realm of criminal investigations is in estimating time since death. After death the body begins to emit gasses that are attractive to insects, especially certain types of flies (phylum Arthropoda, class Insecta, order Diptera). Some of these substances are cadaverine and putrescine. From the behavior of these flies, it seems that production of these substances begins immediately upon death. It takes some time for the amount of these substances produced to be great enough for the human nose to detect, however. The flies that usually arrive first are the blow flies (family Calliphoridae) and the flesh flies (family Sarcophagidae).

Female **blow flies** will lay their eggs on the body, especially around the natural orifices such as the nose, eyes, ears, anus, penis and vagina. If the body has wounds, the eggs are also laid in them. **Flesh flies** do not lay eggs, but instead give birth to live larvae, which are deposited in these same areas. After a short time the blow fly eggs hatch into **larvae**, commonly called maggots. The larvae pass through three larval stages during their development, called **instars**. The instars are defined by the larva shedding its exoskeleton, called **molting**, and emerging as a larger larva. The larvae feed on dead tissue. When a third instar larva reaches a certain point in the fly life cycle, it will move away from the body, build a casing around itself, and become a **pupa** (plural pupae). The pupal casing (puparium) is similar to the chrysalis of a butterfly in that eventually an adult fly will emerge to begin the life cycle over again. Flesh flies have a similar life cycle, but skip the egg stage.

Fly larvae can reduce a body to bones and dessicated tissue in an extremely short period of time, under two weeks in the Southeast U.S. under the most favorable of conditions. The rate of insect activity depends on temperature and other climatic factors, and studies by some of my former graduate students have shown that this process takes at least a year in Montana. Surprisingly, however, recent studies by graduate students here at The University of Montana show that insects can survive freezing conditions and continue to feed, though more slowly, through the winter. Most of these studies have been done using the carcasses of non-human animals, such as pigs or wolves. However, the University of Tennessee, Knoxville, has a facility for doing decay rate studies using donated human cadavers.

If the flies are regarded as the "first wave" of invaders of a corpse, the beetles are the "second wave". The **beetles** (order Coleoptera) are one of the largest groups of animals and some of them eat dead flesh, both as larvae and as adults. Maggots feed on softer moist tissue and are apparently incapable of feeding on dried tissues. Beetles, however, specialize on consuming the drier tissues. The larvae of beetles can be distinguished from maggots due to the fact that they have 3 pairs of legs, while maggots do not have legs. Other insects and other arthropods also feed on a corpse or utilize it in other ways. Taken together these insects provide evidence about time since death because their life cycles are very well known. The theory behind estimating time of death, often referred to as the **post mortem interval** (PMI) is that since insects arrive on the body soon after death, estimating the age of the insects will also lead to an estimation of the time of death.

The most precise way to determine the age of larvae and eggs is by collecting them at the crime scene and rearing them to adulthood in a laboratory, at the same average temperature as the crime scene. When the larvae complete their life cycle and become adults, their species can be identified precisely and the length of that species life cycle can be looked up in the literature on insects. If, for example, the species to which some collected larvae belong has a life cycle of 23 days and it took 14 days for the collected larvae to reach adulthood, then they must have been collected on day 9 of the life cycle. Therefore, the person probably died about 9 days before the larvae were collected.

Insect evidence can sometimes be informative about the cause of death. In fact, it is not unknown for venomous insects such as bees, or other arthropods such as spiders, to be the cause of death. Since flies target wounds when laying their eggs or larvae, an area of especially rich infestation may be the site of a wound. If poison or drugs were involved in the death, it may not be possible to detect them in the remaining tissues of the body, but they may be detectable by analyzing the maggots collected from the area of the liver, or even from shed pupal casings of earlier generations of insects.

29.1.2 Effects of Insects on Blood Evidence

Insects can affect the interpretation of blood spatter patterns. Roaches, flies, and fleas walking through pooled and splattered blood will produce tracks that may complicate the interpretation of the blood spatter. In addition, flies will feed on the blood and then pass the partially digested blood in its feces, which are known as "fly specks". Flies will also regurgitate and possibly disgorge a blood droplet onto a surface some distance away from the place where the blood originated, which may serve to confuse crime scene analysis. Therefore it is important to recognize and properly document the presence, feeding, and activities of insects at the crime scene.

29.2 Forensic Botany

Botany is the study of plants. Forensic botany is the generation and use of evidence from plants and plant parts (wood, seeds, pollen, etc.) in investigations. There are several topics that are considered part of forensic botany. These include: identifying wood, leaves, seeds, or pollen; interpreting the habitats of plants; and matching pieces of wood or other parts of plants. Plants can provide forensic evidence because the components and construction of a plant's body, and its ecological requirements, are unique to the species.

The case that brought forensic botany to the attention of the public was the kidnapping of the Lindbergh baby in the 1930's. A scientist named Arthur Koehler applied his knowledge of wood anatomy to matching some boards, which helped in tying the suspect to the crime scene. Botanical information has been used many times since.

29.2.1 Identifying plants

At the gross (non-microscopic) level, the leaves, flowers, wood grain, stems, and other features of plants differ enough that even an amateur with a good book on plant identification can identify most plants. When dealing with fragments of plants, however, botanists often turn to the microscope to help them make identifications. Plant cells have rigid external walls made of cellulose and the protein lignin. **Cellulose**, though simply a chain of sugars, is not digestible by animals except ruminants, termites, and a few other arthropods, all of which depend on certain bacteria and/or protozoans that reside in their intestines to actually digest the cellulose. The wood protein **lignin** is chemically and structurally more complex than cellulose, and can be broken down only by a few fungi. The manner in which cell walls are laid down and the patterns formed by cells are unique to the wood of each taxonomic group. Although it may not be possible to identify the exact species of oak wood, a botanist has no trouble distinguishing that it is oak wood or some sort. Nonwoody plants also have characteristic cell shapes and arrangements that allow them to be identified. Some plant cells contain crystals of calcium oxalate, starch grains, characteristic vein patterns, or characteristic hair-like structures that aid in their identification.

Since no botanist is familiar with every plant in the world, identification ultimately depends on access to information about plant characteristics. Databases of plant characteristics are available, the largest of which is maintained by the Jodrell Laboratory, which is part of the Royal Botanic Gardens in the city of Kew in the United Kingdom.

29.2.2 Interpretation of Plant Habitats

Different plant species have preferences for particular sets of environmental conditions, under which they grow best, and perhaps not at all elsewhere. Plant species tend to occur in **communities** of plants with similar preferences. Ecological knowledge about the preferred habitats of various species can be useful in forensic contexts because a botanist finding a certain set of plant parts associated with a suspect or a victim can sometimes determine that the person had been to a certain location where that community of plants was growing.

In forensic investigations, plant parts specifically adapted for certain methods of seed dispersal can be particularly useful. Cockleburs and the seeds of beggars' lice, for example, are adapted for sticking to the hair of animals in order to obtain a free ride to a location where they might be able to grow. Not being picky, they also stick to people's clothing, where they can serve as evidence that the person has been to a location where such plants grow.

29.2.3 Forensic Palynology

Palynology is the study of pollen, and forensic palynology is a specialty within forensic botany. Pollen exists everywhere, being shed from plants and often carried by the wind and air currents. The pattern of which species of plants are represented in a sample of pollen is known as a **pollen profile**. Pollen profiles are often distinctive for certain plant communities. Like other plant community evidence, a pollen profile can be used to demonstrate that a person has been to a certain location where that community of plants grows.

Pollen analysis is a technique occasionally used by a trace evidence examiner, who can match pollen profiles without having to delve into the ecological and community aspects of the pollen profile. For more complex cases, however, or when the placement of a suspect in a certain area is important for a case, a forensic botanist can supply expertise that might allow these additional questions to be answered.

29.2.4 Class and Individual Characteristics of Wood

Lets consider wood as an example of plant evidence that might be a candidate for matching. Evidence wood recovered at a crime scene is compared with standard wood items recovered in association with a suspect. Wood has both class and individual characteristics, though the class characteristics predominate.

Class characteristics of wood include its species, its tree ring pattern, and a few other details. The only natural individual characteristic of wood is its grain pattern, but grain patterns are difficult to work with in matching. Most cases of successful positive matching of pieces of wood are made based on fracture pattern matching between broken ends, or on the basis of toolmarks from the same tool being found on both pieces.

29.2.5 Other Applications

There are many cases in which a forensic botanist has been consulted in order to identify plants that contain drugs or which are poisonous. Sometimes these identifications have to be made on partially digested plant remains recovered from the stomach of a victim. A nationwide, computerized database on poisonous species is maintained by the Hunt Institute of Carnegie-Mellon University in Pittsburgh.

Botanists are also called upon to identify noxious or invasive species growing on agricultural lands. Some states have laws that require removal of nuisance plants, with the cost being shared by the landowner and the county. Removal operations of this sort are expensive, and everyone involved is concerned with the correct identification of the plant.

29.3 Forensic Geology

Forensic geology is the application of evidence from soils, rocks, minerals, coal, oil, and other substances obtained from the ground to criminal and civil investigations. Forensic geology emerged in the 1970's, later than many other forensic sciences. One of the pioneers in forensic geology is my friend, Ray Murray, a retired University of Montana vice president, dean, and geology professor, who wrote the first textbook on the subject. Forensic geologists primarily do matching of dusts, muds, soils and hydrocarbons to their sources. As such, they are similar in approach to criminalists, and some of the things they match can also be matched, though with a lesser level of expertise, by trace evidence analysts.

29.3.1 Matching Soils

The most common examination conducted by a forensic geologist is the analysis and comparison of soil samples. The goal of this type of examination is to determine whether two soil samples, an evidence item and a standard item, derive from the same source. Fortunately soils have both class and individual characteristics that allow this form of matching to be done.

Soil consists of a mixture of substances, which is usually unique. It may include rocks, fine particles produced from rocks, minerals, organic material, and various industrial substances such as particles of plastic or metal. Soils are found in some form nearly everywhere and they often cling to shoes or other articles of clothing.

Soils have several properties, some of which change and are not considered in forensic examinations. These unusable characteristics include the amount of water in the soil, its acidity (pH), and its density. Soils dry out, and their acidity changes for several reasons. As discussed in chapter 27, the degree of compaction of soil is altered by the process of digging it from the earth, which changes its density. Further, some characteristics of soil, its organic content for example, vary between different samples collected in the same area and are not suitable for forensic matching. In general, soil has three characteristics suitable for matching: color, mineralogy, and distribution of soil grain size.

As is common in forensic matching, a comparison of evidence and standard samples can result in a positive match, an exclusion, or indeterminate. In the case of comparing soils the result of indeterminate carries the same meaning as a consistent match (*i.e.* the class characteristics match, but it can not be determined whether the individual characteristics match or not), and usually occurs when the soil sample is either too small or is contaminated.

29.3.2 Class and Individual Characteristics of Soils

Soil color is a class characteristic, but one in which there is enough range of variation that it can prove useful for exclusions. No two soils have exactly the same distribution of grain sizes (silt, clay, sand, gravel, etc.), but simple methods for determining the range of grain sizes are not precise enough to allow this characteristic to be used for positive matching. Soil mineralogy, the chemical composition of the soil grains is an individual characteristic.

Color of soil is determined by visually comparing a dried sample to a standard color system called the **Munsell color chart**. It is estimated that several hundreds of types of soils can be distinguished by color alone. Since this method is rapid, it is usually the first test applied during matching of soil samples.

Soil grain size distribution is determined by stacking a set of sieves with differing mesh sizes so that the largest mesh sieve is at the top of the stack and the finest at the bottom. The soil sample is placed at the top of the sieve stack and washed or shaken through. Each sieve will trap soil particles that are too large to pass through its mesh, so that the soil grains are efficiently separated by size. Particles of a certain size category (trapped by a certain sieve) are collected and weighed to determine the size distribution of the soil particles in percents by weight.

If these two tests do not result in an exclusion, the forensic geologist will determine the mineralogy of the sample. The **mineralogy** of a soil sample refers to the chemical composition of the particles of which it is composed. Since soils contain particles from different types of rocks, plastics, metals, and organic materials, the mineralogy is unique to each soil.

29.3.3 Dusts and Muds

Dusts and muds can be regarded as soils with a certain moisture content. Dusts are often so dry that they are easily blown around by the wind, while muds are saturated with water. Therefore, essentially the same procedures are used when comparing dusts or muds as when comparing soils. The main difference is in the preparation necessary, since muds will need to be dried before being examined.

29.3.4 Collecting and Preparing Soil Evidence

The quantity of soil required for matching is surprisingly small, approximately equal to the size of a wooden match head, though some cases require a larger quantity. Soil samples can be obtained from clothing, footwear, or other items by either shaking the item over a large sheet of paper or by scraping a sample off the item. If the items have been shipped or transported it is likely that some of the soil will have fallen off the item and be loose in the bottom of the container or within the packaging material. The preferred method for collecting standards is to obtain them directly from the location in question.

Samples are generally allowed to air dry for around 12 hours, then any clods or lumps are broken up and the sample is oven dried at about 95 degrees Celsius for four hours. This removes any excess moisture, producing samples with the same amount of moisture that can then be compared.

29.3.5 Hydrocarbon Matching

The long association between geology and the petroleum industry has resulted in geologists being authorities on hydrocarbons. Hydrocarbons are molecules formed of carbon and hydrogen, as described in chapter 11, and other than methane are normally produced from petroleum. One common task of forensic geologists, therefore is matching samples of hydrocarbons.

While one batch of a pure hydrocarbon is identical to any other, many products containing hydrocarbons are produced at refineries from petroleum which contains impurities. The types and amounts of impurities vary from one oil field to another. Refineries accept delivery of petroleum from a wide range of sources and each batch of a hydrocarbon product, gasoline, for example, is produced from a unique, random mixture of petroleum from several sources. The resulting product will, therefore, have a unique combination of trace impurities, called a **hydrocarbon fingerprint**.

Matching of samples of a hydrocarbon product, such as gasoline or kerosine is often called hydrocarbon fingerprinting, and uses gas chromatography. The recorder attached to the output of the gas chromatograph produces a chart with many peaks, each corresponding to some substance in the hydrocarbon product. It is a straightforward matter to compare the recorded charts of two samples of a hydrocarbon product to determine whether their trace impurities are the same. Hydrocarbon products that can be matched by hydrocarbon fingerprinting include gasoline, kerosene, naphtha, reformate, jet fuel, diesel, fuel oil, hydraulic oil, lubricating oil, crude oil and a few other refinery products.

One of the most common applications of hydrocarbon fingerprinting is in matching traces of accelerants recovered from the scene of a fire to containers of these products in the possession of an arson suspect. Accelerants are substances such as gasoline or lighter fluid that are used to make a fire burn faster and hotter. Arson investigation will be discussed in more detail in the chapter on forensic engineering. Hydrocarbon matching is also used in cases of oil spills or other environmental contamination to identify the source of the contaminating hydrocarbon product, in order to determine who is responsible.

29.4 Questions for Study and Review

1. Describe the three branches of forensic entomology. Which branch(es) is/are mostly concerned with evidence that might be presented in civil court, and which mostly concerned with evidence for criminal court?
2. Describe the life cycle of a blow fly.
3. How are the life cycles of flies used to estimate time since death?
4. What insects other than flies are useful in estimating time since death?
5. In your own word, define the following terms: larvae, instar, pupa, PMI.
6. Discuss ways in which insects can complicate a forensic investigation.
7. How do forensic botanists identify plant species?
8. In your own word, define the following terms: cellulose, lignin, plant community, palynology.
9. What are the class and individual characteristics of wood?
10. In your own word, define the following terms: grain size distribution, mineralogy, soil, dust, mud, hydrocarbon fingerprint.
11. What are the class and individual characteristics of soils?
12. A trace evidence examiner can match pollen profiles and dust profiles. Under what conditions might a forensic botanist or geologist be able to generate more evidence than a trace evidence examiner?

29.5 Sources

Amendt J, Campobasso CP, Gaudry E, Reiter C, LeBlanc HN, Hall MJR. 2007. Best Practice in Forensic Entomology - Standards and Guidelines. International Journal of Legal Medicine 121 (2): 90-104.

American Board of Forensic Entomology. 2004. American Board of Forensic Entomology. [Internet]. [cited 2007 Oct 16]. Available from: http://web.missouri.edu/entomology/.

Archer MS, Bassed RB, Briggs CA, Lynch ML. 2005. Social Isolation and Delayed Discovery of Bodies in Houses: the Value of Forensic Pathology, Anthropology, Odontology and Entomology in the Medico-legal Investigation. Forensic Science International 151(2-3): p259(7).

Barnes S. 2000. Forensic Entomological Case Study and Comparison of Burned and Unburned Sus Scrofa Specimens in the Biogeoclimatic Zone of Northwestern Montana. Thesis submitted in partial fulfillment of the requirements for the Master of Arts degree in Anthropology, The University of Montana – Missoula.

Benecke M. 2001. Forensic Entomology: the next Step. Forensic Science International 120(1-2): 1-1.

Benecke M. 2001. A Brief History of Forensic Entomology. Forensic Science International 120(1-2): 2-14.

Benecke M, Barksdale L. 2003. Distinction of Bloodstain Patterns from Fly Artifacts. Forensic Science International 137(2-3): p152-159.

Bock JH, Norris DO. 1997. Forensic Botany: an Under-utilized Resource. Journal of Forensic Sciences 42 (3): 364-367.

Bruce LG, Schmidt GW. 1994. Hydrocarbon Fingerprinting for Application in Forensic Geology; Review with Case Studies. AAPG Bulletin 78(11): 1692-1710.

Byrd JH. 2007. Welcome to Forensic Entomology.com, Explore the Science of Forensic Entomology [Internet]. [cited 2007 Oct 24]. Available from: http://www.forensic-entomology.com/.

Coyle HM, Lee C, Lin W, Lee HC, Palmbach TM. 2005. Forensic Botany: Using Plant Evidence to Aid in Forensic Death Investigation. Croatian Medical Journal 46: 606-612.

Gennard DE. 2007. Forensic Entomology: An Introduction. Hoboken (NJ): John Wiley & Sons.

Hayes RA. 2000. Forensic Geologists Uncover Evidence In Soil And Water. [Internet]. [cited 2007 Oct 16]. Available from: http://www.geoforensics.com/geoforensics/art-1101a.html.

Finley JA. 2004. Geologic Material as Physical Evidence [Internet]. FBI Law Enforcement Bulletin 73(3). [cited 2007 Oct 19]. Available from: http://www.fbi.gov/publications/leb/2004/mar2004/march2004.htm#page_2.

Joy JE, Liette NL, Harrah HL. 2006. Carrion Fly (Diptera: Calliphoridae) Larval Colonization of Sunlit and Shaded Pig Carcasses in West Virginia, USA. Forensic Science International 164(2-3): 183(10).

Mildenhall DC, Wiltshire PEJ, Bryant VM. 2006. Forensic Palynology: Why Do it and How it Works. Forensic Science International 163(3): 163(10).

Morgan RM, Wiltshire PEJ, Parker A, Bull PA. 2006. The Role of Forensic Geoscience in Wildlife Crime Detection. Forensic Science International 162(1-3): 152(11).

Morgan RM, Bull PA. 2007. The Philosophy, Nature and Practice of Forensic Sediment Analysis. Progress in Physical Geography 31 (1): 43-58.

Murray RC. 2004 Evidence from the Earth: Forensic Geology and Criminal Investigation. Missoula (MT): Mountain Press.

Murray RC. 2005. Collecting Crime Evidence from Earth [Internet]. [cited 2007 Oct 24]. Available from: http://www.forensicgeology.net/science.htm.

Murray RC, Tedrow JC. 1998. Forensic Geology. New York: Simon & Schuster.

Randerson J. 2002. Unfellable Evidence: Wood "Fingerprints" Will Make it Harder to Smuggle Lumber. New Scientist 174 (2342): 14(1).

Ruffell A, McKinley J. 2005. Forensic Geoscience: Applications of Geology, Geomorphology and Geophysics to Criminal Investigations. Earth-Science Reviews 69 (3-4): 235-247.

Terneny TT. 1997. Estimation of Time since Death in Humans Using Mature Pigs. Thesis submitted in partial fulfillment of the requirements for the Master of Arts degree in Anthropology, The University of Montana – Missoula.

Wagster, L. 2007. Decomposition and the Freeze-Thaw Process in Northwestern Montana: A Preliminary Study. Thesis submitted in partial satisfaction of the requirements for the Master of Arts degree in Anthropology, The University of Montana – Missoula.

Woodall J. 2007. The tales pollen tells. The Scientist 21(5): 67(1).

Chapter 30
FORENSIC ENGINEERING

In this chapter we will examine forensic engineering, a forensic science often overlooked because many of its major applications are to the realm of civil rather than criminal cases. Forensic engineering is the application of the engineering sciences to evidence that might be used in court. Engineering is one of the broadest fields of study in existence, and there are numerous branches, including civil engineering (engineering of public structures), mechanical engineering, chemical engineering, electronic engineering, hydraulic engineering, and aerospace engineering among many specialties. Engineering is about the practical application of sciences, and engineers are the people who take the knowledge that scientists generate in their research and make something practical out of it in the real world.

Forensic Engineering is considered to have four branches: traffic accident investigation, investigation of structure failure, investigation of product failure, and investigation of fires and explosions.

30.1 The History of Forensic Engineering

Some aspects of forensic engineering must go back to ancient times. For example, ancient pyramids, temples, etc. were engineered, and when they fell down, the failure was investigated by somebody. We could call that person a forensic engineer. Questions, such as "Why did this break?" and "How did this fire start?" are probably equally ancient. Traffic accidents are more a phenomenon of modern times, but you can be sure that in ancient times, when two horses collided and one or both riders were hurt, that someone performed an investigation of the accident based on what they knew at that time about the characteristics of horses and horse riding practices.

Therefore, unlike the forensic sciences discussed so far, forensic engineering appears to continue from ancient times to the present, with the major cause of evolution in the field being advances in science, particularly in materials science. **Materials science** is an interdisciplinary field that involves the study of materials (such as building materials) and their properties. The professional society for forensic engineers is the Society of Forensic Engineers and Scientists, founded in 1980.

30.2 Traffic Accident Investigation

Vehicles of all types collide. They collide with other vehicles, with stationary objects such as telephone poles, with pedestrians, and also with domestic or wild animals. In order to keep this chapter's length under control I will focus on automobile accidents, but similar principles and questions apply to other forms of transportation including railroad, airplane, and boating accidents.

There are always several practical questions that need to be answered in the case of a traffic accident. First, was there a crime committed, such as reckless driving, DUI, or speeding? If a death occurred, is the manner of death accidental, or did it involve homicide or suicide? Similar questions apply in cases of injury. Second, are there grounds for a civil suit? Is somebody responsible for damages? Who took the actions that caused the accident? Should an insurance company pay for the damages? If so, which company, and, if not, who should pay?

As we have seen previously for crime scene processing, in traffic accidents cases the evaluation and investigation of the incident is routinely done by a patrol officer. More complex accidents might be handled by a accident investigator, who is often a peace officer with training in the physical and engineering sciences beyond that of most peace officer, and who might even be a forensic engineer. It is only in especially serious and complex cases that a forensic engineer will be consulted. For most cases in which a forensic engineer is consulted, it is to dispute the law enforcement version of the events that happened. The forensic engineer's greater expertise in the physics of motion, properties of materials, the nature of friction against surfaces, and other things may allow them to find something that the peace officers missed or calculated incorrectly.

The steps in a peace officer's investigation of a traffic accident are quite similar to crime scene processing in other types of cases. I will assume here that the accident involves two vehicles. The steps in the investigation consist of scene control, data gathering, and data analysis. The first step is scene control, which means taking every effort to make sure that the accident doesn't get any worse. The people and vehicles already involved need to be protected, ambulances need to be summoned if there are injuries, and towing vehicles need to be summoned if the vehicles involved in the accident are disabled. The safety of citizens and officers is the main concern, and flares, flashing lights, traffic controllers, and/or reflectors will be employed to warn approaching traffic that there is a hazard in their path. The often dark colored uniforms worn by peace officers are detrimental, especially after dark, so the investigating officers will don reflectorized vests. Secondary to safety, but also important, is to protect the evidence at the crime scene.

The second step is data gathering, which corresponds to evidence documentation and collection in a more typical crime scene protocol. Testimonial evidence will be gathered in the form of statements made by the drivers and any witnesses. Important physical evidence present will be collected as well, such as the final stopping place of each vehicle, the direction the vehicles are facing when finally stopped, the paths the vehicles followed from impact to the stopping point, the location and length of skid marks, the location of debris produced during the accident, the nature of the damage to the vehicles, and any road conditions or obstructions that may have contributed to the accident. As with conventional crime scene processing, these items of physical evidence will be documented both photographically and by sketching.

It is during the data analysis phase of the investigation that principles of science and engineering are used to understand the events and forces that occurred leading up to and during the accident. This type of analysis is very complex, and most investigators, engineers or not, use one of many software packages available commercially for exactly this purpose. Through an understanding of the events and forces involved, it may be possible to determine the cause of the accident and who, if anybody, is at fault. Here is where the basis for dispute of the investigator's findings usually arises. There is tremendous pressure on the investigator to assign fault to one of the drivers involved, and this determination often relies of making very fine distinctions, which may turn out to be disputable.

30.2.1 An Example: Speed Estimation From Skid Mark Data

Let's take speed estimation from skid mark data as an example of the data collected and calculations made to determine the facts surrounding a traffic accident. Skid marks are produced by an automobile's tires during hard braking events. When the brakes are applied, the vehicle's weight shifts forward, transferring more weight to the front tires. The kinetic energy of the automobile's motion is dissipated in the form of heat due to the friction between the tires and the roadway surface. Kinetic energy was discussed previously in chapter 12, and involves the velocity and mass of the moving object. On pavement, the heat from the dissipating energy melts the tars and oils on the roadway surface, thus creating the skid mark. On concrete, skid marks are made by the rough concrete surface actually "grinding up" the tire, or melting it, causing fragments of the tire material to adhere to the road surface.

It is possible to estimate the speed at which a vehicle was moving when the brakes were applied from the length of the skid marks produced as it braked to a stop, by applying what is known from engineering science about the friction between a tire and a road surface. This is the formula used, which is actually fairly simple.

$$S = \sqrt{30DFE}$$

In this formula, S is the speed of the vehicle in miles per hour; D is the distance taken to stop, in feet, which is equal to the length of the skid mark. D is measured at the scene. F is the coefficient of friction of the road surface, and is usually considered to be between 0.50 and 0.90 for an asphalt surface, though some authorities believe that F can be as high as 1.2 for modern tires and braking systems. F is determined by consulting a table of coefficients of friction for different surfaces or by doing a "skid test" in which the officer produces a skid mark with their own vehicle traveling at a known speed. E is the efficiency of braking, judged from whether all four tires have left skid marks, or only some of them. Typically, each front tire is thought to contribute 30% of the braking efficiency, and each rear tire 20%. If all four tires have left skid marks, the efficiency of braking is 100%.

For example, let's say that we find skid marks for all four tires, which are 50 feet long on dry asphalt. Let's say that the coefficient of friction is estimated to be 0.80, a commonly used value. Since all four tires have left skid marks the braking efficiency is 100%. Therefore, D = 50, F = 0.80, and E = 1. This give us the following calculation. Therefore, it is likely that the vehicle was traveling at about 35 miles per hour at the time the brakes were applied.

$$S = \sqrt{30DFE} = \sqrt{(30)(50)(0.80)(1)} = \sqrt{1200} = 34.64 mph$$

30.2.2 What an Engineer can Add to the Investigation

Even though forensic engineers becomes involved with only a small percentage of traffic accidents, the sheer volume of traffic accidents assures that there is plenty of work for them. Many forensic engineers are in private practice, and are hired to render a expert opinion on the events of an accident by one party or another in a civil suit. Often, this amounts to rechecking the evidence and verifying the calculations, but there is more that an engineer can contribute as well. Table 30.1 lists some additional things that a forensic engineer may consider that may not be routinely considered by the peace officers who process traffic accident scenes.

Table 30.1: Additional Factors in a Traffic Accident that May be Considered by a Forensic Engineer

Road Issues
- Was the roadway designed, maintained, and signed adequately?
- Did roadway or roadside hazards contribute to the accident?

Visibility Issues
- Were the lighting & visibility adequate?
- What was the visibility? Were there obstructions, such as buildings, parked vehicles, trees, a crest in the roadway, or another moving vehicle that may have obscured a driver's line of sight?

Injury Issues
- What were the collision forces involved, and how did these contribute to any injuries sustained by the drivers?
- How did the the bodies of the vehicle's occupants move during the accident?
- What portions of the interior of the vehicle might the occupants' bodies have struck?
- Working with medical doctors, were there sufficient forces to cause the injuries claimed?
- Were the vehicle's occupants wearing seat belts?

Mechanical Failure Issues
- Did the seat belt system fail, leaving the occupants unprotected?
- Was the accident caused by improper maintenance or an inherent design problem?

Human Factors
- Were there any actions a driver could have taken to avoid the collision or lessen its severity?

30.2.3 Time-Distance Studies

Many times a forensic engineer will tie all of the information together into a time-distance study. A **time-distance study** is a second-by-second reconstruction of the accident showing where the vehicles were in relation to each other, when hazard factors took effect, what effect they had, when the vehicles impacted, when injuries were sustained, and the other events suggested by the nature of the evidence. These studies can be of great help in allowing judges and juries to understand the critical events involved in a traffic accident.

30.3 Structure Failure

Bridges, buildings, roadways, and many other types of structures fail. In some cases this may result in considerable injury and loss of life. Even in cases where no humans were harmed, there is always financial loss. Therefore, there is almost always some form of civil suit involved in the case of a **structure failure**. Structure failure is usually an awe-inspiring event, which leads people to question whether they are really safe as they go about their daily lives. Combining this with the fact the structure failures are fairly uncommon, when such an event occurs it is investigated by the best experts available, and this is usually a team of forensic engineers who can often determine the cause.

There are a set of practical questions involved in the investigation of a structure failure. The first is, of course, what caused the failure? For the purposes of the nearly inevitable civil litigation that follows, other questions are also important. Some of these are as follows. Could it have been prevented? Was there a disaster plan in place? Who is responsible for losses? Should an insurance company pay?

Forensic engineers often use a system of categories into which the cause of a structure failure is assigned. These are reminiscent of the manner of death categories used in death investigation. The categories are: bad design or construction, improper maintenance, natural hazard (earthquake, fire, flood, etc.) including cumulative effects from previous events, deliberate damage (sabotage, arson, etc.), and undetermined.

30.4 Product Failure

Many products also fail. Ultimately, all products fail as they wear out. What is of most concern is when the product fails in an unexpected way. For example, some high profile **product failures** that have occurred in the past include automobiles that explode when struck from behind, cribs that strangle the child they are supposed to keep safe, and household appliances that electrocute the user. Product failure can cause financial loss, injury, or death, and therefore civil suits are common. Forensic engineers can often find the facts in product failure cases.

The practical questions involved in product failure are similar to those involved in structure failure, including: How did it fail?, Why did it fail?, Who is responsible?, Who should pay? Likewise the categories of cause of product failure are similar to those of structure failure, with the added category of improper use.

The actual causes of product failure (similar to cause of death in a death investigation) are many and varied. Electrical systems such as lighting, electrical appliances, motors, and electrical wiring can fail due to irregularities in the power supplied to a building or to the use of the system beyond its rated capacity. Products made of metal can fail due to corrosion, weld failures, and metal fatigue. Plastics, glass, and other non-metals can fail due to stresses placed upon them.

30.5 Other accidents

Forensic engineers can often render an expert opinion in a wide range of common accidents. Some cases in which forensic engineers have testified include falls from ladders, scaffolding collapse, failure of railings, breaking of windows, and "trip and fall" accidents.

30.6 Arson and Explosions

The fourth category of expertise for forensic engineers is arson and explosions. In this chapter I will concentrate on arson. Explosives and explosions are the subject of the next chapter. **Arson** is the crime of deliberately starting a fire that endangers another person or another person's property. In order to be arson, the starting of the fire must be deliberate, but the endangerment of people or property need only be negligent or reckless. It is also arson to burn one's own property with the intent of collecting insurance for the loss. Most arsons are felonies.

Arson is sometimes a difficult crime to investigate for two reasons. First, it is often pre-planned. Thus, the perpetrator has usually taken steps to minimize the amount of incriminating evidence left behind. Only rarely will an arsonist be found at the scene of the crime by the time authorities arrive, though the psychology of arsonists sometimes prompts them to watch the fire and firefighters and, later, investigators, as they work. Second, the destruction caused by the fire may obliterate significant evidence. In fact, a fair number of arsons are committed exactly for the purpose of covering up or destroying evidence. An equally difficult task is determining that the fire was caused by arson. A variety of types of product failure can lead to a fire, and there are a few natural causes, thus the situation is often confusing.

30.6.1 Who Investigates Arson?

Although traditionally considered part of forensic engineering, arson investigation is most often undertaken by a fire investigator associated with a fire department or a state fire marshal's bureau. These officials are recognized as peace officers in most states, and in some cases have training in forensic engineering methods or are forensic engineers. As in traffic accident cases, a forensic engineer is likely to be consulted if the case is particularly complex, or if a defendant wishes to dispute the findings of the initial investigators.

Several other types of forensic scientist will usually become involved in an arson investigation, depending on what evidence is present, such as fingerprints, blood, toolmarks, impressions, or trace evidence. Forensic chemists or forensic geologists may do hydrocarbon matching. If a human body is discovered, forensic pathologists, anthropologists, and odontologists will attempt to identify the person and determine whether they died from the fire or were already dead from another cause before the fire started,

30.6.2 Arson Crime Scenes

Because of the challenges described above, arson investigators need immediate access to the scene of a fire, as soon as it is safe. For this reason, fire investigations are exempt (by a decision of the U.S. Supreme Court) from the requirement to obtain a warrant to search the premises and seize evidence. Usually, the first thing searched for is the point of origin of the fire. The point of origin is one of the best clues as to the cause of a fire; and if it is arson, traces of the materials used to start the fire may still remain. If the point of origin is some electrical wiring, the kitchen stove, or a toaster, for example, then the fire is probably not a case of arson.

It is often difficult to pinpoint the origin of a fire. In general, a fire will burn in an upward direction from the point of origin, but many factors can affect this. This is one of the areas in which the expertise of a forensic engineer, who is familiar with the characteristics of fires, fuels, structural issue and other subjects can be valuable.

Many arsonists use **accelerants** in starting a fire. Accelerants are usually a hydrocarbon product, such as gasoline, which is poured over materials at the location the arsonist wishes to start a fire. Using an accelerant makes the fire start rapidly and burn briskly, thereby assuring that a significant amount of burning has occurred by the time the fire is noticed and reported. Although one would expect that any accelerant used would have burned up by the time the fire is put out, this is rarely the case. Accelerants can soak into wood, brick, carpets, or fabric, and be recovered in amounts sufficient for analysis such as hydrocarbon matching. The spraying of water onto a fire often slows the evaporation of flammable hydrocarbons to the point that the fuel/air mixture will not sustain combustion, and can result in pools of mixed accelerant and water that can be recovered. Vapor detector devices, often called "sniffers" exists and can be of use in detecting residual accelerant. These devices usually consist of a wand that directs air over a heated filament. If there are traces of accelerant in the air, some activity will usually be noticeable around the filament.

Another evidence item that investigators search for is the device that was used to start the fire. A fire starting device is known as an **ignitor**, and can range from a simple paper match to a sophisticated timing device. Other devices that have been used to start a fire include a burning cigarette, firearms, ammunition, electrical sparking devices, and "Molotov cocktails" (a quantity of accelerant in a breakable container to which is attached a burning wick). The metal and glass from more complex ignitors are more likely to survive the fire than are paper or wooden matches. If the ignitor can be recovered, it may yield valuable evidence for locating a suspect. Even in the case of paper matches there have been cases where the unburned remains of the match were recovered and its torn surface matched to a stub in the matchbook from which it was pulled.

Other telltale signs of arson include multiple origins of fires, and evidence of "streamers", which are trails of accelerant designed to spread the fire from one location to another.

As in other crime scenes, it is imperative that any evidence present be protected from damage or tampering. However, no other type of crime scene presents quite the challenge to protecting evidence as an fire scene. The fire burns, the water sprays, and the firemen are moving through the area in their efforts to put out the fire. However, once the fire is out and safe conditions have been established, the next step should be the normal cordoning off of the area in order to protect the crime scene.

Documentation and collection of evidence at an arson scene proceeds in the fashion with which you should now be familiar. Samples of materials suspected to contain traces of accelerant are usually packaged in new, clean paint cans. These types of cans can be sealed to prevent evaporation or contamination of the accelerant, and the metal from which the can is made can not be dissolved by the accelerant, as is the case with plastic. Of course, standard items must be collected from any suspects, including samples of any supplies of substances that could be used as an accelerant, and evidence that might connect an ignitor to the suspect. If the suspect can be apprehended within a few hours, there may even be traces of accelerant remaining on their clothing.

30.7 Questions for Study and Review

1. What are the four branches of forensic engineering?
2. In your own words, define the following terms: materials science, time-distance study, structure failure, product failure, arson, accelerant, ignitor.
3. Discuss the sorts of traffic accident cases that would be investigated by a peace officer and which might be investigated by a forensic engineer.
4. Discuss ways in which a traffic accident investigation differs from normal crime scene processing.
5. Discuss what kinetic energy has to do with traffic accidents and their investigation.
6. Say that a traffic accident occurs between a Ford and a Chevrolet in a 25mph speed zone. The driver of the Ford claims that the driver of the Chevrolet was speeding and that the accident is, therefore, her fault. The investigating officer finds only one skid mark, made by the Chevrolet's left rear tire, and measures it to be 100 feet long. How fast was the driver of the Chevrolet going when she applied the brakes? Use an estimated coefficient of friction of 0.80.
7. Make a list of types of evidence that a forensic engineer can add to a traffic accident investigation that are not normally produced by a peace officer's investigation.
8. List the categories of causes of structure failure.
9. List the categories of causes of product failure.
10. Discuss the process of investigation of an arson crime scene. What special considerations does an arson investigation present?

11. List some types of evidence that, if found at a fire scene, suggest that the fire was caused by arson.

30.8 Sources

Abelson, P. 2007. To Reconstruct an Accident. Public Works 138(8): 21(2)

Bertsch W. 1996. Chemical Analysis of Fire Debris: Was it Arson? Analytical Chemistry 68(17): 541A(5).

Bonadiman JSC. 2003. Causes of Failure: What a Forensic Engineer Looks For. The Forensic Examiner 12(11-12):p23(5).

Borusiewicz R, Zieba-Palus J, Zadora G. 2006. The Influence of the Type of Accelerant, Type of Burned Material, Time of Burning and Availability of Air on the Possibility of Detection of Accelerants Traces. Forensic Science International 160(2-3): p115(12).

Brown S. 2007. Forensic Engineering: Reduction of Risk and Improving Technology (For All Things Great and Small). Engineering Failure Analysis 14 (6): 1019-1037.

Byrd M. n.d. Crash [Internet]. [cited 2007 Oct 23]. Available from: http://www.crime-scene-investigator.net/crash.html.

Deans J. 2006. Recovery of Fingerprints from Fire Scenes and Associated Evidence. Science & Justice 46 (3): 153-168.

Doerzaph Z, Gill R. 2001. Forensic Engineering. Journal of the Idaho Academy of Science 37(1): p22(1).

D'Onofrio JA. 2007. Forensic engineering evaluation of premises maintenance. The Forensic Examiner 16(3): 28(6).

Gagg CR. 2005. Failure of Components and Products by 'Engineered-in' Defects: Case Studies Engineering Failure Analysis 12 (6): 1000-1026.

Harris JO. n.d. Determining Vehicle Speeds for Skid Marks [Internet]. [cited 2007 Oct 4]. Available from: http://www.harristechnical.com/articles/skidmarks.pdf

Insurance Committee for Arson Control. 2002. Investigation [Internet]. [cited 2007 Oct 24]. Available from: http://www.arsoncontrol.org/2002/investigation/investigation.htm.

Kudarauskas N. 2007. Analysis of Emergency Braking of a Vehicle. Transport 12(3): 154–159.

Nokes LDM, Knight BH. 1996. Forensic Biodynamics: a Valid Subsection of Forensic Engineering? Forensic Science International 77 (1-2): 1-2.

Pert AD, Baron MG, Birkett JW. 2006. Review of Analytical Techniques for Arson Residues. Journal of Forensic Sciences 51 (5): 1033-1049.

Sandercock PML, Du Pasquier E. 2003. Chemical Fingerprinting of Unevaporated Auromotive Gasoline Samples. Forensic Science International 134(1): 1-10.

Van Biema D. 1993. Clues in the Ashes. (arson investigation in Southern California). Time 142(20): 58(3).

Yates JK, Lockley EE. 2002. Documenting and Analyzing Construction Failures. Journal of Construction Engineering and Management – ASCE 128 (1): 8-17.

Chapter 31
Advanced Topics
EXPLOSIVES

This chapter will address the topic of explosives. It depends on chapter 11, in which the chemistry of oxidation reactions was presented. Explosives are considered within the expertise of forensic engineering. While not exactly commonplace, thankfully, bombs have always provided a convenient way to cause a large amount of damage. Explosives have perhaps become more relevant in the 21st century due to terrorism, which will be discussed in more detail in the chapter on homeland security.

Some people are surprised at the ready availability of explosives to the criminal element of society. There are two factors involved in this. The first is that some countries do not regulate explosives to the extent that the United States does, or social and political forces have made it difficult to regulate them. This is alleged to have occurred around 1990, after the breakup of the former Soviet Union, when corruption among the authorities charged with guarding supplies of explosives, perhaps even including nuclear explosives, led to their ready availability on the black market. Those with the money to buy them, such as organized crime associations and terrorist groups were able to acquire substantial quantities. The second factor involved is that some potentially explosive substances are widely available and only minimally regulated. Almost any hydrocarbon product can be made into a bomb, including methane, propane, and gasoline. Diesel fuel is a component of one of the most widely abused high explosives, along with ordinary ammonium nitrate fertilizer. Many household products available in aerosol cans can be made to explode. I once saw an episode of one of the classic TV cop shows of my youth – I think it was Beretta – in which an explosives expert claimed to be able to make a bomb out of baking soda and vinegar.

31.1 Explosions

There are three basic types of explosion: mechanical, nuclear, and chemical. A **mechanical explosion** occurs due to a physical reaction such as overloading a container with compressed air. The popping of a balloon is a very small type of mechanical explosion. Mechanically produced explosions are sometimes used in mining, where the air supply is tenuous enough that polluting it with the gas released from a chemical explosion is a concern. A **nuclear explosion** is one in which a nuclear chain reaction occurs, resulting in the nearly instantaneous release of enormous amounts of energy. Nuclear explosives have well known military applications, and more controlled nuclear reactions are used in nuclear power plants. **Chemical explosions** are by far the most common. They arise from a rapidly, exothermic, chemical reaction, very often an oxidation reaction similar to those with which we are familiar from chapter 11. In fact, natural gas (methane) can be considered a low explosive, and methane explosions are far from rare. They occur via the same oxidation reaction as burning of methane. The vast majority of explosives of relevance to the forensic sciences are chemical in nature.

31.2 What Happens During an Explosion?

Confining ourselves to the consideration of chemical explosives, the exothermic chemical reaction that causes the explosion produces gases and heat. This is reminiscent of the oxidation reactions, such as combustion, that we examined in chapter 11. What distinguishes an ordinary reaction such as combustion, from an explosion is the vastly more rapid rate at which the reaction occurs. The rapid reaction causes the rapid release of heat and gasses. It is a principle of physics that whenever a solid or liquid is converted to a gas, the gas occupies more space than the solid or liquid, and therefore expands. This is true whether the conversion to gas is via a chemical reaction or by simply heating the material. For example, when water is boiled in an old fashioned tea kettle, the steam escapes with enough force to make the tea kettle "whistle".

Gasses want to expand and hot gasses want to expand even more and faster. If the explosion reaction occurs inside a container, the expanding gasses will build up tremendous pressure. This pressure is a form of stored energy. Professional tools used in the construction industry are often powered by compressed air, for example. Quite rapidly, the pressure will exceed the ability of the container to hold it and the container will fail suddenly, resulting in the release of the pressure as a shock wave as the air around the explosion is compressed by the gasses. This shock wave is the blast of the explosion. Further, the failure of the container may cause it to break into pieces, which are blown outwards as shrapnel. Both the blast and the shrapnel cause damage to anything in the surrounding area.

31.3 Low and High Explosives

The details above describe the events that happen during the explosion of a low explosive. One of the more practical distinctions between a **low explosive** and a **high explosive** is that while a low explosive usually requires a container for the gasses and energy of the reaction to build to explosive levels, a high explosive does not. Whether in a container or not, a reaction of a high explosive occurs so rapidly that it detonates, producing an immediate blast. Again, the distinction between an ordinary chemical reaction, a low explosive reaction, and a high explosive reaction is all about reaction speed. In an ordinary oxidation, a fuel combusts; in a low explosive reaction, it "deflagrates" at up to 1000 meters per second; and in a high explosive reaction it "detonates" at rates between 1000 and 8000 meters per second. Yet another way to distinguish a low explosive and a high explosive is by the speed of the shock waves produced. Low explosives produce shock waves that are subsonic (less than the speed of sound), while high explosives produce shock waves that are supersonic (greater than the speed of sound).

One thing that the distinction between low and high explosives does not include is the force or lethality of the blast. This is more a function of the amount of explosive than of its type. Low explosives can create just as much devastation as high explosives, though it may take more of it to achieve the same effect.

31.4 Low Explosives

Low explosives are very easy to manufacture. All that is required is a fuel, an oxidizing agent, a container, and some way to supply the ignition energy to start the reaction going. As one example, potassium chlorate is an excellent oxidizing agent, and can be combined with a variety of substances to make a low explosive. The substances (fuels) that potassium chlorate has been combined with include sugar, carbon, sulfur, starch, phosphorous, and magnesium filings, among other things. The ignition energy can be supplied using a fuse, by adding other chemicals to the mixture, or by a high tech sparking device. The container used can be nearly anything, and is only limited by the bomb maker's imagination. Pipes, bottles (metal or glass), metal cans, and many other types of containers have been used.

Low explosives deflagrate at rates that make them ideal for propelling or throwing a projectile. Gunpowders used in ammunition and fireworks are a good example, and are the most widely used explosives. Gunpowders, including smokeless and black powder, are widely available, since they are used by firearm enthusiasts who reload their own ammunition, either to save money or to customize the amount of powder and thus the force and speed of the bullet.

Black powder, the simplest form of gunpowder is often used in the fuses that are lit to set off charges of other explosives, including high explosives such as dynamite. Most of us can picture a stick of dynamite with a fuse protruding from the top of it. The fuse is usually black powder, wrapped in or with plastic or fabric to hold it in place. Unless confined, black powder merely burns, and the rate of burning is such that a fuse of sufficient length can be lit and the person who lit the fuse has time to retreat to a safe location before the detonation occurs.

Smokeless powder is another type of gunpowder. The active ingredient in this low explosive is nitrocellulose or nitroglycerin (or both). It is the gunpowder most often used in ammunition and is considered relative safe, predictable, and powerful.

Hydrocarbons and hydrocarbon products are also low explosives. For example, if enough natural gas escapes into a confined area, such as an oven or even an entire kitchen, it can be explosive if the correct fuel to air mixture occurs (5.3% to 13.9% methane). This is a common type of household accident and has caused many fires and injuries.

31.5 High Explosives

High explosives are those that detonate so rapidly that they do not need to be confined to create a blast that exceeds the speed of sound. High explosives are further divided into initiating and non-initiating types. **Initiating high explosives** are highly unstable, and can be caused to detonate by the application of a moderate amount of heat or pressure. **Non-initiating high explosives** are considerably more stable and take a substantial kick of energy to detonate.

31.5.1 Initiating High Explosives

Initiating explosives are sensitive to heat, shock, and friction. This makes them difficult and dangerous to handle in all but the most minute quantities. However, this property makes them ideal for use for setting off other high and low explosives. Devices that contain a small amount of initiating high explosive and are used to initiate the explosion of more stable explosives are called primers. As one example, ammunition invariably has a primer that is detonated by the action of a firearm's firing hammer or pin, which causes the main charge of gunpowder to explode. Blasting caps, used to set off dynamite and other explosives, work on the same principle.

One initiating high explosive used in homemade bombs is TATP (triacetone triperoxide). TATP is relatively simple, though extremely hazardous, to make from acetone, hydrogen peroxide, and an acid. Not only is TATP hazardous to make, it is also extremely hazardous to work with because it is an initiating high explosive that requires very little energy to detonate.

31.5.2 Non-Initiating High Explosives

Non-initiating high explosives, unlike the initiating type, are not overly sensitive to heat, shock, or friction. These are the type of explosives that are used for most commercial and military applications. Some common examples of noninitiating explosives are nitroglycerine, dynamite, TNT (trinitrotoluene), PETTED (pentaerythritol tetranitrate), RD. (cyclotrimethylenetrinitramine) and tetryl (2,4,6-trinitrophenylmethylnitramine).

The traditional non-initiating high explosive is **dynamite**, though in modern times its has been rendered nearly obsolete by the ammonium nitrate based class of explosives. It is an irony of history that the prize most symbolic of humanity's search for peace, the Nobel Peace Prize, is named for Swedish chemist Alfred Nobel, who invented dynamite in 1867. Dynamite is made of nigroglycerin mixed with wood pulp or another stabilizing substance; sodium nitrate, which serves to furnish oxygen for complete combustion; and a small percentage of a stabilizer such as calcium carbonate. The strength of dynamite is specified by the weight percentage of nitroglycerin in the formula. For example, 60-percent straight dynamite contains 60 percent nitroglycerine. .

The most common explosives in commercial and industrial applications today are based on ammonium nitrate. Ammonium nitrate is a superb oxidizing agent, and when mixed with a fuel forms a stable, non-initiating high explosive. The explosive most commonly used in homemade bombs, including those used by terrorists, is made by soaking an ammonium nitrate based fertilizer in a fuel oil such as diesel fuel. This explosive is usually referred to as **ANFO**, which is an acronym for ammonium nitrate and fuel oil. ANFO was used both in the 1993 bombing of the World Trade Center and the 1995 bombing of the Murrah Federal Building in Oklahoma City. ANFO is easy and relatively safe to handle, but packs a strong explosive punch.

Commercial ammonium nitrate based explosives are commercially available as water gels or emulsions. Water gels are formulated of ammonium nitrate, sodium nitrate, powdered aluminum or another fuel, and a suitable sugar-based chain molecule such as guar gum to hold it all together. The resulting product looks like jello or gel toothpaste. Water gels, get their name from the fact that they are resistant to water and are often used when blasting in wet conditions. Ammonium nitrate emulsions are formulated of two distinct components, an oil based component and water. Since oil and water do not mix, thesr two components must be emulsified, just as olive oil and vinegar are emulsified to form Italian salad dressing. Typically, emulsifying agents will be added, and extremely small spheres of glass, resin, or ceramic will be added as inactive ingredients to control the explosive potential of the product.

31.5.3 Military High Explosives

No discussion of high explosives would be complete without a mention of military high explosives. In many countries outside the United States, the accessibility of these products to terrorist organizations makes them common constituents of homemade bombs.

RD, which stands for rapid detonating, is a popular and powerful military explosive. **Plastic explosive**, which comes as a pliable substance with a consistency similar to bread dough is a formulation of RD. Plastic explosive is also referred to by its military desigation, composition C4.

TNT (trinitrotoluene) is often used in artillery and other applications requiring an explosive charge, such as demolition, bombs, and grenades. TNT is a component of several mixtures and formulations designed to increase its yield. Military "dynamite" is not really dynamite, in that it contains no nitroglycerin but is actually composed of a mixture of RD. and TNT. The military uses a variety of other TNT mixtures as well. Amatol is a mixture of TNT and ammonium nitrate; ammonal is a mixture of powdered aluminum with TNT, charcoal, and ammonium nitrate; RDX is a mixture of hexamine and TNT; and pentolite is a mixture of pentaerythritol tetranitrate and TNT.

PETTED (pentaerythrite tetranitrate) is also known as PETN or PETNA. Often combined with TNT, it is used by the military for small-caliber projectiles and grenades. PETTED is also used commercially in fuses that are designed to interconnect a set of explosive charges that should explode nearly simultaneously, as opposed to fusing made from black powder which is designed to allow the person who lights it enough time to safely leave the area.

31.6 Homemade Bombs

Homemade bombs are used not uncommonly by organized crime and terrorists. The IED's (informal explosive devices) commonly used against U.S. troops during the war in Iraq are essentially homemade bombs. It is fair to say that most of the bombs encountered by peace officers in the U.S. are homemade bombs.

Homemade bombs are often camouflaged or disguised in some manner. Disguising them as packages or suitcases is common, and car bombs can be considered disguised bombs as well. Homemade bombs can also be hidden in a variety of locations as booby traps, or with the intention of blowing up the location or its occupants.

The most common ingredient of homemade bombs is ANFO, though dynamite and TATP power many of them as well. The detonation device can range from none at all to sophisticated timers and switches. If the criminal deploying the bomb believes that some event will occur with enough force to set off the explosive, such as ramming a car into a barricade, no detonator is needed. One of the most ingenious detonators is a simple mercury switch, which is a tube with two electrical contacts and the liquid metal mercury. When held upright the mercury flows to a position away from one of the contacts so that it doesn't form an electrical circuit. When the tube is tilted, however, the mercury covers both of the contacts, completing a circuit that can deliver an electrical charge to the explosive to detonate it. A mercury switch can be placed, for example, in a car's trunk so that when the trunk is opened the mercury closes the circuit and the explosion occurs. For blowing up a car and it's occupant however, using the ignition switch to complete a circuit is a popular option. Other types of switches detect motion or other special condition. Clocks, both low tech and high tech, can also form the basis of a detonation circuit and have given us the common expression "ticking time bomb".

31.7 Investigation of Bombs and Bombings

The investigation of a bomb threat or a bombing is possibly the most dangerous crime scene scenario of all. Therefore, it should only be undertaken by the most highly trained and experienced individuals. As is a common element of movies, disarming a bomb is often extremely intricate, and sometimes only the person who designed the bomb really knows how to do it. For this reason, most bombs are disposed of by the crude process of detonating them under controlled conditions. Once the bomb has been rendered safe, the situation can be treated like an ordinary crime scene and investigation.

If the bomb actually exploded, the crime scene will be even more devastated than occurs in arson cases. The process is, however, similar to that of arson investigation in that one of the first tasks is to identify the origin point of the explosion. The origin point of an explosion will normally exhibit a crater. Traces of the chemicals used in the bomb, and perhaps some of the products of the detonation reaction are likely to still be present in the area where the explosion occurred, and like the accelerants used in arson, these substances may be volatile and should be recovered as quickly as possible. Soil and debris from inside the crater should be collected. As with accelerants, porous materials such as wood, brick, and fabric may hold traces of the chemicals of interest. Again, as in arson investigation, the detonating device is a crucial piece of evidence and it, or its fragments, can sometimes be recovered. If a container was utilized as part of the bomb, fragments of it should be strewn around the location of the blast and easily collectable.

One piece of high tech equipment that is often used in bomb scene investigations is an EGIS system. An **EGIS** is essentially a portable gas chromatograph. The EGIS has a vacuum tube that collects vapors from surfaces suspected to contain explosive residues. The gas chromatograph then separates and identifies the components of the vapor. If residues are detected on the surface of an item, the item can be recovered for further examination.

Particles of unconsumed explosive are often readily identifiable under the microscope. Therefore, once back at the crime lab, the samples collected will be analyzed microscopically. After microscopic examination, the samples will be treated with acetone or another solvent to dissolve any explosive residue and these extracted samples will be analyzed by a forensic chemist.

31.8 Questions for Study and Review

1. In your own words, compare and contrast mechanical, nuclear, and chemical types of explosions.
2. In your own words, describe what happens when an explosive explodes inside a container.
3. List the differences between low explosives and high explosives.
4. What two types of explosives are used in firearms ammunition?
5. Describe the difference between initiating and non-initiating high explosives.
6. Categorize the following explosives as low explosives, initiating high explosive, or non-initiating high explosives: RD (C4), potassium chlorate, TNT, black powder, TAPT, smokeless gunpowder, dynamite (nitroglycerine), hydrocarbons, ANFO.
7. Discuss the investigation of bomb scenes. What special considerations exist for these types of investigations?

31.9 Sources

Brown Gl. 2005. The Big Bang: A History of Explosives. Stroud (Gloucestershire, United Kingdom): Sutton Publishing.
Fuller TC. 1999. Bomb Threat: A Primer for the First Responder [Internet]. FBI Law Enforcement Bulletin 68(3). [cited 2007 Oct 19]. Available from: http://www.fbi.gov/publications/leb/1999/mar99leb.pdf.
Global Security.org. 2007. Explosives [Internet]. [cited 2007 Oct 24]. Available from: http://www.globalsecurity.org/military/systems/munitions/explosives.htm.
Interactive Learning Paradigms Incorporated. 2007. Explosive [Internet]. [cited 2007 Oct 24]. Available from: http://www.ilpi.com/msds/ref/explosive.html.
N.C. Dept of Transportaton, Safety & Loss Control. 2000. The History of Explosives [Internet]. [cited 2007 Oct 24]. Available from: http://siri.uvm.edu/ppt/blast1/index.htm.

Thermo Electron Corporation. n.d. EGIS™ Defender: Portable Desktop Explosives/narcotics Detection System [Internet]. [cited 2007 Oct 24]. Available from: http://www.envimet.com/pdfs/EGIS%20Defender%20Brochure.pdf.

Wikipedia contributors. 2007 Acetone Peroxide [Internet]. Wikipedia, The Free Encyclopedia; 2007 Oct 17, 22:02 UTC [cited 2007 Oct 21]. Available from: http://en.wikipedia.org/w/index.php?title=Acetone_peroxide&oldid=165277527.

Wikipedia contributors. 2007. Explosion [Internet]. Wikipedia, The Free Encyclopedia; 2007 Oct 23, 14:01 UTC [cited 2007 Oct 24]. Available from: http://en.wikipedia.org/w/index.php?title=Explosion&oldid=166510503.

Wikipedia contributors. 2007. Explosive Material [Internet]. Wikipedia, The Free Encyclopedia; 2007 Oct 23, 19:06 UTC [cited 2007 Oct 24]. Available from: http://en.wikipedia.org/w/index.php?title=Explosive_material&oldid=166575247.

Chapter 32
Basic Science
COMPUTERS AND NETWORKS

In this chapter we will focus primarily on the basics of computer science and computer networking. This information will be foundational for the following chapters on forensic computer science, computer networks and security, and to a lesser extent electronic surveillance and biometrics. As in the previous basic science chapters, my goal is to provide you with enough background to understand the methodologies that forensic scientists use, but not to burden you with too much detail. The chapter will start by presenting the basics of computer architecture, with an emphasis on how the various parts of a computer communicate. It will then consider networks, with the emphasis again on how different computers connected to a network communicate. Finally, we will briefly explore two issues of computer science that are also foundational for coming chapters, hashing and cryptograph.

32.1 Computers

A **computer** is an electronic machine that manipulates numbers and other types of data extremely efficiently and rapidly. It consists of millions of transistors, which primarily act as switches, and other electronic components wired together in a configuration that allows for data flow and data manipulation. Data storage is also an essential aspect of modern computers, and magnetic media (hard disks), optical media (CD's and DVD's), and several types of silicon-based memory chips make this possible. As part of their role as data manipulators and processors, computers have taken on parts and functions that allow them to communicate; both with users via keyboards and mice, and with other computers over networks.

32.2 Binary Data

The most important aspect of a computer to a forensic computer scientist is its data storage capabilities. This is because the main thrust of forensic computer science is to recover evidence, in the form of stored documents and records, that reveal some illegal activity.

Numbers and characters are represented within a computer in **binary** form, as strings of 0's and 1's, because in the world of transistors and the circuits built from them there are only two states - on and off (or present/absent, or positive/negative, etc.). Exactly what a particular string of binary digits means depends on the context. For example, the binary string 01000001 can represent the number 65, or it can represent the letter 'a', or it can be an instruction for the computer's central processing unit to do some task (such as an addition), or it can represent something else. The meaning depends on whether the computer is adding numbers, reading a document, or running a program. Data is also stored within a computer's memory, or on an external memory device, using strings of binary digits. A single 1 or 0 is a "**bit**" of data, and 8 bits is a **byte**. A byte can represent up to $2^8 = 256$ different codes, which is enough to have a separate code for all the letters, numbers, punctuation marks, etc. with plenty left over.

The use of binary representations of data is the fundamental distinction between a digital device and an analog device, and computers are absolutely and completely digital devices. For example, let's consider a piece of music recorded in two different ways: as an analog recording on a cassette tape and as a digital audio file on a computer's hard drive.

A note of music has two characteristics: a frequency or set of frequencies that represent the speed of vibration of the air as a person hears it, and the intensity or amplitude at that frequency or frequencies. The cassette tape stores the information that corresponds to the music in analog format. If the note is at a certain frequency, the tape records a wave at that frequency, and records the amplitude as the strength of the wave. As the tape is then read, the information on it is converted back into an electric signal that drives a speaker at the correct frequency and at an amplitude corresponding to the strength of the stored information.

Now, consider the same note of music stored digitally as a digital audio file. As the note is played it is "sampled" by the recording device. **Sampling** consists of measuring the frequencies and their amplitudes at regular intervals, say 30,000 times each second if the absolutely best recording fidelity is desired, and perhaps 5,000 times each second if you can tolerate less than perfect sound reproduction. At each of these intervals, the recording device converts the frequencies and amplitudes of the sounds into numbers (in binary form). These numbers are stored sequentially in the file, along with the sampling rate. During playback of the note, the file is read, and the numbers representing frequency and amplitude are taken at the same rate as when the music was sampled, and converted into signals that drive a speaker with the required frequencies and amplitudes.

32.3 Parts of a Computer

A computer consists of a data processing chip, several types of memory systems, several types of systems to move data into and out of the computer, and all the wiring and components that tie the systems together. A set of wires that deliver data from one part of the computer to another is often called a **data bus**, or simply a "bus".

In the terminology I will use here, the **central processing unit** (CPU) refers to the main computer unit, that contains the most important of the computer's parts. Other people use this term to apply to the main data processing chip that forms the heart of the computer, but I will refer to this device as the **processor chip**. You will find this terminology common in advertisements for computers, but other terminologies are more common in computer science textbooks.

Within the CPU of most computers you will find the processor chip; a hard disk; some silicon-based memory chips that the processor uses to store data as it works on it, called main memory or **RAM**; some input/output systems, several data busses; and all the wiring and miscellaneous components to make all these parts work together. For most computers, the input and output systems are primarily external to the CPU, and the CPU will have several jacks, plugins, and ports to which they are connected. Other input/output systems are located inside the CPU. Notice a couple of things here. First, the combination of RAM and hard disk provides for both short term memory (RAM) and long term memory (the hard disk). A better way to refer to these is volatile and non-volatile memory. RAM is **volatile memory** in that when the system is shut off any data stored in RAM disappears. The hard disk provides **non-volatile memory**, which does not disappear when the system is shut off. Second, it is equally convenient, perhaps more so, to consider the hard drive unit, which contains the hard disk, as an input/output device instead of a memory device. In reality it is both, in that it is a memory device that interacts with the processor in a way similar to an input/output device.

Input/output (I/O) systems provide mechanisms for getting data into and out of a computer. I/O systems that are usually contained within the CPU include a CD or DVD system, which provides for external storage of data on optical disks; the video system, to which an external monitor is usually connected; the systems that read the keyboard and mouse, which are likewise external components; an audio system that usually includes a small internal speaker along with ports to which a microphone and external speakers may be connected; a printer system, which connects to an external printer; a modem for connecting to other computers over the telephone system; one or more types of network devices; and the circuitry for several types of general data ports, such as parallel ports, serial ports, USB ports, and firewire ports.

32.4 Data Handling and Storage

From the point of view of a forensic computer scientist, nothing is as important as data handling and storage. There are three fundamental principles of data handling and storage that you should become familiar with, as they form the basis of some of the commonly used ways in which forensic computer scientists find incriminating evidence on a suspect's computer.

First, all input/output devices send or receive data in packages consisting of a certain number of bytes, and each device has a different number of bytes in its package. The number of bytes per package may differ even between devices of a given type, say hard disk drives, made by different manufacturers. Data busses also have characteristic package sizes that they handle. RAM also has a certain characteristic package size, that depends on the **operating system** (Windows vs Macintosh vs Linux, etc.). Various **controllers**, actually small computers in their own right, within each device, break apart and combine packages from other devices so as to make this all work and the data flow smoothly.

Second, when sending data, each device has to send a full package. There is no provision for sending a package that is, for example, 63% full. When presented with the need to send less than a full package, the controller for the system will get data from somewhere to fill up the package before sending it.

Third, a memory device always stores a full package, even when it receives less than a full package. In this case the controller does not gather up extra bytes to fill up the package, but simply stores what it has, wasting any additional allocated memory storage space.

32.4.1 RAM Space

Here's a situation of practical interest to a forensic computer scientist. When a user is working with a file, the file is stored in RAM. When the file is saved, it is sent to the hard disk drive. According to the second principle discussed above, the RAM has to send a full package, but it is extremely rare for a file to be of exactly the right size to fill a package evenly with no bytes left over. Therefore, in virtually every case, the RAM has to find extra bytes to fill up the package before sending it to the hard disk drive. Now consider where those extra bytes come from. They come from whatever immediately follows the file as it is stored in RAM and may consist of fragments of other documents. When the RAM sends its package to the hard disk drive, the entire package, including the extra bytes, is stored on the hard disk. These extra bytes are called **RAM space**. Figure 32.1 illustrates how RAM space is produced.

As an example of how a forensic computer scientist can take advantage of RAM

Figure 32.1: The Data Movement and Storage Processes That Lead to the Creation of RAM Space and Slack Space

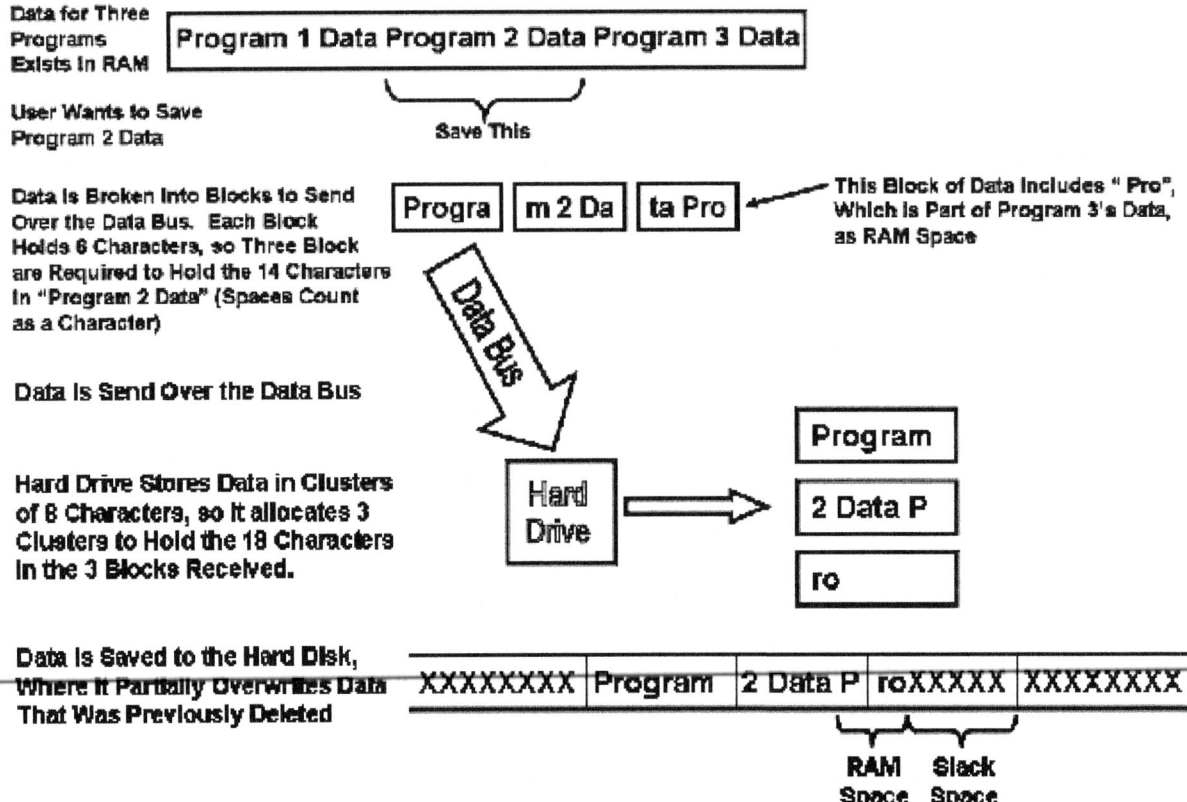

space, imagine that we have a clever drug supplier, who keeps his or her list of contacts on a USB flash drive. A USB flash drive (also called a memory stick, jump drive, and several other names) plugs into the USB port of a computer to serve as external memory, and is small, therefore easily concealable and easily disposed of, for example, by flushing it down a toilet. By keeping his or her contacts on the USB flash drive, the criminal believes that this data is secret and not recorded anywhere on the computer. Now, imagine that our criminal has the file of contacts read into the computer's RAM, and decides to write a letter to his or her mother. When finished, he or she saves the letter to Mom on the hard disk. It's very possible that part of the list of drug contacts is saved along with the letter as RAM space. If this person is arrested and their computer seized, it is well within the ability of a forensic computer scientist to recover the information stored as RAM space.

32.4.2 Slack Space

Let's look at the receiving end of a file sent to the hard disk drive. Unless the package sizes of the RAM, the hard disk drive, and the bus in between are exactly the same, it is unlikely that the hard disk drive will receive a file of exactly the size that its controller has allocated on the hard drive surface for its storage. Therefore, the file will not fill the entire space allocated for storing it and will not overwrite the information that might be present in the unused portion of this space. This part of the hard disk that is not overwritten with new data is called **slack space** and may contain a portion of an older deleted file. Figure 32.1 illustrates how slack space is produced.

To see how slack space can be useful, let's imagine that we have a criminal who has been running a confidence operation (a "scam" in today's street talk) and has defrauded several people and sent the money to an overseas back account. Initially she or he kept records of who their victims were, to avoid trying the scam on them twice. Now our criminal has moved on to a new city and deleted the old file. Over time, most of the old file will be overwritten by new data, but some of it might be preserved in slack space. If this person is arrested and their computer seized, it is within the expertise of a forensic computer scientist to recover this information, thereby gaining the names of a few people who may be willing to serve as witnesses against the criminal.

32.5 Computer Networks

The users of computers often want their computers to communicate directly with other computers for the purposes of sharing information. A connection between two or more computers is called a **network**. There are many different types of networks. Some are informal, and consists of simply wiring together two or more computers in the same home or office so that they can share data. Others, especially in commercial settings, are more formal.

The smallest and simplest type of formal network is the **Local Area Network** (LAN). A LAN connects computers within the same building or set of buildings. LANs can be **peer-to-peer**, in which all the computers communicate as equals, or they can be set up in a **server-client relationship** where one larger computer (the server) stores the information and smaller computers (the clients) retrieve information from the server.

Figure 32.2: The Relationship Between LAN's, Intranets, and the Internet

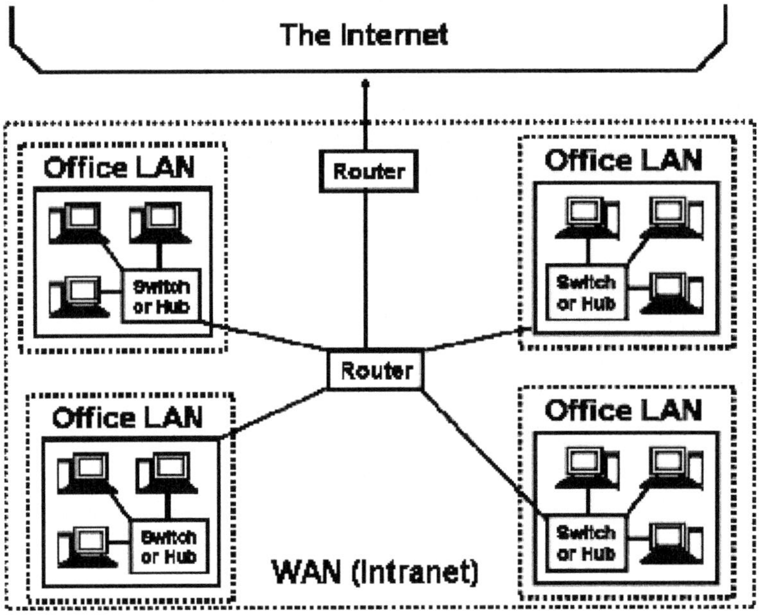

A **wide area network** (WAN) spans a larger area of space, and can be nationwide or even worldwide in scope. Like LANs, WANs can be set up as peer-to-peer or as server-client. There are two types of WANs, intranets and internets. **Intranets** are private networks owned by an organization. For example, a large business might install a nationwide WAN that connects all their local offices and provides employees throughout the country with a common database of information that is stored on a server at company headquarters. This information is private and people who are not employees of the company are not welcome to access it. **Internets** are public WANs that connect intranets and/or LANs together. The largest and most well know internet is "**the internet**", which is a worldwide WAN that allows users to connect peer-to-peer (as in internet chat) or in server-client modes (such as using a web browser to access a web server).

Today's reality is that almost every single computer in use has the potential to connect to the internet, and because the internet is used by many companies to connect together their LANs into an intranet, almost every computer in a business setting in the developed countries of the world is attached to the internet.

32.5.1 Data Handling and Transfer on a Network: Packets and Protocols

Let's look at how information is exchanged over a network, and how this leads to security issues. The information exchanged over a network is referred to as network **traffic**. Like automobile traffic, the information takes up some space as it travels from its origin to its destination. The amount of traffic a network can handle is known as its **bandwidth**. Bandwidth is very analogous to the size of a highway. A four lane highway can carry more traffic than a two lane highway. The bandwidth of a network is a function of the type of medium it is carried over (wire vs fiber optic cable) and the speed of the machines that route the information around the network.

32.5.1.1 Packets

In order to maximize bandwidth, information sent over a network is broken down into chunks. These are very analogous to the "packages" of data that are exchanged inside a computer, and are called **packets**. Packets are often fairly small, a few kilobytes in size. Communicating in this way is like putting the information into a bunch of cars instead of a single freight train. Each car is responsible for delivering its portion of the cargo (information) to the destination, and the different cars may take different routes and may arrive in a different order from the order they started out in. This means that one of the tasks of a network is to fragment the information into packets, then reassemble the packets in the correct order into a complete message once they reach their destination.

32.5.1.2 Protocols

In the practical world of exchanging messages between computers on a network, there needs to be some standard for how the computers and the applications they are running are going to communicate with each other. Let's say that someone is using a web browser to read a document on a web site. There are several different entities involved in this process. First, there is the user's web browser. There is also the computer upon which the browser is running. The computer has a network card or a modem that has to be able to communicate with the phone lines or the network. Then there is the network itself with its cables and routers. At the other end of the network is the web server's network card or modem, the server computer itself, and the web server application that is providing the document.

In order for all of these entities to communicate and work together, a set of instructions and formats in which those instructions are to be provided have been developed. These instructions and their formats are called **protocols**. In reality, several protocols are needed and these are referred to as **protocol suites**. The protocols of a suite work together in a simple way. The traditional way to think of them is as being layered one upon the other, like the layers in an onion. The information packets from a deeper layer are wrapped in a larger packet that is sent up to the next layer. One analogy is to think of the process of sending a message over the network as a chain of delivery people. The first person in the chain hands an envelope to the second person, who stuffs that envelope into a larger envelope and hands it to the next person, who stuffs it into a still larger envelope and hands it on. This process continues to the end of the chain. When the now multiply wrapped package is receive on the other end of the chain, the chain reverses, with each person in the chain opening an envelope, doing what it says to do on the outside of the envelope and passing the contents to the next person in the chain. The final person gets the original message.

The main security concerns arise when the data is actually transmitted over the network. The packet of information has the originating address and the destination address, but it doesn't know how to get where it is going. Therefore, in order to find its way, it has to ask the routers and other machines encountered along the way how to find the path toward the destination.

32.5.2 Old LAN Technology and the Problem of Packet Sniffing

In the days when LAN's were just getting started, there were few enough computers and simple enough networks that all the packets were sent to every computer, and each computer simply listened for its own address. A device called a **hub** connected all the wires leading to individual computers together and sent every packet down every wire. If a computer saw a packet not addressed to itself, it ignored it. If a computer saw a packet addressed to itself, it accepted it.

I think you can see the security issues that arise from this techology. Since every computer on the LAN sees every packet, a snoopy person could tell their computer to accept all packets, regardless of who they are for. This is called packet sniffing. Packet sniffing is a way that a hacker or spy could intercept all the messages and files going over the LAN.

32.5.3 Newer LAN Technology

Fortunately, newer technology arrived several years ago that makes LANs more secure. This technology is called **switching technology**. It relies on intelligence within the devices that connect the individual computers on the LAN. With the older hub technology, all the packets from individual computers go to a central hub, that rebroadcasts each packet to all the other computers connected to the hub. **Switches** are devices similar to hubs, except that they know which connection leads in the direction of each computer in the network. Therefore, a switch sends each packet only along the connection that leads to the destination computer. This makes the packets more secure because somebody sniffing packets has to somehow tap into the more direct connection between the two computers rather than simply being located anywhere on the LAN. Switching technology has the added benefit of reducing traffic on the LAN, because connections that do not contain the destination computer are not sent the packets addressed to it. Most LAN's have now converted to switching technology.

32.5.4 WAN Technology: How the Internet Works

WAN's have never had anything like hub technology, because it has always been impractical to send each and every packet to every computer. Imagine the confusion and loss of bandwidth, for example, if every packet sent by each of the hundreds of millions of computers on the internet was received by every other computer on the internet. Therefore, WAN's have always used a technology similar to that of switches in a LAN. The intelligent data handling machines on a WAN are called **routers**. In a manner similar to a switch, a router knows a set of destinations that can be reached via each wire connected to it. It may not know the exact destination, but it at least knows the correct direction in which to forward the packet most of the time. This results in packets flowing in a relatively blind manner. To envision this manner of packet flow, imagine a person leaving Missoula, Montana on a journey to Albuquerque, New Mexico, but not really knowing how to get there. Now imagine that as he or she comes to each fork in the road they ask a wise old-timer sitting there which fork to take to get to Albuquerque. At first, the old-timer may simply say, "The right fork goes south." As the traveler gets closer to the destination, he or she encounters ever more knowledgeable old-timers, who may say, "The left fork leads to New Mexico," and eventually will encounter an old-timer who says, "take the right fork to get to Albuquerque."

Routers are fairly intelligent devices. They even have the intelligence to know how much traffic is on a certain route and can tell a packet the equivalent of "This route is too busy, take another." This is why the different packets that make up a message may travel to their destination on different routes. However, routers don't know everything, and if they receive a packet for a destination they do not know about, they will simply ignore the packet. This is necessary because it would be detrimental to bandwidth to have packets drifting aimlessly around the WAN. Packets that are lost in this way are noticed to be missing by the receiving computer, which will send a request to the originating computer to resend the packet.

The router system means that not every message can be seen everywhere on the WAN. This makes it unlikely that a hacker, spy, or even a law enforcement agency can intercept a person's data outside of their LAN. The fact that different packets take different routes makes this even less likely. Therefore, packet sniffing is only effective within a LAN.

32.5.5 Wireless Networks

Wireless networks are the least secure of all. In wireless networks, the packets are broadcast as radio signals. Anybody within the range of the broadcast can pick up the packet without even being connected to the LAN. Therefore, if the administrator of a wireless network feels that it needs security, he or she will set up the network to encrypt the data, using an algorithm that allows only the intended recipient computer to read the contents of a packet. We will look at encryption later in this chapter.

32.6 Hashing

One technique that is used in forensic computer science and other applications such as biometrics and forensic databases is hashing. **Hashing** is a way of summarizing the contents of a computer file as a number. In some cases the "hash code", as this number is called, represents a summary of the content of the information present in the file. In other cases, the hash code is simply a virtually unique representation of the file.

Generating a hash code for a file utilizes a mathematical algorithm. **Algorithm** is simply the term for a process to accomplish some task. For example, let's say that we want to organize a large collection of documents so that they could be retrieved quickly. We might do this alphabetically by the title of the document, but what if a document lacks a title? Also, an alphabetical system is sequential, meaning that the documents that start with 'A' come first and those that start with 'Z' come last. In a sequential system it is difficult to retrieve documents from the middle of the sequence, say those that begin with 'K', quickly and efficiently. There are algorithms for searching a sequential list, but they tend to be slow. Here is another problem. What if you wanted to access all documents with a certain content? You would have to look at every document in the list. These are practical problems for information storage. For example, in chapter 16 we looked at the FBI system for archiving fingerprint cards by sorting them into 1032 file drawers based on the fingerprint patterns of the ten fingers. This system quickly proved too cumbersome, and has now been replaced by computerized databases (AFIS systems). Computerized databases have solved this problem, and part of the solution in almost every case involves hashing.

A very simple hashing algorithm for text documents would be to sum the values of all the letters in each document, assigning a = 1, b = 2, etc. This would be extremely tedious for a human to do, but a computer can do it in a flash. The sums resulting from applying this algorithm to a large number of documents would usually differ significantly, but can not be guaranteed to be unique because different combinations of letters could sum to the same number. However, we wouldn't expect this to happen often with documents of normal sizes. Hash codes generated this way would reflect the lengths of documents to the extent that the letters occur with approximately the same frequency in all the documents, but would not be a summary of contents of the documents. For three hypothetical documents we might get hash codes something like 10,031, 12,222, and 13,845. Hash codes can be used in a efficient method of indexing documents, called binary trees, which is beyond the scope of this discussion. Suffice it to say that once a hash code has been entered into a binary tree, searching for the document or item it represents is extremely fast.

Real hash code algorithms are more complex than our example, of course. Those designed to summarize the contents of a document often utilize a process called a fast Fourier transform, the details of which are also beyond this discussion. Hash code algorithms that seek to express the uniqueness of the information in a document, but not its content, use complex mathematical operations on the data in the document to insure that the probability of two documents having the same hash code is extremely low.

32.7 Cryptography

Cryptography is the hiding of information. There are two common forms of cryptography: encryption, in which the information is translated into some form of code; and steganography, which is the art of hiding something in plain sight. It is widely believed that criminals and even terrorist organizations communicate over the internet using both encrypted messages and steganography.

32.7.1 Encryption

Encryption is the process of hiding information by encoding it in some manner so that only the intended recipient of the information can decode it. Converting a message to code is called **encryption** and converting a coded message back to readable form is called **decryption**. There are two commonly encountered forms of encryption – secret key and public key. Of the two, public key encryption is more difficult to break, and is more widely used. The word "key" in this case refers to the method by which the encryption and decryption are accomplished. One example of a key widely used in internet (usenet) newsgroups is called rot-13. In rot-13 encryption, the alphabet is simply rotated 13 positions upward, so that 'm' stands for 'a', 'n' stands for 'b', etc. This is a very simple scheme and easy to crack. It is intended to keep the casual eyes of children from seeing the content of postings in adult newsgroups. Another key is the simple substitution cipher, where 1 = a, 2 = b, etc.

Secret key encryption relies on keeping the key by which the message was encrypted secret. In an ideal situation only the sender and the receiver of the message will know the key. The key must be kept secret, because its use is simple enough that anyone who has the key can encrypt or decrypt a message. The success of this type of encryption depends on the complexity of the encryption process and how secret the key can be kept. One modern version of a secret key encryption scheme is blowfish encryption, which uses a complex mathematical manipulation of a password or number designated by the users.

Public key encryption is more secure. In **public key encryption** each person has two keys, one of which is called their "private key" and is known only to the person themself; and the other of which called their "public key", and is made public to anybody who wants it. Any message encrypted using a person's public key can only be decrypted using that same person's private key, and vice versa. Therefore, a person's public key can be made freely available to anybody and everybody, so long as their private key is kept secret. There is a complex relationship between the two keys that makes it nearly impossible to determine the private key by examining the public key.

Here's how public key encryption works in practice. Let's say that Mary wants to send John a secret message. Mary has her private key and her public key, and John has his private key and his public key. Mary gives John her public key, and John gives Mary his public key. Now, Mary knows her private key, her public key, and John's public key. Likewise, John knows his own private key, his public key, and Mary's public key. In order to send a message to John, Mary's software uses both her private key and John's public key to encode the message. The message is received by John, who uses his private key and Mary's public key to decrypt the message. Since Mary used her private key and John's public key, only someone who knows Mary's public key and John's private key can decrypt the message. If you got lost in this explanation, try again, keeping in mind that the whole idea is this – public and private keys form pairs and something encrypted using a private key can only be decrypted using the corresponding public key. Likewise, something encrypted using a public key can only be decrypted using its corresponding private key.

Public key encryption forms the basis for security of sensitive information that is sent over a network, from a corporate CEO directing an employee at another office to buy certain materials, to the authentication that an ATM machine does of your ATM card.

The security of public key encryption comes from the secrecy of each user's private key and by the size of the numbers used in the keys. Key number sizes in use range from 40 bits (numbers as large as 1,099,511,627,776) to 128 bits (numbers as large as 34,028,236,692,093,846,346,337,460,743,177,000,000). The 40-bit keys are considered weak and breakable, but anything over 56 bits is considered virtually impossible to break. Some files on computers recovered from al-Qaeda strongholds in Afghanistan after the Americans invaded were successfully decrypted because they were encrypted using 40-bit encryption.

32.7.2 Steganography

Steganography is the science and art of hiding a secret message in plain sight, normally by embedding it within another message. The use of steganography goes back hundreds of years. For example, this message was reportedly transmitted by a German spy during World War I.

Apparently neutral's protest is thoroughly discounted and ignored. Isman hard hit. Blockade issue affects pretext for embargo on byproducts, ejecting suets and vegetable oils.

When you take the second letter of each word, it spells out: "Pershing sails from NY June 1".

In modern times, steganography is often done by embedding a message into a digital picture. The distortion caused by this embedded information is usually faint and subtle enough that it goes unnoticed. It is widely believe that terrorist organizations in particular communicate over the internet using steganography, and the volume of traffic on the internet, including pictures, is so enormous that the communication occurs undetected.

32.8 Questions for Study and Review

1. In your own words, define the following terms: computer, binary numbers, bit, byte, sampling (of sounds), data bus, CPU, processor chip, RAM, input/output system, operating system, controller (of a computer-related device), network, LAN, WAN, intranet, internet, traffic (on a network), bandwidth (of a network),

packet (of network data), protocol (for network data), algorithm, encryption and decryption, steganography.
2. Describe the difference between volatile and non-volatile computer memory.
3. In your own words, describe the processes that lead to the formation of RAM space and slack space.
4. Describe the difference between peer-to-peer and server-client type of network configurations.
5. In your own words, define "the internet" and discuss its relevance for forensic science.
6. In your own words, describe how protocol suites work to pass data from a transmitting computer to a receiving computer over a network.
7. Compare and contrast hubs, switches and routers with regard to how they work and the security they provide for a network.
8. In your own words, describe hashing.
9. In your own words, describe public key encryption.
10. Why is public key encryption considered to be more secure than private key encryption?

32.8 Sources

Carrier B. 2005. File System Forensic Analysis. Upper Saddle River (NJ): Pearson Education.

Carvin A. 2001. When a Picture Is Worth a Thousand Secrets: The Debate Over Online Steganography [Internet]. [cited 2007 Oct 24]. Available from: http://www.andycarvin.com/archives/2001/10/when_a_picture.html.

Casey E. 2000. Digital Evidence and Computer Crime : Forensic Science, Computers, and the Internet. Burlington (MA): Elsevier Science & Technology Books.

Elmasri R, Navathe SB. 2006. Fundamentals of Database Systems . Upper Saddle River (NJ): Addison Wesley.

Englander I. 2002. The Architecture of Computer Hardware and Systems Software: An Information Technology Approach. Hoboken (NJ): John Wiley & Sons.

Fletcher C. n.d. Public Key Encryption [Internet]. [cited 2007 Oct 26]. Available from: http://www.krellinst.org/UCES/archive/modules/charlie/pke/.

Jollish B. 2002. The Encrypted Jihad [Internet]. [cited 2007 Oct 26]. Available from: http://dir.salon.com/story/tech/feature/2002/02/04/terror_encryption/index.html.

Kessler GC. 2004. An Overview of Steganography for the Computer Forensics Examiner [Internet]. Forensic Science Communications 6(3). [cited 2007 Oct 21]. Available from: http://www.fbi.gov/hq/lab/fsc/backissu/july2004/research/2004_03_research01.htm.

Knight W. 2002. Weakened Encryption Lays Bare al_Qaeda files [Internet]. New Scientist (17 January, 2002). [cited 2007 Oct 26]. Available from: http://www.newscientist.com/article/dn1804-weakened-encryption-lays-bare-alqaeda-files.html.

Olson D. 2000. Analysis of Criminal Codes and Ciphers [Internet]. Forensic Science Communications 2(1). [cited 2007 Oct 21]. Available from: http://www.fbi.gov/hq/lab/fsc/backissu/jan2000/olson.htm.

Stallings W. 2005. Cryptography and Network Security. Upper Saddle River (NJ): Prentice Hall.

Westphal K. 2003. Steganography Revealed [Internet]. [cited 2007 Oct 26]. Available from: http://www.securityfocus.com/infocus/1684.

Wikipedia contributors. 2007. Hash function [Internet]. Wikipedia, The Free Encyclopedia; 2007 Oct 3, 21:24 UTC [cited 2007 Oct 26]. Available from: http://en.wikipedia.org/w/index.php?title=Hash_function&oldid=162101485.

Chapter 33
FORENSIC COMPUTER SCIENCE

This chapter is the first in a series of five chapters on the more high tech side of forensic science. The series starts with this chapter on forensic computer science, from which we will move on to network security issues such as hacking. Next, we will examine forensic digital image analysis and forensic audio analysis, which are forensic sciences in their own right. A chapter on electronic surveillance and biometrics will follow, and the series will end with a chapter on forensic databases.

The subject introduced in this chapter is forensic computer science. Forensic computer science is the application of computer science to find or develop evidence that might be used in court. There are several branches, as will be discussed below. Topics included in this chapter are some of the non-network oriented parts of forensic computer science: digital evidence, cybercrime, and software/music pirating.

Forensic computer science seems to be one of the most rapidly growing forensic sciences as computers become ever more common. Although it does not have its own branch of the AAFS yet, there has been a movement underway since at least 2001 to create one. Originally, forensic computers science was a university forensic science. Crimes involving computers were rare enough that professors consulted on the few cases that came up. Since the 1990's however, it has been increasingly seen as a law enforcement tool and a concern of corporations. It appears today as if most forensic computer scientists, like forensic artists and photographers, work for a law enforcement agency. Another large contingent works for private corporations, either as in-house specialists seeking to combat network security problems, or as a part of a private force of trained individuals working for a company that provides forensic computer services.

There are two organizations that certify forensic computer scientists: the International Association of Computer Investigative Specialists (IACIS), and the High Tech Crime Network (HTCN). Both require a serious amount of training and experience.

33.1 The History of Forensic Computer Science

Computers have been among the most rapidly developed of all modern technologies. Nearly non-existent before the space race of the 1960's, they emerged in the 1970's and 1980's as toys for hobbyists before becoming mainstream in the 1990's. Given the ubiquity of computers today, forensic computer science is emerging from the university and into both law enforcement agencies and private companies as one of the primary ways in which documents relating to crimes are discovered. A whole new genre of crime that revolves around the internet, such as hacking, has led to the emergence of forensic computer scientists who specialize in network and server security issues.

33.2 The Scope of Forensic Computer Science

Forensic computer science has several branches and an additional interest in many other topics. The branch that other forensic scientists are most likely to encounter is **digital evidence**, which is the recovery of documents, records, and other types of evidence that have been stored on a computer. **Cybercrime** is the use of computers in the commission of traditional crimes such as fraud, theft, terrorism, drug trafficking, child pornography, and even murder on occasion. Most cybercrimes are investigated using traditional methods, but since a computer was involved, there is often digital evidence that can be recovered. **Software piracy** is the illegal copying and/or sale of software. **Hacking** is defined here as gaining unauthorized access to a computer system. It can be used for industrial espionage, various forms of sabotage, or simply to satisfy the hacker's curiosity. Hacking is the special concern of forensic computer scientists who specialize in networking issues. Another task of forensic computer scientists who specialize in network security is the prevention and mitigation of system attacks, such as viruses, email floods, and denial of service attacks. Another topic of interest is the use of computer technology as an aid to other branches of the forensic sciences. One example is the use of computers in digital image analysis and digital audio analysis; and another is the use of computerized databases to store information of value to forensic scientists.

33.3 Digital Evidence

Some law enforcement agencies claim that computers have made their job more difficult in some ways. Even if this is true, however, computers have in other ways become the best friends of law enforcement. The reason for this is that long term memory devices, such as hard disks and DVD's, record information, which may persist even after the user thinks they have gotten rid of it. Hard drives in particular will often preserve evidence that can be retrieved by a forensic computer scientist.

Digital evidence is simply evidence that is stored or transmitted in digital form. As we saw in chapter 32, data is stored digitally on a computer, a network, or one of a computer's peripheral devices. Other forms of digital evidence include images or sound that have been recorded digitally or converted to digital format, and the transmissions of certain types of telephone systems. In the 21st century almost everybody receives and/or stores information digitally, and this includes information about or relating to any crimes they may commit. Much of this information is retrievable, even if it has been deleted.

33.3.1 Histories and Caches

There are some ordinary places to look to find documents that a person has been viewing or working on recently. The most common is, of course, as files that have been saved to the computer's hard drive or a similar storage device. In some cases, however, as when browsing the web, the files being viewed are not normally saved by the user to their computer's hard drive. Even so, it is usually possible to find out what a person has been viewing online recently. In my class I often tell my students that I'm going to show them some "kiddie porn". I then use my web browser to visit the website that has the picture, actually a pen and ink sketch of a baby in a bathtub, which I believe was used several decades ago in advertisements for a line of baby bath products. After everyone quits laughing, I close the browser and have them tell me ways that they could tell that I had viewed that picture. The more internet-savvy students will suggest looking at the browser's history, which is indeed the first place a forensic computer scientist would look for evidence a suspect has visited a questionable web site. In addition to web browsers, many other types of programs, including word processors and spreadsheet programs keep a **history** of at least the last few documents worked on. In fact, the Microsoft Windows operating system, has a "Recent Documents" folder that records the last several documents viewed or worked on by almost any program that runs on a computer with Windows. These history entries are usually tied to dates and times as well, and this information is recorded for each stored file, so it is easy to determine exactly when a document was last accessed.

Back to the "kiddie porn". Some students, who would probably be proud to claim the title of "geek" (I am) will point out that most web browsers will cache pictures from the websites that have been visited. This is a common means of speeding up the apparent speed of a computer or a program. To **cache** information means, in this context, to save it in an easily accessible place, usually the computer's hard disk, for fast retrieval if it is needed again. Once information is cached, the next time it is needed it is rapidly retrieved from the cache. For example, if I wished to view the "kiddie porn" again, the image is likely in my cache and would load immediately without having to download it again from the internet. Caches are often well hidden, and this includes web browser caches, which are often buried in some subfolder of a subfolder of a subfolder of a folder that nobody ever looks at anyway. However, a forensic computer scientist knows where to look for various caches, and can retrieve documents and images from them.

Similar to caching is the use of **virtual memory** in modern computer operating systems. RAM is always finite in size, and for files larger than the size of RAM or if many files are open at the same time, portions of the RAM will be cached on the hard drive. Special files exist to make the caching and retrieval of the data efficient and often unnoticeable by the user. Windows, for example, calls this file a "pagefile". Although RAM is volatile, and the data in it disappears when the system is turned off, this is not necessarily true of the virtual memory file residing on the hard disk. In addition, recall the concept of RAM space from chapter 32, which consists of portions of RAM that are written to the hard disk during the process of saving another file. A forensic computer scientist can examine a hard disk for RAM space that may contain incriminating information.

33.3.2 Deleted Files

In many cases in which a forensic computer scientist becomes involved, evidence is present as ordinary files, produced by some software program, on the criminal's hard disk. Most criminals are either stupid or arrogant, and believe that they will never be caught and that these files of evidence will never be recovered. Others, perhaps more intelligent or more paranoid, are careful to delete files that they don't want law enforcement to have access to.

When using a modern computer with a modern operating system, deleting a file does not actually remove it from the system. In most cases the file's name and its location information, i.e. where it is stored on the hard disk, is simply written to the trash, recycle bin, wastebasket, or whatever it is called on a particular operating system (I will call it a "trash can"). This is done so that the user can quickly undelete the file if they change their mind. For example, in Windows, you simply locate the file in the "Recycle Bin", right click on it, and choose "Restore" from the menu that appears. This places the file's name and location information back into the folder from which it was deleted. A deleted file stays in the trash can until the trash can is emptied. There have been a truly astonishing number of cases in which a criminal has deleted a file, but forgotten to empty the trash can, where the file remains for anybody, professional or not, to find.

Even when the trash can is emptied, the file still remains on the hard disk. Most modern operating systems simply change a couple of bits associated with the file name, and this signals the hard disk drive's controller that the disk space allocated for that file can be used for something else. Unless the hard disk is nearly full, however, it may be a considerable amount of time before this space is actually reused. Therefore, using appropriate software, a forensic computer scientist can simply change the bits that signal the deletion of the file back to their original state and the file may be intact.

Even if the file is overwritten by new information, some or all of it might be recoverable. Remember slack space from chapter 32, which is space at the end of a file's allocated storage area that doesn't actually get written with data, and in which part of a previously stored file's data may still exist. Even if slack space is not helpful, there are sophisticated techniques and devices that can be used to retrieve the overwritten information. As new information is written over old information, the old information is not totally destroyed – just weakened. It still exists as a faint trace called a "**ghost**". Using the appropriate software and devices, the forensic computer scientist can compensate for the effect of the currently written information to recover the ghost. In some cases of multiple overwrites, the ghost under the ghost can be read. The device used to read ghosts is called a magnetic force probe. How many layers deep can a magnetic force probe read? It depends, but the U.S. Department of Defense considers a file to have been securely deleted if it has been overwritten seven times. Programs that will overwrite your data several times with random numbers are available for home use as free downloads from the internet.

33.3.3 Some Practicalities in Discovering Digital Evidence

Given all these ways of recovering digital evidence, the bottom line is that once evidence is in a computer system, at least traces of it are almost always recoverable, and often the entire document or other file can be recovered. It becomes almost impossible to hide visits to questionable web sites or to delete an incriminating file from the system in such a way that it can not be recovered. This means that anytime a criminal uses their computer to communicate something relevant to their crime or works with a document related to their crime, evidence is left in the computer that can be used to link them to the crime. Because of this, one of the most routine evidence items to be collected at a crime scene or during a search for evidence is any computers and/or computer media that are present.

Not all methods of recovering digital evidence are equally easy, however. Some digital evidence recovery methods can take a considerable amount of time. It is relatively easy to check for deleted files, look in the history or cache, or undelete files from the trash can. However, it becomes a real chore to look through a large hard drive for fragments of deleted files, or to probe slack space and RAM space. Therefore, these types of analyses are usually done only when the nature of the crime warrants it. Trying to recover overwritten data using a magnetic force probe is extremely tedious, and is done only in the most serious and critical cases.

33.3.4 Using Hashing to Search for Files

One tool that can help speed up the process of searching an enormous data storage device for either deleted or undeleted files is called MD5 hashing. Hashes and hashing were discussed in chapter 32. This technique is used when there are certain known files that might be present on the data storage device. For example, in child pornography cases there is a certain large set of pictures that most or all of the porn merchants sell and which a consumer of kiddie porn is likely to have had on their hard drive. You can imagine the time it would take for an investigator to find every file or file fragment on a large data storage device and check to see if it's an illegal image. This also raises issues because a search warrant may not give permission for the investigator to snoop into files other than those directly related to the crime being investigated.

MD5 is a hashing technique. It works by taking any file (document, program, image, or whatever) and computing from it a form of hash code called a message digest (the MD in MD5). The message digest is unique to 1 in 340 undecillion (undecillion is the eleventh number in the sequence that begins million, billion, trillion ...). Therefore it is effectively unique and can be considered a fingerprint for a file. MD5 hashing was first developed for security purposes. Because even a tiny alteration in a document will produce a different message digest, the fact the a file yields the same message digest as when it was originally created offers proof that the file was not altered. For forensic purposes, if two files yield the same message digest, then their content must be exactly the same.

In order to detect illegal files, the forensic computer scientist will use software that will scan all the data (deleted and undeleted) on the data storage device and produce a list of each file and its message digest. The message digests of the illegal files are known, and if one of them matches a message digest produced from a file on the suspect's computer, then that file must be a copy of the illegal file. This can be checked quickly by inspecting the suspect file.

33.3.5 Other Ways to Gather Digital Evidence

There are other ways of obtaining digital evidence other than getting it from the computer's hard drive. I would like to discuss two of them: packet sniffers and keystroke recorders. Both of these involve installing a device or program on the suspect's computer or network and require a court order similar to the type of court order required in order to wiretap someone's telephone.

Packet sniffers are also called network listeners, and were introduced in chapter 32. If a particular computer or its user is the target of an investigation, a packet sniffer can be installed on their LAN to make a copy of each packet of information sent to or from that computer over the network. Packet sniffers can also be installed at an internet service provider's facility and used to trap all the communications sent by a user over the internet.

Keystroke monitors are programs that make a record of each keypress that the user makes. The record of the keypresses can be transmitted to authorities over the internet, or can be saved in a file to be retrieved later by the authorities.

33.4 Cybercrime

Cybercrime refers to traditional crimes, such as fraud, murder, or stalking, that are facilitated by the use of a computer or network. Computers facilitate certain types of communications and other activities that can be exploited by criminals.

33.4.1 Things that Facilitate Cybercrime

There are several ways in which the modern computer-enhanced world makes fraud and theft easier. First, computers allow a form of communication that is much more anonymous than more traditional forms of communication. If you go buy something at a department store, such as Walmart, you know three things. First, you know that the company exists, because you see their building. Second, you know that the goods you are buying exist, because you can see them and hold them. Third, you know that there is someone to complain to if there is a problem and ultimately there is someone to send the police to if the problem is severe enough. If you buy something from a web site on the internet there is a higher potential for fraud because you don't know that the business exists, that it actually has inventory, and that there is someone to direct complaints to.

The other facet of the modern world that facilitates fraud and theft is the wide availability of personal information. Much of this information is available over the internet. All forms of public information, such as names, addresses, court processes, marriage dates, birthdates, deaths, college attendances, and other things you might read in a phonebook or newspaper are easily accessible. Even private information such as social security numbers, credit histories, purchasing histories, school grades, and more are available, though you may have to pay a small price to get it. One of the most problematical things this encourages is identity theft, which occurs when someone gets your personal information and poses as you to obtain credit cards and other privileges. There are at least 50,000 cases of identity theft reported each year.

The anonymity of internet communications makes it easier for terrorists, drug traffickers, child pornographers, and other criminals to communicate with each other.

33.4.2 Enforcement of Cybercrime

A few of these cybercriminals are prosecuted each year, but the reality is that there is too much communication occurring over the internet for authorities to keep track of it all and watch for suspicious communications. Therefore, most cybercrime enforcement occurs when someone makes a complaint. There are computers in place on the internet that scan communications for suspicious content, but they are fairly ineffective. The programs that these computers run are usually limited to looking for certain key words (such as "bomb") or looking for messages directed to certain sites or individuals. More intelligent content analysis tools are being developed, but the more intelligent the tool the longer it takes to analyze a packet. Therefore, it become virtually impossible to monitor all packets that travel over the internet.

Another problem that hinders the authorities in policing the internet is the use of encryption. For this reason, the government is very interested in cryptography legislation. For example, under the Clinton administration, the government tried to enact a law that required everyone using encrypted messages to make the encryption key available to the government. This legislation failed to pass congress because of concerns about privacy, but the issue has not gone away and many people feel that it's only a matter of time before such legislation is enacted for the sake of national security.

Certain types of crimes against persons are becoming more common because of the anonymity of the internet and the wide availability of personal information. There have been cases in which a criminal has obtained personal information that has allowed them to stalk and eventually kill a victim. There have also been cases of people, even children, raped or killed by someone they met over the internet and decided to rendezvous with in person. There have been cases in which people have been frightened by threats that have been sent anonymously by email. These have included both threats against businesses or government agencies and threats against individuals.

These sorts of traditional crimes are not new – only the use of computers in committing them is new. Therefore, traditional methods of tracking and prosecuting the criminals are fairly effective. Usually it simply takes good investigative work by someone who is computer literate. Forensic computer scientists get involved in the analysis of evidence for some of these crimes, however. In particular, forensic computer scientists will usually be called upon in cases that involve encryption, or in cases where the origin or location of email or web sites involved in a crime needs to be tracked down. At the present time, the FBI's National Computer Crime Squad (NCCS) is the main police force for the internet in the U.S. The crimes investigated by the NCCS include: computer network breakins, industrial espionage, software piracy, child pornography, email attacks, password sniffers, and credit card fraud.

33.5 Software Piracy and Counterfeiting

The software industry claims that it loses billions of dollars each year to illegal copying of software. In part these claims are exaggerated, but there is no doubt that they lose a substantial amount of money in this way. There are several aspects to the problem. The newest crackdown is on music sharing over the internet, with the same arguments presented. This discussion will focus on software piracy, but nearly everything said herein applies to music piracy as well. The unique feature of software and music that makes it a target of illegal copying is that the copy works as well as the original.

33.5.1 Software Sharing among Individuals

As you are probably all aware, it is illegal to copy and use someone else's commercial software. This is a very common practice, and it's rare to find anybody who is so honest that they don't have or use any illegally obtained software. In fact, there's an old joke that goes something like this: "There are two types of people in the world, those who admit to pirating software and liars." This is where the software industry gets a large amount of their claim of huge losses due to software piracy. They figure that if everyone who had pirated a copy of their software actually bought the software, they would take in large amounts of additional money. In reality, studies and experience have shown that most people who use pirated software would do without it before they would pay for it. Still, being paid by even a tiny fraction of the people illegally using the software would generate a lot of money, so the software publishers actively pursue strategies designed to get people to buy the software. These strategies usually focus on scare tactics wherein the companies claim that people caught with pirated software are subject to huge fines.

In reality it costs more to take someone to court for software piracy that what would be made if the case were won, even if judges or juries were willing to impose punitive fines upon the offender. Police agencies are usually unwilling to investigate or prosecute these cases, so when they are prosecuted it is usually in civil court. Therefore, most software piracy on the individual level is ignored. However, there are occasional cases in which the Software Publisher's Association (SPA) will make an example of someone by taking them to court. They hope that this scares people into buying the software rather than pirating it.

Rather than trying to prosecute individual software pirates, the software industry has focused on copy protection schemes. Of course, as soon as a copy protection scheme is developed, some clever computer programmer will figure out a way to defeat it. A good programmer can defeat most copy protection schemes by simply rewriting the program instructions so that the copy protection scheme is avoided. This process is called **cracking** the software.

There are many, usually informal, groups of people within the computing community whose hobby it is to crack software and redistribute the cracked version to the other people in their group, or even to anybody who wants it. These cracked or otherwise illegal copies of software are referred to as **warez**. Warez are usually made available on a usenet newsgroup or on a web server somewhere, often without the operator of the server even knowing about it. For example, when I first set up the anthropology department's web server, I was naive and left a certain area on the server open to the public so that other faculty could upload lecture notes and similar documents that I would then make available on the web. However, within less than two weeks I found that somebody had discovered this area and was using it to store warez and porn.

Warez sites are considered a problem by the software industry, which tries its best to see that they are shut down, usually by threatening the owner of the system that has the warez on it with a lawsuit. Posting or hosting warez is also a violation of criminal law in the U.S., so there might be criminal penalties that apply as well.

33.5.2 Software Sharing Within an Organization

Most businesses and other organizations rely on a set of software programs to compose documents, keep their books, and in other ways facilitate the business. In some organizations or businesses one copy of the software is bought, which is then copied and given to several employees without paying for separate copies of it. This type of situation is more often prosecuted because an organization or business is often seen as having deep pockets (i.e. a lot of money) and able to pay large fines. The claim can also be made that some part of the organization's revenue was earned by the use of the software. Unlike for individuals who receive copied software, the employees of a business that engages in this practice can not say that they would do without the software before buying it, because the business depends on its use and normally the business is expected to supply employees with those things they need to do their job, including software.

The SPA has teams of people who go to organizations and businesses and do software audits. A **software audit** consists of making sure that each copy of software in use has been purchased or a license to use it has been obtained under some agreement. Of course there is probably no organization or business in existence wherein every computer is completely free of pirated software. Therefore, it is relatively easy for a software audit to turn up something. Historically, large fines have been imposed on some organizations, so it is cost effective for the Software Publishers Association to do this.

33.5.3 Software Counterfeiting

A commercial form of software piracy, in which someone other than the publisher of a software product makes illegal copies of the product and sells them as if they were legitimate copies, is called **software counterfeiting**. Often the copies are dressed up to look real, and a copy works as well as the original, so often the buyer isn't even aware that they are buying an illegal counterfeit product. Software counterfeiting is not tolerated in the U.S. and so only occurs on a small scale. However, counterfeiting is common in some countries, where copies are made and sold over the counter or over the internet (even to Americans). Like counterfeit money, it's the person who is holding the counterfeit item when it's detected that loses their investment.

Software counterfeiting is one problem that does cost the software industry a lot of money. If the software companies were able to receive the money that people spend on counterfeit software, they would gain more profit. It can't be argued that people buying counterfeit software would not have bought it anyway, because they in fact did buy it.

Although the enforcement of software piracy doesn't usually require specialized knowledge of computer science, forensic computer scientists may be called upon to verify that a certain copy of a piece of software is authentic or counterfeit or to determine the process by which a particular piece of software was cracked or copied.

33.6 Questions for Study and Review

1. In your own words define each of the following terms: digital evidence, cybercrime, software piracy.
2. List the places on a computer where a forensic computer scientist might look to find digital evidence relating to a crime.
3. Explain how a forensic computer scientist can often retrieve a deleted document. How might this document be retrieved even if the "trash can" has been emptied? How might this document be retrieved even if it has been overwritten by another file?
4. In your own words explain how hashing can be used to find some forms of incriminating files.
5. Describe how packet sniffers and keyboard monitors can be used to collect digital evidence.
6. What features of the internet make committing certain types of crimes easier than at a traditional "bricks and mortar" store?
7. Who is in charge of investigating cybercrime?
8. Explain the difference between software piracy and software counterfeiting.
9. What is the most common response to software piracy of the sort that involved software sharing between individuals?
10. What is being done to combat software sharing within organization?
11. What is being done to combat software counterfeiting?

33.7 Sources

Access Data Cor. 1999. How secure are you? [Internet]. [cited 2007 Oct 26].
    Available from: http://www.diament.pl/diament/ad/secureclean/howsecure.htm.
Aeilts T. 2005. Defending Against Cybercrime and Terrorism [Internet]. FBI Law
    Enforcement Bulletin 74(1). [cited 2007 Oct 19]. Available from:
    http://www.fbi.gov/publications/leb/2005/jan2005/jan2005.htm.

[Anonymous]. 2004. Summary of the Organized Crime Situation Report 2004: Focus on the Threat of Cybercrime. Trends in Organized Crime 8(3): p41-50.

Ballezza RA. 2007. Identity and Credit Card Fraud Issues [Internet]. FBI Law Enforcement Bulletin 76(6). [cited 2007 Oct 19]. Available from: http://www.fbi.gov/publications/leb/2007/may2007/may2007leb.htm.

Casey E. 2000. Digital Evidence and Computer Crime : Forensic Science, Computers, and the Internet. Burlington (MA): Elsevier Science & Technology Books.

Clark K. 2002. .Data Recovery: the Forensics Wave of the Future. The Forensic Examiner 11(5-6): 36(2).

Cogar SW. 2003. Obtaining Admissible Evidence from Computers and Internet Service Providers [Internet]. FBI Law Enforcement Bulletin 72(7). [cited 2007 Oct 19]. Available from:
http://www.fbi.gov/publications/leb/2003/july2003/july03leb.htm#page_12.

Gillespie H. 1997. Computers and the Science of Solving Crime. Today's Chemist at Work, 6(9), 40-44.

Goodman M. 2001. Making Computer Crime Count [Internet]. FBI Law Enforcement Bulletin 70(8). [cited 2007 Oct 19]. Available from:
http://www.fbi.gov/publications/leb/2001/august2001/aug01p10.htm.

Hardy, R, Kreston SS. 2002. Computers are like Filing Cabinets...: Using Analogy to Explain Computer Forensics [Internet]. American Prosecutors Research Institute Update Volume 15, Number 9. [Internet]. [cited 2007 Oct 26]. Available from: http://www.ndaa-apri.org/publications/newsletters/update_volume_15_number_9_2002.html.

High Tech Crime Network. n.d. Certification Requirements [Internet]. [cited 2007 Oct 26]. Available from: http://www.htcn.org/cert.htm.

Hosmer C. 2002. Proving the Integrity of Digital Evidence with Time. International Journal of Digital Evidence 1(1). [cited 2007 Oct 26]. Available from: http://www.utica.edu/academic/institutes/ecii/ijde/articles.cfm?action=article&id=9C4EBC25-B4A3-6584-C38C511467A6B862.

hternational Association of Computer Investigative Specialists. 2007. IACIS Certification Types [Internet]. [cited 2007 Oct 26]. Available from: http://www.cops.org/certifications.

Mercer LD. 2004. Computer Forensics: Characteristics and Preservation of Digital Evidence [Internet]. FBI Law Enforcement Bulletin 73(3). [cited 2007 Oct 19]. Available from:
http://www.fbi.gov/publications/leb/2004/mar2004/march2004.htm#page_29.

New Technologies Inc. 2005. Software Suites [Internet]. [cited 2007 Oct 26]. Available from: http://www.forensics-intl.com/tools.html.

Noblett MG, Pollitt MM, Presley LA. 2000. Recovering and Examining Computer Forensic Evidence [Internet]. Forensic Science Communications 2(4). [cited 2007 Oct 21]. Available from:
http://www.fbi.gov/hq/lab/fsc/backissu/oct2000/computer.htm.

Palmer, Gary, L., 2002. Forensic Analysis in a Digital World [Internet]. International Journal of Digital Evidence 1(1). [cited 2007 Oct 26]. Available from: http://www.utica.edu/academic/institutes/ecii/ijde/articles.cfm?action=article&id=9C4E938F-E3BE-8D16-45D0BAD68CDBE77.

Panda B, Giordano J, Kalil D. 2006. Next-generation Cyber Forensics. Communications of the ACM 49(2): 44(4).

Robbins J. n.d., The Devil"s Advocate: Computer Forensics Can Support Both Sides of Computer Litigation [Internet]. [cited 2007 Oct 26]. Available from: <http://www.expertnetwork.com/computer_expert.htm>.

Samuel DV. 2006. Code Breaking in Law Enforcement: A 400-Year History [Internet]. Forensic Science Communications 8(2). [cited 2007 Oct 21]. Available from: http://www.fbi.gov/hq/lab/fsc/backissu/april2006/research/2006_04_research01.htm.

Scientific Working Group on Digital Evidence (SWGDE), International Organization on Digital Evidence (IODE). 2000. Digital Evidence: Standards and Principles [Internet]. Forensic Science Communications 2(2) [cited 2007 Oct 26]. Available from: .http://www.fbi.gov/hq/lab/fsc/backissu/april2000/swgde.htm.

Simons EB. 2005. Forensic Computer Investigation Brings Notorious Serial Killer Btk to Justice. The Forensic Examiner 14(4): 55(3).

Whitcomb CM. 2002. An Historical Perspective of Digital Evidence: A Forensic Scientist's View [Internet]. International Journal of Digital Evidence 1(1). [cited 2007 Oct 26]. Available from: http://www.utica.edu/academic/institutes/ecii/publications/articles/9C4E695B-0B78-1059-3432402909E27BB4.pdf

Wise B. 1995. Catching Crooks With Computers [Internet]. American City & County (May, 1995). [cited 2007 Oct 26]. Available from: http://www.clickit.com/touch/arg/crooks.htm.

Chapter 34
COMPUTER NETWORK SECURITY

In this chapter we will explore the parts of forensic computer science that focus on networks, including the internet. The relevant technology was presented in chapter 32. The chapter will begin with a consideration of general network security, then examine threats to a computer system such as hacking, viruses, and system attacks. The chapter will conclude with a look at cyberstalking as an example of a crime made easier by the presence of the internet.

34.1 Network Security

There exist many computer systems that store private or sensitive information. We don't have to look far for examples, ranging from a company's customer database to military computers that we think might store the launch codes for nuclear missiles. In fact, there are a lot of files on my own computer, such as income tax records, bank statements, and research notes that I don't want anyone else to have access to.

The internet, and networks in general, are never completely secure, in that somebody who knows enough can gain access to the data in a computer, and perhaps even gain access to the ability to run programs on the computer. The fact that almost all computers are connected to the internet and/or other networks on at least some occasions, creates a high potential for security breaches. This is a problem for everybody, but a bigger problem for government, business, and industry than for individuals.

34.2 Hacking

Illegally gaining access to a computer system is popularly called **hacking**. There are debates among computer scientists over whether "hacking" is the correct term. Traditionally, hacking has meant mastering the intricacies of computers and making them do what you want – a noble goal; and the term hacker has been a complimentary term applied to a proficient computer user. Traditionalists insist that the proper term for illegally gaining access to a system is "cracking", but this term is more commonly applied to breaking copy protection on software, so using this term causes confusion.

There are also debates over whether hacking (the illegal form) is a legitimate pursuit or not. Some people believe that there are "white hats" (good hackers) and "black hats" (evil hackers), with the good ones being people who violate computer security simply for the learning experience and with no malevolent intent. Fortunately or not, the law doesn't distinguish between the motivations for hacking and in many states it's a felony. There have also been attempts to classify hacking as a terrorist act, and therefore subject to even harsher penalties.

One of the most high profile cases involving the prosecution of a hacker was the Kevin Mitnick case. Mitnick is considered a "white hat" by almost everyone in the online community, including myself, and is known to hackers around the planet as "Condor", a name taken from the Robert Redford movie "Three Days of the Condor". Mitnick was arrested on February 15, 1995, for allegedly breaking into the home computer of Tsutomu Shimomura, a well-respected member of the computer security world. He was suspected of breaking into Shimomura's system and stealing computer security tools to distribute over the Internet. Mitnick eventually accepted a plea bargain in which he plead guilty to "possessing unauthorized access devices". For this crime he received the maximum prison sentence of eight months. His sentence reached farther than this however, in that part of his parole conditions was to not touch a computer for ten years. I happened to be watching a TV show called "The Screen Savers", on the now defunct Tech TV cable channel, when they invited Mitnick onto the show to access the internet for the first time since 1995. Needless to say, it was an interesting episode.

34.2.1 Network Security

There are an overwhelming number of factors that affect network and computer security. Some of these include: features of the networking software, the computer's operating system, the network hardware, and the programming language in which the software is written. The use to which the computer is put also affects its security. For example, it is much easier to secure a computer that a company's employees connect to using custom software than to secure a web site that has material that is intended to be readable by everyone in the world. The greater the number of people given any type of access, the greater the security risk. The best way to insure complete security is to not connect the computer to any network. There are a host of people-oriented security issues as well, particularly people's choices of passwords, their gullibility when presented with some trick designed to get them to reveal their password, and any malice they may bear toward the owners or administrators of the computer system.

There are a variety of strategies for improving network security. As discussed in chapter 32, older LAN hardware presents opportunities for packet sniffing, and should be replaced with modern switching systems. Unfortunately, since Microsoft Corporation is the producer of the most popular networking systems and computer operating systems, their products are the targets of most hackers, rendering them almost continually vulnerable to some sort of attack. Any of you who use Microsoft Windows, for example, are probably aware of the huge number of software updates that Microsoft issues in an attempt to keep up with the hackers. In my university, these updates are automatically "pushed" to our computers so that they install automatically whenever we turn on our computers each morning. There are times when these updates arrive literally every day. Other operating systems don't have such severe problems, and some have been making security improvements for so many years that they are extremely difficult targets for hackers. If I were going to run a sophisticated or sensitive server, I would use the Unix operating system, or one of its descendants such as Linux, which are regarded by computer science professionals as more secure than the rest.

Something as seemingly trivial as the programming language used to develop the software is also critically important. The main principle here is to avoid any software written in C or C++. Although these languages can be used to program software as secure as any in existence, some of their commands are used naively by programmers. For example, C and C++ have commands to read data and write it to a **buffer**. A buffer is memory space set aside by a program to hold input or output data as it is being received or sent. If more data is read than the space allowed in the buffer, a **buffer overrun** will result, in which data is written to memory areas beyond the end of the buffer. The memory overwritten by a buffer overrun usually contains other program code. The data that overwrites this program code due to a buffer overrun can consist of anything, including program codes that will be run whenever the program tries to execute the program instructions that were originally there. The new program instructions can do anything, including giving control of the computer to some outsider. C and C++ have programming commands that avoid this problem, but they are somewhat longer and more cumbersome, so sometimes are not used by the person programming the software.

Data encryption can be used to improve network security. With a data encryption strategy the data in the packets is encrypted so that they can only be decrypted by the destination computer. Especially in server/client type systems it is common to use a public key encryption scheme, as discussed in chapter 32, in which the server knows every client computer's public key and every client computer knows the server's public key. This is the strategy used in applications like online shopping and online banking, and it's very secure.

34.2.2 Account and Password Security

Basic network security strategies are based on user accounts and passwords. A particular user will have an account on a system. That account will have certain privileges with regard to certain data on the system. For example, a certain account may allow the ability to read, write, and delete documents in that person's personal data storage area, be able to read documents in a public area but not alter them, and have no access at all to other people's private areas. Each account requires that a password be entered in order to gain access.

The most obvious strategy for breaking into a computer or network system is to obtain the name of an account and its password. Many organizations assign account names using some logical method so they are easy to find. For example, my user name on the University of Montana's main network is "randy.skelton". In addition, some systems have system commands, such as "finger", that allow a person to find or verify the account name for a user of a system. Therefore, the main task is finding the correct password to use with a given account. There are a lot of strategies for doing this.

Often the easiest and most common strategy for obtaining a user's password is simply to guess what it is. Many people use ridiculously simple passwords, like "computer". Many people use passwords that have some meaning to them personally, such as the name of their wife or their cat. The more you know about a person, the easier it is to guess their password. Since there is so much information about a person available on the internet today, it is often fairly simple to guess a person's password. In the world of hacking this is just a statistical game. Perhaps only one out of every 100 users has a ridiculously easy password, but all the hacker needs is one password to get into the system. Therefore, they try repeatedly until they finally succeed, perhaps after hundreds of attempts. If this gets boring, the hacker may use software to automate the process.

Another hacker strategy is to use a password cracking program. There are programs that automatically try a large number of passwords. Often, these programs have a dictionary and simply start at the a's and work toward the z's. Since most people use fairly simple passwords based on English words this approach is fairly effective.

Each networking system and each computer operating system has **specific vulnerabilities**, such as buffer overruns as discussed above. As also discussed above, Microsoft products are considered weak. Unix systems and their descendants are stronger, but no system is perfect and all have certain specific vulnerabilities. Often it is a race for the operating system programmers to fix a newly discovered specific vulnerability before enough hackers become aware of it to cause a problem. When the operating system programmers find a solution, they then have to distribute it to all the users of that operating system. Microsoft, as one example, has an update utility built into their newer operating systems that automatically check with the Microsoft support site to see if new patches for specific vulnerabilities are available. Many people turn this feature off, however, or don't download the patch because it's too big, so there are millions of computers with known specific vulnerabilities. Often, automated programs or sets of program instructions called **scripts** will be developed to automatically scan systems for these specific vulnerabilities. One form of hacker, called a **script kiddie**, specializes in using these programs and scripts to gain access to systems.

A **backdoor** is a special set of instructions built into a network system that are intended to allow legitimate maintenance personnel and system administrators to carry out their tasks even if they don't know a user's password. Often, a backdoor is simply an account and password that only the suppliers of the software are supposed to know about. For example, it is well known that Oracle systems used to ship with an account name "scott" with a password of "tiger" enabled. The users of the system are either not supposed to know about this or are supposed to inactivate these back door accounts. Many users don't deactivate them, however, leaving a vulnerability for a clever hacker to exploit.

A **Trojan horse**, as the name implies, is a gift that contains a nasty surprise. In this case, a Trojan horse is a program that opens a backdoor for a hacker to gain entrance to the system. Hackers often distribute Trojan horses via mass emailings with the promise that something good or interesting will happen if you run the program. Don't be fooled.

Social engineering is the term used to refer to obtaining a user's password using some sort of scam. For example, a hacker might call a user claiming to be from the computer support department of the company the user works for and ask for the user's password in order to do some "testing" of the system. Many people will give up their password right away. There are many strategies for doing this. The rule is that you should never give your password to anybody. None of the maintenance or support people who work on your system will ever need your password in order to do their work.

34.2.3 Beefing Up Security

The best way to defeat hackers is to make the system as tough to get into as possible. No system is perfect, but hackers find it a lot more rewarding to break into the less secure systems. If a system is tough to break into, they will probably give up soon, because there are a lot of systems out there that are easier to break into. Guidelines for tightening system security are simple. Some of them follow.
- Make passwords hard to guess. Make them longer, change them more often, require them to contain digits and punctuation along with letters, forbid any words that are in any dictionary, and forbid the use of the names of spouses and pets.
- Download and install every system patch or enhancement as soon as it comes out.
- Deactivate all the backdoors.
- Have company-wide training in security issues in which it is made clear that nobody will ask for your password for a legitimate purpose, and that employees should never give it to anybody, ever.
- Institute the policy that only software that has been checked by the technical staff for the presence of Trojan horses may be run on company computers.

34.2.4 Industrial Espionage

Industrial espionage seems like something out of a low budget movie. However, authorities assure us that it is extremely widespread in the real world. If one company can gain access to a competitor's trade secrets then it can obtain a big advantage over the competitor. Since most companies' documents and records are stored on computer systems, they can be accessed by a hacker if he or she can gain access to the system. There are unscrupulous companies that hire hackers or which are willing to pay hackers for secrets they may discover.

In addition to industrial espionage, there are also cases of industrial sabotage, where a hacker damages or destroys information to give a competitive advantage to one company over another. Since modern industrial espionage and industrial sabotage often involve hacking, they are considered network security or computer security issues.

The first high profile case of industrial espionage occurred in 1989, when system administrator Clifford Stoll tracked down an international industrial espionage ring that was spying on American companies. Stoll, who was working at the University of California, Berkeley computer lab at the time, first noticed a small discrepancy in an accounting log. He became intrigued with tracking down these people who had invaded the university computer system he was in charge of. He eventually tracked them to Europe. From his experiences he wrote the best-seller "The Cuckoo's Egg."

Evidence of break-ins is sometimes difficult to find and often overlooked. It really is often just a small discrepancy in some log file. **Log files** are files that record accesses to specific parts of the system. Log files can be examined for suspicious activity. There are several types of system monitor programs that can spot inappropriate activity or unauthorized use of an account. Often, however, it's difficult to distinguish appropriate activity from inappropriate activity. Therefore, many system administrators learn about a break-in only when damaged files or other malicious acts are discovered, and sometimes they never know that the system has been hacked.

34.2.5 Catching Hackers

Even when a hacker's activity is detected, it may be difficult to trace them. Hackers are clever and sophisticated. They may break into one system and use it as a base of operations to break into other systems. Tracing internet connections through a chain of systems is difficult at best, but this is often what has to be done. There are software tools available that can help, but it takes a lot of time and effort.

Once a person is suspected of hacking, however, it then becomes fairly easy to catch them. One way is to obtain a court order and use a packet sniffer to monitor the hacker's activities. Having packets arrive from a computer system to which the hacker does not have authorized access is proof of illegal access. Of course, any files that the hacker illegally downloaded or scripts they used to gain access will probably still be on their computer and recoverable using digital evidence methods. Depending on the operating system, the hacker's computer may actually have logged all of their network activity to a log file.

Penalties for hacking have been fairly light in the past, and the few people who were prosecuted received minimal sentences. That has changed in the 21st century as more organizations are beginning to take hacking seriously and legislation is being passed that provides stiffer penalties for hackers.

There are many ways in which forensic computer scientists become involved with network security, and employment opportunities in this field are good. Forensic computer scientists can help determine the extent of invasion or damage caused by a hacker, help determine how the hacker gained entrance to the system (and eliminate the vulnerability), help track and identify the hacker, and recover proof of the hacker's activities in the form of digital evidence.

34.3 Viruses, Email Floods, and other System Attacks

In this section we look at the activities of truly malevolent computer users – those who seek to damage people's computer systems and/or documents; or who seek to interfere with the intended use of a system. There are many ways of causing deliberate damage. Since not all of these types of malicious activites can be covered in this chapter (whole books have been written on the topic), I will focus on viruses, email floods, and a type of system attack called a denial of service (DOS) attack.

34.3.1 Viruses

A **virus** is a computer program that seeks to propagate itself, meaning that it seeks to make as many copies of itself on as many different computers as possible. It is truly amazing how many computer viruses there are – thousands of them at this point in time. Viruses are fascinating to computer scientists on many levels. Viruses mimic life, especially disease organisms, in that they reproduce and spread. Viruses also represent an interesting technical and social challenge in that they must be given a mechanism for spreading from user to user and they need to be able to avoid existing mechanisms for detecting them. Because of these interesting features there are a lot of people writing viruses. Most viruses are actually not harmful to the system or its data, though they do consume system resources. Some, however, are malicious, and will delete files or cause other severe problems.

Because of the prevalence of viruses, there is a whole industry devoted to producing software that detects and eliminates them. There are many computer scientists whose job it is to find viruses, figure out how they work, and figure out how to remove them. Forensic computer scientists become involved with the task of trying to catch the writer of a virus. Creating or propagating any type of virus is illegal but it is fairly easy to avoid being caught. Only an extremely stupid virus writer would use their own system as the original point of introduction of the virus. Therefore, attempts to trace a virus back to its writer usually fail. Also, many viruses originate in other countries, which makes it especially difficult to trace the writer. Therefore, the best way to catch a virus writer is by traditional methods of investigation. Once a virus writer has been identified, evidence against them can be developed using the methods of digital evidence collection.

34.3.2 Email Floods

Email floods are a form of **system attack**, in which the effect is to interfere with the normal operation of a computer, server, or system by swamping it with more things to do than it can handle. In this case, the system is swamped with more email messages than it can handle. One of the interesting features of email floods, is that the person or people who start them often conceive of it as an attack against an individual, rather than against a system. Email floods are a traditional form of punishment used against people who act badly on newsgroups, email lists, or chat rooms. The idea is that if all the people offended by the boorish user sent many emails to that user's account, the user's mailbox space with their internet service provider (ISP) would become full, which will hopefully result in the ISP shutting off the account and speaking harshly with the offensive user. This is occasionally the outcome, and no permanent damage is done, but in the meantime the system that received the email flood and its other users were inconvenienced.

Email floods can occur when many users all email a certain person or system, or when a single user uses a program or script to send out many emails. Systems are better designed these days, and an email flood is usually recognized and responded to, often by the system simply refusing further emails from an account that has already sent several. Still, email floods waste resources, not the least of which is internet bandwidth.

34.3.3 Denial of Service Attacks

Denial of service (DOS) attacks are another form of system attack. In this case, however, the target is not some offensive user, but a server or the company that runs the server. DOS attacks are similar to email floods in that an overwhelming number of service requests are sent to some server, but these are rarely emails since emails are more traceable that some other forms of service request, such as a request to send a web page. The intention is to so overwhelm the server with requests for service that legitimate users are denied use of the service or server (hence the term "denial of service").

Servers usually provide several services. These services are of many different types, including the correct time, telnet, ftp, and web services. Each service is accessed through a certain **port** or "socket", which is nothing more than an electronic address, no matter how mechanical it sounds. Each service has its own protocol, or set of actions and a format for requesting them. A person wanting, for example, SMTP service (the service used for sending emails), would use software that sent the code word "HELO" (short for hello) to port 25 of a server. The details of what each service does and the protocols that are used are beyond the scope of this book, but are easily found on the internet. A DOS attack involves sending an overwhelming number of service requests to one or more of these services.

System attacks are illegal, but as with many other types of crime that originate over the internet, it is often very difficult to track and identify the person(s) making the attack. Often, a malicious person will hack into one or more systems and launch the attack from there. I this case, the attack can only be traced back to the hacked system, the owners of which may know nothing at all about the attack.

A forensic computer scientist may become involved in the investigation of a system attack by determining from which system(s) the attack was launched (these systems need to improve their security). A forensic computer scientist can also determine how the attack was made, which is valuable information in eventually linking a suspect to the crime. If a suspect is identified, the techniques of digital evidence can be used to help build a case against the suspect.

34.4 Cyberstalking

I would like to look a brief look at one form of crime that has been made easier by the internet – cyberstalking. There is no universally accepted definition of **cyberstalking**, or probably of ordinary stalking for that matter. The idea of stalking involves use of private or semi-private knowledge about a person, such as their address, phone number, place of employment, etc., to follow or keep track of the victim in a way that the victim feels is inappropriate or threatening. Another idea underlying the concept of stalking is that it is a repeated behavior, that the criminal won't stop doing when asked. Most laws that cover stalking require that the stalker actually makes a credible threat of violence against the victim, their family, or their property. Cyberstalking goes beyond ordinary stalking by using the internet, e-mail, or some other form of electronic communications to stalk someone. There is an incredible amount of information about all of us online, as discussed previously, and the ready availability of this information makes any form of stalking even easier.

In 1999, former U.S. Vice President Al Gore asked the U.S. Attorney General to prepare a report on the nature and extent of cyberstalking. This report was produced, and showed that the problem was larger than most people realized. Further, it stated that the ability to law enforcement agencies to respond to cyberstalking is limited. Little has changed since. The bottom line is that the responsibility for protection against and enforcement of cyberstalking is the responsibility of local law enforcement agencies, which vary greatly in their willingness and ability to deal with it. The first traditional stalking law was only passed as recently as 1990 (in California). Cyberstalking laws are still lagging behind.

Law enforcement agencies have been more aggressive in responding to cyberstalking issues when children have been the victims. They have historically been less aggressive when adults are the victim. This seems to be partly because they don't often see cyberstalking as posing a "credible threat" of violence, since for all they know the cyberstalker may live completely across the country or across the world, making actual physical contact with the victim unlikely to occur. There are enough examples at this point in time to show that this attitude is not warranted. Even if it were true, there are indirect ways in which a cyberstalker can cause harm. For example, if two people have some sort of online romance (very common) that breaks up, they can spread damaging gossip or rumors about each other. For example, if one of them is a woman, her former online friend can go to a chat room or a newsgroup and post a notice in her name saying she wants to be raped. This has happened to several women, who have had men show up at their door wanting to rape them. The same sort of thing can be done to a male victim with a little creativity.

All the evidence about cyberstalking suggests that it is a growing problem. What can a person do about it? Many authorities believe that it is too easy to obtain personal information about someone online. However, no legislation has emerged to address this issue so far. At the personal level, it pays to guard one's personal information. In ordinary (not online) stalking cases, the victim usually has to move, change their phone number, and in extreme cases even change their name and place of employment. Online it's a little easier to change your email address or internet service provider. There are also things that internet service providers and internet access programs (such as web browsers) can do, such as offer the ability to selectively block instant messenger alerts or to filter out unwanted emails. These abilities are built into some software, though not many people know how to use them.

It is probably fairly apparent that most of the physical evidence in cyberstalking cases is going to be digital evidence collected from a suspect's computer or the logs and records of an internet service provider. Therefore, successful prosecution of cyberstalkers will likely require the services of forensic computer scientists.

34.5 Questions for Study and Review

1. List as many different types of data that need to be kept secret as you can.
2. Define the following terms: hacking, buffer, buffer overrun, specific vulnerabilities, script kiddie, back door, trojan horse, social engineering, industrial sabotage, log files, (computer) virus, email flood.
3. Explain the difference between "white hat" and "black hat" hackers as they are perceived within the hacker community. Discuss whether any form of hacking is legal or ethical.
4. Explain how each of the following things has an impact on network security (you may need to consult chapter 32 as well): the type of network hardware used, the user account and password policy used, the type of network software used, the operating system used, the programming language used to develop network or operating system software.
5. List the ways in which network security can be improved.
6. Is industrial espionage a concern in the real world, or is it only a common plot device for action/adventure movies?
7. Is it easy or difficult to identify a hacker?
8. Once a person is suspected of hacking is it easy or difficult to find evidence of their illegal activities?
9. Explain how a denial of service attack works.
10. Explain how system attacks, such as denial of service attacks, are investigated.
11. Discuss cyberstalking. How does it differ from traditional stalking? What features of the internet make cyberstalking (or traditional stalking) easier? Who investigates cases of cyberstalking and how are they investigated?

34.6 Sources

Attorney General of the United States. 1999. 1999 Report on Cyberstalking: a New Challenge for Law Enforcement and Industry [Internet]. [cited 2007 Oct 26]. Available from: http://www.usdoj.gov/criminal/cybercrime/cyberstalking.htm.

Ballezza RA. 2007. Identity and Credit Card Fraud Issues [Internet]. FBI Law Enforcement Bulletin 76(6). [cited 2007 Oct 19]. Available from: http://www.fbi.gov/publications/leb/2007/may2007/may2007leb.htm.

Berghel H. 2003. The Discipline of Internet Forensics; a Well-defined Field of Study and Practice Has Evolved as a Result of Network Hacker Activity. Communications of the ACM 46(8): p15(6).

Cogar SW. 2003. Obtaining Admissible Evidence from Computers and Internet Service Providers [Internet]. FBI Law Enforcement Bulletin 72(7). [cited 2007 Oct 19]. Available from: http://www.fbi.gov/publications/leb/2003/july2003/july03leb.htm#page_12.

Davis JB. 2001. Computer Intrusion Investigation Guidelines [Internet]. FBI Law Enforcement Bulletin 70(1). [cited 2007 Oct 19]. Available from: http://www.fbi.gov/publications/leb/2001/jan01leb.pdf

D'Ovidio R, Doyle J. 2003. A Study on Cyberstalking: Understanding Investigative Hurdles [Internet]. FBI Law Enforcement Bulletin 72(3). [cited 2007 Oct 19]. Available from: http://www.fbi.gov/publications/leb/2003/mar2003/mar03leb.htm#page_11.

Granger S. 2001. Social Engineering Fundamentals, Part I: Hacker Tactics [Internet]. [cited 2007 Oct 26]. Available from: http://www.securityfocus.com/infocus/1527.

Kelley PW. 1997. The Economic Espionage Act of 1996 [Internet]. FBI Law Enforcement Bulletin 66(7). [cited 2007 Oct 19]. Available from: http://www.fbi.gov/publications/leb/1997/july976.htm.

Koestner LG. 2006. Law Enforcement Online [Internet]. FBI Law Enforcement Bulletin 75(2). [cited 2007 Oct 19]. Available from: http://www.fbi.gov/publications/leb/2006/feb2006/feb2006leb.htm.

Lease ML, Burke TW. 2000. Identity Theft [Internet]. FBI Law Enforcement Bulletin 69(8). [cited 2007 Oct 19]. Available from: http://www.fbi.gov/publications/leb/2000/aug00leb.pdf.

Luders W. 2007. Child Pornography Web Sites [Internet]. FBI Law Enforcement Bulletin 76(7). [cited 2007 Oct 19]. Available from: http://www.fbi.gov/publications/leb/2007/july2007/july2007leb.htm#page17.

Perry S. 2006. Network Forensics and the Inside Job. Network Security 2006(12): 11(3).

Petherick W. 2007. Cyberstalking: Obsessional Pursuit and the Digital Criminal [Internet]. [cited 2007 Oct 28]. Available from: http://www.crimelibrary.com/criminal_mind/psychology/cyberstalking/1.html.

Rugala E, McNamara J, Wattendorg G. 2004. Expert Testimony and Risk Assessment in Stalking Cases: The FBI's NCAVC as a Resource [Internet]. FBI Law Enforcement Bulletin 73(11). [cited 2007 Oct 19]. Available from: http://www.fbi.gov/publications/leb/2004/nov2004/nov04leb.htm#page_9.

Stallings W. 2005. Cryptography and Network Security. Upper Saddle River (NJ): Prentice Hall.

Stutler TR. 2000. Stealing Secrets Solved [Internet]. FBI Law Enforcement Bulletin 69(11). [cited 2007 Oct 19]. Available from: http://www.fbi.gov/publications/leb/2000/nov00leb.pdf.

Sullivan S. 1999. Policing the Internet [Internet]. FBI Law Enforcement Bulletin 68(6). [cited 2007 Oct 19]. Available from: http://www.fbi.gov/publications/leb/1999/jun99leb.pdf.

Wattendorf GE. 2000. Stalking Investigation Strategies [Internet]. FBI Law Enforcement Bulletin 69(3). [cited 2007 Oct 19]. Available from: http://www.fbi.gov/publications/leb/2000/mar00leb.pdf.

Wikipedia contributors. 2007. Internet crime [Internet]. Wikipedia, The Free Encyclopedia; 2007 Oct 23, 04:21 UTC [cited 2007 Oct 26]. Available from: http://en.wikipedia.org/w/index.php?title=Internet_crime&oldid=166446023.

Wood RA, Wood NL. 2002. Stalking the Stalker: A Profile of Offenders [Internet]. FBI Law Enforcement Bulletin 71(12). [cited 2007 Oct 19]. Available from: http://www.fbi.gov/publications/leb/2002/dec2002/dec02leb.htm#stalking

Chapter 35
FORENSIC IMAGE AND AUDIO ANALYSIS

In this chapter we will examine two of the more high tech forensic sciences – forensic digital image analysis and forensic audio analysis; plus one related lower tech forensic science, forensic linguistics. Image and audio analysis are involved with clarifying and enhancing evidence obtained during an investigation. In the case of image analysis the evidence needing enhancement is usually some form of recorded image, and in the case of audio analysis it is usually some form of recorded sound. Even in routine cases, images and audio recordings may need to be enlarged, amplified, cropped, expletives deleted, or some other adjustment to prepare them for presentation to a jury. Forensic linguistics will also be discussed briefly in this chapter, since its subject matter overlaps to a considerable extent with forensic audio analysis. This chapter will be foundational for the chapter on electronic surveillance and biometrics.

35.1 Forensic Digital Image Analysis

This field does not need the term digital in its title, since an analog image, such as a photograph, can be easily converted to a digital image with modern technology, such a a scanner. Therefore, I will henceforth drop the word when referring to this field. Forensic image analysis involves taking images from cameras, video recorders, and other devices and improving their quality. In many cases images are of low quality, with relevant details not easily visible. This occurs when the device used to make the image is of low quality, or when the user of the device that made the image was unfamiliar with its operation, or when the image has to be made from a long distance away or under challenging conditions. The task of the image analyst is to enhance the image so that relevant details that it recorded can be seen. A second task is to determine whether a photograph or other image has been tampered with.

The technology of image enhancement has emerged from several roots, and its present status within the forensic sciences is unclear. The basic imaging methods, irrespective of the device used, are the province of photography. The instruments, such as cameras, seem to have arisen partly from photography and partly from computer science. The enhancement methods and software have emerged from art, computer science, and, perhaps surprisingly, from medicine. Medicine was one of the first fields to seek to enhance and clarify images such as radiographs and MRI scans, and provided a great deal of the impetus for the development of the methods and software now used.

At present, I am uncertain where to place this specialty. There are digital image analysts who consider themselves photographers. Others consider themselves computer scientists, and yet others do not describe themselves as either and approach the subject as a stand-alone field that is not part of any other forensic science. Going somewhat out on a limb, I'm going to bet that forensic image analysis will settle as a part of forensic photography. Already, much of the training of forensic photographers focuses on image analysis, and it seems a natural extension of photography to include other image sources such as video recorders.

35.1.1 Image Enhancement

Image enhancement refers to clarifying details of an image that are unclear in the original. Those people I know who do image analysis use Adobe Photoshop software, most often on an Apple Macintosh computer. Those I have queried on the subject are convinced that this combination of software and computer provides the ultimate set of tools for the job, and this may indeed be true.

Image enhancement is an endeavor in which forensic scientists have to walk a fine line. The goal is to enhance details already present in an image, but this must stop short of adding details to the image that were not already present. The former is the desired result, but the latter is flatly unethical. Judges will normally insist that when enhanced evidence is used, the process of enhancement be shown to her or him (and the jury if relevant) to show that the enhancement process really just clarified details already present in the image. Figure 35.1 shows the results of image enhancement of a digital photograph taken under low light conditions. This enhancement was done using my computer and a simple freeware program called Digital Camera Enhancer 1.1.

Figure 35.1: Image Enhancement of a Digital Photograph Taken Under Low Light Conditions

 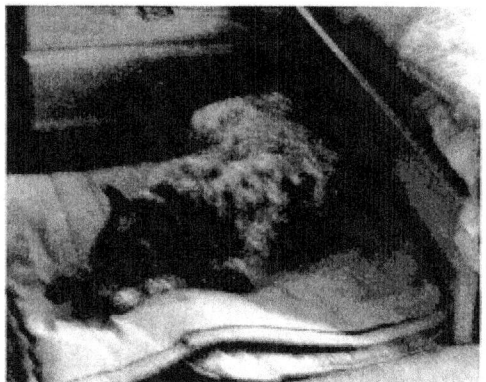

Before Enhancement   **After Enhancement**

35.1.2 Sources of Images

The sources of images to be enhanced are many and varied. Some of them are ordinary photographs or video recordings, digital or otherwise, taken by peace officers or witnesses to an event. Another source of images is various forms of surveillance. Many locations are now videorecorded, either 24/7 or at least during the hours of operation. These include gas stations, both the cash register and the pumps;

convenience stores; banks and ATM locations; several types of government offices; and now even some fast food vendors. Indeed it is difficult to go about one's daily routine without being videorecorded at least once during the day. Once of the challenges of the videorecording systems used is that most of them record to a VHS video tape at an extremely slow speed. Those of you old enough to have used video tapes to record a TV show may remember that there is usually at least a standard recording speed and an slower speed designed to record up to 6 hours of programs on a tape. The quality of the video recorded at the slower speed is poorer than the quality of video recorded at the higher speed. Surveillance recording apparatus often records 24 hours of video on a standard tape, with a great deal less quality. Sometimes the quality of surveillance video is good enough to work with, and sometimes it needs to be enhanced.

Other surveillance methods include taking photographs from a long distance away. Even with the best telephoto lenses, these images are not always clear and may need enhancement. Also, there are military applications in which satellite imagery needs to be enhanced. Although not a common tool for forensic science yet, some authorities believe that the use of satellite imagery will become routine in the near future.

The crime lab is another source of images for enhancement. Sometimes the evidence being worked with can benefit from some enhancement. For example, I once saw a presentation by a fingerprint examiner, who discussed her efforts to recover detail from a fingerprint on a piece of rough fabric. The texture of the fabric was such that parts of the print was preserved on the fabrics ridges and the other parts lost in its grooves. When shown a slide of the unenhanced photograph of this print it was clearly impossible to see any details. However, this clever forensic scientist was able to use Adobe Photoshop to subtract the texture of the fabric from a scanned image of the photograph, and the result was stunning. The print was still only partial, but there were clearly details of minutia present and visible. I do not know whether this effort led to solving a case, but it illustrates the utility of image enhancement even for the routine cases at a crime lab.

35.1.3 Authenticating Images

An **authentic image** is one that shows what it is purported to show in the way it purports to show it. An image that is not authentic may have been tampered with, or it may have been staged. Tampering with an image is almost always easy to spot because it's difficult to get the details of shading, color, etc. correct. When, for example, someone grafts the image of Brittney Spears' face onto the image of Arnold Schwartzenegger's body, nobody is ever fooled into thinking that this is an authentic image of a woman who has overdone the body building. The same is true of pictures of "jackalopes" (a mythical creature half jackrabbit and half pronghorn antelope), and similar chimeras (a chimera is a creature with a mixture of characters of other creatures). The way to detect tampering in an image is to enlarge it to the level where you can see the actual **pixels** (small spots of color) that make up the image. Under these conditions any alterations will be visible as inconsistencies between the patterns of pixels in one part of the image and those in the rest of the image.

It is more difficult to detect a staged image. A **staged image** is one in which subjects are placed specifically to suggest that some event occurred, which did not actually occur. Detecting staged images relies primarily on common sense and the knowledge of the examiner. For example, people who are staging a scene often have body postures that are not natural, and someone with expertise in acting can possibly detect this.

35.2 Forensic Audio Analysis

Forensic audio analysis shares many characteristics with forensic image analysis, with the main difference being the obvious one that image analysis works with images and audio analysis works with recorded sounds. The main task of audio analysis is to enhance recorded sounds, usually voices, to the point that the details of the sounds recording are clear. Forensic audio analysis includes authenticating sounds, a pursuit usually called sound identification. As for forensic image analysis, forensic audio analysts need to make sure that enhancing a recording does not result in adding information that was not originally present.

Audio analysis has been around longer than image analysis, since easily manipulable and analyzable sound recording devices and media, such as tape recorders, is several decades older than comparable methods of working with images. Therefore, audio analysis includes several types of tasks that have no counterpart in image analysis. These include voice (speaker) identification; audibility analysis, i.e. could a certain sound be heard at a certain location; and event sequencing, which refers to figuring out the sequence of events captured by a sound recording.

35.2.1 The Origin of Forensic Audio Analysis

Forensic audio analysis is a moderately new technology, though unlike forensic image analysis, it does not have roots in a wide variety of other areas – only two that I know of. The first root of audio analysis is the commercial sound recording industry. These are the people who bring us music CD's, which emerge highly edited and manipulated from the original recordings of musicians in a studio. The other root is computer science, which has long been interested in audio recordings, leading to the development of a huge amount of software for analysis and manipulations of digitally recorded sound. The forensic aspects of audio analysis emerged during World War II, due to the need to identify the voices of people in recorded radio broadcasts. One of the first forensic audio techniques, voiceprints, was developed in 1941. Since that time, forensic audio analysis has been used in several high profile cases, such as the assassination of John F. Kennedy, to answer questions about certain sounds in the recording of the event.

35.2.2 General Methods

Most of the techniques of forensic audio analysis utilize digitized versions of the audio recording, which is manipulated using software. The results of an analysis are often presented in the form of a spectrogram (no relation to spectrograms produced by chromatography or mass spectrography as discussed in chapter 18). A **spectrogram** is a two dimensional diagram that presents two or three types of information that are occurring simultaneously. Usually, either the frequencies involved in the sound are displayed as a function of time, to produce a T-F spectrogram (T for time, F for frequency); or the amplitude of the sound is displayed as a function of time to produce a T-A spectrogram (T for time, A for amplitude). In certain applications, frequency and time are plotted relative to each other to produce an F-A spectrogram. For complex tasks, such as voiceprints, both frequency and amplitude may be displayed as a function of time to produce a T-F-A spectrogram. Figure 35.2 shows a T-A spectrogram, a T-F spectrogram, and a T-F-A spectrogram of a 100Hz tone. Note that the T-A spectrogram shows about 20 wave cycles in 0.2 second, which corresponds to 100 wave cycles in one second. The T-F spectrogram shows a straight line at 100 Hz because the frequency of the tone does not vary. Hertz (Hz) is the metric measure of frequency and is equivalent to cycles per second. The T-F-A spectrogram plots time vs frequency and shows amplitude as color, where darker color corresponds with greater amplitude. This spectrogram shows that while the frequency remains constant, the amplitude at that frequency varies, with peaks of loudness occurring 100 times each second.

35.2.3 Methods for Improving Listenability

Listenability refers to the level of comfort of the person listening to a recording. If the recording contains hum or noise, and especially if it contains loud pops and clicks, listening to it for a long period of time causes considerable discomfort. Listenability enhancement is the process of reducing these annoying, distracting, or painful extraneous sounds so that the recording is more comfortable to listen to.

Fortunately, methods exist to improve the listenability of a recording. One method is equalization, in which the amplitude at certain frequencies is adjusted to achieve a balance between treble, bass, and frequencies in between. Clicks and pops are annoying because they often occur at amplitudes many times greater than those of the recorded voices or other information. Amplitude limiting, also called "clipping" is used to reduce the volume of clicks and pops by cutting off amplitudes greater than a certain value. The clicks and pops remain, but their amplitude is reduced to where they do not produce discomfort when they occur. Hums and other types of interference can often be reduced using filtering. Filtering limits the amplitude at certain frequencies. Many forms of interference occur at frequencies either below or above those of speech. High pass filters can reduce low frequency sounds, such as most hums. Low pass filters can likewise reduce high pitched interference.

Figure 35.2: T-A, T-F, and T-F-A Spectrograms

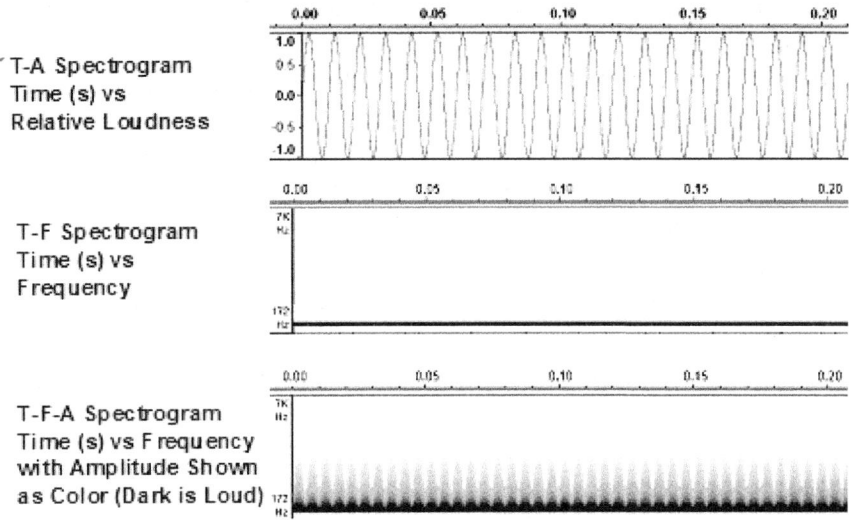

Three Type of Spectrograms of a 100 Hz Tone

35.2.4 Methods for Improving Intelligibility

Intelligibility refers to the ability of the listener to understand the content of an audio recording. An intelligible recording is one for which the words in the dialog can clearly be understood. The first step in enhancing intelligibility is usually to improve listenability. Other techniques for improving intelligibility include simply turning up the volume of the recording and perhaps turning down its speed. If slowing the speed of a recording makes the speaker's pitch too low, this can be corrected for by adjusting the overall pitch of the recording.

Another useful technique is compression. Most speech sounds occur within a limited range of frequencies. Compression removes all but this range of frequencies. This often increases the clarity and the perceived volume of the audio. Television commercials almost universally use compression. They remove the high and low ends of the frequency range which allows the portion of the volume accounted for by sounds in these ranges to be used in the portion of the frequency range in which speech normally occurs. Therefore, the commercials sound louder, even though the advertising agencies and TV stations claim that the volume of a commercial is no louder than that of the regular programming. This claim is true if the volume measured is average volume over the entire frequency range.

Another approach to intelligibility, especially in the case where a certain critical word is difficult to understand, is **dialog decoding**. If a case hinges on whether a suspect said "gun" or "gum", dialog decoding may be able to determine which word was actually spoken (sorry, no method exists to reliably distinguish "bomb" from "balm"). Each speech sound (phoneme) has distinct characteristics, otherwise we would not be able to distinguish "gun" from "gum" under any conditions. These characteristics may be obscured in a low quality recording. Dialog decoding uses a T-F spectrogram of the speech sound in question, which is often sufficient to determine which sound it is, in our example an 'n' or an 'm'. In especially difficult cases, if a T-F spectrogram is insufficient, a T-F-A spectrogram can be used. These methods will not work in all cases, however.

Several other advanced techniques, beyond the scope of this chapter can also be applied, often with near magical effect on the intelligibility. The names of some of these methods are: sequential mirroring, deconvolution filtering, and wideband and multiband sidechain compression.

35.2.5 Authenticity Analysis

Authenticity analysis is the process of determining whether a recording was actually made at the time and place, and under the circumstances claimed. One common question is whether the recording has been edited to either add or remove something. The method for making this determination is similar to that used in image analysis. Under normal conditions, and using common equipment, it is very difficult to edit a recording in such a way that continuity is maintained throughout the edited area. However, sound studio engineers do this all the time, and the results are not noticeable on a CD. It is likely, though, that professional quality sound editing equipment is required to make seamless transitions between edited sections of a recording, and that the portions edited together must have otherwise identical characteristics, as is normally only obtained when recordings are made in a professional studio.

As discussed in chapter 17 with regards to writing, human events do not ever occur the same way twice. Therefore, one way to demonstrate tampering with a recording is if it can be shown that the events recorded occur with the exact same timing and sequence (event sequence analysis) more than once.

Other things important for determining authenticity include the nature of background sounds, which should be consistent with the place and time claimed for the recording; and the physical nature of the recording method and recording media. For example, an authentic recording of Elvis Presley's last concert would not be made by a digital recorder onto a flash memory card.

35.2.6 Audibility Analysis

Audibility analysis seeks to answer the question of whether a certain sound could be heard under a specific set of conditions. For example, could a person sitting in a car outside a building have heard a shot fired inside the building. The answer to an audibility question often depends on the conditions, especially the presence of other noise in the environment, often referred to as **masking noise**. In our example, it would be much more difficult for the person sitting in the car to have heard the shot if their motor was running. The loudness of the sound in question compared to any masking noise present is often referred to as the **signal to noise ratio**.

The process of doing an audibility analysis normally involves reconstructing the acoustic event, perhaps several times as someone listens or records the sound level at the location where the audibility is in question. The more the conditions of the original event can be duplicated, the more accurate the determination will be. In our example, one forensic audio analyst could duplicate the sound of the gun firing (hopefully not by actually shooting it) while their partner sat in the car under conditions as similar to those of the original event as possible (the motor running or not, the windows up or not, heavy traffic on the street or not, etc.). The results of this re-enactment may reveal whether the sound in question was audible at a certain place or by a certain person.

35.2.7 Sound Identification

Sound identification seeks to answer the question, "What is that sound?" For example, was a bang heard in a recording the sound of a gun being fired, the sound of a chair falling over, a car backfiring in the street outside, a "pop" in the recording device, or any one of a number of events that can produce a similar sound. Certain sounds, especially those that begin and end quickly, are difficult to distinguish by the unaided human ear. However, similar sounds from two sources are not likely to be identical.

If the nature of a sound is difficult to determine simply because it happens too quickly, slowing down the recording may offer some insight. A more reliable analysis can be done using an F-A spectrogram of the sound. Similar sounds with different causes should have different profiles of amplitude at certain frequencies, which can be detected using this method.

Taking the time to identify the nature of background noises can sometimes prove to be time well spent. Cases abound in which a background noise was identified as a type of sound produced and audible only in a specific area, which led to quick identification of a suspect. There are also several cases in which a conversation occurring in the background has provided valuable evidence.

35.2.8 Event Sequence Analysis

Event sequence analysis is determining the order of events that are present in a recording, and the exact times at which they occur. A famous example is the chain of events that occurred during the assassination of U.S. President John F. Kennedy. A recording of the event made by a bystander, Abraham Zapruder, furnished crucial evidence for the investigation, including an exact event sequence. In more prosaic, but no less critical situations, the 911 recording of an event can be used to make an event sequence of what happened to a person after they dialed for help.

35.2.9 Voice Identification

Voice identification seeks to determine the identity of a certain speaker, recorded or otherwise. This is one of the oldest uses of forensic audio analysis, as discussed previously, and dates to World War II.

This is a form of matching, and as such requires both an evidence item (the recording of the questioned voice) and one or more standard items, which consist of recordings of voices of known speakers. The more closely the recording devices and other conditions can be matched between the evidence and standard voice recordings, the more successful voice identification is likely to be. The standard method relies on the fact that each of us has a unique way of speaking. The exact details of the pronunciation of certain words will vary between individuals. These details can be captured to a certain extent in a T-F-A spectrogram, also known as a **voiceprint**. The limiting factor when working with voiceprints is that they can only represent a speaker's pronunciation of a certain word or phrase. It is not possible, for example, to match a person's pronunciation of "monkey" with their pronunciation of "chicken". The voiceprints of the same word or phrase must be used. Figure 35.3 shows a voiceprint of the Japanese word "minato" as spoken by a native female speaker from Shinagawa, Tokyo.

Figure 35.3: Voiceprint of a Japanese Woman Saying the Word "Minato"

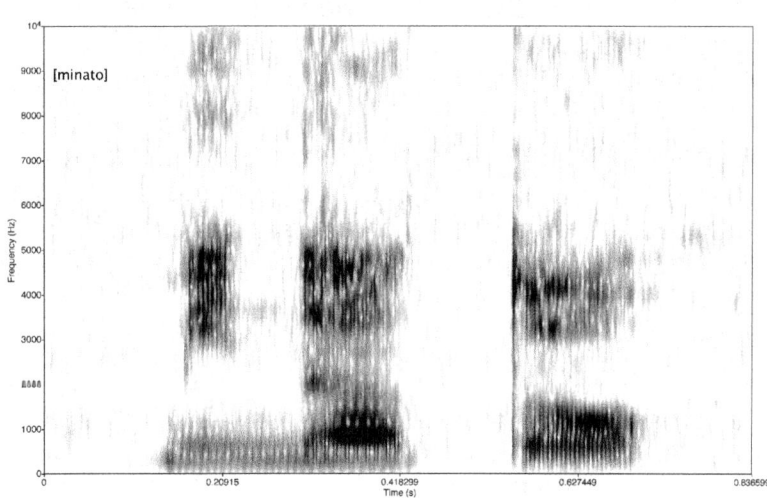

The process of making a match is reminiscent of matching handwriting. A person never says the same word exactly the same way twice, yet two utterances of the same word by the same person are usually very similar. Not unlike handwriting, speaking is an activity that originates in the brain as a desire to speak a certain word. The brain sends signals along nerves to the lungs, vocal folds, throat muscles, tongue, lips, and mouth, which all work together to accomplish the utterance. A person's speech is therefore a product of their mental template of how a word should be pronounced, and their vocal apparatus. The uniqueness of each person's speech comes from the interaction of these anatomical and mental factors. Yet another similarity to handwriting is that a person's speech is influenced by their mental and physiological state, and speech produced while excited, fatigued, intoxicated, etc., will differ.

The TFA spectrogram captures the details of how a word or phrase is spoken. The forensic audio analyst has to judge both whether the expected amount of natural variability is present, and whether the voiceprint is consistent with a certain person's normal range of speaking variation. Voice recognition from voiceprints is challenging. This is one area where computerized methods can help, though not replace a trained forensic audio analyst. Some computerized methods reduce the information in a voiceprint to a hash number, which is then compared. Possibly the biggest challenge in voice recognition is that background and extraneous noises have an effect on the voiceprint. Voice recognition also suffers from some of the same problems as biometrics, which will be discussed in a coming chapter. The acceptability of voiceprint evidence in court has not been fully worked out. It is allowed in some cases, but not in others.

35.3 Forensic Linguistics

Linguistics is the study of languages and speech, and forensic linguistics is the application of knowledge from this field to evidence in investigations. It has two basic applications: speaker identification and detecting lies, along with several other applications. Forensic linguistics is an established field and there is even a professional journal, titled "Forensic Linguistics".

The task of **speaker identification** is essentially the same task as the forensic audio analyst's voice recognition – the identification of who is speaking. However, forensic linguists approach it from a different direction. Instead of focusing on matching voice prints, the approach that forensic linguists use focuses on listening to the speech and applying what linguists know about the characteristics of language. Things that forensic linguists listen for include pronunciation of vowels and consonants (phonetics), and the overall pitch and pattern of the speech.

A second application of forensic linguistics is detecting when someone is lying by listening to their speech. It is observed that many people hesitate in their speaking just before telling a lie. Another technique used in lie detection relies on the observation that many people use more words in telling a lie than in telling the truth. Further, when telling a lie, people tend to be more creative in their use of words and do not repeat the same word as often. The creativity in word use can be quantified by dividing the number of different words in an utterance by the total number of words used. A higher ratio may reflect a bigger lie.

35.4 Questions for Study and Review

1. Explain what forensic image analysis is.
2. Define image enhancement.
3. What must an image analyst be very careful about when doing image enhancement?
4. List as many sources of images or video as you can.
5. Explain how an image analysis can show that a photograph of Mount Rushmore with president Bush's head instead of Abraham Lincoln's head is a fake (by looking at the photograph itself).
6. What is a staged image and why are staged images difficult to detect?

7. Define forensic audio analysis.
8. Discuss the different types of spectrograms and for each list the common purpose or purposes for which it is used.
9. Discuss how listenability and intelligibility are different.
10. Discuss methods for improving listenability.
11. Discuss methods for improving intelligibility.
12. Explain what authenticity analysis is when the term is applied to forensic audio analysis, and discuss some of the approaches to doing it.
13. Discuss audibility analysis and how it is done.
14. Discuss sound identification and how it is done.
15. In your own words, define event sequence analysis.
16. Describe how voice identification differs from speaker identification.
17. Define forensic linguistics and list the common tasks of forensic linguists.

35.5 Sources

[Anonymous]. 2007. Restoring Video Evidence [Internet]. Government Security 6(4). [cited 2007 Oct 26]. Available from: http://govtsecurity.com/state_local_security/restoring_video_evidence/.

Berg E. 1996. Forensic Image Processing: An Introduction to Image Enhancement [Internet]. [cited 2007 Oct 26]. Available from: http://www.imagingforensics.com/forensic.pdf.

Berreby David. 1992. Back talk. Discover 13(n7): 76(4).

Blades HB. 2003. Tom Owen Voice Identification & Forensic Audio & Video Analysis (Interview). The Forensic Examiner 12(9-10): 15(7).

Bramble S, Compton D, Klasén L. 2001. Forensic Image Analysis [Internet]. Paper presented at 13th Interpol Forensic Science Symposium, Lyon, France. [cited 2007 Oct 26]. Available from: http://www.aeicomputertech.com/_downloadz/forensics_resources/forensic_image_analysis.pdf.

Computer Audio Engineering. 2007. Forensic Audio: An Introduction [Internet]. [cited 2007 Oct 26]. Available from: http://CAEaudio.com/forensicintro.html.

Furui, Sadaoki, n.d. Speaker recognition [Internet]. [cited 2007 Oct 26]. Available from: http://cslu.cse.ogi.edu/HLTsurvey/ch1node9.html.

Gauvain, J. L., Lamel, L. F. and Prouts, B., 1995. Speaker Identification and Verification [Internet]. LIMSI Scientific Report March 1995. [cited 2007 Oct 26]. Available from: http://www.limsi.fr/Recherche/TLP/reco/2pg95-sv/2pg95-sv.html.

Koenig BE, Lacey DS, Herold N. 2003. Equipping the Modern Audio-Video Forensic Laboratory [Internet]. Forensic Science Communications 5(2). [cited 2007 Oct 21]. Available from: http://www.fbi.gov/hq/lab/fsc/backissu/april2003/lacey.htm.

Koenig BE, Lacey DS, Herold N. 2006. Video and Audio Characteristics in VHS Overrecordings [Internet]. Forensic Science Communications 8(3). [cited 2007 Oct 21]. Available from: http://www.fbi.gov/hq/lab/fsc/backissu/july2006/technical/2006_07_technical01.htm.

Koenig BE, Lacey DS, Killion SA. 2007. Forensic Enhancement of Digital Audio Recordings. Journal of the Audio Engineering Society 55 (5): 352-371.

Nolan F. 2001. Speaker Identification Evidence: its Forms, Limitations, and Roles [Internet]. Proceedings of the Conference 'Law and language: Prospect and Retrospect', December 12-15 2001, Levi (Finnish Lapland). [cited 2007 Oct 26]. Available from: http://www.ling.cam.ac.uk/francis/LawLang.doc.

Scientific Working Group on Imaging Technologies. 2003. Recommendations and Guidelines for the Use of Digital Image Processing in the Criminal Justice System [Internet]. Forensic Science Communications 5(1). [cited 2007 Oct 21]. Available from: http://www.fbi.gov/hq/lab/fsc/backissu/oct2003/2003_10_guide02.htm.

Scientific Working Group on Imaging Technology. 2005. Best Practices for Documenting Image Enhancement [Internet]. Forensic Science Communications 7(3). [cited 2007 Oct 21]. Available from: http://www.fbi.gov/hq/lab/fsc/backissu/july2005/standards/2005_07_standards01.htm.

Scientific Working Group on Imaging Technology. 2005. Best Practices for Forensic Image Analysis [Internet]. Forensic Science Communications 7(4). [cited 2007 Oct 21]. Available from: http://www.fbi.gov/hq/lab/fsc/backissu/oct2005/standards/2005_10_standards01.htm.

Smith SS, Shuy RW. 2002. Forensic Psycholinguistics Using Language Analysis for Identifying and Assessing Offenders [Internet]. FBI Law Enforcement Bulletin 71(4). [cited 2007 Oct 19]. Available from: http://www.fbi.gov/publications/leb/2002/april2002/april02leb.htm#page_17.

35.6 Acknowledgments

Digital Camera Enhance, also known as Digital Camera Enhancer, Free DCE or Free Digital Camera Enhancer is available for download from Mediachance, www.mediachance.com. Mediachance also markets a much more fully featured digital enhancement program called DCE Autoenhance for about $40, as well as other image enhancement software for home users.

The graphic used in Figure 35.3 is from Wikipedia Commons at this URL: http://en.wikipedia.org/wiki/Image:Spectrogram_-minato-.png. It was created in 2005, by User:ish ishwar, who makes it available under the Creative Commons Attribution 2.0 license. See http://creativecommons.org/licenses/by/2.0/ for the details of this license.

Chapter 36
ELECTRONIC SURVEILLANCE AND BIOMETRICS

The focus of this chapter is electronic surveillance and biometrics. Electronic surveillance involves the use of electronic devices to gather evidence about the activities of suspects. Biometrics is a relatively new field that uses computerized methods to identify individuals for a variety of purposes. Although electronic surveillance, especially wiretaps, is widely used in gathering evidence, biometrics is not widely used in the forensic sciences yet. However, it includes a set of technologies and methods that will probably become useful in forensic science in the near future. In the meantime it has application in corporate and individual security, military issues, and homeland security. These are also interesting topics because of their treatment in movies and other media, which portrays their use in ways that are still science fiction.

The chapter begins with an examination of electronic surveillance, including its history, some of the technologies involved, and the roles of forensic scientists in analyzing evidence generated by it. Next, the subject of biometrics is considered, including its main goals and some of the technologies involved.

36.1 Electronic Surveillance

Although not a forensic science, electronic surveillance is a useful forensic technology, and is becoming more important as time goes on. Electronic surveillance is a growth industry, both for those who do the surveillance and those who try to prevent or counteract it. In a previous chapter I discussed industrial espionage, which is of great concern in the corporate world, and which relies heavily on electronic surveillance.

Electronic surveillance has long been used as a tool for law enforcement as well. There are some people who believe that if law enforcement is going to keep up with criminals, given limited budgets and staffing, they are going to have to do it by focusing more on using high tech methods such as electronic surveillance.

There are three important branches of the electronic surveillance field: bugs and cameras, wiretapping, and satellite surveillance.

36.1.1 The History of Electronic Surveillance in the U.S.

The first recorded instance of wiretapping dates to the U.S. Civil War, during which both the Union and the Confederacy tapped the other side's telegraph wires in order to intercept communications. During the late 1800's and early 1900's, audio recording devices and still cameras began to be used for surveillance, particularly of foreign immigrants. From the 1920's through the middle 1960's electronic surveillance became commonplace, both as a means of gathering evidence on criminals, and also, unfortunately, to gather information about political and civil rights activists. During the 1960's and 1970's, the use of electronic surveillance without a warrant became restricted. The Safe Streets Act of 1968 set specific procedures for the use of electronic surveillance, including the requirement to obtain a warrant in most cases. Since the middle 1980's, and particularly since passage of the Patriot Act of 2001, the federal government has been seeking to ease restrictions on the use of electronic surveillance, with some success.

36.1.2 Bugs and Cameras

No longer confined to international spy operations, bugs (electronic listening devices) are in widespread use, especially in the realm of industrial espionage. Statistics are hard to come by, but most authorities agree that the problem of illegal use of bugs and cameras is large – much larger than is commonly imagined. Law enforcement agencies, especially those at the federal level, also employ these devices. Video cameras are now so small that they are being integrated into bugs along with their traditional audio capabilities, with the result commonly called a "video bug".

Video recorders are now installed at a wide variety of locations including convenience stores, ATM locations, gas pumps, government buildings, commercial buildings, and even on street lights in some cities. These cameras are placed in the hope of recording a crime as it occurs so that the nature of the crime and the identity of the criminal are easier to determine. Peace officers are not actually supposed to examine these recordings unless a crime is known to have been committed, or they obtain a court order (warrant) for surveillance of a suspect. There is even an increasing number of them being installed at stop lights and in speed zones for the purpose of detecting traffic violations. Video cameras are also routinely installed on police patrol cars in most jurisdictions, so that they can tape the interactions between peace officers and citizens or suspects.

36.1.3 Wiretapping

Wiretapping is one of the most widespread forms of electronic surveillance, both for industrial espionage and for law enforcement. Wiretapping is traditionally thought of as listening in on telephone lines, but it now includes monitoring of cell phones, cordless phones, pagers and other types of wireless communications as well.

The federal government and 37 states allow wiretap evidence in court. Montana is one of the states that does not allow it. In fact, I once testified (about human bones) at a murder trial in Montana in which the suspects were convicted, but the case was eventually ruled a mistrial because wiretap evidence, which had been obtained legally in another state, was used in the case. When the state trial was thrown out, the case was moved to federal court, where wiretap evidence is admissible. Seeing the handwriting on the wall, the suspects plea bargained.

Normally, a wiretap can not be installed without a warrant, though the Patriot Act has modified this requirement to a certain extent in cases involving national security.

The warrant must state exactly what type of evidence can be gathered, and is not a license for the law enforcement agencies to collect information not related to the crime being investigated.

Having obtained a court order, the law enforcement agency will generally require the suspect's telephone provider to help in installing the wiretap. There are also cases in which the court will authorize a roving wiretap, which means that the wiretap focuses on a suspect, whatever telephone they use. In many cases the communications are recorded and examined later. Sometimes the recorded information is of low quality and needs to be enhanced by a forensic audio analyst. There is a multitude of different locations in a telephone system where a wiretap can be installed, which is one of the reasons wiretapping is such a common method of industrial espionage. Each of these locations requires certain types of devices, but this is mature technology that is usually not difficult to obtain.

36.1.4 Satellite Surveillance

Satellite surveillance is not yet important in routine law enforcement. Despite what you see in movies like "Enemy of the State", even the most sophisticated satellites are only capable of resolving features 10 centimeters or more in size at the present time. Thus a human being would show up on these most sophisticated imaging systems, but it would be impossible to recognize their face. An expert on satellite surveillance stated on TechTV (a now defunct cable TV channel) in 2002 that even if Osama bin Laden were lying on the ground looking up and smiling, the satellites wouldn't be able to see him. This is, however, an area in which rapid progress is being made.

I have some acquaintances, some of them present or former members of the military, who claim that the newest satellites would be able to detect Osama if he were lying on the ground smiling up at them. I'm skeptical, because in order to have a reasonable chance of recognizing a face, the resolution required would need tp be around one milliimeter, which is about 100 times better than the resolution of any satellite surveillance technology admitted to by the military. The jump in resolution from 10cm to 1mm is a large obstacle. Normally, electronic technology of all sorts follows a rule of thumb referred to as "Moore's law", which says that the rate of advancement in electronic technology is a doubling of capability every 18 months. In reality, this is an extremely rapid rate of technological advancement, and I would not believe someone who said that satellite technology improves at a rate greater than Moore's law. Assuming that Moore's law applies, it would take 18 months to improve from 10cm resolution to 5cm, 18 months more for 2.5mm resolution, and so one. Altogether, it would take more than six 18-month intervals, or approximately 10 years to reach 1mm resolution. So, we can expect to be able to recognize faces from space by about 2017 – not such a long time away, but still definitely in the future.

Even though faces are practically invisible, satellite surveillance can and is used to track movements of people and vehicles. Nobody actually knows whether or not satellite surveillance has been used in this way outside of the military arena, but it seems doubtful.

36.1.5 The Role of Forensic Scientists in Electronic Surveillance

Wiretapping and videorecording technologies provide good examples of how forensic scientists become involved with electronic surveillance and how they may become involved more with it in the future. From these examples it is clear that if forensic scientists become involved, it is with the processing and enhancement of the evidence collected. The fields of forensic image analysis and forensic audio analysis have obvious applications to the information recorded by video cameras and wiretap devices, and likely will be the ones involved with other forms of electronic surveillance evidence as they becomes available.

36.2 Biometrics

In this context, the term **biometrics** refers to the emerging field of technology devoted to identification of individuals using their anatomical characteristics. Existing approaches are based on retinal or iris scanning, fingerprints, face recognition, voice recognition, and even ear recognition.

Biometrics is a poorly chosen name for this field. The terms "biometrics" and "biometry" have been used since early in the 20th century to refer to a different field, the statistical and mathematical methods applicable to data analysis problems in the biological sciences. Also, the root of the word "biometrics" is "bio", which is normally applied to non-humans. For a field that is only applicable to humans the root word "anthropo" is a more appropriate choice. One name that could have been coined, and which would have been etymologically and historically appropriate is "digital anthropometry". However misguided, the term "biometrics" has become the accepted name for this field, and we are stuck with it.

36.2.1 The History of Biometrics

Biometrics is actually a form of **anthropometry**, which is a term that refers to the measurement of characteristics of living people. As you who have read the previous chapters of this textbook realize, the use of human body measurements and other characteristics in identification is more than 100 years old, going back to the work of Alfonse Bertillon in the late 1800's and early 1900's. The use of computer technology to accomplish identification of deceased individuals through anthropometry has been practiced by forensic anthropologists since the 1960's.

In its present incarnation, biometrics arises from the efforts of computer scientists to develop systems that will recognize individuals for security purposes. The focus is on achieving this recognition quickly, based on measurements that are gathered by a camera or scanner, analyzed immediately, and compared with standards that are archived in a computerized database.

36.2.2 General Methods

The general approach to biometrics is to utilize an electronic device to capture data about a person's body measurements in such a way that the data can be processed using computer software. The data obtained is usually converted to a hash code that summarizes the information in the data. This hash code can then be indexed and stored, and ultimately compared to other hash codes in a matching process.

Computerizing this process increases the speed at which data can be stored and retrieved. It does not, however, increase the accuracy of matching. Most people assume, incorrectly, that a computer can do matching more accurately than a human can. So many of the forensic science TV shows and movies show someone sitting in front of a computer on which is displayed two fingerprints, two faces, or a similar comparison. The person presses a key and interesting things happen on the computer screen followed by flashes or beeps indicating a match. This is pure science fiction, despite the efforts of some of the best scientists to design such a system. In fact, it is well known that computers do not do matching very well, but humans do it with incredible speed and accuracy. Computers can, however, find potential matches efficiently, due presumably to similarities in hash codes. The potential matches are then supplied to a human for the actual matching.

There are two main uses of biometrics. One is authentication for security purposes, called **verification**; and the other is scanning and monitoring large numbers of people in the hope of detecting someone of interest, called **identification**. At the present state of the art, verification works well and identification works poorly.

36.2.2.1 Verification Applications

The use of biometrics for verification, *i.e.* establishing a certain person's identity, is designed to replace a security guard. I do not believe that anybody in the field would claim that biometric methods are superior to a security guard for recognizing people – this is just another expression of the science fictionish idea that computers do a good job of matching things. The goal is to do the job cheaper, while not tying up the valuable resource of a human mind for the essentially mindless task of guarding something.

In verification applications, a scanner scans some part of the person's body. Next a hash code is computed, which extracts the information in the scan and reduces it to a number. The resulting hash code (essentially an evidence item) is then compared against a small database of hash codes (standard items) derived from scans of authorized persons. If the hash codes are similar enough, then the person is granted access to the area that the biometric scanner guards.

This application partially avoids the relatively poor performance of computers on matching problems by keeping the database of possible matches small. When this is the case, the matching criteria can be kept very specific. Also, this is a situation in which the computer can focus on excluding a match, which is a task they do much better than actually making a match. The challenge is to balance the sensitivity of the matching criteria so that they are not so narrow that they often exclude the authorized persons, yet not so broad that they include unauthorized persons.

The probability of not recognizing an authorized person is called the **false reject rate** (FRR). The probability of recognizing an unauthorized person is called the **false accept rate** (FAR). In most cases the system is "tuned" so that it has a miniscule FAR, while allowing a fairly high FRR. The manufacturers of scanning devices and software are quite positive that their systems work with a high degree of accuracy. Advertised FAR's are usually in the range of one in a million. However, actual scientific studies of success and failure rates are rarely made available, so it is difficult to say for sure. Some independent authorities report FRR's as high as 10%. This is acceptable, however, if whatever is being protected by the system is critical and the penalty for a false rejection is not too severe. For example, if the authorized person whose access has been rejected can simply try again, or even if they have to call a guard or their supervisor to get access, this might be an acceptable price to pay for the security.

36.2.2.2 Identification Applications

The second major application of biometrics is identification, which amounts to scanning and monitoring a large number of people in the hope of recognizing someone of interest – usually a suspect. For example, after the terrorist attack of September 11, 2001, monitors were installed in airports that captured images of people's faces in order to try to screen out or capture people who were known terrorists.

The main issue with identification, privacy and freedom issues aside, is that it works very poorly. If you followed the news about airport monitoring, you know that the program was discontinued after a short period of time. The problem is that the systems did not do a good job at this application. The main problem was false positive matches. Completely random people were identified as potential terrorists and led away for questioning. I have not seen even a single report of these systems correctly identifying a terrorist.

This technology probably does has some potential. At present it is an immature technology and the future will bring improvements that may make it more useful. The ultimate problem, however, is the fact that computers do not do matching well. For now we are left to rely on humans who have memorized the appearance of possible terrorists and other suspects, and whose brains are capable of doing efficient matching, at least under good conditions (which may not be the case at a busy location such as an airport).

36.2.3 Body Parts Used in Biometrics

Biometrics has been most successful, so far, using parts of the hands and the eyes. On the hands, fingerprints and palm prints have proven useful. Both the iris and the retina of the eye are unique to each individual, a fact that has been exploited in biometrics. Although each person's face is unique, face recognition has so far proven unreliable. Voice recognition was discussed in chapter 35, where it was pointed out that its reliability is uncertain. A new direction, so far not actually applied for verification or identification, focuses on the ears.

36.2.3.1 Finger and Palm Prints

For verification applications, fingerprints can't be beat. Since fingerprints have been databased for many years using AFIS systems, the technology for scanning, hashing, and computerized matching of fingerprints is as mature as any in the field. The same technology can also be applied to palm prints.

In verification applications, it is not unreasonable to ask someone to place their fingertips or palms on a scanning device. In fact, the computer I am using to type this text has a fingerprint scanning device built in, that I use to login to the computer instead of typing in a user account name and a password. Figure 36.1 shows the fingerprint scanner on my laptop computer.

Figure 36.1: The Fingerprint Scanner on my Laptop Computer

For identification applications, however, fingerprint and palmprint methods are not often considered because of the feeling that few people at an airport or similar location would be willing to routinely submit their finger or palm for a scan.

As discussed in chapter 16, AFIS fingerprint databases use a method based on detection of three types of minutia. One approach to the biometrics of fingerprints uses these same minutia, and is called the **AFIS method**. This method offers the advantage of being fairly standard and of utilizing the most mature technology available. Another method is called the **correlation-based approach**. In this method, several hundred separate small areas of the two prints are scanned and compared. Both methods work well.

36.2.3.2 Eyes

The eye contains at least two sources of unique individual characteristics. One is the distribution of pigment in the iris of the eye. The second is the distribution of blood vessels over the retina of the eye. Iris and retina scanning are useful for verification, though less convenient than fingerprint scanning since the eye has to be held close to the scanner for some period of time – within two inches of the scanner for several seconds in the case of retina scans, and up to 36 inches away for a few seconds in the case of iris scanning. It is estimated that 5% to 10% of the population can not hold still long enough for a retina scan, even if they are giving it voluntarily. Retina scanning presents a special problem in that the retina is inside the eyeball and in order to scan it, the scanning device must somehow peer through the pupil of the eye to gather its information.

Figure 36.2: Use of a Retina Scanner

Eye scanning, especially iris scanning, is seen by some authorities as being useful for identification because, unlike fingerprint scanning, it is conceivable that a scanner could be constructed that could scan the eye from enough of a distance away that the person being scanned would not be aware of it. At this point, only the time required for an iris scan is a limiting factor in this application, and the most innovative manufacturer in the field claims scan times as short as one second. I have seen news articles claiming that some airports have already installed iris scanners.

Retina and iris scanning both use the correlations-based approach to generate data. The data is then hashed using one of the algorithms that generate a number that reflects the information present in the data. One commercial retina scanner uses a laser to record the characteristics of the retina at over 400 points. This method yields a very low FAR, claimed to be in the neighborhood of 1 in a million by the manufacturer. However, as in fingerprint scanning, the FRR can be quite high, often as high as 10%. One commercial iris scanner uses a video camera to record characteristics of the iris at 266 locations. As for retina scanning, a low FAR and high FRR are reported.

36.2.3.3 Face Recognition

Face recognition is the holy grail of biometrics. I can only imagine the frustration of scientists trying to computerize the recognition of faces, a task that humans do with extraordinary skill and accuracy even under conditions of variable lighting and orientation. Although a promising technology for either verification or identification, it is the most desired technology for identification, since it is a simple matter to capture the faces of everybody in a crowd using a video camera.

There are many obstacles to reliable face recognition, however. The steps in doing a face recognition begin with capturing the image of a face. Next the software has to decide if the image really is of a face, and this turns out to be one of the greatest challenges. One approach to face detection is called "eigenfaces", and involves analyzing the grey values in a grey scale version of the image to estimate how face-like the image is. Once a face is detected, the pose must be estimated. Pose refers to the orientation of the face. For example, is the person looking straight ahead, slightly downward, or to the side. Once the pose is estimated, the image can be corrected to simulate a full frontal image (as if the person were looking straight ahead). The next step is feature detection, wherein the software attempts to recognize the characteristics of a face in the image, such as the edges of the face, the eyes, the nose, and the mouth. Having determined the locations of these features, the relationships between them are hashed to produce a number that summarizes the characteristics of the face. The hash code is then compared with hash codes of known faces in the hope of making a match.

Face recognition is quite challenging because face detection, pose estimation, and feature detection are difficult tasks with current technology. Another problem is changes in the face. What happens if the person shaves their beard, or uses different makeup, or wears glasses? These changes tend to confuse the face recognition system. Variation in lighting conditions can also have an effect, as can the expression on a person's face. Therefore, the accuracy of face recognition is quite low compared to fingerprint-based or eye-based systems. It is perhaps paradoxical that one of the very easiest things for a human to do is to recognize a face, while this task remains one of the most difficult things for a computer to do.

36.2.3.4 Voice Identification

Voice identification also has potential, especially for use in verification systems. I would love to be like Captain Picard of Star Trek and ask the computer to identify my voice, and have it respond, "Skelton ... Randall R ... Professor". Of course, I could program my computer to respond with this phrase when it hears a sound, but teaching it to actually recognize my voice as distinct from anyone else's is a much greater challenge. We examined voice recognition technology in chapter 35, including its challenges, and this information does not need to be repeated here. The accuracy of voice recognition is thought to be greater than that of face recognition, but less than that of eye or fingerprint recognition.

36.2.3.5 Ears

The ear is not often recognized as a unique part of the body, but it is. Characteristics of the ears were part of Bertillon's method for identifying people from their bodies. Some modern work has been done on the biometrics of ears, especially by the Dutch.

Ear identification has occasionally been used in traditional forensic investigations. For example, when earprint evidence is left on a window by someone who pressed their ear against it to listen to what was going on inside. There have been cases, at least in the Netherlands, where criminals have been convicted primarily on the basis of earprint evidence.

It has also been suggested that ear identification could be used for verification or even for crowd monitoring. It has many similarities to face recognition, but might be more useful because it is thought that an ear might be easier for a computer to recognize than a whole face. So far, I don't know of any companies that have developed this technology commercially. If they do, I think we can expect it to operate in a manner similar to face recognition, and to be plagued by approximately the same problems.

36.2.4 The Future of Biometrics

The bottom line for biometrics is that it is a promising emerging set of technologies, based on historically important principles, that function acceptably for verification purposes. Identification applications are lagging behind because computerized methods for making matches form a large database of standards are not yet mature technologies. As forensic scientists, our interest is in the fact that any technology that can be used for these purposes can also be used in the identification and apprehension of suspects.

36.3 Questions for Study and Review

1. For what purposes, legitimate and illegitimate, are bugs used?
2. What sorts of communications devices are subject to wiretapping?
3. Is wiretapping evidence allowed in all states?
4. How do forensic scientists become involved with evidence obtained by electronic surveillance?
5. What is "biometrics"?
6. In your own words, explain the difference between biometric verification and biometric identification.
7. Define FRR and FAR.
8. What are some applications of biometric verification?
9. Has it proven easy to determine a person's identity using computerized biometrics (biometric identification)? What is the main problem in achieving this goal?

10. What parts of the body are commonly used for biometric verification and identification?
11. Why is it that most commercial verification applications, such as access to laptop computers, involve the use of fingerprints or palmprints?
12. What is the individual characteristic of irises that biometric iris scanners seek to capture?
13. What is the individual characteristic of retinas that biometric retina scanners seek to capture?
14. Discuss how face recognition works and why it is very difficult to achieve a working face recognition system.

36.4 Sources

Abut H. 2003. Digitized and Digital Signatures for Personal Identification [Internet]. [cited 2007 Oct 29]. Available from: http://72.14.253.104/search?q=cache:x4KuRJzJsx0J:anadolu.sdsu.edu/abut/dllecture-may_2003c.pdf+AFIS+hashing+algorithm&hl=en&ct=clnk&cd=15&gl=us.

Atkinson JM. 2001. Wiretapping and Outside Plant Security - Wiretapping 101 [Internet]. [cited 2007 Oct 26]. Available from: http://www.tscm.com/outsideplant.html.

Center for Democracy and Technology, 2006. The Nature and Scope of Governmental Electronic Surveillance Activity [Internet]. [cited 2007 Oct 26]. Available from: http://www.cdt.org/wiretap/wiretap_overview.html.

Colbridge TD. 2000. Electronic Surveillance: A Matter of Necessity [Internet]. FBI Law Enforcement Bulletin 69(2). [cited 2007 Oct 19]. Available from: http://www.fbi.gov/publications/leb/2000/feb00leb.pdf.

Coleman S. 2000. Biometrics [Internet]. FBI Law Enforcement Bulletin 69(6). [cited 2007 Oct 19]. Available from: http://www.fbi.gov/publications/leb/2000/jun00leb.pdf.

Delaney DP, Denning DE, Kaye J, MacDonald AR. 1993. Wiretap Laws And Procedures: What Happens When the Us Government Taps a Line [Internet]. [cited 2007 Oct 26]. Available from: http://www.spybusters.com/government_taps.html.

Elder, WJ. 2007. Electronic Surveillance: Unlawful Invasion of Privacy or Justifiable Law Enforcement [Internet]. [cited 2007 Oct 7]. Available from: http://www.yale.edu/ynhti/curriculum/units/1983/4/83.04.07.x.html>.

Hodges K. 2007. Legal Brief: New Federal Rule of Criminal Procedure Addresses Warrants for Tracking Devices [Internet]. FBI Law Enforcement Bulletin 76(2). [cited 2007 Oct 19]. Available from: http://www.fbi.gov/publications/leb/2007/feb2007/feb2007leb.htm#page15.

Kleinberg KF, Vanezis P, Burton AM. 2007. Failure of Anthropometry as a Facial Identification Technique Using High-quality Photographs. Journal of Forensic Sciences 52 (4): 779-783.

Meehan, Michael, 2000. Iris Scans Take off at Airports [Internet]. ComputerWorld July 17, 2000. [cited 2007 Oct 26]. Available from: http://archives.cnn.com/2000/TECH/com

MIT Media Laboratory Vision and Modeling Group, 2002. Face Recognition Demo Page [Internet]. [cited 2007 Oct 26]. Available from: http://vismod.media.mit.edu/vismod/demos/facerec/.

Murray KD. 2006. Describe the Murray Inspection Process [Internet]. [cited 2007 Oct 26]. Available from: http://www.spybusters.com/the_inspection_process.html.

National Center for State Courts. 2002. Retinal Scan [Internet]. [cited 2007 Oct 26]. Available from: http://ctl.ncsc.dni.us/biomet%20web/BMRetinal.html.

O'Neal CW. 1998. Surreptitious Audio Surveillance: The Unknown Danger to Law Enforcement [Internet]. FBI Law Enforcement Bulletin 67(6). [cited 2007 Oct 19]. Available from: http://www.fbi.gov/publications/leb/1998/juneleb.pdf.

Prabhakar S, Jain A. n.d. Fingerprint Identification [Internet]. [cited 2007 Oct 26]. Available from: http://biometrics.cse.msu.edu/fingerprint.html.

Ratha NK, Senior, A. 2001. ICAPR Tutorial on Automated Biometrics [Internet]. [cited 2007 Oct 26]. Available from: http://www.research.ibm.com/people/a/aws/icapr.html.

van der Lugt C. 1997. Ear-Identification State of the Art [Internet]. [cited 2007 Oct 26]. Available from: http://www.crimeandclues.com/earprint.htm.

Yoshino M, Matsuda H, Kubota S, Imaizumi K, Miyasaka S. 2001. Computer-Assisted Facial Image Identification System [Internet]. Forensic Science Communications 3(1). [cited 2007 Oct 21]. Available from: http://www.fbi.gov/hq/lab/fsc/backissu/jan2001/yoshino.htm.

36.5 Acknowledgments

The graphic used in Figure 36.2 is from the U.S. Department of Defense web site, and as such is in the public domain. It was found at http://www.defenselink.mil/photos/NewsPhoto.aspx?NewsPhotoID=8775 on November 13, 2007.

Chapter 37
FORENSIC DATABASES

The subject of this chapter is the databases used by forensic scientists. One of the branches of forensic computer science is the use of forensic databases and other software by forensic scientists and law enforcement. This chapter will conclude the series of chapters about forensic computer science and related technologies.

I belong to a Saturday breakfast group that consists mostly of retired peace officers, who among them represent an incredible amount of practical experience. During one conversation over our scrambled eggs and pancakes we came to the group opinion that law enforcement must ultimately triumph over criminals; not because there are more law enforcement officers than crooks, and certainly not because of superior weapons, but because law enforcement is better organized and has better access to communication and information. Nowhere are the twin heros of organization and information more evident than in the numerous databases of information that exist to aid peace officers and forensic scientists.

The chapter begins with an introduction to databases. It continues with an overview of what types of databases exist and who maintains them. The remainder of the chapter will focus on briefly describing the databases commonly used by forensic scientists, some of which were mentioned or introduced in previous chapters and some of which are presented here for the first time.

37.1 What is a Database?

A **database** is defined as a set of related files that is created and managed by a database management system (DBMS). A **DBMS** is software, such as Microsoft Access or Oracle, to name just two of the most popular. The main value of a DBMS over more informal ways of managing data, such as simply keeping a list, is that the DBMS can manage any form of data including text, images, sound and video. A DBMS also uses advanced algorithms for information retrieval from the database and can rapidly find and retrieve one record out of millions.

The information in a database is usually organized as **records**, each of which is a collection of all the stored information about one particular person or item. Records are indexed by **key** values, which are the items of information that are normally used to search for a particular record. The key value for a person is usually their name or perhaps their social security number. Multiple keys may be defined for a database, allowing records to be looked up using several different search criteria. Records consist of a set of **fields**, where each field holds and presents a certain item of information about the person or item that the record concerns. For example, a person's address would be stored in a field within a record.

Searching for information within a database is designed to be simple. Although the records are indexed by their key values, any information may be searched for, albeit more slowly. For example, our database of people could be searched using a telephone number, and the result would be the record for which that telephone number was recorded in the telephone number field. If the person's name was a key field and their telephone number was not, then it would take longer to retrieve this record by searching for a telephone number than it would by searching for the person's name, but the time the search took would likely still be acceptable.

37.2 Types of Forensic Databases

There are three types of forensic databases – really four, because NCIC is unique and deserves a category of its own. One type of forensic databases is databases of characteristics of items – for example, a database of characteristics of firearms manufactured in the United States. These databases are often primary references for forensic scientists. Another type is databases of evidence from past and present crimes – for example, a database that includes characteristics of firearms evidence collected at crime scenes. These databases are used both by forensic scientists and by law enforcement agencies. The third type is databases of crimes and information about criminals – for example, a database of firearms that have been stolen. These are most often used by law enforcement, but occasionally used by forensic scientists. NCIC combines many types of information from the other types of database, plus other information that is archived nowhere else.

37.3 Who Maintains Forensic Databases

Databases of characteristics of items need only one organization to maintain them. The organization that maintains the most of these, by far, is the FBI. We will see below, in table 37.1, a partial list of databases of this type that are maintained by the FBI. In other cases this type of database is maintained by an organization of manufacturers of a type of item, or a society of scientists (forensic or otherwise) who are interested in that type of item.

Databases of evidence from past and present crimes are usually maintained by several separate organizations that then share their information with other such organizations. These systems are typically arranged in multiple tiers. The base of a multiple tier system is each local law enforcement agency maintaining a local database. These local databases are usually integrated for each state by some branch of that state's Department of Justice. Finally, the state databases are tied together at the national level by one of the federal agencies. For example, each local jurisdiction maintains their own database of fingerprints, which is called an AFIS (automated fingerprint identification system). The local AFIS databases are brought together at the state level to form a statewide AFIS database, and ultimately the state AFIS databases are brought together at the national level by IAFIS (integrated automated fingerprint identification system) which is administered by the FBI.

Databases of crime and criminals use a system similar to that of databases of evidence from past and present crimes, with local databases at the level of the jurisdiction and the state, and a national database that integrates them all. For example, each jurisdiction maintains its own SOR (sex offender registry) and the FBI maintains NSOR (national sex offender registry).

37.4 Databases of Item Characteristics

There is an amazing number of databases of item characteristics that are available to forensic scientists. I will not pretend to give a complete list here, but only to give some representative examples and enough of a list that you can see how numerous and varied they are. The two main categories of these are databases of genetic information and the vastly larger category of databases of other types of information.

37.4.1 Databases of Genetic Information

There are many databases of genetic information available to forensic scientists as well as to geneticists. Some, such as the one generated by the ongoing Human Genome Project, maintain the information known about the genetics of the 24 distinct chromosomes that are part of the human genome. As you remember from chapter 15 these 24 distinct chromosomes are chromosomes 1 through 22, X and Y. Other databases maintain information about mitochondrial DNA and the distribution of MtDNA haplotypes in the world's populations, one of which is called the MtDNA Population Database. The YHRD (Y chromosome haplotype reference database) maintains information about the distribution of Y chromosome haplotypes in the world's populations.

37.4.2 Other Databases

Some databases of characteristics of items are maintained by organizations other than the FBI. Some of these are: SoilFit, a database of characteristics of soils; FORS, a database of forensic literature; a forensic hash database, which contains hash codes of files – not information about the food or the drug; a forensic glass database; and an ignitable liquids reference collection and database.

The number of databases of item characteristics maintained by the FBI is enormous. Table 37.1 presents a partial list.

Table 37.1: Type of Items for Which the FBI Maintains a Database of Item Characteristics

Abrasives
Adhesives, caulks, sealants
Ammunition characteristics
Bird feathers
Building materials

Controlled substances (drugs)
Electronic devices
Explosives
Fiber samples
Firearms characteristics
Glass refractive indices
Lubricants
Paints
Papers and inks
Polymers and plastics
Ropes and cords
Insulations used in safes
Shoe tread patterns
Tire tread patterns
Typewriters, printers, copiers, etc.
X-ray spectra of many substances

37.5 Databases of Evidence From Past & Present Crimes

There are three important database systems that I will address in this section: the NDIS/CODIS system for DNA, the IAFIS/AFIS system for fingerprints, and the NIBIN/IBIS system for firearms evidence. NDIS/CODIS and IAFIS/AFIS use the mulititiered system described above. NIBIN/IBIS is unusual in that its multitiered system focuses on regions rather than local and state jurisdictions. The first acronym in each pair refers to the national level database and the second acronym refers to the local level or region databases.

37.5.1 NDIS/CODIS

NDIS is the acronym for National DNA Index System, and CODIS stands for combined DNA index system. As implied in these names, these databases constitute a system for maintaining records of DNA profiles. The CODIS system is usually two tiered, with a local jurisdiction maintaining a CODIS system, with the various CODIS systems in a state being linked together. NDIS ties all of these state systems together.

The NDIS/CODIS system of databases includes information on DNA profiles of offenders, DNA evidence from crime scenes, DNA profiles of missing persons, and DNA profiles of unidentified deceased persons. Each state has their own rules for which offenders must contribute a DNA sample. For example, the Montana State Legislature passed a bill in 2001 that requires DNA profiles from all felons and sexual offenders to be entered into CODIS, and gives judges the discretion to order a person convicted of any crime to contribute a DNA sample to CODIS.

Data from CODIS systems may be used in a variety of ways. The most common, of course, is to search for a match to DNA evidence recovered at a crime scene. When DNA evidence is recovered it is entered into CODIS, which generates a hash code for fast indexing and searching. The forensic scientist has a choice of searching the local CODIS, the state CODIS, or NDIS, and may search all three. The advantage of searching NDIS, even for a clearly local crime is that a match may be made to evidence from another crime in another state. The database will return a list of possible matches for a DNA analyst to examine and make a match if appropriate. In addition to use in criminal investigations, the Montana Legislature has authorized the use of CODIS for identification of missing persons and for research (with the names of the person omitted).

37.5.2 IAFIS/AFIS

IAFIS stands for integrated automated fingerprint identification system and AFIS for automated fingerprint identification system. The organization is similar to NDIS/CODIS in that local and state AFIS systems are integrated by the national level IAFIS system, which is maintained by the FBI.

AFIS systems include the prints and criminal histories of offenders (there are over 47 million people in the IAFIS database), prints from crime scenes, prints of missing persons, and prints of unidentified deceased persons. As for CODIS systems, each state has its own rules for which offenders are included, and the rules for inclusion are usually less strict than for CODIS. For example, in Montana, offenders' prints can be entered into AFIS if the person is convicted of any crime except traffic, regulatory, or fish and game offenses, including misdemeanors.

Fingerprints are collected from people is two ways – livescan and deadscan. The "live" and "dead" here do not refer to the state of the person in question at the time. A livescan uses laser technology to directly scan the fingers of a person. The person simply places their hand on the device for a few seconds and the prints are collected. Deadscan refers to scanning fingerprint cards. The images of the fingerprints are compressed using an algorithm called wavelet/scalar quantization, which reduces the size of the digital file needed to store the image. The information about the location and nature of minutia in the scan is used to derive a hash code for indexing and retrieval. The operator may search local, state, or national databases. The system will return a list of possible matches, and the actual determination of a match is made by a fingerprint examiner.

37.5.3 NIBIN/IBIS

NIBIN is the acronym for the National Integrated Ballistic Information Network, which is administered by the Bureau of Alcohol, Tobacco, and Firearms (ATF). IBIS is the acronym for integrated ballistics identification system. The NIBIS/IBIS system divides the U.S. into 16 regions, each with a number of IBIS sites. When the system is complete, ATF plans that there will be 235 IBIS sites. A system based on regions was chosen so that each region is about equal is population. For example, Montana is in region 15, along with Washington, Idaho, Oregon, Alaska, and Hawaii; whereas

California is divided into two separate regions because of its large population. Regional IBIS sites are integrated into the nation NIBIN system.

IBIS systems collect characteristics of weapons known to have been used in committing a crime, and characteristics of ammunition from a crime scene where the weapon is unknown. NIBIN currently has information on almost one million bullets and casings. In contrast to CODIS and AFIS systems, there is almost no regulation at all of what firearms related information goes into an IBIS database, and ATF encourages local agencies to submit all their firearms related evidence. For example, there are no regulations in Montana State Law about what is or is not to to be included in IBIS, and the policy of the Montana State Laboratory of Criminal Investigation is to include all evidence.

Firearms examiners use high-precision imaging devices to scan the details of firearms evidence into IBIS, where the information in the image is converted to a hash code. Both regional, and national databases may be searched. A list of possible matches is returned and a firearms examiner makes a match if appropriate.

37.6 Databases of Crime and Criminal Information

Databases of crime and criminal information are often, but not always, organized along similar lines as the databases of evidence from past and present crimes, with local, sometimes state or regional, and national level databases that are integrated for wide-ranging information retrieval. In this section, I will focus on two systems, the NICS/CCH system, and the NSOR/SOR system.

37.6.1 NICS/CCH

Each jurisdiction, or at least each state, has a CCH, which stands for computerized criminal history, in which the details of criminals and their crimes are maintained. The details of crimes are included both for solved and unsolved crimes. NICS, the National Instant Criminal Background Check System, is the national level database that integrates the CCH's.

The sorts of details concerning criminals included in a CCH are pretty much what you would expect. The person's name, date of birth, identifying characteristics, detailed criminal history, most recently known address and telephone number, and links to fingerprint records, DNA records, and firearms records if applicable. Details of crimes include the location, date, and nature of the crime in enough detail that the characteristic methods used by a certain criminal (modus operandi or MO) can be discerned.

Since the information in a CCH is not based on an image, but instead on textual records, hash codes are not used except in the case of some DBMS's for indexing the records.

37.6.2 NSOR/SOR

NSOR is the acronym for the National Sex Offender Registry, while SOR simply stands for sex offender registry. Local and state level SOR's are integrated into NSOR. These databases manage information about sexual offenders. The information submitted to a SOR includes the predictable details about the offender, and also the types of crimes and victims preferred by the offender. Again, this is textual information and hashing is not necessary.

37.7 NCIC

NCIC is the acronym for the National Crime Information Center, which is administered by the FBI. It can be considered the nation's master database of crime and criminal information. It includes a wide variety of information and therefore can not be cleanly placed in any of the previous three database categories.

One type of data that NCIC maintains is records on people. Criminals of various sorts in the database include federal wanted persons, juvenile delinquents escaped from custody, persons wanted in foreign countries, people in IAFIS and NDIS, violent felons, serious drug offenders, individuals who pose a threat to the President of the U.S., members of violent gangs, and members of terrorist organizations. In addition, NCIC maintains records on certain missing persons, including those who are incompetent (i.e. they can't take care of themselves), kidnapped or endangered, are missing after a catastrophe, or are children. A few types of unidentified person are included, such as those who are amnesic, infants, deceased, only known from body parts, or are victims of a catastrophe.

NCIC also maintains records on stolen items. Some of the types of stolen items maintained in the database are vehicles of all types (cars, boat, airplanes, etc.), vehicle license plates, firearms, securities (bank notes, bonds, stocks), and stolen items that have been recovered from criminals but their true ownership is not known.

There is some regulation of who can use the information in NCIC. Any information can be provided to U.S. federal, state, or local law enforcement agencies. Data on firearms may be provided to foreign law enforcement agencies, such as the Royal Canadian Mounted Police. Criminal history data may be provided to a variety of agencies for use in connection with licensing or employment.
Data on vehicles may be released to motor vehicle or driver's license registries and to the National Automobile Theft Bureau. Data on missing persons may be provided to the National Center for Missing and Exploited Children. Any data can be released to Members of Congress and to the National Archives and Records Administration.

37.8 Questions for Study and Review

1. In your own words, define the following terms: database, database management system, record, key value, field, livescan fingerprint collection, deadscan fingerprint collection.

2. In your own words, describe how local, state or regional, and national level databases interact in forming a network of forensic and law enforcement database systems.
3. What is the relationship between AFIS and IAFIS? Is this the same relationship as exists between NIBIN and IBIS or CODIS and NDIS?
4. What single agency maintains most of the forensic and law enforcement databases in the United States?
5. For each of the following types of data give the name of at least one database that includes it: mitochondrial DNA from people around the world who are not necessarily criminals, DNA from criminals, fingerprints, firearms and ammunition, criminal histories of offenders, sex offenders.
6. How is a complex item, such as a fingerprint, processed for storage in a database and fast retrieval by searching the database?
7. What authority determines whose evidence will be databased in a system such as AFIS, CODIS, or NIBIN?
8. What is NCIC, and why is it often considered to be a database category of its own?
9. What sorts of records are databased in NCIC?
10. In general, is it possible for anybody to access the data in forensic or law enforcement databases? What sorts of people is access to these databases restricted to?

37.9 Sources

Allen SP, Williams MR, Bryant C, Byron D, Cerven J, Cooper BD, Hilliard DC, Hoffmann J, Kwast J, Thomas SA, Whitcomb CM. 2006. The National Center for Forensic Science Ignitable Liquids Reference Collection and Database [Internet]. Forensic Science Communications 8(2). [cited 2007 Oct 21]. Available from: http://www.fbi.gov/hq/lab/fsc/backissu/april2006/standards/2006_04_standards01.htm.

American Society of Questioned Document Examiners. 2002. American Society of Questioned Document Examiners Forensic Databases and on-line References [Internet]. [cited 2007 Aug 4]. Available from: http://www.asqde.org/datapg1.htm.

Brislawn C. 2002. The FBI Fingerprint Image Compression Standard [Internet]. [cited 2007 Oct 29]. Available from: http://www.ccs3.lanl.gov/~brislawn/FBI/FBI.html.

Daukantas, Patricia, 2001. Database Sheds Light on Crime [Internet]. Government Computer News 20(3). [cited 2007 Oct 26]. Available from: http://www.gcn.com/print/vol20_no3/3649-1.html.

Federal Bureau of Investigation. n.d. CODIS [Internet]. [cited 2007 Aug 4]. Available from: http://www.fbi.gov/hq/lab/codis/index1.htm.

Federal Bureau of Investigation. n.d. National/State Sex Offender Registry [Internet]. [cited 2007 Oct 26]. Available from: http://www.fbi.gov/hq/cid/cac/registry.htm.

Federal Bureau of Investigation. n.d. National Instant Criminal Background Check System [Internet]. [cited 2007 Oct 26]. Available from: http://www.fbi.gov/hq/cjisd/nics/nicsindex.htm.

Federal Bureau of Investigation. 2003. Handbook of Forensic Services, 2003 [Internet]. [cited 2007 Oct 26]. Available from: http://www.fbi.gov/hq/lab/handbook/forensics.pdf.

Federal Bureau of Investigation. 2005. Integrated Automated Fingerprint Identification System or IAFIS [Internet]. [cited 2007 Oct 26]. Available from: http://www.fbi.gov/hq/cjisd/iafis.htm.

Forensic Science Service Ltd. 2007. FORS Forensic Bibliographic Database. [Internet]. [cited 2007 Oct 26]. Available from: http://www.forensic.gov.uk/forensic_t/inside/products/fors/fors.htm.

Hitt SL. 2000. NCIC 2000. [Internet]. FBI Law Enforcement Bulletin 69(7). [cited 2007 Oct 19]. Available from: http://www.fbi.gov/publications/leb/2000/jul00leb.pdf.

Huguley M. 2001. Wanted and Arrested Person Records [Internet]. FBI Law Enforcement Bulletin 70(9). [cited 2007 Oct 19]. Available from: http://www.fbi.gov/publications/leb/2001/september2001/sept01p20.htm.

Macaulay Institute. 2007. Integration of Soil Fingerprinting Techniques for Forensic Application [Internet]. [cited 2007 Oct 26]. Available from: http://www.macaulay.ac.uk/soilfit/.

Pike J, Aftergood S. 2006. National Crime Information Center (NCIC) [Internet]. [cited 2007 Oct 26]. Available from: http://www.fas.org/irp/agency/doj/fbi/is/ncic.htm.

Wikipedia contributors. 2007. Automated Fingerprint Identification System [Internet]. Wikipedia, The Free Encyclopedia; 2007 Oct 23, 19:34 UTC [cited 2007 Oct 26]. Available from: http://en.wikipedia.org/w/index.php?title=Automated_Fingerprint_Identification_System&oldid=166581766.

Wikipedia contributors. 2007. Automated Firearms Identification [Internet]. Wikipedia, The Free Encyclopedia; 2007 Aug 18, 02:37 UTC [cited 2007 Oct 26]. Available from: http://en.wikipedia.org/w/index.php?title=Automated_firearms_identification&oldid=151956984.

Chapter 38
FORENSIC PSYCHOLOGY

This chapter is the first in a series of three in which we will examine topics related to forensic psychology. In this chapter I will introduce the subject. The next chapter is an advanced topics chapter on lie detection and forensic hypnosis, both of which are historically associated with forensic psychology. The final chapter of the series will focus on psychological profiling.

Forensic psychology forms a surprisingly large branch of the forensic sciences. Forensic psychologists participate in the pursuit of justice from near the beginning of an investigation, through courtroom proceedings, and on into the realm of corrections. A definition of forensic psychology might be the application of psychological knowledge to the justice system. Forensic psychologists also adopt a variety of alternative names, including criminal psychologists, investigative psychologists, correctional psychologists, police psychologists, and social-legal psychologists. The differences between these types of forensic psychologists are mainly related to their roles. For example, an investigative psychologist is more often involved with the investigation of crimes than with testifying about a person's competency to stand trial. It is important to realize that the number of roles psychologists have in the justice system is too large for any one person to be able to do them all well.

38.1 What is Forensic Psychology?

Another definitional issue is my use of the term "psychology" in this book. Herein I am using the terms psychology and psychologist to apply to any of several types of behavioral scientists whose primary expertise is in the workings of the human mind. Therefore, my use of the term includes psychiatrists, who for the most part would probably prefer to be considered distinct from psychologists. The main difference between a psychologist and a psychiatrist is that a **psychologist** usually has a Ph.D. type of doctorate in psychology, whereas a **psychiatrist** has an M.D. type of doctorate in medicine followed by a residency in psychiatry. Forensic psychologists are certified by obtaining a Diploma in Forensic Psychology from the American Board of Professional Psychology, based in their qualifications. A forensic psychiatrist normally undertakes specialized training in forensic psychiatry similar to that described in chapter 25 for forensic pathologists. Psychiatrists can prescribe drugs and psychologists can not. Psychologists focus primarily on behavior and how personality, motivation, experience, and conditioning shape this behavior. A psychiatrist, on the other hand, is more often interested in the biological and physiological bases of mental illness, which is usually viewed as a disease of the brain that should be treated using drugs.

This distinction between the fields is far from perfect. Both contain a large number of scientists whose theories and methods quite often overlap. Despite the characterization in the previous paragraph, there are many psychiatrists who subscribe to behavioral models, and many psychologists who subscribe to a biological model. In this chapter I will refer to them all as forensic psychologists, even if they are really psychiatrists or some other type of behavioral scientist, such as a sociologist or anthropologist.

38.2 The History of Forensic Psychology

The history of forensic psychology begins in the early 1900's with the work of William Stern of Germany on memory, specifically how well people could remember events that they had witnessed. This was one of the first scientific studies showing that eyewitness testimony is unreliable. The first book on forensic psychology was published in 1908 by Hugo Munsterberg. Also during the early 1900's the use of psychological tests for evaluating offenders and applicants for law enforcement jobs became established.

Since the middle 1900's forensic psychologists of all sorts have become accepted as experts with a great deal to contribute to many aspects of the process of justice. The largest association of forensic psychologists in the U.S. is the American Psychology and Law Society (APLS), a division of the American Psychological Association. In 2005, the APLS had over 2000 members.

38.3 Forensic Psychologists in the Criminal Investigation

Forensic Psychologists can play a number of key roles in a criminal investigation. Following the discovery of a crime, a forensic psychologist may be asked to act as a criminal profiler, generating clues about the suspect from the behavioral evidence observed at a crime scene. Psychological profiling will be discussed in greater detail in a coming chapter.

A psychologist may participate in interviewing a suspect. A psychologist's knowledge of human mental states may allow them to suggest a line of questioning or manner of questioning that may result in more truthful or revealing statements from the suspect.

One type of forensic psychologist common in larger jurisdictions is the **police psychologist**. A police psychologist works for a law enforcement agency, and has often taken basic or even advanced peace officer training. These hardworking professionals provide a wide range of services. They may assist in suspect interviews as described above. They often provide counseling, both to victims of crimes, and to peace officers who have survived or witnessed a particularly nasty experience. They also commonly become involved with the process of hiring new peace officers, which usually includes administering a psychological test to make sure that applicants are not hired if they exhibit certain types of personality disorders that would make them incompetent or even dangerous.

38.4 Forensic Psychologists in Court

There are many ways in which forensic psychologists become involved with court proceedings, both criminal and civil. Most of these situations will involve the psychologist giving testimony, but in some situations they serve as advisors to attorneys. Forensic psychologists who specialize in courtroom issues are sometimes referred to as **court psychologists**.

One common issue in both criminal and civil courts is an individual's mental **competency**. There are many shades of meaning to competency, so let me provide some examples. First, a defendant may be determined to be incompetent to stand trial. This means that the defendant's mental state, temporarily or permanently, is such that they can not understand the nature and importance of the trial. In other cases, competency refers to a person's ability to make legal decisions on their own behalf. This occurs in criminal and civil contexts. Montanans, and many other people no doubt remember the "Unabomber", Ted Kazynski. Kazynski wished to act as his own attorney, but he was determined to be mentally incompetent to do so. Some civil cases involving disputed wills, for example, hinge on whether the deceased person was mentally competent at the time they made or changed their will.

Another psychological issue, especially in the criminal realm, is a defendant's mental state at the time they committed the crime. This sets the stage for the insanity defense, which will be discussed in more detail below. A psychologist, or often psychologists on both sides of the issue, will testify as to their opinion of the defendant's mental state at the time.

In the civil realm, a forensic psychologist may evaluate and testify concerning the mental trauma or disability suffered by a plaintiff or a defendant. This determination is often important in influencing the amount of damages awarded to a litigant. Forensic Psychologists are almost always involved in certain types of civil cases, such as child custody suits. Many other types of civil cases, such as divorces and custodial arrangements for the elderly may also benefit from the expertise of a forensic psychologist.

Another role of forensic psychologists in court is to give advice to attorneys. One example is in evaluating potential jury members. Attorneys always want to avoid jury members who are **prejudiced** (meaning that they have pre-judged the case without hearing it), or at least jurors who are prejudiced against their side. Many studies by psychologists and sociologists have revealed characteristics, including characteristic answers to certain questions, that reveal a prejudice. This is carried to the point that some people believe that the way the jury will decide in a case can be determined before the trial begins based on interviews of jurors. Others, including myself, are skeptical of this belief. In any case, the jury selection process allows both sides to rule out jurors who are prejudiced against them, which hopefully results in a balanced jury. Difficulties arise when a case has received so much publicity that few or no unprejudiced jurors can be found among the panel of potential jurors. In this case, one or the other of the attorneys (usually the defense attorney) will ask for a **change of venue**, which involves moving the case to a court in another location where, hopefully, potential jurors are not as uniformly prejudiced.

In other cases, and here we enter an ethically gray area, forensic psychologists who are experts in the psychology of juries can give attorneys advice about how to influence the jury. This can include any number of things from the defendant's clothing to the lawyers' mannerisms. Some forensic psychologists are also experts at how to psychologically motivate witnesses, and can suggest ways in which an attorney can pose questions so that the witness will be cooperative, or perhaps become confused or angry. Types of witnesses will be discussed later in this chapter.

38.5 Forensic Psychologists in Prisons and Treatment Centers

A large number of forensic psychologists work either in a correctional institution (i.e., prison, jail, or juvenile hall), in a psychiatric hospital, or in a facility for the mentally ill offenders. These forensic psychologists may be referred to as **correctional psychologists**. In the corrections environment they fill many significant roles, but by far the most common and important is providing treatment to offenders who are mentally ill. It is believed by many people that certain mental illnesses, such as obsessive/compulsive disorder, bipolar disorder, and schizophrenia can actually lead to criminal behavior. The first step in rehabilitation of offenders with these problems is to treat their illness.

There may also be some personality disorders, such as sociopathy and psychopathy, that predispose people who have them to behave in a criminal manner. Sociopathy and psychopathy are similar versions of **antisocial personality disorder**, a condition in which the affected person is not capable of understanding the feelings of others. These are not illnesses in the traditional sense, and are much more resistant to treatment. Nonetheless, people so afflicted deserve a chance for rehabilitation and forensic psychologists will attempt interventions that may make the problem less severe.

The diagnosis of a mental problem and the evaluation of progress made usually depends on tests that are administered to inmates. These **psychological tests** have been demonstrated to give a reliable assessment of a person's mental state, though there is always concern that a clever offender may be able to score better on such a test than their mental state would normally warrant, either through familiarity with the tests or sheer cleverness.

Forensic psychologists also have a role to play in parole evaluations. If a convict is to be paroled (*i.e.* set free before serving their full sentence) then the risk that the person will offend again has to be judged to be very low. Forensic psychologists can often contribute to this assessment, based on tests or other observations of the convict.

38.6 The Insanity Defense

The history of the insanity defense goes back to 1843 in England, when Daniel M'Naghten shot and killed the secretary of the Prime Minister of England, mistaking the secretary for the Prime Minister himself. M'Naghten was determined to be mentally ill, and was sent to a hospital for the mentally ill rather than to prison. The insanity defense is a controversial issue within the justice system and to the general public. It is widely misunderstood, and there are large numbers of people, both inside and outside the justice system who see the insanity defense as a way in which criminals get away with heinous crimes. The insanity defense came under intense scrutiny in the 1980's when John Hinckley shot U.S. President Ronald Reagan in what he testified was an act designed to impress the actress Jody Foster. Hinkley's defense team tried the insanity defense, to the outrage of a large segment of the American public.

Although the insanity defense is commonly seen as a tactic used by fakers to avoid the consequences of their actions, there is no evidence that anybody is "getting away" with anything. A verdict of "not guilty by reason of insanity" or "guilty, but mentally ill", does not result in a defendant being set free, but instead in their being routed to a psychological treatment facility rather than a traditional prison. In many cases in which the insanity defense was successful, the offender spent more years in the treatment facility than they would have if they had simply served their sentence in prison, because the treatment facilities do not release patients who have not shown significant improvement to the point where they are able to function in society at large. It is worth noting that approximately half the cases in which the insanity defense is invoked are not violent crimes.

The details vary from state to state, but in general the insanity defense rest on proving that at the time the defendant committed the crime, he or she did not do so either intentionally or negligently, due to the presence of a mental illness or mental defect (personality issues usually don't count). This seems fair. If a person was so mentally ill or defective that they did not perceive their actions as criminal, then what they need is treatment, not punishment. A person so mentally ill or defective would likely not understand the punishment or its reason anyway. Let me present an example, the "robot bitch" case (not my choice of name). Some years ago in Montana, a man killed his wife, under the delusion that she was an alien robot sent from outer space to kill him. The fact that the man was suffering under this delusion at the time was apparently never disputed. I'm a normally gentle and peaceful person, but if I was confronted with an alien robot bent on my personal destruction, I would do my best to kill it before it could kill me, and every state allows self defense. I believe that most people would do the same. However, the jury in this case found the man guilty, and so far as I know he is still serving his sentence in a traditional prison, where I can only hope that he is receiving some form of treatment for his apparently profound mental illness.

38.6.1 Verdicts in the Insanity Defense

One of the current issues in the insanity defense is the nature of the verdict reached in such a case. Traditionally the verdict in a successful insanity defense was "not guilty by reason of insanity". This set of terms raises several points of contention. First, what does "not guilty" mean here? It implies that the person did not commit the crime, but the person really did commit the crime. This terminology obscures the fact that the offender is not held to be responsible for the crime due to their mental illness or defect – not because they didn't commit the crime.

Second, what does "insanity" mean. Surprisingly, there is no medical definition of "sane" or "insane". Instead, it is a legal opinion, intended to be rendered by a group of jurors who may know nothing about the medical or psychological issues involved.

Therefore, the nature of the verdict in these cases has undergone change in many states to the verdict "guilty, but mentally ill" as the result of a successful insanity defense case. This newer verdict clarifies some of the issues raised in the paragraphs above.

38.6.2 Success Rate of the Insanity Defense

The insanity defense is actually invoked quite rarely. A study in the early 1990's showed that it was invoked in less than one percent of cases, and most authorities believe that it is invoked less now than in the past. The study also showed that in 90% of the cases where the insanity defense was invoked, the defendant was diagnosed as having a mental illness or mental defect. Even so, the insanity defense was successful in only about a quarter of the cases, presumably the ones in which the offender's mental illness was most severe. The study further reported that in 80% of the cases in which the insanity defense was used succesfully, its appropriateness in the case was agreed to by both sides before the trial began.

From the results of this study we can conclude that the insanity defense is not very successful, except in cases where the defendant's mental incompetence is clear and obvious. It is simply not the case that a significant number of criminals are faking a mental illness and thereby avoiding prison time.

38.7 Reliability of Diagnoses

Discussion of the insanity defense always raises questions of the reliability of diagnoses of mental illness. Indeed, it is often the case that an insanity defense trial can devolve to a "war of the psychologists", where one argues that the defendant is mentally ill and the other argues the opposite. The impression one gets is that psychological diagnoses are not reliable. However, the reason for a "war of the psychologists" is less about the reliability of diagnoses and more about the nature of courtroom proceedings.

The court is adversarial by nature. This aspect of it goes back to ancient times and evolved from debates (forensics) in which people representing two sides of an issue each attempted to convince a judge or the onlookers that their point of view was the correct one. The actual correctness of the points of view were not as important as whether the arguments made were convincing. The modern American court system has inherited this approach, for better or for worse. Therefore, the use of an opposing psychologist is usually for the purpose of making sure that the defense team and their psychologist are able to make a convincing argument. This allows the evidence upon which a diagnosis of mental illness is made to be brought out and subjected to scrutiny by the judge and jury, who then may form an opinion about it. The same situation occurs in many other contexts, and is by no means a feature exclusive to the insanity defense.

The reliability of diagnoses of mental illness, when made by a psychiatrist, is about 80%, which is very nearly the same as the reliability of a diagnosis of any form of illness made by a medical doctor.

38.8 Perception, Memory, and Leading Questions

Perception is a tricky thing, and memory is even trickier. Sometimes the way in which interviews, questioning, and interrogation are carried out lead someone to give an incorrect statement of events. With some people, especially children, the person testifying may not even know that their statement is false. Sometimes a false memory or perception can be stronger than a true one. Since the study of perception and memory is part of the field of psychology, some forensic psychologists specialize in these problems.

One of the more common and serious issues that arise out of the trickiness of perception and memory is leading questions. **Leading questions** are questions in which the desired answer is implied in the question itself. For example: "You shot Bob, didn't you?" is a leading question because it implies the person asking the question is expecting to hear "yes" as the answer. The question "Did you shoot Bob?" is not a leading question, because it does not anticipate any specific answer. Most people can be described as both willing to please and reluctant to contradict. These traits arise from our basic social conditioning. Leading questions are considered unfair, because they predispose the person answering the question to respond to it in the way that the question implies. If the person answering the question answers it in a manner contrary to its implication, they might actually be perceived by the jury as being contradictory or uncooperative.

At an even deeper level, leading questions can actually have an effect on a person's memory of events. The implied answer may be interpreted as reflecting the truth, causing the person to whom it is asked to consciously or unconsciously doubt their memory if it conflicts with the implied answer. For these reasons, leading questions are objectionable, meaning that an attorney can object to the question, and the judge will usually require the questioner to restate the question in a manner free of implications.

38.8.1 Leading Questions and Child Psychology

Leading questions are problematical for adults, but especially so for children. Child psychology is not the same as adult psychology, and children's memories are much easier to influence with leading questions. Repeated studies and actual cases have shown that young children are eager to please and have a less firm grip on reality than adults (some actually do believe in Santa Claus). Children also have a tendency to perceive anything an adult says as truth, and if their perception or memory disagrees with it, then they are the one that is wrong. This may cause children to pick up on what adults want them to say and actually believe that events that never happened did happen. This effect is most pronounced with younger children, and fades as they mature.

What we now know is that when interviewing a child, the interviewers have to be extremely careful not to lead the child or to seem to prefer one answer over another. Asking a child leading questions can result in the formation of a false memory, and child psychology is such that it may be indistinguishable from a true memory.

The false memory effect in children has been most publicized in child sexual abuse cases. If investigators assume that the abuse has occurred and frame questions under that assumption, they will often get responses that agree with their assumption, but which are not true. For example, if the investigator asks "Did he touch your privates?" or "He touched your privates, isn't that right?", most young children will answer yes, regardless of whether or not any "privates" were touched, because it's obvious that the investigator wants to hear the answer "yes". A better way to ask is "Where did he touch you?"

Following the lead of the New Jersey Supreme Court, most judges are now allowing "taint" hearings. These are hearings designed to determine whether the testimony of a child has been influenced by the interview procedure itself.

38.9 Types of Witnesses from the Psychological Point of View

As one example of the ways that psychologists can advise attorneys about jurors and witnesses, let's look at a categorization of types of witnesses from the psychological point of view. Forensic psychologists may be consulted in the evaluation of witnesses, or even to participate in the interview of a witness as part of the discovery pretrial phase. By categorizing a particular witness one can predetermine, to an extent, what to expect from that individual in future encounters, such as their courtroom testimony. A number of different types of witnesses exist. Here is a widely accepted taxonomy.

- The Distraught Witness. A distraught witness exhibits emotional distress. This distress may be the direct result of witnessing the crime or may arise indirectly from their relationship with the victim.
- The Fraudulent Witness. A fraudulent witness actually knows nothing about the crime, but is seeking attention or notoriety by pretending to know something.
- The Hostile Witness. A hostile witness is deliberately antagonistic and/or noncompliant. They can refuse to answer directly, or even invent misleading facts. Some hostile witnesses are antagonistic toward the law or the justice system, while others have a relationship to the offender, whom they are trying to protect. In some courts, special rules apply to hostile witnesses.
- The Intimidated Witness. An intimidated witness is afraid of retaliation from the person they are testifying about or that person's associates.
- The Inventive Witness. Unlike a fraudulent witness, an inventive witness did witness something. However, the testimony they give is exaggerated

or embellished with made-up details. Like a fraudulent witness, the inventive witness may be seeking attention. In other cases the inventive witness simply has a hard time distinguishing reality from imagination.
- The Reluctant Witness. Some people find it difficult to speak freely, or to answer questions, especially when they are the center of attention. I find this true of myself in court. No matter my many years of experience lecturing to college classes, when I take the witness stand, I feel as if I am in 4th grade and standing In front of the principal's desk. While I can deal with this, not everyone can. A reluctant witness finds being the subject of scrutiny on the witness stand to be disconcerting and will be hesitant to answer. Another type of reluctant witness feels that the process of justice is none of their business, and are therefore reticent about answering questions.

38.10 Questions for Study and Review

1. Describe the difference between a psychologist and a psychiatrist.
2. List three tasks that "police psychologists" commonly perform.
3. How does a "court psychologist" differ from a "police psychologist?"
4. What is "mental competency" as applied in court proceedings?
5. What does it mean if a potential juror is described as "prejudiced?"
6. Describe "correctional psychologists" and the roles they perform.
7. Many people believe that criminals often invoke the insanity defense in order to escape punishment for their crimes. Does the true nature of the use of the insanity defense support this belief? Why or why not?
8. Describe the differences you perceive as a citizen between the verdict of "not guilty by reason of insanity" and the verdict of "guilty, but mentally ill."
9. In about what percent of cases is the insanity defense invoked?
10. What does the phrase "the court system is adversarial" mean?
11. What is a leading question and why are leading questions problematical in suspect interviews and courtroom examination of witnesses?
12. Why are leading question especially problematical in questioning children?
13. What are the six types of "interesting" witnesses from the point of view of forensic psychology and what are the characteristics of these types of witnesses?

38.11 Sources

American Board of Forensic Psychology 2007. Brochure [Internet]. [cited 2007 Oct 16]. Available from: . [Online]. Available from: http://www.abfp.com/brochure.asp.
American Psychiatric Association. 1985. Insanity Defense: A Position Statement [Internet]. APA Document Reference No. 85003. [cited 2007 Oct 28]. Available from: http://www.psych.org/edu/other_res/lib_archives/archives/198503.pdf.

[Anonymous]. n.d. Forensic Psych: Just the FAQ's [Internet]. [cited 2007 Oct 28]. Available from: http://forensics.thedatabase.org/faq.htm.

Decaire M. n.d. Forensic Psychology: the Misunderstood Beast (Part 1) [Internet]. [cited 2007 Oct 28]. Available from: http://www.suite101.com/article.cfm/forensic_psychology/15697.

Decaire M. n.d. Forensic Psychology: the Misunderstood Beast (Part 2) [Internet]. [cited 2007 Oct 28]. Available from: http://www.suite101.com/article.cfm/forensic_psychology/16054.

Decaire M., n.d. Forensic Psychologists Vs. Forensic Psychiatrists [Internet]. [cited 2007 Oct 28]. Available from: http://www.suite101.com/article.cfm/forensic_psychology/20603.

Decaire, M. 1992. Witness Evaluation Typologies [Internet]. [cited 2007 Oct 28]. Available from: http://forensics.thedatabase.org/wit_type.htm.

Gado M. 2007. The Insanity Defense [Internet]. [cited 2007 Oct 28]. Available from: http://www.crimelibrary.com/criminal_mind/psychology/insanity/1.html.

Gallagher MP. 2007. State alleges psychologist coached child on sex abuse in custody case. New Jersey Law Journal (Oct 5, 2007). Available from: Expanded Academic ASAP. Gale. Univ of Montana. 22 Oct. 2007 http://find.galegroup.com.weblib.lib.umt.edu:8080/itx/start.do?prodId=EAIM. Gale Document Number:A169542615.

Garvey DV., 2001. Frequently Asked Questions [Internet]. [cited 2007 Oct 28]. Available from: http://www.geocities.com/CapitolHill/Lobby/6027/faq.htm.

Jones R. n.d. The Concept of Criminal Responsibility [Internet]. [cited 2007 Oct 8]. Available from: http://www.forensicmed.co.uk/forensic_psychiatry.htm.

Miller L. 2007. The Psychological Fitness for Duty Evaluation [Internet]. FBI Law Enforcement Bulletin 76(8). [cited 2007 Oct 19]. Available from: http://www.fbi.gov/publications/leb/2007/aug07leb.pdf

Northeastern University Psychology Department. 2002. General Information on Forensic Psychology [Internet]. [cited 2007 Oct 8]. Available from: http://www.psych.neu.edu/academics/forms/forensic.pdf.

Ramsland K. 2007. The Art of Forensic Psychology [Internet]. [cited 2007 Oct 28]. Available from: http://www.crimelibrary.com/criminal_mind/forensics/forensic_psychology/index.html.

Ramsland K. 2007. The McMartin Nightmare and the Hysteria Puppeteers [Internet]. [cited 2007 Oct 28]. Available from: http://www.crimelibrary.com/criminal_mind/psychology/mcmartin_daycare/1.html.

Sharps MJ, Barber T, Stahl H, Villegas AB. 2003. Eyewitness Memory for Weapons. The Forensic Examiner 12(9-10): 34(4).

Sharps MJ, Hess AB, Casner H, Jones J. 2007. Eyewitness memory in context: toward a systematic understanding of eyewitness evidence. The Forensic Examiner 16(3): 20(8).

Underwager R, Wakefield H. 1997. The Taint Hearing: Issues for Forensic Psychologists [Internet]. Presented at the 13th Annual Symposium in Forensic Psychology, Vancouver, British Columbia (April 17, 1997). [cited 2007 Oct 28]. Available from: http://www.tc.umn.edu/~under006/Library/Taint_hrg.html.

Van Dorsten B (Editor). 2002. Forensic Psychology: From Classroom to Courtroom. New York: Kluwer Academic Press.

Wikipedia contributors. 2007. Insanity defense [Internet]. Wikipedia, The Free Encyclopedia; 2007 Oct 26, 15:15 UTC [cited 2007 Oct 28]. Available from: http://en.wikipedia.org/w/index.php?title=Insanity_defense&oldid=167229228.

Wikipedia contributors. 2007. Leading question [Internet]. Wikipedia, The Free Encyclopedia; 2007 Oct 12, 04:10 UTC [cited 2007 Oct 28]. Available from: http://en.wikipedia.org/w/index.php?title=Leading_question&oldid=163974588.

Chapter 39
Advanced Topics
LIE DETECTION AND HYPNOSIS

In this chapter we will explore two subjects that are often associated with forensic psychology, lie detection and forensic hypnosis. Both can be described as on the edge of credibility, in that they definitely contribute to investigations in some cases, but rarely is the evidence they generate accepted in court. This chapter will examine lie detectors and lie detection first, then move on to explore forensic hypnosis.

39.1 Lie Detectors and Lie Detection

The use of a lie detector, more precisely referred to as a **polygraph device**, is fairly common in investigations. A polygraph device measures and records several physiological variables of an individual while she or he is being asked questions. The physiological variables measured are usually blood pressure, pulse, respiration rate, and skin conductivity.

The theory underlying lie detection is that it is stressful to lie, as compared to telling the truth. For one thing, lying involves being creative, which is a difficult mental task for some people. Many people were also brought up being taught not to lie and therefore have inhibitions against lying that have to be broken. When lying, there is also the stress caused by the possibility that the lie will be discovered, leading to embarrassment or worse. Therefore, certain physiological reactions usually occur when lying, including increased muscle tension, more rapid heart rate, increased sweating, and perhaps others. Polygraphs are designed to detect these physiological changes.

39.1.1 The History of Lie Detection

The idea that lying provokes anxiety is ancient. Peoples in various parts of the world have devised rudimentary lie detection tests based on this principle. In West Africa, for example, one group of people believes that the anxiety involved in lying causes a person to be jittery and have trouble with dexterity. They have devise a test that consists of passing an egg back and forth, and if the accused drops the egg, they are held to be lying. The ancient Chinese believed that the anxiety of lying caused a person to have a dry mouth, incapable of producing significant amounts of saliva. The lie detection test they devised consisted of the accused attempting to swallow a mouthful of raw rice while a prosecutor read the statement of the charges that they had claimed to be innocent of. The accused person was considered guilty if they were not able to work up enough saliva to swallow the rice. In the U.S., psychologists began doing significant work on lie detection during the 1920's. Polygraph devices were popularized by psychologists in late 1930's, and the field of lie detection continues to be closely associated with forensic psychology.

39.1.2 How a Lie Detector Test is Conducted

A lie detector test is more formally called a **psychophysiological detection of deception** (PDD) examination. Electrodes are placed in contact with the skin of subject (the possible liar) in order to measure heart rate and skin conductivity. Skin conductivity gives a good indicator of sweating. The proper contacts for measuring blood pressure and respiration rate are also attached. The subject is then asked questions while their physiological data is measured and recorded.

Three types of questions are asked of the subject. One type is called **control questions**, and consist of questions to which the subject will probably give a lie as an answer. For example, "Have you ever stolen anything?" The second type of question asked is **irrelevant questions**, which should provoke no anxiety, such as "What is your dog's name?" The final group of questions is **relevant questions**, which are those that are relevant to the crime, such as "Did you shoot Bob?".

Interpreting the PDD results is done by comparing the physiological reactions to the irrelevant, non-anxiety-provoking questions with the reactions to the questions that are intended to receive a lie in response. This establishes that person's characteristic physiological response to truths and lies. The reactions to the questions relevant to the crime are then compared to these standards to determine whether the individual answers are more likely truths or lies.

The results obtained are usually expressed as "test failed", meaning that the subject was lying in their responses to the relevant questions, or "test passed" if the subject was likely not lying in their responses to the relevant questions. An ambiguous result is occasionally obtained.

39.1.3 Problems with Lie Detection

It has been replicably demonstrated that lying does cause anxiety in most people. For those people it is very difficult to lie in such as way that a polygraph won't detect it. Yet, it is also clear that there are some people who either don't experience anxiety when lying, are capable of controlling their physiological reactions while lying, or are able to convince themselves that the lie is true to the extent that they believe it as if it were the truth. Regardless of the mechanism, several known criminals have passed the PDD.

What is even more common is for the polygraph to flag a truthful statement as a lie. If there are emotional attachments to an event or an action, it may be stressful to think or talk about it, especially when connected to a polygraph machine.

Lie detection remains controversial. Advocates claim that it is accurate 70% to 90% of time. Others say that it's just a fancy method of interrogation, and that all forms of interrogation are inherently unreliable. Arguably the best investigation into the reliability of lie detection was made by the National Research Council (NRC) in 2003.

The NRC found that the majority of polygraph research is of low quality. The few good studies they identified concluded that lie detection is better than chance, but not perfectly accurate. Overall, the NRC recommended against the use of lie detection in most circumstances.

39.1.4 The Use of Lie Detector Evidence

Because of the problems with lie detection, the results of PDD's are only rarely considered acceptable evidence in court. In a variety of court rulings there have been severe restrictions placed on the use of polygraph evidence. One of the most thorny questions is whether failing a lie detector test amounts to testifying against oneself. There are constitutional protections against self-incrimination that might be violated by lie detector testing. Therefore, when polygraph evidence is used in court, it is more often used to exonerate, rather than to incriminate. In other words, it is more common to say that the person passed the lie detector test and must be innocent than it is to say that the person failed the test and must be guilty. The fact that a person failed a polygraph test can sometimes be stated in court, but only when qualified with the information that polygraph results are not scientifically certain. The main exception is the use of PDD's with sex offenders on probation or parole, who are routinely polygraphed in many states. Surprisingly, state and federal courts have upheld this practice.

In many cases, polygraph tests are used generate leads that are then investigated further in order to verify their accuracy. It is common for a polygraph session to produce new facts that can be used in an investigation. Polygraph officers and investigators know that the results of a PDD are rarely acceptable as evidence in court. Therefore, if the subject fails the test, the examiner or a peace officer will often use this result as leverage to obtain an admission of guilt. For example, they may say, "We know you are lying, so you might as well tell the truth."

Although polygraph operators were originally psychologists, all sorts of people operate them nowadays, including criminalists, police psychologists, and peace officers. There is a set of training courses for polygraph operators and a certification process, which is overseen by the American Polygraph Association.

39.2 Forensic Hypnosis

Forensic hypnosis is the use of hypnosis to acquire evidence for an investigation. In most cases this consists of hypnotizing a subject in an attempt to enhance their memory. In other words, a victim or witness is hypnotized and asked to try to remember details of a crime. Under hypnosis, they may remember details that are otherwise lost to them. Hypnosis is sometimes even used on a suspect who claims not to remember having done the crime.

39.2.1 The History of Forensic Hypnosis

Hypnosis emerged out of the mysticism movement of the 1700's and 1800's. The first recorded use of hypnosis in a criminal investigation was in 1845, when a theft victim was able to identify the thief during hypnosis. Initially, courts were skeptical of forensic hypnosis and in 1897 the California Supreme Court ruled that hypnotic evidence was not admissible. During the 1900's as the work psychologists have done with hypnosis has improved its use, it has become more acceptable. In 1927 the U.S. Supreme Court ruled that hypnotically refreshed testimony is admissible, and in 1968 the Maryland Supreme court ruled that hypnosis is just like any other memory aid.

39.2.2 Memory Theory

Psychologists tell us that there are two kinds of memory, short term memory and long term memory. Most psychologists believe that we use **short term memory** for things that we consider unimportant and **long term memory** for things we consider important. So, for example, I can remember a lot of details about forensic anthropology, but I can never remember where I parked my darn car.

Apparently, the categorization of information as "important" or "unimportant" is mostly unconscious. So, even if I tell myself to remember where I parked the car, I may not be able to remember when it comes time to drive home, because my unconscious mind doesn't recognize today's parking spot as important information. Applying these ideas to witnessing a crime is straightforward. If a witness doesn't perceive a detail as important, it goes in short term memory and is soon lost. For example, lets say that I am walking down the sidewalk and a car is coming toward me on the street. I see the front of the vehicle, so I see the license plate and the driver, but since it seems unimportant at the time, I take no notice of either. Now, if after passing me, the driver runs over a pedestrian and then speeds away, it is extremely unlikely that I will remember the number on the license plate or the face of the driver. However, I might be able to remember these details under hypnosis.

Victims of crimes often forget details as well. This can occur for several reasons. One reason is that the detail seems unimportant within the context of what is happening to them. Another reason is that some victims cope with the psychological aftereffects of their experience by trivializing it, or even forgetting it completely. An extreme form of this is repression, where the victim's mind blocks the memory of the experience from them because it is too traumatic to think about. Hypnosis can sometimes free up these memories so the victim can access them.

39.2.3 Hypnosis and Its Effects

Hypnosis is an altered state of awareness. Nobody completely understands what hypnosis is, but some people think of it as a state of deep relaxation in which the mind becomes more open to suggestion. We have all heard stories of people being hypnotized and made to act like a chicken. Occasionally, in some people, such deep states of hypnosis can occur. Further, it seems as if the barriers between the conscious and subconscious portions of the mind, including normally inaccessible portions of memory, are lowered during hypnosis.

Increased suggestibility is a known consequence of hypnosis. **Suggestibility** is a mental state in which a person becomes more likely to believe or do what someone else tells them. Therefore, great care must be taken when questioning a hypnotized person to not frame questions in such a way that the hypnotized person takes the question as a suggestion. The hypnotist definitely has to avoid leading questions.

Another phenomenon associated with hypnosis is **hypercompliance**, wherein the hypnotized person may seek to please the hypnotist, police, or prosecutor by "doing the right thing" and providing erroneous but well meant information.

Memory distortion, also known as "confabulation" may occur during or after hypnosis and consists of remembering something in such a way that the facts are distorted. There is no such thing as a perfect recollection. All memories are a combination of factual observation with false information that results from the mind filling in the gaps in our perceptions. One common problem is **time distortion**. When a person is hypnotized, their sense of time is distorted – usually accelerated by a quarter to a third. Therefore, the timing of events may be substantially off. When the perception of an event is marginal in the first place, there is a large chance for it to be distorted by purely fictitious information supplied by the person's mind or by the questions of the hypnotist. Therefore, all information obtained by hypnosis needs to be verified by independent investigation.

It is also possible that the witness or victim doesn't actually have any memories to remember. Trying to recall the number on the license plate will fail if the witness happened to be looking in a different direction as the car passed, and no amount of hypnosis can produce a true memory of what was not actually experienced. In this type of situation, the probability of inducing a false memory is high.

39.2.4 Memory Refreshment

Memory refreshment consists of hypnotizing a person, then asking them to relive an experience in their memory, so that forgotten details may be recalled. **Age regression**, or time regression, is a closely related form of hypnosis in which the hypnotized person is returned to a younger age. With help from the hypnotist a person can sometimes approach the memory of events without the accompanying emotions that were felt at the time and which may be blocking the memory of the event. However, during a memory refreshment or age regression the hypnotized person may experience **revivification**, which is experiencing the event over again, including the emotional and physical sensations experienced when it occurred the first time.

39.2.5 The Reliability of Forensic Hypnosis

Many of the potential problems with forensic hypnosis were described above. Does forensic hypnosis work, despite these problems? Forensic hypnosis is considered fairly marginal by most forensic scientists. The problem of false memories is severe. Perhaps the best test is to summarize the cases in which forensic hypnosis has contributed or attempted to contribute. What do the cases tell us?

First there are many cases in which a subject remembers crucial information under hypnosis. This information has turned out to be accurate and has led to the identification of a suspect. Second, there are also many cases in which a subject remembers events and detail inaccurately, including remembering the most minute details of events that never happened. Putting these together we have to say that the results are mixed. Therefore, information obtained under hypnosis can only be considered reliable if it can be verified by independent investigation.

In an attempt to salvage forensic hypnosis, the FBI has recommended a three-part procedure that if followed will result in reliable, perhaps even court admissible, evidence. This procedure minimizes the problems of forensic hypnosis. The first part of the procedure is a case review, to determine whether there is likely to be some information gained by hypnotic refreshment of someone's memory. The second step is the hypnotic interview, in which the person is hypnotized and questioned. The third part is the most critical and is independent corroboration of facts and details that emerge from the hypnotic interview. It is this third step that is most important, and gives us our bottom line on the reliability of forensic hypnosis – don't trust hypnotically derived evidence unless it can be independently verified.

39.2.6 The Use of Hypnotically Obtained Evidence

The highest and best use of forensic hypnosis is to generate possible leads that can be followed up by further investigations. The use of hypnotic evidence in court is undergoing renewed scrutiny and courts are becoming more reluctant to allow it. When it is allowed, it becomes the burden of the prosecution to show that the remembered details are true. The State v Hurd case is perhaps informative. In this case, a victim of a knife assault was unable to identify her attacker until she underwent hypnotic refreshment, during which she was able to identify her attacker as her ex-husband. However, the state was not able to prove that the memory was genuine by independent investigation, and the ex-husband was found innocent. The ruling opinion in this case placed several restrictions on the collection and use of hypnotic evidence. One such restriction, for example, was that the hypnotist not be a part of the prosecution team. The State v Hurd opinion established guidelines for how a hypnotic interview should be conducted, and established a requirement to record the hypnosis interview so that the possibility of introducing a false memory can be tested by reviewing the record of the session.

39.3 Questions for Study and Review

1. What is the basic theory for how a polygraph device can be used to detect when a person is lying?
2. What is a PDD examination?
3. What three types of questions are asked during a PDD examination? Why are these three different types of questions asked?

4. What does "test failed" mean when it is given as the result of a PDD examination?
5. What is the main problem with lie detection which prevents its use as evidence against a suspect?
6. What is the most common use of lie detector evidence in court today?
7. How do short term and long term memory relate to whether a person considers information important or unimportant?
8. What exactly is a hypnotic state?
9. Define the terms suggestibility, hypercompliance, memory distortion memory refreshment, and age regression; and explain how they apply to forensic hypnosis.
10. What is the FBI's recommended procedure for obtaining and using evidence obtained via hypnosis? Why is it important to follow this procedure?

39.4 Sources

American Polygraph Association. 2002. The American Polygraph Association [Internet]. [cited 2007 Oct 28]. Available from: http://www.polygraph.org/.

Frazier K. 2007. Energy Department will end most polygraph testing. Skeptical Inquirer 31(1): 8(1).

Gilbert S. n.d. Forensic Hypnosis A Guide for the Perplexed [Internet]. [cited 2007 Oct 28]. Available from: http://members.aol.com/SVG2254/hypno.htm.

Lykken D. 2004. Nothing like the Truth: the Polygraph Is Still the Most Popular Tool for Ferreting out the Guilty. New Scientist 183(2460): p17(1).

National Research Council. 2003. The Polygraph and Lie Detection. Washington (DC): National Academies Press.

Newman AW, Thompson JW. 2001. The Rise and Fall of Forensic Hypnosis in Criminal Investigation. Journal of the American Academy of Psychiatry and the Law 29 (1): 75-84 2001.

Ruscio J. 2005. Exploring Controversies in the Art and Science of Polygraph Testing. Skeptical Inquirer 29(1): 34(6).

Scheflin AW. 2001. Investigative Forensic Hypnosis. American Journal of Clinical Hypnosis 44 (2): 159-160.

Scheflin AW, Frischholz EJ. 1999. Significant Dates in the History of Forensic Hypnosis. American Journal of Clinical Hypnosis 42 (2): 84-107.

Thompson SK. 2007. A brave new world of interrogation jurisprudence? American Journal of Law & Medicine 33(2-3): 341(17).

Warner WJ. 2005. Polygraph Testing A Utilitarian Tool [Internet]. FBI Law Enforcement Bulletin 74(4). [cited 2007 Oct 19]. Available from: http://www.fbi.gov/publications/leb/2005/apr2005/april2005leb.htm#page10.

Wikipedia contributors. 2007. Memory [Internet]. Wikipedia, The Free Encyclopedia; 2007 Oct 26, 13:30 UTC [cited 2007 Oct 28]. Available from: http://en.wikipedia.org/w/index.php?title=Memory&oldid=167209603.

Chapter 40
PSYCHOLOGICAL PROFILING

This chapter completes the series on forensic psychology and begins a series of two chapters on criminal profiling. I will start with an introduction to the subject of criminal profiling, then focus more closely on psychological profiling. Psychological studies of aggression, murders, and serial murders will be discussed briefly before considering some of the types of behavioral evidence a psychological profiler would consider. The chapter will conclude with a retrospective look at the Jack the Ripper case.

40.1 Approaches to Criminal Profiling

Criminal profiling is a set of methods that allow some of the characteristics of the criminal to be deduced or inferred by examination of the behavioral evidence surrounding the crime. Criminal profiling is both a science and an art, which requires considerable experience and training. This and the following chapter provide a brief introductory overview, nothing more.

There are three main approaches to criminal profiling: statistical, psychological, and geographic. Although DNA profiling, discussed in chapter 21, sounds as if it belongs in this group, it actually does not because it is not based on behavioral evidence.

Statistical profiling, also known as inductive profiling, has been developed primarily by criminologists (sociologists) and psychologists by correlating characteristics that certain types of criminals have in common. When this information is analyzed it becomes possible to say that a certain percentage of people who commit a certain type of crime have a given characteristic. We will explore statistical profiling in the next chapter.

Psychological profiling has been developed by psychologists, and relies on a psychologist's deep knowledge of the human mind, personality, and motivation. In this approach, a psychologist looks at the nature of the crime, the nature of the victim, and the way the crime was committed to learn about the psychological characteristics of the criminal. Knowing some psychological characteristics of a criminal may give insight into the criminal's background, how they think, and who they are. We will look at some aspects of psychological profiling in this chapter.

Geographic profiling is the newest type of criminal profiling. It has been developed using concepts from geography, mathematics, and ecology. It seeks to determine where the criminal lives (or otherwise considers "home") by analyzing the pattern of where the crimes occur. Geographic profiling will be explored in the next chapter.

The **behavioral sciences** includes several academic disciplines, including at least psychology, sociology, and (sociocultural) anthropology. People with education in different disciplines approach profiling with different assumptions. Criminologists and other sociologists, tend to focus on environmental and family factors in criminality and criminal profiling. Here, the term "environmental" means non-genetic as described in chapter 15. Sociologists make heavy use of statistics in profiling. Psychologists, are more convinced that genetics plays a role in forming a person's mind and in criminality. They make heavier use of psychological testing analyzed by statistical methods. Anthropologists are a strange motley bunch who have a tendency to fall between sociologists and psychologists in theoretical approach. Anthropologists tend to believe that both genes and environment are important. Anthropologists are also skeptical of models that see humans behavior as being so simple that it can be explained as anything less than a phenomenon that results from the combination of environment, upbringing, and genetics; plus additional individual factors. Therefore, anthropologists tend to be skeptical of profiling, considering it unscientific and unreliable. It is indisputable, however that there are indeed cases in which profiling of all sorts has made a contribution.

Nobody who is familiar with profiling believes that profiling alone can adequately describe a criminal. The fact that a person fits a profile is not considered acceptable evidence in any court. Profiling should be used in a manner discussed in the previous chapter for lie detection and hypnosis – to generate leads that are then followed up by traditional investigation. Profiling should not and can not take the place of investigative work. Cases are not solved by profiling or profilers – they are solved by investigators who use all resources available to them, including profiling.

Profiling of all types is tricky. Different profilers will often interpret the evidence differently. For example, there was a case in which a victim was fatally stabbed with a pencil. One profiler suggested that the perpetrator was most likely a person who had been in prison where a pencil would be considered a weapon, and was likely an experienced criminal. After this lead did not produce a suspect, investigators called in another profiler, the renowned John E. Douglas (not the University of Montana archaeology professor), who examined the same evidence but came to a dramatically different conclusion. He suggested that the offender was a young person who had simply grabbed the first thing he saw. This new profile led to the arrest of a 16 year-old boy.

40.2 Types of Cases for Which Profiling is Useful

Not all cases, even serious ones such as sexual assaults or murders, are suitable for profiling. In order to be a candidate for profiling a case has to be unusual in some way. This could include being extremely violent, having an unusual type of victim, or being a set of serial crimes. Profiles are often developed in cases involving abducted children if the abductor is not known to be a family member. The term **serial crime** refers to crimes that are committed repeatedly by the same person or persons. We often think of serial murderers and serial rapists, but many other types of criminals commit their crimes serially, including robbers, burglars, con artists, and many others.

The most useful contribution of profiling is typically to narrow down the range of suspects in cases where there is not clearly one or a small number of suspects. The profiler can often give an idea of what type of person the perpetrator is likely to be, and in the case of geographic profiling, a place to start looking.

40.3 Psychological Studies of Aggression

One of the more fascinating and relevant topics that have been researched by forensic psychologists is the question of why people act aggressively. Psychologists traditionally define **aggression** as behavior toward another person that intentionally inflicts harm. Other authorities broaden this definition to include the threat of harm as well as the actual infliction of harm.

Psychologists identify eight distinct kinds of aggression that can be found in some form in virtually all species, including humans. These types of aggression are predatory, inter-male, fear induced, territorial, maternal, irritable, sex-induced, and instrumental. Profilers find it useful in certain cases to determine the type of aggression that led the perpetrator to commit the crime, because this often gives direct insight into the motivations and perceptions of the perpetrator.

Predatory aggression is displayed by a predator toward their prey. Humans are a predatory species. In the criminal realm, predatory aggression is usually displayed by criminals who see other humans as lesser beings. These criminals tend to fall into two groups. **Narcissists** are those people whose only concern is themselves and their own needs and desires. This may lead them to consider the welfare of others as less important, and therefore not to be concerned about harming them.
Psychopaths lack empathy, and therefore may not ascribe to others the same feelings and desires that they themselves have, leading them to consider others as not human to the same extent they are.

Intermale aggression consists of violent or intimidating behaviors displayed by males towards other males as a form of competition for mates, resources, and status. This is seen every time men get into a dispute over a woman, and when they compete in employment settings and sports. Statistics uniformly indicate that most violence is perpetrated by males, and the largest portion of it is directed toward other males.

Fear-Induced aggression is an apparently instinctive inclination to respond aggressively in response to a threat. It is likely to be related to the well-known fight or flight response to a threatening situation. Interestingly, psychologists have found that forced confinement also elicits this form of aggression. Forensic implications include the possibility that such aggression could be a major issue in prison environments. If already aggressive individuals are placed into an environment that fosters aggression via confinement, the risk of violence may be increased.

Territorial aggression is displayed in the form of threats or attacks in response to an invasion of one's territory. There is also a predisposition for intruders to behave submissively and retreat when confronted. The concept of "territory" may extend beyond an actual physical space and include a person's possessions, job, or professional status.

Maternal aggression is displayed by females in response to a threat to their offspring. This response is also common in males of some species, and some authorities include humans in this group.

Irritable aggression is threatening or harmful behavior directed towards an object when the aggressor is frustrated, hurt, deprived, or stressed. This is what students take the brunt of when they drop by my office unannounced (just kidding!). While aggression toward inanimate objects, such as a punching bag, is usually seen as an acceptable outlet for stress, aggression toward people for this reason is not.

Sex related aggression is perhaps most commonly seen in the form of jealousy. It is a fact that both sexual desire and some forms of aggression are elicited by the same sets of hormones and brain chemicals. Therefore, it is possibility for some individuals to associate or actually confuse sexual desire with domination and violence.

Instrumental aggression is a form of learned behavior. The person who displays it has learned that they can get what they want by behaving aggressively. Much of human aggression, especially aggression that is not easily explained as one of the seven types discussed above, seems to be of this type.

40.4 Studies of Murderers

Murder is an aggressive act, often with a motive that places it within one of the types of aggression described above. Psychologists have studied those situations and motivations that provoke a person to murder. They find that the motivation for murder generally falls into one of two categories. The first, and probably most common category is when the murder is the result of a certain event, such as a quarrel, a reaction to an insult, or a case of jealousy. In these cases it is likely that the situation has provoked one of the forms of aggression described above, and it is more likely that the murder is spontaneous and not premeditated. The second category of motive is the "for gain" type. Murders committed for the desire to gain wealth, revenge, sex, power, or status fall into this category. Protection of one's life, family, and goods also falls into this category more often than not. Murders that occur when gain is the motive are more likely to be premeditated.

When a murder falls into one of these categories, we can understand the motive and the situation that led to the murder. These types of murders we can, perhaps hesitantly, call "normal". Once we go beyond these more typical situations, however, the motivation becomes much less clear, and the cause less understandable. It is in these sorts of cases that psychological profiling can be of the greatest use.

40.5 Studies of Multiple Murderers

Some murderers offend more than once in their lives. They may, for example, be sent to prison for a murder and while there murder a fellow inmate. A person who does this is usually called a repeat murderer. The distinguishing feature is that there is nothing connecting the murders except that the same person committed them. A multiple murderer, in contrast, is someone who kills more than one person as part of the same event, though the event in this case may encompass a lengthy amount of time.

A **mass murderer** is someone who kills several people at one time. For example, someone who set off a bomb in a shopping mall is a mass murderer. A **spree**

killer is someone who decides to kill everyone associated with a certain place or event, one at a time. Those who seemingly insanely go about shooting their fellow employees or fellow students are of this type. Occasionally, a spree killer will not be so focused and actually try to kill anybody they can. A "killing spree" of this sort can last for a considerable period of time and sometimes cover a considerable amount of territory. Distinct from these other types of multiple murderers are the serial killers. A **serial killer** kills several people over a period of days, weeks, months, even years. There is usually a cooling off period between the killings, and the killer goes through phases, or cycles. Although all of these types of multiple murderers may be profiled, psychological profiling is best noted for cases of serial killers, which will be explored in more detail below.

40.6 Serial Killers

Serial killers produce complex murders. These murders often have several unusual features. The most recognizable factor is multiple occurrences – at least two, and in many cases the serial nature of the murders is not recognized until three or more victims have been identified. Second, a serial killer will normally never stop killing until they are stopped somehow, by being apprehended or becoming so ill or elderly that it is not possible for them to continue. Third, the interval between killings is variable. Sometimes they will kill two or three times in quick succession, then go through a cooling down period in which they do not kill for days, months, or even years. For a true serial killer, however, the urge to kill will invariable arise again.

The fourth unusual feature of serial killings is that the victim is usually a stranger to the killer. In most "normal" murders the victim is acquainted with the perpetrator, and may even know them well, as in the case of murder of a spouse. In a serial killing on the other hand, the killer and their victim usually have not had any previous contact. Rather than choosing the victim as a person, the serial killer will usually choose them as a symbol of some person or group of people against whom the serial killer has hostility. Fifth, serial killers almost always work alone. There are some cases where two or more people have carried out a series of killings as a team, but this is very rare. Sixth, the motivation for the murder is neither gain nor the direct out-come of an event. Instead, a serial killer simply wants to kill. It seems to be almost a physical need for them. This need may arise out of what we might consider a normal desire, such as for power or sex, or it may be entirely fantastical, arising from the imagination of the killer.

It is these unusual features of a serial killer that makes them candidates for psychological profiling as well as so darned interesting. Motivations so far outside our normal understanding capture our imaginations and we seek to understand how such abnormal thought processes can come about.

40.6.1 Characteristics of Serial Killers

At the risk of trampling on the subject matter of the next chapter, here are the statistics of serial killers. Most serial killers are male, between the ages of 25 and 35 when they begin, and Caucasian. There is no pattern to their level of intelligence, social class, whether they are married or the equivalent, whether they have children, their occupation, whether they can hold a job, or whether they seem normal or strange to their acquaintances.

The victims of a serial killer are usually of the same race as the killer (even when the killer is not Caucasian), and more often female than male. There is no pattern to the age of the victim, either by itself or in relation to the age of the killer. Nor is there a pattern to the victim's social class, or any of the victim's other characteristics. As discussed above, the victim is almost always a stranger who is in some way symbolic to the killer.

40.6.2 Motivations of Serial Killers

Psychologists have identified four categories of motivation for serial killers. These differ from the two categories of motivation for normal murders described above. Though some authorities identify a "for gain" type of serial killer, of which an example would be a "hit man" for an organized crime organization, it seems that these serial killers fit the characteristics of "normal" murderers more closely than they do the characteristics of other types of serial killers. It is likely that not all serial killer can be placed into one of these categories, but at least the majority can.

The **visionary motive** refers to the situation where the directive to kill is perceived by the serial killer as coming from some external force or entity. The killer may in fact hear a voice or voices that tell him to commit the crime, and perhaps the type of victim to kill. Clearly, this reflects a severe mental illness.

The **missionary-oriented motive** was captured extremely well in the movie "Seven", which featured Kevin Spacey as the serial killer. Operating under this motive, the serial killer is compelled to rid the world of something he considers evil or immoral. For example, a serial killer operating under this motive may target prostitutes.

The **hedonistic,** or thrill-oriented **motive** is displayed by those serial killers who do it for the thrill. They are in it for what is to them the ultimate in fun and excitement – killing a human being. Thrill oriented serial killers genuinely enjoy the killing. The murders committed under this motivation may be especially brutal or sadistic.

The **power and control motive** is the most common, and may be found along with one of the other three motives in a "mixed motive" serial killer. This type of serial killer seeks to exercise power over his victims. This is closely related to the **lust motive**, because it is believed by many psychologists that sexual assault is often committed more as a way to exercise control and power over a victim than for purposes of sexual gratification. Another related observation is that a serial killer may receive sexual gratification from the killing, even if there was no overt sex act involved. Power and control or lust motivated serial killers are apt to torture their victim. The more pain and domination they can inflict, the more gratification they receive.

40.6.3 Organized vs Disorganized Serial Killers

Serial killers can be categorized as organized or disorganized, or a combination of the two. **Organized serial killers** are usually indistinguishable from ordinary citizens. They dress well, may be married or have a significant relationship, they are friendly, they can hold a job, and so on. One characteristics of the way organized serial killer operate is that they often trick their victims into accompanying them by seeming friendly or even by impersonating a police officer.

The **disorganized serial killer** is similar to what we tend to think a serial killer should be like. These people are usually loners, and their appearance is often typical of what we might consider a "loser". He usually has trouble with relationships and trouble keeping a job. He is often sloppy and personally unkempt. Disorganized serial killers are less intelligent as a group than are the organized type.

Of these two types of serial killers, the disorganized type is much easier to catch. His sloppiness will often lead to his leaving a greater amount of evidence for authorities to recover. Besides, they often look like they could be serial killers, which prompts citizens to report them to authorities. Organized serial killers are much more difficult to apprehend, and may continue for years before finally making a mistake that allows them to be identified. While there might be slightly more disorganized that organized serial killers, the disproportion is not very great, and most authorities consider them to occur with about equal frequency.

40.7 Behavioral Evidence

Now, let us explore the types of evidence that psychological profilers work with. A great deal of the results of psychological profiling are, in my opinion, the result of simple deductions, not requiring any special expertise in order to draw a inference about the perpetrator or their motivation. The rest, as a package, is referred to as behavioral evidence. **Behavioral evidence** consists of inferences drawn from consideration of how the crime was committed and what else was done along with the crime. The behavioral evidence may reveal something about the perpetrator's psychology.

Let's look at some forms of behavioral evidence in murder cases. There are several ways in which to kill a person. Stabbing and strangling are examples of personal methods of killing someone. If one of these methods is used, there is little doubt that the murderer intends to kill the victim. Further, the murderer is killing in a way that shows their strength and power over the victim, and leaves no doubt in the victim's mind that they are being killed and who is killing them. This reveals that there is antipathy, perhaps only symbolic, toward the victim. In contrast, using a bomb, a fire, or a rifle from a distance is an impersonal method of killing. The victim will generally not know the identity of their killer, and, until the final moment, even that they are being targeted. Impersonal methods indicate that there is antipathy (again, possibly symbolic) toward some category of people, or perhaps some ideal or principle, rather than toward any of the individual victims.

The way in which the murder was committed also provides behavioral evidence. Multiple small injuries may indicate a desire to make the victim suffer. Multiple major or fatal injuries probably reflects deep rage toward the victim. The nature of the implement used in the murder may also be important. For example, if the implement used was brought from outside the crime scene, the murder is likely to have been premeditated. If, however, the implement was something that was at the crime scene before the murder it may reflect a spontaneous act.

Other actions taken by the murderer may be informative. If the body is mutilated, this may reflect the desire to completely obliterate the person or their identity. If part of the body was eaten, this may reflect the murder's desire to "own" the victim or incorporate some of the victim's characteristics into themselves. Sexual assault often reflects a lust motive. Clues left behind, such as writing on the wall or a letter, may reflect a desire for notoriety. Alternatively, it may reflect the desire of the criminal to engage law enforcement in a perverted form of "catch me if you can" game.

40.8 Victim Psychology

Attempting to gain knowledge about a perpetrator by studying the characteristics of their victims is known as **victimology**. Victimology will be discussed further in the next chapter, but it is necessary to mention here that victimology provides an essential perspective on a murderer, especially a serial murderer. The characteristics of the victim may reveal the motive behind the killing as well as something about the personal history of the murderer. For serial killers, psychological profilers assume that the victim is symbolic in some way of some things or events in the killer's life.

40.9 An Example: Jack the Ripper

Seemingly no discussion of psychological profiling is complete without a consideration of Jack the Ripper (JTR). Every single source I read in my research on psychological profiling used the JTR case as an example. I am not sure of the reason for this. Following this tradition, and fully aware that this chapter is getting rather long, I will present the major facts of the case.

JTR was a serial killer who was active for a brief period during the late 1800's in London, England. JTR killed five women, possibly more, all prostitutes. In most of the cases, the murders were performed quickly, in a location that was at least partly in the open. In all cases a knife was used, and the bodies were mutilated to one degree or another, including removing the uterus from two victims. The murders escalated in brutality, with each somewhat more savage than the previous one. JTR sent at least one letter, and possibly more, to police authorities bragging about what he had done, including eating parts of at least one body, and challenging them to catch him.

At the time, psychological profiling was in its infancy, if it can be said to have existed at all. Dr. Thomas Bond, the surgeon who performed autopsies on some of the victims, inferred the following characteristics of JTR, based on the behavioral evidence. First, all five murders had been committed by one person, working alone, who was physically strong, cool, and daring. He further commented that JTR was likely quiet and inoffensive in appearance, middle-aged, and neatly attired. Bond's analysis of some of the characteristics of JTR reveal the misconception, still common today, that all serial killers are of the disorganized type. He characterized JTR as a loner, without a real occupation, eccentric, and mentally unstable. Further, he speculated that JTR might suffer from satyriasis, the condition in which a man is considered to be overly interested in sex. In the 1800's satyriasis was considered a sexual deviancy, though today psychologists are reluctant to say that there is anything deviant about an interest in sex – even a strong interest in sex. Finally, Bond suggested that those who knew JTR would be aware that he was not right in his mind.

Psychological profiling has matured considerably since the late 1800's. Today, psychological profilers would infer several things differently. One observation, from the victimology, is that all the victims were prostitutes. This may reflect a missionary motive, as in trying to rid the world of this form of immorality. From the mutilations and nature of the stabbings we would infer rage toward the victims and a desire to obliterate them. The method was by stabbing, which is a personal method, indicating that JTR had a personal connection to the victims symbolically. We do not really know whether his antipathy was toward women in general, toward female sexuality, or specifically toward prostitutes. If he ate parts of the bodies this may reflect a desire to own them, or even to be more like them, perhaps even to the extent of wishing to be a woman. Some psychological profilers see this as evidence of a lust motive, but since none of the women seemed to have been overtly sexually assaulted, caution is advised in this assessment. The letters to the police seem to be a classic case of inviting them to participate in what to JTR must have been an exciting intellectual game.

Today we would probably assess JTR's personal characteristics somewhat differently from Bond's profile. The fact that the murders were quickly accomplished in partly open areas allows us to logically agree with Bond that JTR was strong and daring. Psychological profilers today would be more likely to regard him as a person with a history of failed relationships with women, perhaps indicating that he was divorced and perhaps middle aged. His actions seem to speak of a person who received little respect, due possibly to being of lower social class; and who envied women the ability to be paid (a form of respect) for sex. The fact that he was never apprehended suggests to most psychological profilers that he was an organized type of serial killer, who, in contrast to Bond's opinion, was probably not noticeably different from the rest of the citizens of London at the time.

40.10 Questions for Study and Review

1. What are the three types of criminal profiling?
2. How is a criminal's profile used in a forensic investigation?
3. Is it possible for evidence to be interpreted differently by different profilers?
4. For which types of cases is profiling most useful?
5. What is the definition of aggression?
6. What distinct types of aggression do psychologists recognize?
7. What are the two categories of "normal" motives for murder? What sorts of motives fall into the "for gain" category?

8. Describe the differences between mass murderers, spree killers, and serial killers.
9. List the unusual features of serial killings and serial killers.
10. I am a 53 year-old Caucasian male, who is highly educated and therefore presumably quite intelligent, divorced with a (former) stepdaughter, and a professor who has had the same job for 17 years at the time these words are being written. Perceptions of me differ, in that some people consider me normal and others consider me completely weird (I agree with the latter group). Is it possible that I am a serial killer? Why or why not?
11. If I were a serial killer, what would be the most likely profile of my victims?
12. What are the four common types of motives for serial killings?
13. Describe the differences between an organized and a disorganized serial killer?
14. What is behavioral evidence? Can you give some simple examples?
15. What does the statement that a serial killer's victim is often "symbolic" mean?
16. In looking at the Jack the Ripper case, do you see anything that needs to be added to JTR's profile because Dr. Bond or more modern psychological profilers missed it?

40.11 Sources

[Anonymous]. n.d. The Importance of Victimology in Criminal Profiling [Internet]. [cited 2007 Oct 28]. Available from:
http://www.angelfire.com/weird/flash333/killers/victimology.html.

[Anonymous]. n.d. Serial Killers [Internet]. [cited 2007 Oct 28]. Available from:
http://www.macalester.edu/psychology/whathap/UBNRP/serialkillers/serialkillers.html.

[Anonymous]. 2007. Inside the mind of the mind hunter: an interview with legendary FBI agent John Douglas.(Federal Bureau of Investigation)(Interview). The Forensic Examiner 16(1): 10(4).

Brantley AC, Ochberg FM. 2003. Lethal Predators and Future Dangerousness [Internet]. FBI Law Enforcement Bulletin 72(4). [cited 2007 Oct 19]. Available from: http://www.fbi.gov/publications/leb/2003/apr2003/april03leb.htm#page_17.

Britton V. n.d. Jack the Ripper: Case Study [Internet]. [cited 2007 Oct 28]. Available from: http://www.suite101.com/lesson.cfm/18593/1952/4.

Decaire M. n.d. Aggression Types and Criminal Behavior [Internet]. [cited 2007 Oct 28]. Available from:
http://www.suite101.com/article.cfm/forensic_psychology/17707.

Gado M. 2007. Bad to the Bone [Internet]. [cited 2007 Oct 28]. Available from:
http://www.crimelibrary.com/criminal_mind/psychology/crime_motivation/1.html.

Hickey EW. n.d. Serial Killers: Defining Serial Murder [Internet]. [cited 2007 Oct 28]. Available from: http://www.serialhomicide.com/serial-killers.htm.

Kelly D. 2006. Antisocial Personality Disorder [Internet]. [cited 2007 Oct 28]. Available from: http://www.ptypes.com/antisocialpd.html.

Kocsis RN, Cooksey RW, Irwin HJ. 2002. Psychological Profiling of Offender Characteristics from Crime Behaviors in Serial Rape Offences. International Journal of Offender Therapy and Comparative Criminology 46 (2): 144-169/

Scott SL. 2007. What Makes Serial Killers Tick? [Internet]. [cited 2007 Oct 28]. Available from: http://www.crimelibrary.com/serial_killers/notorious/tick/victims_1.html.

Turvey BE. 1998. Deductive Criminal Profiling: Comparing Apllied Methodologies between Inductive and Deductive Profiling Techniques [Internet]. [cited 2007 Oct 29]. Available from: http://www.criminalprofiling.ch/article2.html.

Turvey BE. 2002. Criminal Profiling : an Introduction to Behavioral Evidence Analysis. Burlington (MA): Elsevier Science & Technology Books.

Vaisman-Tzachor R. 2006. Psychological Profiles of Terrorists. The Forensic Examiner 15(2): 6(12).

Wikipedia contributors. 2007. Jack the Ripper [Internet]. Wikipedia, The Free Encyclopedia; 2007 Oct 26, 16:24 UTC [cited 2007 Oct 28]. Available from: http://en.wikipedia.org/w/index.php?title=Jack_the_Ripper&oldid=167241341.

Wikipedia contributors. 2007. Serial killer [Internet]. Wikipedia, The Free Encyclopedia; 2007 Oct 28, 15:34 UTC [cited 2007 Oct 28]. Available from: http://en.wikipedia.org/w/index.php?title=Serial_killer&oldid=167644409.

Chapter 41
STATISTICAL AND GEOGRAPHIC PROFILING

This chapter will complete the series on criminal profiling by presenting statistical and geographic profiling. The chapter will begin with an overview of statistical profiling, in which the statistics of various types of offenders are compiled and analyzed. Next, we will take up the subject of victimology, to see how the characteristics of the victim can prove to be useful knowledge in an investigation. The chapter will conclude with a discussion of geographic profiling, which uses theories and models primarily from ecological studies of predator/prey relationship to narrow down what a serial criminal considers to be their home territory.

41.1 Statistical Profiling

Statistical profiling is referred to as inductive profiling by some authorities, because it is based on a form of the scientific method called induction. Induction is arguing from general observations to a specific case. For example, if we know that, in general, most serial killers are male, then it makes sense that the serial killer in a specific case is more likely to be male than female. Statistical profiling is done by compiling statistics on offenders. These statistics are then analyzed for patterns. When a certain type of offense is committed, these statistics are consulted to build a statistical profile. A statistical profile may suggest, for example, that the offender in a particular case is 80% likely to be male, 70% likely to be a high school dropout, 50% likely to be under 25 years old, etc. Note that this approach makes less use of drawing inferences from the details of behavioral evidence, though certain methods of committing a crime are candidates for statistical profiling as well. The FBI prefers this method of profiling.

The National Crime Prevention Council of Canada did a study of statistics of certain offenses, which they have made available to the public, and the following results are from their study, published in 1996. Statistics for the United States are likely to be slightly different, but probably not profoundly so.

41.1.1 Statistics of Violent Offenders

These statistics are from studies involving Canadian federal male offenders who were convicted and sentenced to two or more years of incarceration.
- 1/3 had a history of committing family violence (spouse abuse, child abuse, etc.).
- Almost ½ had been a victim of child abuse (physical, sexual, psychological, neglect) or had witnessed family violence as a child or adolescent.
- 69% had some degree of hearing loss. This is nearly 10 times the rate (7%) of hearing loss in the general population.

- The prevalence of major mental disorders (i.e. schizophrenia, major depression, bipolar disorder) within these populations considerably exceeds that in the general population. Only 48% of inmates with a major mental disorder had reported their symptoms to a physician or a mental health professional.
    - Psychotic disorders - 10.4%
    - Psychosexual disorders - 24.5%
    - Depressive disorders - 29.8%
    - Anxiety disorders - 55.0%
- Approximately 75% to 80% were persistent offenders in their youth.
- Most violent offenders were undereducated, although the I.Q. distribution among the inmate population is not significantly different from that of the general population. Approximately 65% of offenders tested at lower than a Grade 8 completion level and 82% tested lower than Grade 10.
- 55% of offenders reported that they were under the influence of alcohol, drugs or both on the day they committed the offence(s) for which they were incarcerated.
- Approximately 50% of the offender population suffered from some type of substance abuse problem.

41.1.2 Statistics of Young Offenders

It has been clearly demonstrated that there is a solid relationship between school experience during early and late adolescence and criminality. A 1993 Canadian study found school performance to be the best and most stable predictor of adult offending. The authors of this study theorize that poor school performance and a weak attachment to school will increase the probability of misbehavior in school that, in turn, provokes disciplinary reactions. This escalates through elementary and secondary school, leading to a higher level of adolescent delinquency and, eventually, to adult offending.

The 1996 study from which the statistics of violent offenders presented above are drawn found that the younger the age of the criminal career's onset and the more serious and extensive the offender's juvenile crime record, the greater the likelihood that the offender will remain criminally active as an adult. Further this study found that offenders who first tried alcohol in their preteens became involved in illegal activities at a significantly younger average age (15.8 years) than those who first tried alcohol as teenagers (18.8 years).

All studies of youthful offenders agree that there are two categories of crimes in which youthful offenders engage. One category can be considered a "normal" part of adolescence, and consists of occasional minor crimes. It is estimated that about 80% of young people have committed these types of crimes as I did occasionally in my youth, such as shoplifting a candy bar from a grocery store. The second category is much more worrisome, and consists of crimes that are habitual and often harmful to someone. One example is common bullying behavior. It appears that these crimes often continue through into adulthood.

A study of 10,000 boys born in Philadelphia in 1945 found that less than 7% of the sample were responsible for nearly 70% of all crimes attributed to the 10,000. This supports the often asserted idea that there is a small element of youthful offenders who are the ones who commit the majority of crimes attributed to young people.

Here are some statistics that apply to youthful offenders.
- Neglect (low levels of parental involvement and supervision of child) is strongly linked to criminality.
- The probability of a boy becoming a childhood offender more than doubles if he has an older member of the family who has been convicted of a criminal offence.
- Among the strongest predictors of delinquency in boys are aggression, drug use and stealing.
- Even though homeless youths constitute a relatively small proportion of all adolescents, they are involved in a substantial and disproportionate share of crime.
- The deeper the level of poverty, the higher the incidence of violence among children. 14% of the poorest boys were violent, compared to 5% of boys who lived in the wealthiest areas. 5% of very poor girls and 1% of the most well-to-do girls, committed acts of violence.
- Some newer studies point a finger toward substance abuse by the mother while pregnant with the child being strongly correlated with the child offending when older.

41.1.3 Statistics of Female Offenders

Statistically, fewer women than men commit crimes, and the crimes women commit are less likely to be serious. For example, a 1992 Canadian study found that women committed only 16.4% of all crimes committed by adults. Here are some other statistics of female offenders.
- About 55% committed the crimes of petty theft, fraud and prostitution.
- 75% had a middle school level education or below, and 40% were classified as functionally illiterate.
- 2/3 had children. Most of these women were the primary caregivers for their children prior to their incarceration, and consequently many of these children ended up in state care.
- 43% had substance abuse or addiction problems and 69% had indicated that drugs and/or alcohol played a major part in their offence and/or their criminal history.
- Approximately 80% had experienced physical and/or sexual abuse.
- 59% exhibited self-injury (such as slashing herself).

41.1.4 Statistics of Sex Offenders

Sex offenders exhibit some statistics that can lead to a statistical profile.
- Over 40% had a history of sex-related offenses as juveniles.
- 80% had less than a high school education, and ½ had less than grade 10.

- More than 50% were found to be unstable in their employment pattern. 65% were unskilled laborers.
- 60% were separated from their biological parents before age 16.
- More than 1/3 had been abused (physically or emotionally) by their parent(s). This is less than for violent criminals in general.
- 1/3 had been sexually abused before the age of 16.
- 1/3 had suffered severe emotional problems prior to the current offence.
- 3/4 had an adult history of alcohol abuse, 2/3 had a history of drug abuse.

41.1.5 Statistics of Robbers

Robbery is the form of theft in which the offender confronts the victim, demanding money or possessions. This differs from other forms of theft, such as larceny or burglary in which the victim is not directly confronted. The following statistics come from Canadian studies, but the statistics for the U.S. are stated to be approximately the same.
- Nearly 100% of robbers are male.
- 2/3 are younger than 25 and virtually none are older than 50.
- There is no more likely to be a background of violence than for the general population.

41.1.6 Some Summary Comments

One common statistic for all of these offenders, adult, youthful, male, or female, is a high percentage who are undereducated. The studies discussed above, which link criminality with a poor experience in school are undoubtedly correct. One wonders whether more might be done in the realm of intervention when children are clearly disconnected from their school and performing poorly. It may be that a small investment of time and money early in the lives of these children may save them from a miserable adult life, while improving society in general.

Another common statistics for many of the types of offenders examined is poverty. Poverty often arises, of course, from a poor educational background. Again, the potential for improvement of these conditions seems obvious.

41.2 Victimology

In the last chapter I briefly introduced **victimology**, the examination of the victims of a crime in order to gain information that may reveal something about the offender. The theory underlying victimology is that criminals (not just serial killers as discussed in chapter 40) choose their victim for a reason. The reason may be psychological, circumstantial (the victim was in the wrong place at the wrong time), or may actually be something to do with the characteristics of the victim.

The characteristics of a victim that lead an offender to choose them are usually neither obvious nor often under the control of the victim. Victimology does not blame the victim, nor does it focus exclusively on obvious things like rapists preferring young women who park their cars in dark alleys. Much of the time, the characteristics are more subtle. Here are some characteristics that victimologists find useful in at least some cases.

- Family Background;
- Reputation;
- Likes and dislikes in food, entertainment, clothing, music, etc.;
- Drug or Alcohol abuse;
- Financial difficulties;
- Routines, such as being at a certain location at a certain time regularly.

In the case of a murder, investigators and victimologists will want to know who was the last person(s) the victim spoke to, and the circumstances in which they did so; and whether the victim had any known enemies or there exists any known reason why someone would have wanted to kill the victim.

Victimology is helpful to profilers and investigators, even beyond the information it provides about the offender. Victimology can help investigators to devise strategies for drawing the offender out into the open. For example, if a certain robber chooses middle aged, overweight males, the police might use one of their officers who fits this description as "bait". Further, victimology may provide some insights that are useful when interviewing potential suspects. For example, if the police know that a certain robber chooses middle aged, overweight, male victims, then they might ask a question such as, "Do you think your father owes you something?"

Victimology also provides a way to predict who the next victim of a serial criminal might be, so that appropriate warnings can be issued to law enforcement agencies and citizens. Again, this goes beyond the obvious, such as petite blondes should stay home at night, and might equally likely be something like, businessmen should not carry a briefcase on Wall Street until we catch this crook.

41.3 Geographic Profiling

Geographic profiling is the newest branch of profiling. It seeks to determine where the "home" of the criminal is, by examining the pattern of where the crimes were committed. The term "home" in this context, may be where the criminal actually lives, or it might be where they work or otherwise feel secure. The important point is that it is the place where the criminal feels most comfortable. Geographic profiling relies on knowing the locations of several crimes scenes, therefore it is only useful in cases of serial crimes.

41.3.1 The Theoretical Basis of Geographic Profiling

The underlying theory of geographic profiling treats criminals as intelligent predators. As such, their behavior can be modeled using formulas that ecologists and animal behaviorists have devised to describe and model predator-prey relationships. In particular, a certain family of models called optimal foraging models, seem to work fairly well. **Optimal foraging models** seek to explain regularities in how certain species or individuals forage (obtain food). They assume that the animal is capable of making choices based on perceived costs and benefits.. Optimal foraging models begin with the assumption that predation has both costs and benefits, that depend in part on circumstances such as the location of the prey. A predator has to balance the costs and benefits of hunting a certain prey in a certain location. The predator then chooses to hunt in a certain location based on the balance of costs and benefits.

For a human criminal, say a serial killer, the benefit is in getting to carry out of the act of killing. The costs are more complex and involve several factors that influence how far from their "home" the criminal will look for their victim. The primary cost for the criminal is the possibility of getting caught. The farther from "home" the greater the cost, not only in terms of the time to get there, but also in terms of being in unfamiliar territory where the criminal feels less safe and perceives the chance of getting caught as greater. In familiar territory, the criminal will know good places to do the killing and perhaps good places to dispose of the body, but these will not be known in unfamiliar territory.

On the other hand, doing the crime too close to home is costly because it may focus law enforcement attention on that area. Therefore, the criminal balances the cost of being in unfamiliar territory against the cost of being too close to home, and reaches a compromise that dictates a set of areas they will focus on.

41.3.2 How Geographic Profiling Works

Geographic profiling methods look at where the crimes were committed, apply the statistical and mathematical models, and come up with a probable location for the criminal's home. In the simplest case, a map of the locations of the crimes will look like a donut. The criminal's home will be located within the donut hole, an area in which no crimes have been committed, whihc is at the center of a ring of locations where crimes have been committed. In other cases the pattern of where the crimes have been committed is more complex than a simple donut, and the models may suggest two or more locations that might be the criminal's home territory.

The end result of geographic profiling is a map, for which areas are coded as having certain levels of probability of containing the criminal's home. This map can then be used by law enforcement agencies to focus their attention on certain areas.

It is also possible to use geographic profiling in reverse, to predict where the next incident might occur. This method also generates a map, in this case showing probabilities of where the serial criminal is likely to strike next.

41.3.3 Utility of Geographic Profiling

The greatest utility of geographic profiling is in sorting between suspects. Often, police will have many suspects in a crime. Using geographic profiling, they can focus on suspects that live or work within an area identified by the geographic profile and spend less time on suspects that live or work outside that area. Geographic profiling can also be another tool for preventing crimes, in those cases where the probability of the criminal striking next in certain locations can be estimated.

41.4 Questions for Study and Review

1. What is the main idea of statistical profiling?
2. Sociologists and anthropologists often argue that education prevents crime in that better educated people are less likely to be criminals. Do the statistics on the several types of offender presented in this chapter offer any support for this idea?
3. In contrast to sociologists and anthropologists, psychologists sometimes argue that the issue for certain types of criminals is that they fail to become engaged with school and with other students at school, which leads them to be more likely to engage in criminal activities. Is there any support for this idea in the theories presented in this chapter?
4. Compare and contrast the statistics of female offenders with those of (male) violent offenders.
5. Explain what victimology is and how it is used in an investigation.
6. What costs and benefits should a criminal weigh in deciding where to commit a crime?
7. Geographic profiling usually produces a map. What does this map show, and how is this information used in an investigation?
8. How can geographic profiling be used to predict where a serial criminal will strike next?

41.5 Sources

[Anonymous]. n.d. The Importance of Victimology in Criminal Profiling [Internet]. [cited 2007 Oct 28]. Available from: http://www.angelfire.com/weird/flash333/killers/victimology.html.

Bureau of Justice Statistics. n.d. Sourcebook of Criminal Justice Statistics [Internet]. [cited 2007 Oct 29]. Available from: http://www.albany.edu/sourcebook/.

Bureau of Justice Statistics. 2007. Criminal Offenders Statistics [Internet]. [cited 2007 Oct 29]. Available from: http://www.ojp.usdoj.gov/bjs/crimoff.htm.

Environmental Criminology Research Inc., 2001. What is geographic Profiling. Online at <http://www.ecricanada.com/geopro/>

Laukkanen M. Santtila P. 2006. Predicting the Residential Location of a Serial Commercial Robber. Forensic Science International 157(1): 71(12).

Le Comber SC, Nicholls B, Rossmo DK, Racey PA. 2006. Geographic Profiling and Animal Foraging. Journal of Theoretical Biology 240 (2): 233-240.

National Crime Prevention Council of Canada, 1995. Offender Profiles [Internet]. [cited 2000 Jan 7]. Available from: http://www.crime-prevention.org/ncpc/publications/children/ offpro_e.htm.

Petherick W. 2007. Criminal Profiling: How it Got Started and How it Is Used [Internet]. [cited 2007 Oct 29]. Available from: http://www.crimelibrary.com/criminal_mind/profiling/profiling2/1.html.

Pritchard J. 2001. Statistics Just One Aspect of Profiling [Internet]. Cincinnati, The Enquirer, Sunday January 14, 2001. [cited 2007 Oct 29]. Available from: http://www.en

Kocsis RN. 2006. Criminal Profiling: Principles and Practice. Totowa (NJ): Humana Press.

Chapter 42
FORENSIC ACCOUNTING

This chapter will explore a traditional forensic science that is proving to have new relevance in the modern era, forensic accounting. Forensic accounting applies the accounting, auditing, and investigative skills of the accounting profession to investigations of financial crimes. It has long been associated with the enforcement of laws regarding "white collar crime", more accurately known as occupational fraud, and the subject of a coming chapter. Forensic accountants contribute to both criminal and civil cases. In recent times, forensic accounting has emerged to provide expertise in spotting financial irregularities and tracing the flow of money, which are among the more successful tools in the investigation of organized crime and terrorists. There's an old saying, "Follow the money," which is as relevant today as it ever was.

The bulk of the chapter will focus on the traditional practice of forensic accounting, and the chapter will conclude with a brief look at some of the newer areas to which this specialty contributes. This chapter is foundational for the chapter on occupational fraud, and is also relevant to the chapters on organized crime and homeland security.

42.1 The History of Forensic Accounting

Forensic accounting is another field that could be argued to have ancient roots. Surely, as soon as business and investing appeared in the early civilizations, records were kept, and somebody emerged with the skills to examine the records and detect evidence of wrongdoing. For example, we know that the scribes of ancient Egypt kept track of the Pharaoh's gold and other possessions. In its modern sense, forensic accounting emerged out of corporate and governmental financial investigative fields in the early 1800's.

The first "official" case involving forensic accounting occurred in 1817, when an accountant was charged with the valuation of an estate in a bankruptcy case. Known more commonly as "auditors", forensic accountants have been prominent in the U.S. in the corporate and governmental spheres since the 1940's.

42.2 The Practice of Forensic Accounting

Forensic accountants examine financial records for evidence of irregularities. If the situation warrants, they may take stock of inventories as well to see if they match the financial records. The unit of importance in accounting is the transaction. A **transaction** is an event that results in the movement of money or goods from one account or status to another. For example, selling a widget is a transaction. It results in the movement of a widget from the seller's inventory to the buyer's inventory, and the movement of money from the buyer's account to the seller's account. Transactions of all types are recorded, and form the fundamental financial records that forensic accountants work with. Transactions need to be recorded accurately, and the arithmetic involved with the movement of items and money needs to be done accurately.

Organizations that earn or handle money, which includes businesses, corporations, not-for-profit entities, and even ordinary people, are required to publish statements of their earnings or losses with one or more government agencies. We are all familiar with income tax, and the fact that some version of a 1040 form needs to be filed with the Internal Revenue Service each year. For organizations of all types there are other agencies to which reports of gains, losses, and other information must be made as well. These **statements** are required to be accurate, and are another fundamental type of financial record that forensic accountants work with.

Forensic accounting is normally done in one of two possible contexts. The first is **investigative accounting**, in which the financial records of some organization are examined to make sure that everything has been done correctly. The other context is called **litigation support**, and applies when there is already a criminal or civil case in progress and the questions to be answered are about the nature of the wrongdoing and the amount of money involved.

When an examination of financial records is completed, there are three possible outcomes. An outcome of **normal** means that all of the financial records are accurate and in order, and that all required statements have been filed accurately. Alternatively, the forensic accountant may determine that **mistakes** have been made in the records of transactions or in the filing of statements, but that there is no evidence of criminal intent involved. A finding of **abnormal** means that something in the transaction records or statements is intentionally or deliberately inaccurate or improper, which usually implies that there has been some criminal activity or that there are grounds for civil action. Usually, the difference between an outcome of "mistakes" and an outcome of "abnormal" is the pattern and context of the inaccuracies. Unintentional errors are not a criminal offense, though the mistake has to be corrected, which might result in owing more (or less) money or even in a penalty. Systematic errors in transaction records, however, or long term trends involving the inaccurate reporting of data to regulatory agencies are likely to be evidence of a crime.

42.3 Things Forensic Accountants Look For

An examination of the sorts of things forensic accountants look for will be informative. As usual, I am not attempting to be complete or thorough here, but to simply to impart the general idea of what forensic accountants consider "interesting".

42.3.1 Discrepancies in Financial Records

There are many types of discrepancies in financial records. Some of those that are most indicative of possible criminal activity include those shown in table 42.1.

Table 42.1: Common Discrepancies in Financial Records

Account balances that are significantly overstated or understated
Transactions not recorded in a complete or timely manner
Transactions improperly recorded as to amount
Transaction assigned to an incorrect accounting period
Unsupported or unauthorized records, balances, or transactions
Conflicting transaction records or statements
Missing documents
Only the photocopies of a document are available
Missing inventory or physical assets
Excessive voids or credits
Alterations on documents (e.g. back dating)
Duplicate transactions

42.3.2 Unusual Behavior

All investigators know that one of the biggest clues that some criminal activity has occurred is unusual or unexpected behavior when the investigator shows up. In the realm of forensic accounting, unusual behavior is typically behavior designed to prevent or delay access of the investigator to information of some form. Some of the more common forms of unusual behavior that are encountered include inconsistent or vague responses to questions; denying or delaying access to records or facilities; denying or delaying access to certain employees, customers, or vendors; and unusually long delays in providing requested information.

42.3.3 A Few "Red Flags" (Symptoms of Fraud)

Forensic accountants have identified a wide range of features of financial records that are not discrepancies, but are unusual. The presence of one of these features may indicate the fraud has occurred. These features are often referred to as **red flags** and are of significance if found repeatedly in financial records. A few red flags are: recurring transactions for identical amounts from the same vendor or customer; unusual purchases not consistent with prior trends; an increase in labor force overtime not justified by production or sales volume; consistent shortages in cash on hand; and frequent undocumented and/or unapproved adjustments (credits, writeoffs, etc.) to bills or sales slips.

42.4 Investigative Skills of Forensic Accountants

One thing that is often overlooked about forensic accountants is their need for and use of general investigative skills. In many cases, the difference between determining whether discrepancies in financial records are mistakes or evidence of criminal activity; or whether the existence of red flags is symptomatic of fraud or simply due to circumstances, rests on evidence external to the financial records themselves. We have seen this pattern in other forensic sciences, ranging from death investigation to forensic hypnosis. In particular, forensic accountants need the skills to find out information about companies, the employees of companies, and the companies' customers and suppliers.

In addition, good forensic accountants need to have an understanding of human psychology, in order to understand what situations may prompt someone to commit fraud. This allows them to focus their investigations on those people who are most likely to be doing something improper.

Sometimes, general investigation will reveal a close association between suppliers or customers and key people in the company. If, for example, supplier A and employee B are friends, and it is discovered that supplier A's bills to the company are consistently adjusted, then a forensic accountant may become suspicious that A and B are perpetrating a fraud.

It is common for companies that are trying to hide financial wrongdoing to use a large number of different banks, so that none of them can see the company's entire financial picture. It is no surprise that an employee who is having financial difficulties, is a drug user, or has a gambling problem is more likely to be defrauding the company than an employee free of these problems. In other cases, investigation may reveal that the company is deeply in dept or is under pressure to merge, which may set the stage for the company to engage in shady practices.

In addition to these investigative skills, as is true for all forensic scientists, forensic accountants need to develop the skill of simplifying their evidence, financial information in this case, in order to make it understandable to lawyers, judges, and juries.

42.5 Specialties in Forensic Accounting

As is the case for some of the other forensic sciences, forensic accounting is extremely broad, and individual forensic accountants will often develop a specialty in a certain type of investigation. Some of these specialties are insurance claims fraud, occupational fraud (about which I will discuss more in a coming chapter), construction fraud, or investment fraud. Other forensic accountants will specialize in a certain segment of the financial world, such as automotive sales and repairs, real estate, or even environmental organizations.

42.6 Trends and New Directions in Forensic Accounting

The practice of forensic accounting extends beyond enforcement of laws related to companies and finances, to include a host of other interesting applications. One trend in recent times is for a company or concerned individual to hire a forensic accountant to do a financial investigation of a company. For example, if a person is considering making a large investment in a company, she or he may hire a forensic accountant to do an investigation to make sure that the company's financial statements (which are likely what attracted the investor to the company) are accurate and that the company is not engaging in any fraudulent practices. When one company is considering a merger or buyout of another company, they may hire a forensic accountant to verify the financial status of the company under consideration. A company may even hire a forensic accountant to do an investigation of another company whose merchandise they are considering buying.

A concerned individual or group of them in a company may hire a forensic accountant if the company has fallen into financial difficulties, to see if some fraud lies behind the trouble. It is apparently also customary to have a forensic accounting of a company before it issues stock. In cases of company partner and shareholder dissents, a forensic accountant may also be consulted.

There are even some individuals who consider marriage a financial arrangement and may hire a forensic accountant to find out their potential spouse's true financial status before saying, "I do." Divorces, likewise have profound financial implications, and it is typical for one or both of the parties to attempt to hide financial assets in order to prevent them from being divided. A forensic accountant can usually easily locate these.

Some new directions in forensic accounting include the investigation of terrorist and organized crime organizations. Terrorist organizations can not operate without money, and one of the best tools in combating them is to identify their sources of funding and shut them off. Also, tracing sources of funding may reveal the full extent of the organization and its supporters. In the case of organized crime, the questions more often concern how the money they acquire is being hidden, since it can not simply be reported to IRS as extortion income or the like. Making "dirty" money look "clean", i.e. as if it is from a legitimate source, is called **money laundering** and will be discussed further in the chapter on occupational fraud. Again, tracing the flow of money may reveal the full extent of the crime organization and its supporters.

Another new direction for forensic accountants is in patent and intellectual property infringement cases. If someone or some company makes money using materials patented or copyrighted by someone else, the amount of money they made directly from this source needs to be separated from the money made from other sources, and a forensic accountant can figure it out. Yet another new direction is related to the aging and eventual mortality of the baby boomer generation, which is resulting in increased rates of probate litigation in which the extent of the deceased's assets needs to be determined.

42.7 Another "Golden Team": Forensic Accountants and Forensic Computer Scientists

I discussed the "golden team" of forensic anthropologists, odontologists, pathologists, and DNA analysts in sections of previous chapters about death investigation and identification of deceased persons. I described this group as a "golden team" because of their effectiveness in accomplishing the job at hand. Now, let me introduce a new "golden team", which consists of forensic accountants paired with forensic computer scientists. In the modern era almost all financial records are stored on computers, including the records necessary to demonstrate that fraud has occurred and the manner in which it has occurred. Forensic computer scientists recover evidence from a computer system, using the methods of digital evidence recovery if necessary, and the forensic accountants are able to analyze the evidence so recovered. Their efficiency at recovering and interpreting evidence, when cooperating in this manner, is spectacular.

42.7.1 An Example: Data Mining for Evidence of Fraud

As an example of the power of the golden team, let me present one example – data mining for evidence of fraud. In general, **data mining** is the application of methods from computer science and mathematics to find patterns in massive data sets. Data mining is commonly used to identify trends in consumer behavior and preferences, which is why many supermarkets now have customer cards. With a customer card, the supermarket is able to track the items you buy, and even more importantly relationships between the things you buy. You can observe this anytime you use one of the major online shopping services, such as Amazon, which is always ready to inform you that people who bought the item you just ordered also bought several other specific items.

Many types of financial data and records can be considered massive data sets, and can be analyzed using data mining techniques. Some of these are stock market data, a company's total set of financial data, and bank transactions data. Using data mining techniques it is possible to spot patterns that are similar to the red flags described earlier. In this sort of investigation, the forensic computer scientist identifies patterns, and the forensic accountant interprets these patterns as evidence of fraud or as being suspicious.

42.8 Certifications for Forensic Accountants

As one might expect in such a robust and varied field as forensic accounting, there are a variety of certifications that a forensic accountant may, should, or in some cases is required to obtain. These certifications start with a basic Certified Public Accountant (CPA) license, which is administered by each state, and which is considered the minimum qualification for any accountant.

The Association of Certified Fraud Examiners offers the most common certification for forensic accountants, the certified fraud examiner (CFE) certification. The National Association of Certified Valuation Analysts administers some additional certifications that a forensic accountant may earn. These include CFFA: Certified Forensic Financial Analyst, CFD: Certified Fraud Deterrence Analyst, ABV: Accredited in Business Valuation, CVA: Certified Valuation Analyst, and ASA: Accredited Senior Appraiser.

42.9 The Future of Forensic Accounting

Forensic accounting is definitely experiencing growth and an abundance of employment opportunities. Accounting journals already identify a shortage of qualified forensic accountants, and some authorities predict exponential growth in the need for forensic accountants in the near future. Despite the clear job opportunities, I find that many of my students are reluctant to consider a career in forensic accounting. Some feel that it lacks the adventure of the more well known forensic sciences, though I believe that this is an inaccurate perception and that forensic accounting is as exciting as, for example, serology (and a lot less messy). Yet others fear that they lack the mathematical skills to succeed in accounting. I hope that this chapter has adequately conveyed the variety, excitement, and cutting edge nature of forensic accounting.

Another factor in the looming shortage of forensic accountants is a lack of college programs specifically in forensic accounting, and perhaps lack of awareness on the part of accounting and other business school faculty that this is an interesting specialty. I have spoken with one of my colleagues, a professor in the accounting department, who believes that all accounting is forensic accounting. Perhaps. There are, however, a growing number of colleges and universities that are offering specialized programs in forensic accounting, including master's degrees in the subject.

42.10 Questions for Study and Review

1. What is a "transaction", as the term is used by forensic accountants?
2. How does investigative accounting differ from litigation support accounting?
3. If a forensic accountant examined some financial records and found evidence of deliberate deception or fraud, what one word would they report in their findings to describe this situation?
4. How many of the common discrepancies in financial records can you list from memory?
5. What sorts of behaviors on the part of company representatives might rouse make a forensic accountant suspicious that fraud may be occurring?
6. What is the meaning of "red flag" in forensic accounting? What are some examples of red flags?
7. Besides their skills in accounting, what other skills are useful for forensic accountant?
8. Under what conditions might an individual who is not an agent of a law enforcement or regulatory agency hire a forensic accountant to investigate a company or a person?
9. How can forensic accounting contribute to the investigation of terrorism and organized crime?
10. Why is the combination of forensic accountant and forensic computer scientist described as a "golden team?"
11. Describe data mining.
12. Does it look like the demand for forensic accountants in the future will be less than at present, greater than at present, or about the same as at present?

42.11 Sources

[Anonymous]. 1996. Double-Entry Autopsy. The Economist (US) 340(7983): 77(1).

[Anonymous]. 2004. The Increasing Demand for Forensic Accountants: an Interview with Certified Forensic Accountant Joseph Siget, Jr. The Forensic Examiner 13(3): 51(1).

Apostolou N, Crumbley DL. 2005. Financial Statement Fraud: a New Ball Game. The Forensic Examiner 14(1): 39(4)

Bukics RML. 1996. Fraud Auditing and Forensic Accounting. Internal Auditor 53(3): 18(2).

Cauthen J. 2001. Investment Fraud [Internet]. FBI Law Enforcement Bulletin 70(5). [cited 2007 Oct 19]. Available from: http://www.fbi.gov/publications/leb/2001/may01leb.pdf.

Crumbley DL, Apostolou NG. 2005. The Expanding Role of the Forensic Accountant. The Forensic Examiner 14(3): 38(6).

Friedman D. 1997. Financial sleuth hot on the trail. U.S. News & World Report 122(5): 19(1).

Gold L. 2007. Forensic Accounting: CPA Gumshoes: Accounting Veterans and New Grads Alike Join Forensics Crusade. Accounting Today 21(8): 1.

Goodman B. 1995. Sleuthing Miami Accountant Pioneers Forensic frontier. Accounting Today 9(9): 14(2).

Kruchten GJ. 1999. The Bank Secrecy Act: A Powerful Weapon for Law Enforcement [Internet]. FBI Law Enforcement Bulletin 68(8). [cited 2007 Oct 19]. Available from: http://www.fbi.gov/publications/leb/1999/aug99leb.pdf.

Lewis PG, Gray B. 2006. Understanding Data Forensics: Electronic Evidence Encompasses a Vast Universe of Data. Bank Accounting & Finance 19(6):p36(4).

Martin J. 1998. Dissecting the books. Management Review 87(6):47(5).

McCue C, Stone ES, Gooch TP. 2003. Data Mining and Value-Added Analysis [Internet]. FBI Law Enforcement Bulletin 72(11). [cited 2007 Oct 19]. Available from: http://www.fbi.gov/publications/leb/2003/nov2003/nov03leb.htm#page_2.

Michaelson WM. 1996. Divorce: a Game of Hide and Seek? Journal of Accountancy 181(3): 67(3).

Miller JN, Jenkins HG, Houser JD. 2005. The Use of Forensic Accounting Techniques in the Determination of Intellectual Property Damages. The Forensic Examiner 14(4): 40(5).

Rezaee Z, Crumbley L. 2007. The role of forensic auditing techniques in restoring public trust and investor confidence in financial information. The Forensic Examiner 16(1): 44(6).

Schroeder WR. 2001. Money Laundering [Internet]. FBI Law Enforcement Bulletin 70(5). [cited 2007 Oct 19]. Available from: http://www.fbi.gov/publications/leb/2001/may01leb.pdf.

Schwartz ND. 1999. If the Numbers Look Fishy, Here's the Man to Call: Forensic Accountancy. Fortune 139(8): 466+(1).

Silverstone H, Sheetz M. 2007. Accounting and Fraud Investigation for Non-Experts. 2nd Edition. Hoboken (NJ): John Wiley & Sons.

Slotter K. 1999. The CPA's Role in Detecting and Preventing Fraud [Internet]. FBI Law Enforcement Bulletin 68(7). [cited 2007 Oct 19]. Available from: http://www.fbi.gov/publications/leb/1999/jul99leb.pdf.

Todd KJ. 2004. Using Digital Evidence to Ferret out the Dishonest Employee. Employee Relations Law Journal 30(2): p13(10).

Yockey DW. 1988. So You Want to Be a Forensic Accountant: Accountants Are Taking on a New Role: Expert Witness. Management Accounting (USA) 70(5): 19(3).

Zysman A. 1998. Forensic Accounting Demystified [Internet]. [cited 2007 Jul 13]. Available from: http://www.forensicaccounting.com/cphome.htm.

Chapter 43
Advanced Topics
OCCUPATIONAL FRAUD

The subject of this chapter is occupational fraud. I have designated it as an advanced topics chapter because it is not actually about one of the forensic sciences, but instead about an application of the forensic sciences, among which forensic accounting is the most obvious. Occupational fraud is crime, and one of the most common in our modern era. The chapter will begin with an overview of occupational fraud and its many manifestations. We will take a look at the most common types of occupational fraud before considering how forensic scientists become involved with its investigation. The chapter will continue with a brief look at the problem of counterfeiting and conclude with an introduction to money laundering that will be foundational to the chapters on organized crime and homeland security.

43.1 What is Occupational Fraud

Occupational fraud includes a large number of crimes that all involve stealing from a person or an organization or misusing the organization's resources for personal gain. Other names for occupational fraud are "white collar crime", which is more accurately one specific type of occupational fraud, "financial crime", and "economic crime". I have chosen to use the term occupational fraud because it is the term preferred by the Association of Certified Fraud Examiners (ACFE) that we encountered in chapter 42 as the organization that certifies most forensic accountants.

Within the category of occupational fraud there are some broad subcategories that are commonly discussed. What distinguishes these subcategories is who the perpetrator is and who the victim or victims are. **White collar crime** is the term used to describe the situation where the perpetrator is an employee or other member of an organization and the victim is either the organization or the organization's customers. The "payoff" for the crime, the illegal gains in other words, goes to the employee. **Corporate crime** is a term used to describe the situation where the perpetrator is a company or employees of a company and the victims are the company's customers or employees. In this case the "payoff" for the crime goes to the company. Government agencies are, unfortunately, not immune to occupational fraud, and **Governmental crime** is the term used for the equivalent of corporate crime when the organization is the government or its employees. In this case the "payoff" goes to the government agency.

Occupational fraud includes a huge number of misdeeds of various sorts. The ACFE has an enormous list which can be simplified into three categories: corruption, asset misappropriation, and fraudulent statements.

43.1.1 Corruption

Corruption is a large category of offenses. The simplest definition of **corruption** is any situation in which someone receives money that they did not earn in the traditional manner of buying or selling legitimate goods or services. This can be extended to unfair competitive practices as well. Let me provide some common examples of corruption.

Conflict of interest is a form of corruption in which an employee or organization buys from or sells to an organization in which they have a financial interest. Normally, conflict of interest is not illegal if two conditions are met. First, the conflict of interest must be revealed to regulatory agencies, and second the prices of goods and services must be the same as for transactions with other organizations with which no conflict of interest exists. If these two conditions are not met, then a crime has been committed.

Bribery, kickbacks, and illegal gratuities are forms of corruption in which the organization or its employees give money or take money for some special consideration. Economic extortion is a form of corruption in which the organization or its employees use some form of economic threat to get the victim to give them money. Threat of physical harm or property damage constitutes plain extortion, a serious offense. If instead the threat is economic, such as "We won't supply this critical part anymore," or "We will increase the fees we charge you," then the offense is economic extortion.

43.1.2 Asset Misappropriation

Asset misappropriation is nothing more nor less than stealing. There are many forms of this offense. Larceny and skimming refer to taking money or supplies from the organization or from the money that passes through the organization's hands. Fraudulent disbursements refers to Inflated bills, inflated payrolls, inflated expense reimbursements, "ghost" employees (people on the payroll who do not actually exist), claiming hours not worked, check tampering, false refunds, etc., in which money is being given for something other than legitimate purchases, sales, or employee hours worked.

43.1.3 Fraudulent Statements

Recall from chapter 42 that companies are required to make statements of the financial status to various regulatory and tax agencies. These statements include tax documents, audit reports, and earnings or loss statements. Fraudulent statements refers to cases in which these statements are intentionally incorrect or inaccurate. Some common instances of this type of offense include overstating or understating the amounts of assets or revenue, overstating or understating the amounts of liabilities or debts, and misrepresenting the company's abilities or expertise. It is also considered fraudulent to withhold information until after some detrimental or beneficial event occurs.

43.2 The Scope of Occupational Fraud

Authorities agree that occupational fraud is a large and growing problem. The National Consumer Law Center (Ralph Nader's organization) reports that financial crime is flourishing. It seems as if this is one area in which there is a definite distinction between the rich and the poor. The more affluent a person is, the more likely they are to be a perpetrator of occupational fraud, while the less affluent a person is the more likely they are to be a victim of it.

The punishments for occupational fraud tend to be quite light. There is a general feeling that occupational fraud is non-violent and, therefore, less serious. However, the victim of the crime is not always convinced that it's less serious. Let's say that a robber pulled a gun on you and forced you to give him $100, and let's say that a corrupt salesperson intentionally overcharged you by $100. First, the police are not going to be nearly as excited about tracking down the criminal salesperson as they are in tracking down the robber. So, there is less chance that anything would be done at all. Second, it is more difficult to prove that the criminal salesperson was actually intending to defraud you, since he or she could claim that it was just a mistake. Third, it would be much more difficult to obtain a conviction of the criminal salesperson. Fourth, if the criminal salesperson was convicted, chances are he or she would go to a minimum security facility, some of which have been described as "glorified summer camps". Fifth, the criminal salesperson's sentence is likely to be drastically shorter than the robber's. In reality, the harshest thing likely to happen to the criminal salesperson is that he or she will lose their job.

Putting together the potential for a big payoff, the low probability of getting caught, and the low probability of being seriously punished, occupational fraud seems like a good career choice. People who are reasonably intelligent but lack moral fiber notice this quickly.

43.3 The Five Most Common Occupational Frauds

By far the most common type of occupational fraud is asset misappropriation by an employee, a crime also called embezzlement. This single offense accounts for more than 80% of occupational frauds. In the vast majority of cases, around 90%, the targeted asset is cash. Employers are correct to be wary of their employees stealing cash. The second most common type of occupational fraud is management fraud, usually by top management of the company and by deceptive manipulation of financial statements.

The number three occupational fraud is investment fraud, in which worthless investments are sold to unsuspecting investors. An example of this is telemarketing fraud, for which it is estimated that hapless victims lost more than five billion dollars in 2000. The fourth most common fraud is vendor fraud, in which vendors, acting alone or in collusion with buyers, overcharge for purchased goods or fail to ship goods that have been paid for. In fifth place is customer fraud, in which customers, perhaps in collusion with the seller, don't pay the full price for goods or services.

43.4 The Role of Forensic Scientists in Investigating Occupational Fraud

Several types of forensic scientists get involved with the investigation of occupational fraud cases. Document examiners apply their skills to detecting alterations to documents, including checks and bills. They can also sometimes help pinpoint who made the alteration. Forensic accountants apply their skills to determine whether a financial crime has occurred, and, if so, by whom and for how much money. They can examine accounting systems for weaknesses, design internal controls to help prevent fraud, determine the degree of organizational fraud risk, interpret financial data for unusual trends, and investigate any indications of fraud.

Forensic economists can apply their skills to estimating the amount of loss to an individual or an organization due to an incident of fraud. This loss may extend beyond the strict dollar value involved in the crime, due to missed opportunities and damage to the company's reputation. Forensic computer scientists can apply their skills in a variety of ways. First, they can collect digital evidence of occupational fraud from the computer of the perpetrator. Second, they can help figure out how an incident of fraud was accomplished. In modern times many types of occupational fraud involve tweaking a computer or a program in some way. For example, an employee at a bank may crack into the part of the bank's software that handles fractional cents on interest charges and have all the fractions of cents sent to his or her own account. Fractional cents don't even show up on a customer's account statement, but if the fractional cents are collected for thousands of customers, the criminal can accumulate a substantial amount of money. Third, they can examine computer and network systems for weaknesses. They can also examine credit card readers and bar code scanners, and their associated software, for signs of tampering.

Traditional investigation is also important in cases of occupational fraud. If someone is suspected of fraud, their bank records can be examined to see whether all the money passing through their account is actually from their salary, and, if not, where it is from. Investigation may also reveal whether a person has debts or other problems that make them more likely to commit an act of occupational fraud.

43.5 Counterfeiting

Counterfeiting is related to occupational fraud in that both are financial crimes, and both involve a form of fraud. Counterfeiting is the unauthorized reproduction of an item for sale or use in a financial transaction. We usually think of counterfeiting as applying to money, but in chapter 33 we looked at software counterfeiting. Counterfeiting can apply to many kinds of consumer goods ranging from Beanie Babies to food and liquor. The principle is that something worthless or worth a lesser amount is passed off as being worth a greater amount. Here I will only look at the counterfeiting of money, but it is important to note that these other forms of counterfeiting exist as well.

The concept of counterfeiting money is similar to that of forgery. Forgery, however, technically only applies to documents that have no inherent value in themselves, such as identification documents. Counterfeiting, on the other hand, can apply to a much wider range of things, in which the genuine article has inherent value.

43.5.1 Counterfeiting Money

Counterfeiting was once the domain of extremely skilled crooks who needed expensive engraving and printing equipment. As the price of desktop-publishing systems has dropped, however, counterfeiting has become something of a home industry. Personal computers with the graphics needed for counterfeiting are now available for a few hundred dollars. Color photocopiers also contribute to the problem, because many of them can produce reproductions that are very similar to real money, especially if the correct paper is used. The counterfeit money produced by a simple scanner, computer, and color printer system can sometimes fool an inexperienced store clerk, apparently often enough to make the effort worthwhile. These lower quality counterfeits almost always get detected by bank employees, however, because they have training in how to distinguish authentic money from counterfeit. In order to make counterfeits that might pass a bank teller's inspection, the highest quality equipment and skills are necessary, and the single largest problem is in matching the paper upon which currency is printed.

The question of who gets stuck with the loss when a counterfeit bill is detected is that it's the person who had it in their possession when the conterfeiting was identified. Often, this is a merchant or individual who included the bill in a bank deposit.

Because U.S. currency is widely accepted, it is widely counterfeited. Unfortunately, U.S. currency is not only the most desirable currency in the world. It is also one of the most easily counterfeited. Most counterfeiters operating in the U.S. duplicate the $20 bill because it is in such widespread use in everyday transactions. Less often the $100 bill is counterfeited.

Outside the U.S. however, the $100 bill is the most commonly counterfeited bill. One authority has estimated that a group of highly-skilled counterfeiters backed by Iran and Syria have produced as much as $1 billion in superb reproductions of the old U.S. $100 bill. As a point of comparison, the U.S. Bureau of Engraving and Printing makes about $9 billion in bank notes each year. There is also a $100 bill counterfeit produced at an unknown location (or perhaps known but not disclosed) that is such a good reproduction that officials are calling it the "superdollar". A raid in the Philippines in 1999 revealed a counterfeiter with more than $50 billion in U.S. currency and treasury notes. Some authorities estimate that in 1989, more than 80% of the U.S. hundred-dollar bills circulating in Europe were counterfeits. Counterfeiters in Colombia are suspected of manufacturing more than a third of the counterfeit notes seized in the U.S. in 1999. Clearly, this is a problem of considerable magnitude and significance.

The U.S. Treasury Department has responded to the problem of counterfeiting by recently redesigning the currency. The new money has a lot of security features such as optically variable ink. The Treasury Department believes that these features will make U.S. currency virtually impossible to counterfeit. However, even the new bills are being faked, with varying degrees of success.

43.5.2 The Role of Forensic Scientists in Investigating Counterfeiting

Most of the involvement of forensic scientists in enforcing counterfeiting laws is support of traditional law enforcement investigations, as would be true in the enforcement of most crimes. Sometimes document examiners are asked to apply their knowledge of papers and inks to helping solve a case. Since much counterfeiting is done using computerized equipment, forensic computer scientists may be consulted. Forensic accountants can often spot influxes of money that have no legitimate explanation and may be due to counterfeiting.

43.6 Money Laundering

Money laundering is any process of making "dirty" money (money from illegal activities or for illegal purposes) look "clean" (as if it came from or is being used for a legitimate source). Money laundering is a need shared by any person or organization that makes money by illegal means, including organized crime groups and terrorist groups.

43.6.1 Common Money Laundering Schemes

The schemes devised to launder money are both numerous and clever. Here are some of the methods of money laundering that have been used.

- **Front Companies**. These are companies that carry on legitimate business where illegal profits can be mixed with revenues derived from legitimate activities. Front companies can be traditional businesses, non-profit organizations, or even charitable organizations.
- **Shell Companies**. These are businesses that exist, but don't actually do anything. Illegally acquired money is passed through these companies in such a way as to appear to be money acquired through the activity of a legitimate business. Like front companies, shell companies can be traditional businesses, non-profits, or charities.
- **Money or Value Transfer Systems**. There are two forms of these, "normal" and gray/black market money transfer systems. The principle is the same in either case, once the money or goods have been transferred to another person or another form, it is difficult to trace its true origin. An example of a "normal" money transfer system is wiring money via Western Union or other wire transfer services. Gray/black market systems include such things as non-official currency exchanges or money transfer systems.
- **Smurfing**. Smurfs are little blue people who inhabit (or at least used to inhabit) a children's TV show. Smurfing, also called structuring, is a method for depositing money in a bank in small enough amounts that the systems in place for reporting large deposits to authorities are avoided. Traditionally, deposits of over $10,000 were reported to regulatory

authorities, and recently the trigger amount has been lowered to $5,000 in at least some cases. Smurfing simply involves dividing a large amount of money into deposits of less than the trigger amount and having a different person make each deposit. Once money is in a bank it is difficult to trace its true origin.
- **Credit card front-end loading**. This is a strategy wherein a person pre-pays cash to their credit card to build up credit that can then be used for purchases. Once cash has been transferred to a credit card account, it is difficult to trace its source.
- **Currency Smuggling**. This is the physical movement of cash from one location to another to disguise its source and ownership.
- Converting the money to Gold or another **commodity item**. The commodity can then be resold legitimately.

43.6.2 Responses to Money Laundering

Money laundering is a difficult problem to deal with. There are some approaches to combating it, none of which is completely effective. Suggestions for improving the response to money laundering mostly focus on doing what is already being done more effectively.

Some authorities suggest that increased governmental oversight of money transfer systems, including licensing and regulation, could have an impact on money laundering. Although there are already reporting requirements for suspicious deposits or withdrawals of money from financial institutions, it has been suggested that the amounts considered "suspicious" be lowered. Another suggestion is routine investigation of businesses to make sure that they actually manufacture or deliver a product or service and aren't a shell company. A similar suggestion is to increase oversight of accounting practices for businesses, non-profit organizations, and charitable organizations to detect cases in which the amount of money made does not match the amount of goods and services bought or sold.

43.6.3 The Role of Forensic Scientists in Investigating Money Laundering

The ways that forensic scientists become involved in investigations of money laundering parallel the ways that they become involved in enforcing occupational fraud. Mainly, this involves many types of forensic scientists supporting traditional law enforcement investigations into cases involving money laundering. The golden team of forensic accountants and forensic computer scientists is especially valuable in this type of investigation.

43.7 Questions for Study and Review

1. In your own words, define occupational fraud.

2. Compare and contrast white collar crime, corporate crime, and governmental crime in terms of who is the perpetrator, who is the victim, and who gets the illegal benefit of the crime.
3. What is the meaning of "corruption" in the context of occupational fraud? What are some common forms of corruption?
4. Define "asset misappropriation" and give some examples?
5. Why is it more difficult to identify and prosecute occupational fraud that some other forms of theft?
6. What is the single most common form of occupational fraud?
7. What types of forensic scientists are most commonly involved in investigations of occupational fraud?
8. How does counterfeiting differ from forgery?
9. What technological advances in recent times has made it easier to counterfeit money?
10. In your own words, define money laundering.
11. How does a front company differ from a shell company?
12. In your own words, describe what "smurfing" is used for and how it works.
13. In what ways do forensic scientists become involved in investigations of money laundering?

43.8 Sources

[Anonymous]. n.d. Counterfeiting [Internet]. [cited 2007 Oct 30]. Available from: http://www.sniggle.net/counterfeit.php.

[Anonymous]. 2003. Qui Tam Whistleblower Statute - Report Governmental Fraud [Internet]. [cited 2007 Oct 30]. Available from: http://www.personal-injury.com/practice_areas/Governmental_Fraud.asp.

Anti-Counterfeiting Group. 2004. The Anti-Counterfeiting Group: Campaigning Against the Trade in Fakes [Internet]. [cited 2007 Oct 30]. Available from: http://www.a-cg.com/guest_frames.html.

Association of Certified Fraud Examiners. 2004. 2004 Report to the Nation on Occupational Fraud and Abuse [Internet]. [cited 2007 Oct 30]. Available from: http://www.acfe.com/documents/2004RttN.pdf.

Burns KS. 2004. White Collar Crime [Internet]. [cited 2007 Oct 30]. Available from: http://www.karisable.com/crwc.htm.

Coenen TL. 2007. The Fraud Files: Fraud in Government Versus Private Industry [Internet]. [cited 2007 Oct 30]. Available from: http://www.wislawjournal.com/archive/2007/0709/coenen-070907.html.

Coffin B. 2003. Trends in Corporate Fraud. Risk Management 50(5): p9(1).

Hendrie E. 2006. Breaking the Bank [Internet]. FBI Law Enforcement Bulletin 75(7). [cited 2007 Oct 19]. Available from: http://www.fbi.gov/publications/leb/2006/july2006/july06leb.htm.

Holtfreter K. 2005. Is Ocupational Fraud "Typical" White-collar Crime? A Comparison of Individual and Organizational Characteristics. Journal of Criminal Justice 33(4): p353(13).

Mokhiber R. 1994. Soft on Crime. Multinational Monitor 16(5): pp25(3).

Salierno D. 2005. The Fight Against Fraud. Internal Auditor 62(1): p62(5).
Scott A. 2002. The High Cost of Occupational Fraud. Internal Auditor 59(5): p13(2).
United States General Accounting Office. 1996. Counterfeit U.S. Currency Abroad: Issues and U.S. Deterrence Efforts [Internet]. [cited 2007 Oct 30]. Available from: http://www.fas.org/irp/gao/ggd96011.htm.
Wells JT. 2001. Why employees commit fraud. Journal of Accountancy 191(2): p89.
Williams HE. 1997. Investigating White-Collar Crime: Embezzlement and Financial Fraud. Springfield (IL): Charles C. Thomas.

Chapter 44
Advanced Topics
ORGANIZED CRIME

In this chapter we will explore the problem of organized crime. I have designated this chapter as an advanced topic because it does not discuss a forensic science, but rather an application of the forensic sciences. The chapter will begin with a look at the definition, characteristics, and scope of organized crime. Next it will discuss strategies for combating organized crime, followed by an examination of known crime organizations and their activities. It will briefly examine the roles of forensic scientists in policing organized crime and conclude with an exploration of the use of money laundering by organized crime.

44.1 Organized Crime

Organized crime is the illegal activity of people and organizations whose purpose is profit through illegitimate business enterprise. Put more simply, an organized crime enterprise (hereafter a **crime organization**) is a business that makes their money through illegal activities. Organized crime involves marketing techniques that are illegal, such as extortion and smuggling; products that are illegal, such as drugs and prostitution; or services that are illegal, such as gambling and loan-sharking. Rarely does organized crime operate in the open. Most often crime organizations pretend to be legitimate businesses, and in some cases infiltrate legitimate businesses and other organizations as cover for their illegal activities.

Often, a crime organization resembles a legitimate business, and may even have the structure of a normal corporation. It is when one examines the products and services offered or the methods used in dealing with their customers does it become apparent that it is not a legitimate business.

44.2 Characteristics of Organized Crime

Crime organizations have a set of characteristics, some of which may be shared by legitimate businesses. It is the combination of these characteristics that reveals the criminal nature of the organization. Here is a list of characteristics of crime organizations.
- Hierarchy. Crime organizations have a command structure consisting of three or more permanent levels of rank, each with their own duties. For example, as anyone familiar with "The Godfather" books or movies knows, the Italian mafia as a hierarchy consisting of a boss (don), with an advisor (consigliere), and caporegimas (lieutenants or captains) that are chiefs to soldati (soldiers).

- Ideology. Organized criminals subscribe to the philosophy that their activities and tactics are "just business, nothing personal".
- Perpetuity. The organization is designed to last through time, beyond the lifespan of current members.
- Monopoly. The organization has unique and total control over the illegal activities in a certain location, known as their "turf"; or over a certain product or service in their "turf".
- Discipline. The group controls its members, and often nonmembers, with prompt, often deadly responses. There is a code of silence, an expectation of complete secrecy, and extensive rules and regulations.
- Restricted membership. Membership is based on ethnicity, kinship, criminal record, or other strict grounds. These organizations don't accept employment applications – potential new members need a sponsor, or are recruited by someone who is already a member of the crime organization.
- Corruption. The group is uses corruption as a primary tool in their dealings with customers and the justice system. Corruption was defined in chapter 43 and includes bribery, kickbacks, extortion, and other similar illegal incentives. Sometimes this includes corruption of peace officers, attorneys, and even judges.

44.3 The Scope of Organized Crime

Organized crime is thought to account for somewhere between 1% and 2% of the U.S. gross domestic product (GDP). This sounds trivial, but when you consider that the GDP is now in the trillions of dollars, the amount of money involved is huge. The annual income of organized crime is larger than that of most industries in the United States.

Where does all this money come from. The largest single source overall is probably the sales of drugs. Another historically large source is thought to be infiltration of labor unions, where the organized crime element takes control of pension funds and dues. Other large sources include thefts and sale of stolen goods, prostitution and pornography, and gambling.

Organized crime is known to infiltrate legitimate businesses. They especially target businesses that are low tech, have uniform products, and either have a monopoly or exist in a type of market where price increases will not reduce demand. Other businesses are infiltrated simply to provide a "front" for the illegal activities, or for the purpose of money laundering.

44.4 Responding to the Problem of Organized Crime

Organized crime has always been difficult to prosecute. The main reason for this is public apathy and lack of awareness. Many people think of organized crime as something they see on TV, and not something that exists in their neighborhood. A second reason is that the tactics of corruption often used by organized criminals are effective. Organized crime also has political connections, and some of the most wealthy and politically influential people have connections to organized crime, or even originally made their family fortune through organized crime (for example by smuggling liquor during the prohibition era).

Because of these issues, normal police tactics are often ineffective. For example, drug users and occasional sellers are arrested, but the underlying structure that stretches from the production of drugs through distribution and eventually to the street seller remains intact. A few types of other tactics are more successful, including the use of informants, surveillance, undercover operations, witness protection programs, and immunity for those willing to testify. All of these tactics raise problems and issues.

Informants are extremely difficult to deal with, but are absolutely essential in order to gain information about crime organizations. This is especially true for newly emerging crime organizations, the leaders and activities of which are not previously known. Informants are quite unreliable, and this is yet another situation in which any leads generated need to be independently verified by investigation or at least by the statements of other informants. In some jurisdictions, each informant is rated on a scale of reliability.

Undercover operations are another staple of crime dramas. However, the danger involved for undercover officers is so high that this tactic is only used as a last resort, when it is not possible to get crucial evidence about a crime organization by any other method. The fact that new members join a crime organization through recruitment or sponsorship make infiltration by undercover peace officers difficult.

We haven't looked at the topic of immunity before, although we have looked at the related issue of plea bargaining. **Immunity** is an agreement between prosecutors and a witness that the witness won't be prosecuted for a crime, in exchange for their testimony against other people involved in the crime. There are two forms of immunity. The first is **transactional immunity**, which gives complete protection from prosecution for crimes about which the witness is asked to testify. The second is **use immunity**, which is a less complete form of immunity and only protects the witness from being prosecuted based on the information in any testimony they provide. With use immunity the witness can still be prosecuted using evidence obtained independently. The "code of silence" required of members in a crime organization makes obtaining witnesses difficult, even with the promise of immunity. Some authorities believe, however, that the current generation of organized criminals is more willing to break the code of silence and become a witness in exchange for immunity and/or favorable plea bargaining arrangements.

Currently, about 200 to 500 medium and high level organized crime members are arrested each year, which is considered a large number by some authorities. The fact that this number seems relatively small to many of us is a testament to the difficulty of gathering enough evidence to build a case against organized criminals. The most common reasons for arrest of organized crime members are racketeering, drugs, stolen property, prostitution, gambling, and murder.

Racketeering is a term that comes from the federal Organized Crime Control Act, passed in 1970. Title IX of that act is called **RICO** (Racketeer Influenced and Corrupt Organization). It listed 24 different activities (such as bribery, extortion, and murder) associated with organized criminal activity. To be deemed a racketeer, a suspect must have committed at least two of the 24 different activities on the list within a ten-year period, and be shown to be a member of an organization that supports these activities. Participants are then found guilty for the acts of the criminal organization regardless of their personal participation. RICO penalties are severe, and include prison sentences that are double or triple the usual amount. RICO also gave law enforcement the tool of asset forfeiture, which is the seizure of the criminal's assets that were presumably gained illegally.

No single law enforcement agency has sole responsibility for policing organized crime. The responsibility is shared by the FBI, the Drug Enforcement Administration, ATF, the Department of Labor, the Internal Revenue Service, and local law enforcement agencies.

44.5 New Ideas for Combating Organized Crime

The inability to effectively control organized crime using traditional methods has prompted various authorities to suggest new tactics. Some of these are being applied, but others require a restructuring of laws and are slower to be tested. Let's examine a few of these suggestions.

One suggestion is **decriminalization** of some of the activities and products from which organized crime makes the most money. The idea here is that by lessening the penalties associated with traditional vice activities, organized criminals are not able to charge as much for the product or service. This hurts the crime organizations financially, and may cripple them. This strategy has not been widely tried, but one instance known historically is when liquor was decriminalized after prohibition, with regulation and taxes enforced by the federal government and the individual states. Several of the crime organizations based on smuggling liquor found themselves out of a source of income. This idea could be applied, for example, to decriminalization of certain drugs, gambling, and even prostitution. Precedents for the decriminalization of these vices exist in certain states. For example, certain forms of gambling are legal, though highly regulated, in Montana. The oversight of the legal forms of gambling by state authorities makes it unlikely that organized crime makes much money from gambling activities in Montana.

Another idea is new forms of regulation that involve the use of zoning and licensing ordinances to control vice activities. This essentially creates "red-light" districts in which organized crime is allowed to operate, but with the government exercising control over the situations and receiving a portion of the revenues.

Other ideas for reducing the profitability of organized crime involves capping certain types of payments, such as on insurance claims for buildings damaged by arson. At a certain point in a building's lifespan it may be worth less than the amount it is insured for, which tempts organized criminals to burn it down in order to collect on the insurance. Capping the insurance payment at the amount of the building's current value limits this temptation. Other types of compensatory payments that could be capped exist as well.

On older tactic, which is still effective and has advocates is simple harassment of the crime organization and its members. This involves disrupting distribution networks for illegal goods, electronic surveillance of members of the crime organization, special grand juries that are devoted to cases of organized crime, and especial vigilance of suspected criminals by peace officers. Some of these tactics are controversial, however, because they single out certain segments of the populace for unfavorable treatment.

44.6 Known Crime Organizations

The crime organizations known to exists at the present time are almost all based on ethnicity. A person's **ethnicity** is the cultural group to which each person believes themself to belong. Nationality and race are aspects of ethnicity. A partial listing of those known to operate in the United States follows.
- Aboriginal. There are several crime organizations around the world where the criminals are members of the native or indigenous peoples of the country. One country that has a large problem with aboriginal organized crime is Canada, where several "First Nations" control the illegal tobacco, weapons, and gaming markets. Since some Native American tribes have traditional homelands in both the U.S. and Canada, some of this activity is also encountered in the U.S.
- Chinese. The Tongs and Triad (Chinese Mafia) have been involved in business extortion, smuggling aliens into the U.S. (mostly from Mexico), software piracy, and of course, the more traditional drugs, gambling, and prostitution in Chinatowns across the U.S. and the rests of the world. The Tongs are an old system of secret societies, going back to the middle 1800s. They operate behind immigrant protection associations, but have many of the characteristics of traditional organized gangs.
- Colombian. The Colombian drug cartels have been well studied. A large amount of effort, both in the U.S. and in Colombia, has led to the arrest or removal of the leadership of these crime organizations on occasion, but the groups simply reappear under new leadership. Part of the problem is that there is considerable popular support for these organized criminals in Colombia, where they are seen as common people who have made it big. The different cartels have complex relationships with each other and differences in the ways in which they operate. The crime organizations in the region of Mendellin, Colombia are described as acting more like petty street hoodlums, while those in the region of Cali are described as operating more like legitimate businessmen. The Colombian cartels tend to have many sympathizers, and large organizations employing as many as 24,000 people in a typical operation.
- Italian. More has been written about Italian mafia than any other brand of organized crime in the U.S. They are the inspiration for "The Godfather", "The Sopranos", etc. and are relatively well known in popular culture. Perhaps the mafia can be regarded as the "traditional" American organized crime organization.
- Jamaican. The Jamaican posse crime organizations are best known for having developed the market for crack cocain in the U.S., where they control about 40% of it, according to authorities. They also engage in kidnaping for profit, murder, robbery, and auto theft. Authorities believe

that there are about 40 known posses operating throughout the U.S., with a total of about 22,000 members.
- Japanese. The yakuza traces its origins back to the 600's C.E., and has always been a major part of Asian organized crime. Authorities believe that the yakuza is gradually expanding into the U.S. where they now have a presence in California and Hawaii. The traditional activities of the yakuza are primarily money laundering and weapons smuggling. Their legitimate activities include banking and real estate.
- Jewish. Jewish organized crime seems to have gotten its inspiration from the mafia, and has existed in the U.S., primarily on the East Coast, since the 1930's.
- Mexican. Smuggling illegal aliens, drug trafficking, and money laundering are the primary activities of the la eme, also known as the "Mexican Mafia".
- Russian. Russian organized crime dates back to the days of the Czars, and thrived under the corrupt Soviet Union. Since the collapse of the Soviet Union, Russia has become even more corrupt, and crime organizations have become big business through smuggling military equipment and weapons (including nuclear), loansharking, smuggling fuel, and protection rackets. Authorities believe that there are about 5,000 different gangs in Russia, with an average of 20 members each; and that about 25 of these gangs operate outside of Russia, in the U.S. or Europe. Those in the U.S. seem most active around New York City, but some are springing up in places where immigrants from the former Soviet Union have settled.
- Vietnamese. Crime organizations of Vietnamese origin are of growing concern to U.S. authorities. Considered by many to be the most ruthless of the Asian gangs, they are among the most secretive and least well known. It is likely that most of their income comes from the drug trade.
- Outlaw Motorcycle Gangs. The only major group of crime organizations in the U.S. not based on ethnicity is the outlaw motorcycle gangs. Bikers are notorious for being rowdy, but some of the gangs have gone beyond ordinary rowdiness to become crime organizations, primarily involved with drugs and prostitution. Law enforcement agencies are wary of them, because they have occasionally conducted "wars" on police and even disrupted court proceedings. Some of the gangs have been known to infiltrate legitimate businesses, such as strip clubs.

44.7 Organized Crime Activities

Typical activities of organized crime groups include the following: alcohol & tobacco smuggling, arms trafficking, arson, corruption (bribery), counterfeiting, drug trafficking, extortion, fraud, gambling, hijacking, home invasion, labor racketeering, loan sharking, murder for profit, protection rackets, robbery, sex slavery, smuggling of humans and human organs, smuggling of animal parts, and theft. Most of these are familiar, but some students may not have previously heard the term "protection racket."

A **protection racket** is a form of extortion in which businesses or individuals are asked for money by the crime organization, in return for protection from crime, perhaps from the actions of other crime organizations. If the person or business refuses, the organized crime organization may retaliate by beating the person or business owner or causing property damage.

As one example of an organized crime activity, let's focus on illegal gambling. There are two major forms of illegal gambling: numbers and bookmaking. **Numbers** is a form of lottery in which a person places a small bet with a "numbers runner", and chooses 3 numbers between 000 and 999. The bettor then receives a receipt. The runner, who has several copies of the receipt, passes one to a "pickup" who carries it to the "bank" or the accounting room of the numbers operation. If the customer wins, they get a payoff that is usually 100 to 500 times the amount of the bet. There are different ways of determining the winning number. The "Brooklyn method" is based on the last three digits of the total amount of money a particular racetrack handles on a specific day. Runners make a 25% commission on their receipts, plus tips. The banker usually pays off the police to avoid interference in the operation.

Bookmaking is a form of sports betting. For basketball and football, one team is usually given a handicap of a certain number of points, called a "spread". If the Green Bay Packers are handicapped by 8 points against the Chicago Bears, a person betting on the Packers wins if the Packers win by more than 8 points. A person betting on the Bears wins if the Bears win or lose by less than 8 points. If the Packers win by exactly 8 points, all bets are returned. To prevent returns, most bookmakers express the spread in half-point values, such as 8.5 points. A bookmaking operation requires a "bookie", who sets limits on the size of bets; a runner, who transfers money; and a clerk, who handles the records. Some bookies also have a "tabber", who keeps tabs on the changing spreads.

44.8 The Role of Forensic Scientists in Investigating Organized Crime

Forensic scientists can become involved in the investigation of organized crime in a variety of ways. The most common is through providing traditional forensic examinations of evidence in cases being investigated by traditional methods. Surveillance of organized criminals is a common law enforcement technique. Wiretapping, bugging, video surveillance, and other strategies are used. For reasons we discussed in chapter 35, these forms of evidence often have to be processed by a forensic audio analyst or forensic image analyst in order to make the information understandable or visible.

Forensic computer scientists may become involved with recovering digital evidence from the computers of organized criminals. Forensic accountants may work with this evidence or other financial data to identify instances of fraud or other crimes. Forensic linguists may become involved with interpreting information recovered from surveillance. For example, what might "pushing up daisies" mean? This is an American expression for being dead, so we American know it, but what about similar expressions used by Chinese, Russian, or Mexican organized criminals? A forensic linguist can help interpret such phrases.

44.9 Use of Money Laundering by Organized Crime

Crime organizations have a clear need to launder the money they acquire. For the most part they use the mechanisms described in chapter 43 to accomplish this. For organized crime there is less emphasis on value transfer systems and more emphasis on front companies and shell companies. The widely known tendency for organized crime to infiltrate legitimate businesses, labor unions, and the like is primarily due to the need to launder their illegally acquired money. After being infiltrated, these legitimate companies or organizations essentially function as front companies for the crime organization.

44.10 Questions for Study and Review

1. In your own words, define organized crime.
2. Explain what is meant when a crime organization is described as having a hierarchy?
3. What is the typical ideology of crime organizations?
4. Are crime organizations open to anybody who wants to join? If not, what are some common membership criteria?
5. What is thought to be the single largest source of income for organized crime?
6. What is the difference between transactional immunity and use immunity?
7. What is "racketeering" and what is "RICO?"
8. Explain how decriminalizing certain drugs or activities might be a way to fight organized crime.
9. What is a person's ethnicity?
10. In your own words, define the following activities of crime organizations: protection racket, numbers, bookmaking.
11. How do forensic scientists become involved in investigations of organized crime?

44.11 Sources

Dijk J. 2007. Mafia Markers: Assessing Organized Crime and its Impact upon Societies. Trends in Organized Crime 10(4): 39-56.

Hagan F. 2006. "Organized Crime" and "organized crime": Indeterminate Problems of Definition. Trends in Organized Crime 9(4): 127-137.

Hendrie E. 2003. When an Informant's Tip Gives Officers Probable Cause to Arrest Drug Traffickers [Internet]. FBI Law Enforcement Bulletin 72(12). [cited 2007 Oct 19]. Available from: http://www.fbi.gov/publications/leb/2003/dec2003/dec03leb.htm#page_9.

Hight JE. 2000. Working with Informants: Operational Recommendations [Internet]. FBI Law Enforcement Bulletin 69(5). [cited 2007 Oct 19]. Available from: http://www.fbi.gov/publications/leb/2000/may00leb.pdf.

Kleinknecht W. 1996. The New Ethnic Mobs: The Changing Face of Organized Crime in America. New York: The Free Press (Simon and Schuster).

Kugler M, Verdier T, Zenou Y. 2005. Organized Crime, Corruption and Punishment. The Journal of Public Economics 89(9-10): p1639(25).

Levi M, Maguire M. 2004. Reducing and Preventing Organised Crime: an Evidence-based Critique. Crime, Law and Social Change 41(5): 397(73).

McFeely RA. 2001. Enterprise Theory of Investigation [Internet]. FBI Law Enforcement Bulletin 70(5). [cited 2007 Oct 19]. Available from: http://www.fbi.gov/publications/leb/2001/may01leb.pdf.

Nelen H. 2004. Hit Them Where it Hurts Most? The Proceeds-of-Crime Approach in the Netherlands. Crime, Law and Social Change 41(5): p517(18).

O'Connor, T., 2004. Organized Crime [Internet]. [cited 2007 Oct 30]. Available from: http://www.apsu.edu/oconnort/3220/3220lect07a.htm.

O'Neal S. 2000. Russian Organized Crime [Internet]. FBI Law Enforcement Bulletin 69(5). [cited 2007 Oct 19]. Available from: http://www.fbi.gov/publications/leb/2000/may00leb.pdf.

Shelley LI, Picarelli JT. 2005. Methods and Motives: Exploring Links Between Transnational Organized Crime and International Terrorism. Trends in Organized Crime 9(2): p52(16).

Stewart RC. 2006. Reflections on Labor Racketeering and Interdisciplinary Enforcement. Trends in Organized Crime 9(4): 60-101.

Sukharenko A. 2004. The use of corruption by 'Russian' organized crime in the United States. Trends in Organized Crime 8(2): p118-129.

von Lampe K. 2006. The Interdisciplinary Dimensions of the Study of Organized Crime 1. Trends in Organized Crime 9(3):77-95.

Chapter 45
Advanced Topics
HOMELAND SECURITY

In this chapter we will take a look at the topic of homeland security, especially the problem of terrorism, and what a forensic scientist needs to know about it. Since this is an application of the forensic sciences rather than a forensic science in itself I have designated it an advanced topic.

Homeland security is the newest hot topic since the terrorist attacks of 9/11/2001. Personally, I am not a big fan of strong homeland security because I perceive it as an erosion of constitutionally guaranteed freedoms and protections – though I would certainly agree that it provides a needed function. However, with the new Department of Homeland Security and the increased investment of money in federal law enforcement agencies, this is one of the areas where forensic scientists may find themselves working.

This chapter will begin with a consideration of terrorism, the greatest current concern of homeland security. I will consider Al Qaeda as an example of a foreign terrorist organization, then direct the focus toward domestic terrorists. We will look at how terrorists raise and launder money, and how forensic scientists become involved in the investigation of terrorism.

45.1 Terrorism

Terrorism is widely misunderstood. It is nothing new, nor is it a uniquely Middle Eastern phenomenon. **Terrorism** can be defined as any act designed to cause fear in a population. The fear that it is seeking to inspire could range from fear of injury or death, to fear of economic loss, or simply fear of being inconvenienced. Fear is a potent political tool. When a populace feels fear, it loses confidence in its government and in the law enforcement organizations that are there to protect it.

Usually terrorism is perpetrated by a small marginalized group that sees itself as oppressed, and is carried out against a larger more mainstream group that the marginalized group sees as being the oppressors. Whether it's a good act or a bad act depends on which of these groups one belongs to. For example, the Boston Tea Party, which occurred on the eve of the American Revolution, is thought of as a great moment in American history, but it was in fact an act of terrorism against the commercial monopoly of the British Empire.

Terrorism almost always causes some financial loss, usually from damage to property. It often causes injury or death to people as well. It can have a specific target, either real or symbolic, or it can be random in nature where those people impacted may simply be in the wrong place at the wrong time and may even be sympathetic to the terrorists' cause.

45.2 The Organization of Terrorist Groups as Cells

Most terrorist groups are organized as cells. This means that the organization is broken down into small groups, **cells**, of perhaps five to twenty people. Cells do not communicate with each other, do not know what other cells are doing, and may not even know of the existence of other cells. If interaction between cells is necessary, members of the terrorist organization can recognize each other using passwords or some similar method. Usually, new members are recruited and sponsored by old members.

The cell organization serves two purposes. First, it makes it difficult to determine the extent of the terrorist organization, and nearly impossible to apprehend all the members of the group. Even if one cell is eliminated, others remain. If one cell is captured, its members can not serve as informants against other cells because they don't know about them or their activities. Second, it makes the terrorist group difficult to infiltrate. If an informer or undercover agent manages to become a member of a cell, it is hard to get any information other than what is known by that particular cell.

45.2.1 Leadered and Leaderless Cells

There are two main ways in which cells are organized: leadered and leaderless. The **leadered** type of cell organization was invented in France in the early 1800's (possibly much earlier), and has been used by various revolutionaries in Russia, Ireland, Germany, and the U.S. In this type of cell organization, most members of the cell are followers who don't know anything about other cells or the overall structure of the organization. However, each cell has a leader, who communicates with those higher up in the organization.

The **leaderless** type of cell organization was invented in Egypt in the early 1900's, although some authorities trace it back to the assassins of the 1100's. The assassins were a Shi'ite Muslim organization whose members targeted Sunni leaders for execution. Contrary to what the name implies, this type of cell organization usually designates a leader for each cell. The difference is that the leader does not communicate with higher-ups. Instead, the cell follows a master timeline, or waits for some communication from the heads of the organization, or simply makes its own plans. Actual operations or attacks may be led by operation commanders who meet with the cell only in the final days before an attack. The commanders may be the only link between local cells and the larger umbrella organization. The commander may not even participate in the operation himself, often leaving the country before the attack occurs.

Another feature of leaderless cells is that local cells recruit their own members. In leadered cell organizations recruitment is typically done by specialized recruiters so as not to endanger the cell. As a leaderless cell grows larger, it may split into two or more separate cells. This makes it even more difficult to destroy the entire terrorist organization because if even one cell is left it can continue to propagate new cells.

45.2.2 Types of Cells

Most terrorist cells fall into one of a few distinct types, although this organization is flexible. Cells of the leaderless type are especially flexible and may function as any of these types, depending on the situation.

Planning or support cells usually consist of a small number of often local residents, and are responsible primarily for fund raising. They may also be responsible for providing members of other types of cells with drivers' licenses, cash, credit cards, or lodging, as well as procuring weapons or materials for bomb construction.

Sleeper or submarine cells merely wait for the right time to do something. The members of these cells may live in the target country or community for years, doing nothing until the cell is activated. When the time comes, or upon receiving order, they are activated as one of the other types of cells.

Execution cells are the cells that actually carry out an operation or attack. They will utilize resources supplied by support cells, and may previously have been sleeper cells.

45.3 An Example: Al Qaeda

Perhaps no foreign terrorist organization is as well known to Americans as al Qaeda, because of its successful attacks on the World Trade Center in New York City, and nearly successful attack on the Pentagon. In Arabic, "al Qaeda", also written al-Qaeda, means "the base". Its leader is, or was, Osama bin Laden, who was a Saudi Arabian until Saudi Arabia revoked his citizenship. It is not known with certainty whether bin Laden is alive or dead. Rumors of his death, most recently that he died from dysentery while in hiding near the border between Afghanistan and Pakistan have not been confirmed. Whether alive or dead, he has not been visibly active for some time, yet al Qaeda continues to be a nearly worldwide threat.

The principal stated aims of al Qaeda are to drive Americans and American influence out of all Muslim nations, especially Saudi Arabia; destroy Israel; and topple pro-Western dictatorships around the Middle East. Bin Laden has also said that he wishes to unite all Muslims and establish, by force if necessary, an Islamic nation adhering to the rule of the first Caliphs.

Unlike many terrorist groups, Al Qaeda does not depend on the sponsorship of a country, and is not defined by a local conflict. Instead, al Qaeda operates more like a business franchise. It provides support and name recognition to terrorist groups operating in a variety of countries. If there is any common denominator to al Qaeda members it is that they are Sunni Muslims.

In 1998, bin Laden's issued a fatwa (religious decree), in which he declared that it is the duty of Muslims around the world to wage holy war on the U.S., American citizens, and Jews. Muslims who do not heed this call are considered apostates (people who have forsaken their faith) by al Qaeda. This fatwa has been accepted by some Muslims, even though bin Laden has no formal religious credentials. Al Qaeda's ideology, often referred to as "jihadism," is marked by a willingness to kill "apostate" and Shiite Muslims, and an emphasis on religious war (jihad).

Al Qaeda is organized as leaderless cells. Al Qaeda cells have been identified or suspected, or have carried out attacks, in the following countries: Afghanistan, Albania, Algeria, Australia, Azerbaijan, Bali, Bangladesh, Bosnia, Canada, Chechnya, Egypt, Ecuador, Eritrea, Ethiopia, Jordan, Kenya, Kosovo, Lebanon, Libya, Malaysia, Mauritania, Morocco, Pakistan, the Philippines, Qatar, Saudi Arabia, Somalia, Spain, Sudan, Tajikistan, Tanzania, Tunisia, Uganda, the United Kingdom, the United States, Uruguay, Uzbekistan, and Yemen.

Al Qaeda first emerged in Afghanistan in the 1980's out of the Mujahideen, who were fighters recruited from around the Islamic world to come to Afghanistan and help the Afghans fight the Soviet Union's invasion of their country. The war in Afghanistan was the Soviet Union's equivalent of America's Vietnam War – a war that couldn't be won, despite superior firepower.

Osama bin Laden emerged as a financial backer for the organization that recruited Mujahideen from mosques throughout the world. After the Soviets withdrew from Afghanistan, bin Laden returned to Saudi Arabia, where he founded an organization to help veterans of the Afghan war. These veterans are the core of al Qaeda. Bin Laden may have already been organizing al Qaeda when Iraq invaded Kuwait in 1990. Bin Laden was outraged when the Saudi government allowed U.S. troops to be stationed in Saudi Arabia, the birthplace of Islam. In 1991 he was expelled from Saudi Arabia for anti-government activities, and fled to Khartoum, Sudan, where he established the first formal headquarters for al Qaeda.

During the middle 1990's al Qaeda directed several attacks against U.S. interests. In 1994, Sudan expelled bin Laden and the request of Saudi Arabia and the U.S. Bin Laden then moved his base of operations to Afghanistan, where he stayed as a "guest" of the Taliban government of Afghanistan until the U.S. drove them from power in Nov. 2001. In August 1996 bin Laden issued a "Declaration of War" against the U.S. In 1998, bin Laden announced an alliance of terrorist organizations, the "International Islamic Front for Jihad Against the Jews and Crusaders", which united several previously independent terrorist organizations.

Since 2001, bin Laden has been (possibly literally) underground. He has occasionally released video tapes to Islamic news services, but their authenticity is questionable. Since that time, the most prominent leader of al Qaeda has been Dr. Ayman al-Zawahiri, an Egyptian surgeon from an upper-class family. Al-Zawahiri is suspected of masterminding several operations during the late 1990's. He is wanted by the FBI and has been sentenced to death by Egypt in absentia. In March 2004 the Pakistani military engaged in a clash with resistance troops along the Pakistan-Afghanistan border who may have been defending al-Zawahiri, but neither he nor his body were recovered. This was not the last time that operations have come close to apprehending or killing al-Zawahiri, but as of January 2007 he is known to still be alive and at large.

Despite the U.S. "War on Terror," al Qaeda continues to be a threat world-wide. Some authorities believe that it is actually growing due to widespread opposition in Islamic countries to U.S. involvement in Iraq. There have been about a dozen major attacks by al Qaeda terrorists since September 11, 2001. Authorities are divided on whether or not al Qaeda is a major threat inside the U.S. Some believe that we are safe from a major attack and others believe that a major attack is probable. What is clear is that al Qaeda does have cells in the U.S. and as long as this is true the possibility of some form of attack remains.

45.4 Domestic Terrorists

The United States hardly needs foreign terrorists, because we have our own home-grown terrorist organizations. Experts divide these into three groups: right wing, left wing, and special interest. It is important to say that not every member of a group that is considered to be a domestic terrorist organization is actually a terrorist. The terrorist acts are usually committed by a small minority of extreme members.

45.4.1 Right Wing Domestic Terrorists

Right wing terrorist groups are opposed to the authority of the U.S. government and suspicious of government activities. They are especially opposed to taxes and to regulation of firearms. Many right wing groups are opposed to U.S. involvement in the United Nations and other international organizations. Many of them are characterized by hatred of people of races or religious groups other than their own (overwhelmingly White and fundamentalist conservative Christian). Right wing terrorist groups have their roots in White supremacist groups such as the Ku Klux Klan and in various anti-government groups. They have been prominent since the 1970's and FBI officials consider them the most serious current domestic terrorist threat. Examples include skinheads, neo-Nazis, and groups that refer to themselves as militias or have the word "free" in their name. Examples of right-wing domestic terrorism include:
- the April 1995 bombing of the Murrah federal building in Oklahoma City, which killed and injured many people;
- the July 1996 bombing at Centennial Park during the Atlanta Olympics, which killed one person and injured more than 100;
- the summer 1999 shooting sprees by lone gunmen targeting minorities in the Chicago and Los Angeles metropolitan areas, which left three people dead.

45.4.2 Left Wing Domestic Terrorists

Left wing terrorist groups are based on philosophies of opposition to capitalism and American business and industrial practices. They can be communist oriented, anarchist oriented, or have a poorly defined orientation. Left wing terrorists are responsible for the first wave of domestic terrorism in the U.S., which was the anarchist movement of the late 1800's. Anarchists were thought to be responsible for the bomb that exploded in Chicago's Haymarket Square during a labor riot in 1886. In 1892 anarchists attempted to assassinate the steel tycoon Henry Clay Frick, and in 1901, an anarchist sympathizer named Leon Czolgosz assassinated U.S. President William McKinley in Buffalo, New York.

Another wave of left-wing terrorist activity began in the 1960s. The theme that connected the various terrorist groups was opposition to the war in Vietnam, which the U.S. government portrayed as a war against communism. Far-left groups such as the Weathermen Underground and the Symbionese Liberation Army used bombings and kidnapings to draw attention to their causes. Left wing terrorist groups have pretty much disappeared since the middle 1980's. According to authorities, the only such group still active is the Puerto Rican separatists. However, the FBI warns that left wing terrorism may be experiencing a revival in the wake of the World Trade Organization meeting of 1999, which was highly protested by Americans opposed to the practices of American businesses in other countries.

45.4.3 Special Interest Domestic Terrorists

Special interest domestic terrorists, unlike left-wing and right-wing groups, which have broad revolutionary agendas, focus on single issues such as abortion, the environment, or animal rights. For example, eco-terrorism, motivated by support for animal rights and environmentalism, began appearing in the U.S. during the early 1990s. Most of the incidents of domestic terrorism since the year 2000 have been perpetrated by these special interest terrorists. Typical examples include attacks on abortion providers, attacks on lumbering operations, and attacks on laboratories that conduct research using animals. The Earth Liberation Front (ELF) and the Animal Liberation Front (ALF) have been responsible for attacks that have caused multiple millions of dollars in property damage, and these groups continue to be monitored by the FBI. The various anti-abortion groups are difficult to distinguish from right wing terrorists in many cases, and can be included with them in being considered among the most dangerous of domestic terrorist groups.

45.5 Terrorist Tools

Terrorists use a variety of tools in planning and executing their attacks. Possibly the most obvious terrorist tool is explosives. Most terrorist attacks include the use of some sort of explosive, most often ANFO. Explosives were discussed in chapter 31.

Cryptography and steganography are also tools used by terrorists. It is widely thought that terrorist organizations communicate over the internet using encrypted messages. Al Qaeda is known to use encrypted messages, and believed to use steganography. Cryptography and steganography were discussed in chapter 32.

Many terrorist organizations rely heavily on press releases, including video footage. Terrorists are eager to take credit for the attacks they perpetrate, and eager to promulgate their message. The leaders of many terrorist organizations, including al Qaeda, release statements or even full videos to the press. These have at least three purposes. The first is to terrorize a populace. Terrorists have found that you can make a group of people afraid just by saying that you might carry out an attack against them. One example is that an al Qaeda leader is quoted as saying that al Qaeda is continually planning at least 100 attacks at any given time, which is a number beyond the ability of any terrorist organization. The second purpose of press releases is to recruit and encourage members. Members of terrorist cells are encouraged to know that their leaders are still alive, free, and functioning. The rhetoric used may also be effective in recruiting new members. The third purpose is to pass messages. For example, it is widely thought that some of the videos released by al Qaeda contain messages, either in the words the speaker says, the way the words are said, the speaker's gestures, or by steganography.

Cyberattack is a tool that many experts consider to be the next major issue in terrorist activities. A cyberattack is an attack against a government or organization by sabotaging or damaging its computer systems. Cyberattacks can be carried out by hacking, releasing viruses, and by system attacks, as described in chapter 34. Some experts have gone so far as to say that it's not a question of "if" a cyberattack will occur, but only a question of "when" and "to what extent". Cyberattacks could be launched against government networks or commercial enterprises. Some people believe that Microsoft, the prominent software company, is a prime target because in addition to being an American company it has links to Israel and India, which are both countries to which Islamic nations are hostile.

Infrastructure attack is another looming terrorist issue. So far, most terrorist attacks have been showy spectacles directed against people. However, some experts point out that terrorists could actually do more damage by attacking infrastructure. **Infrastructure** refers to things like utility and communications systems. Some systems that present possible targets are those that provide electricity, natural gas, telephone service, satellite TV, and waste treatment. Infrastructure also includes transportation systems (roads, train tracks, airports, etc.) and large public works such as dams. Some experts warn that the next major attack on the U.S. could be against a dam or a nuclear power generating facility. Experts believe that of all the infrastructure of the U.S., the most vulnerable is the electrical system.

Attacks against soft targets are also possible. Soft targets are areas that are not well protected. For example, government buildings are fairly heavily protected these days, but elementary schools are not. A terrorist group would find it hard to make an attack on a federal building, but would find it easy to attack an elementary school. One press release from al Qaeda stated that they would be starting to attack soft targets.

One of the greatest fears is that terrorists will acquire and use **unconventional weapons**, also called CBN's (Chemical, Biological, Nuclear) and WMD's (Weapons of Mass Destruction). The supposed existence of weapons of this type was the stated reason for the U.S.'s invasion of Iraq in 2003. In fact, no such weapons have been found by coalition forces, with the possible exception of some weak chemical weapons. Unconventional weapons have been used in the past. Here are some examples.

- The Aum Shinrikyo (Aum, Aum Supreme Truth, Aleph) group, which hopes to one day take over the world, used sarin nerve gas to attack sixteen subway trains in Tokyo, Japan in 1995. This attack killed twelve people and injured thousands more.
- In 1984, followers of the Indian-born guru Bhagwan Shree Rajneesh in Oregon tried to disrupt a local election by poisoning salad bars with salmonella bacteria.
- In 2001, right after the terrorist attacks on the WTC, someone mailed anthrax cultures to several U.S. government officials. That person has never been identified.

So far, nuclear weapons have not been used by terrorists. It is unlikely that terrorists could mount a nuclear missile strike, but what many people worry about is a dirty bomb. A **dirty bomb** is one that actually doesn't do much damage in its initial explosion, but releases a large amount of radioactive material that affects people and the environment for a long period of time afterward. Officials also worry about the possibility that terrorists could plant a nuclear weapon on a boat and sail it into a harbor to detonate it there.

45.6 Terrorist Fund Raising and Money Laundering Strategies

Terrorist strategies for fund raising and money laundering are broadly similar to those of organized crime. Terrorist groups derive income from criminal acts in much the same way that crime organizations do. Some groups derive a considerable amount of money from kidnaping people and demanding ransom. Other groups use extortion in the form of "revolutionary tax" or "protection money". This type of extortion occurs when locals pay the terrorist group to not terrorize them. Yet other groups raise money by smuggling or selling drugs.

Unlike crime organizations, however, terrorist groups may also rely on apparently legal sources of income. Such methods as solicitation of donations, sales of publications, collection of membership dues or subscriptions, and charging fees for cultural or social events are all examples of legitimate sources of funding that may be used by terrorist groups. In some instances, these individual donors may not be aware that their contributions are being used by terrorist groups. In other cases, however, it may be that the donors are aware of the ultimate use that their donations will serve. It's not unusual for a sympathetic person to make a donation to a terrorist cause. For example, many people of Irish ancestry in the city of Butte, Montana contributed money to the Irish Republican Army during their dispute with the United Kingdom over the destiny of Ireland. In a case of this sort what needs to be hidden is the purpose of the money, since it is illegal to financially support terrorist activities.

Once having acquired funds, the terrorist group needs to launder them, so that they appear legitimate and can be moved and used as needed. General methods of money laundering were discussed in chapter 43. One type of money laundering scheme that is used more commonly by terrorists than by other types of criminals is the use of nominees. **Nominees** are family, friends or associates who are trusted within the community, and who will not attract attention. They are used to conduct transactions on the behalf of the terrorist group and to disguise the source and ownership of funds.

Terrorist groups are also known to use front companies and shell companies. In the case of terrorists, however, the front or shell is more likely to be a non-profit or charitable organization. As was true for organized crime, these organizations can be infiltrated by terrorists and used as a front for terrorist money laundering without the legitimate members of the organization knowing it.

45.7 The Roles of Forensic Scientists in Homeland Security

There are many ways in which forensic scientists become involved in issues of homeland security. Because of the slippery nature of terrorism and terrorists, the investigation of terrorism is relatively difficult.

One way that forensic scientists contribute to homeland security is in identification of terrorists. It is important to identify who is responsible for an act of terrorism, just as it is for any crime. Therefore, fingerprint examiners and DNA analysts become involved in identifying the person who left the relevant types of evidence at the location of a terrorist attack. Trace evidence examiners, firearms examiners, experts on explosives, and other criminalists also become involved in the investigation that follows a terrorist attack when the types of evidence they work with are recovered. One focus of new efforts is on identifying terrorists as they enter the U.S. This involves use of traditional forensic sciences such as fingerprint analysis, as well as newer areas such as biometrics.

Another important contribution is in tracing the sources of funding for terrorists. A terrorist attack takes money, often a lot of money, and this money must come from somewhere. Therefore, forensic accountants can be instrumental in tracking down the source of the funding for a terrorist attack.

Forensic computer scientists are the people who know how to deal with a cyberattacks and, to the extent possible, track down the perpetrators. If or when such an attack occurs they will surely be involved in the investigation. Many government and corporate entities are already taking steps to make their systems more secure from such attack, mostly as a response to ordinary hacking and cyberattacks. These efforts may go a long way toward preventing terrorist cyberattacks and minimizing the damage if they occur.

Another job for forensic computer scientists, and one to which forensic mathematicians can also contribute is dealing with encryption and steganography. Although the best encryption schemes (such as 128-bit public key) can't realistically be broken, many other schemes can.

Forensic mathematicians, which will be discussed in more detail in chapter 47, can also make some additional contributions. First they can assess the likelihood of a threat, if they are provided with accurate information. Second, they can model terrorist cells and run simulations of what would happen if certain cell members or even entire cells were removed. For example, I have seen one analysis that asks how you can tell whether enough members of a terrorist cell have been killed or captured so that there is a high probability that the cell can no longer function.

Analysis of press releases is the province of document examiners and forensic linguists, who may be able to generate leads from the evidence contained in the document. Audio and video tapes are the province of forensic audio analysts, and forensic image analysts.

If an unconventional weapon attack occurs it may be a forensic pathologist or forensic toxicologists who first figures out that something is wrong, from the pattern of deaths they are called upon to investigate.

45.8 Lists of Terrorist Organizations

The U.S. Department of State has compiled a list of foreign terrorist organizations. The list includes the name of the organization, its base of operations, a brief description of the organization, a brief description of the organization's targets and goals, an estimate of the number of current members, the year founded, and as least partial lists of the groups' activities. As of April 30, 2003, there were 36 organizations on this list. This list is far too long to reproduce here, but is easily available on the internet. By design, there is no comparable list of domestic terrorist organizations, because the U.S. government has traditionally preferred to label the activities of its citizens as crimes rather than terrorist acts. This attitude may be changing, and the Patriot Act enables several activities of domestic terrorists to be investigated and sentenced to the same standards as attacks by foreign terrorists.

45.9 Questions for Study and Review

1. What is terrorism, what types of people most often become terrorists, and who is most commonly the target of these people?
2. Describe the basic structure of a terrorist cell.
3. What advantages does organization into cells offer?
4. Compare and contrast leadered vs leaderless types of cell organization.
5. Compare and contrast planning, sleeper, and execution cells.
6. Give a brief summary of the history and goals of al Qaeda.
7. Compare and contrast right wing, left wing, and special interest types of domestic terrorists.
8. Explain how terrorist organizations use cryptography, press releases, and cyberattacks.
9. What is an infrastructure attack?
10. What is a "soft target?"
11. What are unconventional weapons? What are some other names for this type of weapon?
12. Describe the strategies terrorists use to raise money for their cause. What sorts of money raising strategies to terrorist organizations use that crime organizations do not use?
13. How do forensic scientists become involved with combating terrorism?

45.10 Sources

Aeilts T. 2005. Defending Against Cybercrime and Terrorism [Internet]. FBI Law Enforcement Bulletin 74(1). [cited 2007 Oct 19]. Available from: http://www.fbi.gov/publications/leb/2005/jan2005/jan2005.htm.

[Anonymous]. 2004. Eco-terrorists Take Toll. State Legislatures 30(2): p9(1).

Belcher J. 2004. Destroying Terrorist Cells [Internet]. [cited 2007 Oct 30]. Available from: http://www.airpower.maxwell.af.mil/airchronicles/cc/belcher.html.

Blades HB. 2003. Nuclear, Biological and Chemical Weapons and Improvised Explosive Devices: an Interview with Paul Errico. The Forensic Examiner 12(7-8): 12(5).

Bodrero DD. 1999. Confronting Terrorism on the State and Local Level [Internet]. FBI Law Enforcement Bulletin 68(3). [cited 2007 Oct 19]. Available from: http://www.fbi.gov/publications/leb/1999/mar99leb.pdf.

Borum R. 2003. Understanding the Terrorist Mind-Set [Internet]. FBI Law Enforcement Bulletin 72(7). [cited 2007 Oct 19]. Available from: http://www.fbi.gov/publications/leb/2003/july2003/july03leb.htm#page_8.

Carvin A. 2001. When a Picture Is Worth a Thousand Secrets: The Debate Over OnlineSteganography [Internet]. [cited 2007 Oct 30]. Available from: http://www.andycarvin.com/archives/2001/10/when_a_picture.html.

Crowe K. 2007. Salad bar Salmonella. The Forensic Examiner 16(2): 24(3).

Dickey BG. 2007. In the Pursuit of Justice: Audio and Video Recordings – Weapons of Terrorism. The Forensic Examiner 16(1): p50(5).

Downey DB. 2000. Domestic Terrorism: The Enemy Within. Current History 99(636): p169(5).

Duffy JE. 1997. Militias: Initiating Contact [Internet]. FBI Law Enforcement Bulletin 66(7). [cited 2007 Oct 19]. Available from: http://www.fbi.gov/publications/leb/1997/july975.htm.

Gilmartin KM. 1996. The Lethal Triad: Understanding the Nature of Isolated Extremist Groups [Internet]. FBI Law Enforcement Bulletin 65(9). [cited 2007 Oct 19]. Available from: http://www.fbi.gov/publications/leb/1996/sept961.txt.

Hayes L, Brunner B. 2004. Al-Qaeda: Osama bin Laden's Network of Terror [Internet]. [cited 2007 Oct 30]. Available from: http://www.infoplease.com/spot/terror-qaeda.html.

Hosenball M. 2002. Islamic Cyberterror: Not a Matter of If but of When. Newsweek (May 20, 2002): 10.

Houghton BK, Schachter JM. 2005. Coordinated Terrorist Attacks [Internet]. FBI Law Enforcement Bulletin 74(5). [cited 2007 Oct 19]. Available from: http://www.fbi.gov/publications/leb/2005/may2005/may05leb.htm.

Information Please Database. 2007. U.S. Designated Foreign Terrorist Organizations [Internet]. [cited 2007 Oct 30]. Available from: http://www.infoplease.com/ipa/A0908746.html.

Jacquard R. 2001. The Guidebook of Jihad. Time 158(19): 58.

Jollish B. 2004. The Encrypted Jihad [Internet]. [cited 2007 Oct 30]. Available from: http://archive.salon.com/tech/feature/2002/02/04/terror_encryption/index.html.

Knight W. 2002. Weakened Encryption Lays Bare al_Qaeda files [Internet]. New Scientist (17 January, 2004). [cited 2007 Oct 30]. Available from: http://www.newscientist.com/article/dn1804.html.

Lewis JF. 1999. Fighting Terrorism in the 21st Century [Internet]. FBI Law Enforcement Bulletin 68(3). [cited 2007 Oct 19]. Available from: http://www.fbi.gov/publications/leb/1999/mar99leb.pdf.

Lithwick D. 2001. How do Terrorist Cells Work [Internet]. [cited 2007 Oct 30]. Available from: http://slate.com/id/1008311/.

Marris E. 2007. Terror Terms for Arsonists. Nature 447(7145): p624(2).

Martin RA. 1999. The Joint Terrorism Task Force [Internet]. FBI Law Enforcement Bulletin 68(3). [cited 2007 Oct 19]. Available from: http://www.fbi.gov/publications/leb/1999/mar99leb.pdf.

Martosco D. 2002. Financing Domestic Terrorism [Internet]. [cited 2007 Oct 30]. Available from: http://www.consumerfreedom.com/oped_detail.cfm?oped=145.

Naval Postgraduate School. n.d. Terrorist Group Profiles [Internet]. [cited 2007 Oct 30]. Available from: http://www.nps.edu/Library/Research/SubjectGuides/SpecialTopics/TerroristProfile/TerroristGroupProfiles.html.

Nelson KR. 1999. Mass Transit: Target of Terror [Internet]. FBI Law Enforcement Bulletin 69(1). [cited 2007 Oct 19]. Available from: http://www.fbi.gov/publications/leb/1999/jan99leb.pdf.

North G. 2003. The Terrorist Cell Group [Internet]. [cited 2007 Oct 30]. Available from: http://www.lewrockwell.com/north/north234.html.

Olson DT. 2007. Financing Terror [Internet]. FBI Law Enforcement Bulletin 76(2). [cited 2007 Oct 19]. Available from: http://www.fbi.gov/publications/leb/2007/feb2007/feb2007leb.htm.

Peterson I. 2004. Splitting Terrorist Cells [Internet]. Science News Online (January 10, 2004) 165(2). [cited 2007 Oct 30]. Available from: http://www.sciencenews.org/articles/20040110/mathtrek.asp.

Raffel R. 2003. Weapons of Mass Destruction and Civil Aviation Preparedness [Internet]. FBI Law Enforcement Bulletin 72(5). [cited 2007 Oct 19]. Available from: http://www.fbi.gov/publications/leb/2003/may2003/may03leb.htm#page_2.

Sharpe TT. 2000. The Identity Christian Movement: Ideology of Domestic Terrorism. Journal of Black Studies 30(4): p604(2).

Sherer JL. 2003. Is Terrorism's Threat Overblown? USA Today 131(2692):14(2).

Simons E. 2006. Faith, Fanaticism, and Fear: Aum Shinrikyo--the Birth and Death of a Terrorist Organization. The Forensic Examiner 15(1): 37(9).

Vaisman-Tzachor R. 2006. Psychological profiles of terrorists. The Forensic Examiner 15(2): 6(12).

Verton, Dan, 2002. Experts Predict Major Cyberattack Coming: Corporate Icons Could Be Targets; Trillions of Dollars Worth of Damage Possible. Computerworld 36(28): 8.

Vice President's Task Force On Combatting Terrorism. 1986. Public Report of the Vice President's Task Force on Combatting Terrorism [Internet]. [cited 2007 Oct 30]. Available from: http://www.population-security.org/bush_and_terror.pdf.

Woolsey RJ. 2002. How Safe Are We? The U.S. Remains Haunted by the Threat of Terrorism. People Weekley 57(20): 105.

Chapter 46
JURISPRUDENCE

The subject of this chapter is jurisprudence. Although there are several common meanings of **jurisprudence**, within the context of forensic sciences it refers to the study of how evidence is collected, preserved, admitted, and presented in court. Jurisprudence shares many concerns in common with criminology, the sociological study of justice and the justice system. This is the area where forensic science and criminology most closely converge.

The chapter will begin with an examination of evidence collection issues. Next, we will look at the conduct of a criminal trial, followed by a consideration of experts and being an expert witness. From there, the chapter will continue with a brief examination of the ethics of forensic scientists and an investigation of the issue of wrongful convictions. Finally, we will look at the findings of the 2009 National Academy of Sciences report that calls for validation of forensic methods.

46.1 Evidence Collection Issues

There are three broad issues regarding evidence in the forensic sciences, some of which we have touched on briefly in previous chapters. The first involves search warrants. The second is chain of custody. Both of these topics were introduced in chapter 4. The third issue is rules of evidence, which vary between jurisdictions, but have some general similarities.

46.1.1 Search Warrants

The Fourth Amendment to the Constitution of the United States says: *"The right of the people to be secure in their persons, houses, papers and effects, against unreasonable searches and seizures, shall not be violated, and no warrants shall issue, but upon probable cause, supported by oath or affirmation, and particularly describing the place to be searched, and the person or things to be seized."*

Many states also have constitutional protections against unreasonable searches and seizures. For example, the Constitution of the State of Montana declares: *"The people shall be secure in their persons, papers, homes and effects from unreasonable searches and seizures. No warrant to search any place, or seize any person or thing shall issue without describing the place to be searched or the person or thing to be seized, or without probable cause, supported by oath or affirmation reduced to writing."*

The requirement for both search warrants and arrest warrants comes from these short constitutional passages. The U.S. version and the state version are very similar, which is is case in most states. With respect to search warrants, both constitutions agree that (1) there must be probable cause, (2) the place to be searched must be described, (3) the evidence being searched for must be described, and (4) there must be an oath or affirmation that the details of the probable cause, the location, and the items are true (and in Montana this must be in writing). Both federal and state code provide some exceptions to the requirement for a warrant, such as in cases of arson as discussed in chapter 30. As an example, Montana allows exceptions if the evidence was discovered during an arrest; if the suspect waives their right to a search warrant; or if the rights of the defendant are not infringed by the search (this is difficult to prove). In many situations, peace officers will try to get a suspect to waive their right to a search warrant, usually by simply asking if it's alright to look around.

It is not widely realized that one exception to the requirement for a search warrant is for motor vehicles. The U.S. Supreme Court has ruled that an officer who has probable cause to suspect the presence of evidence or contraband in a motor vehicle may search to the same extent as if he or she had a warrant, including opening the trunk. This ruling was based on the justices' opinion that people have a lesser expectation of privacy in a motor vehicle due to the intense regulation of motor vehicles; and because traffic stops present an urgent situation ("inherent exigency" in legalese) due to the hazardous location along a roadway, the fact that the driver would not wish to be delayed while a search warrant is obtained, and the fact that once the driver drives off any evidence may be quickly disposed of.

It behooves the person conducting the search for evidence to have their search warrant(s) in order. Many cases are lost due to issues with search warrants. The defense attorney in any case that makes it to court will certainly scrutinize any search warrant(s) for irregularities. When you hear of a suspect being "released on a technicality" this is usually the reason.

Probable cause is normally shown by presenting evidence that a crime has occurred, and that evidence related to that crime is present at a certain location. In the vast majority of cases, this has to be presented to a judge, who will issue the search warrant if convinced that the probable cause exists.

Search warrants are not general "hunting licenses" for evidence of wrongdoing. The specific location and the specific evidence to be searched for must be approved by the judge who issues the warrant. This normally means that evidence of other crimes must be ignored.

Swearing and affirming an oath are two ways of affirming its truth. In a **sworn oath** or statement, the person makes a reference to God that the statement it true. An **affirmation** is a similar solemn statement without reference to God. Often, a written statement, sworn or affirmed to be true under penalty of perjury is called an affidavit.

In most jurisdictions a peace officer may ask a judge for a search warrant over the telephone. When this is the case, some strict guidelines have to be followed. First, the officer must explain why the warrant should be issued immediately without the normally required appearance before the judge. Second, all proceedings that occur as part of the request for the warrant must be electronically recorded by the judge.

46.1.2 Chain of Custody

As discussed in chapter 4, the concept of "chain of custody" refers to the record of who had official custody of a piece of evidence between the time it was collected and the time it appears in court. The main purpose of this is to protect the evidence from tampering.

Forensic scientists have to be sure to carefully record how a piece of evidence was stored, who had access to it, how it was analyzed and what procedures were performed on it. Anything that might change the nature of the evidence has to be recorded and explained.

46.1.3 Rules of Evidence for Admission in Court

In order for a piece of evidence, physical or testimonial, to be used in court it must first be admitted by the judge in the case. Unfortunately, admission of evidence is more like a game than a scientific exercise. The prosecution tries to get any form of evidence supporting the possible guilt of the suspect into court, whether the evidence is fair or even relevant. At the same time they try to prevent the defense from being able to present evidence supporting the possible innocence of the suspect. Likewise, the defense tries to present any evidence supporting the innocence of the suspect, while trying to suppress evidence supporting the guilt of the suspect.

The judge is the arbiter of what evidence is presented at trial and which evidence is not presented. The judge has the final say over what evidence is allowed. There are a few basic principles involved in the judge's decision, though there still remains a large subjective element and one that varies between jurisdictions. For example, Montana does not allow evidence from wiretaps to be presented as evidence, even if the evidence was collected in a state that allows it or under federal authority, which also allows it. So, if a case in Montana rests heavily on wiretap evidence, then it has to be tried in Federal court, where that type of evidence is allowed.

When making a decision about a particular piece of evidence judges try to distinguish whether it is **relevant and fair**. Irrelevant evidence is not allowed, and unfair evidence is also normally excluded. Here's an example of irrelevant and unfair evidence. One common prosecution tactic is to introduce a suspect's record of prior arrests and convictions as evidence showing that the person is already known to be a criminal or to have committed similar crimes. However, this type of evidence is neither relevant nor fair. Arrests that did not result in convictions are irrelevant under all conditions. Prior convictions in which the suspect was found guilty might be relevant during sentencing, but is not relevant or fair to present at a trial. The fact that a person committed a crime in the past is not evidence that they committed the crime with which they are charged, unless some details of the way the crime was committed can be shown to link the crimes together.

Another general principle of admissibility of evidence is that evidence which purports to be scientific, must be shown to have been obtained by the scientific method. As was discussed in chapter 3, this is required by the U.S. Supreme Court decision in the case Daubert vs Merrell Dow.

Judges also attempt to screen out evidence that is unreliable. As discussed in chapter 4 and elsewhere, forensic scientists generally believe that all testimonial evidence is unreliable, but it is difficult to conceive of how a trial could be conducted at all if all testimonial evidence was disallowed. Some types of testimonial evidence are especially unreliable, however. For example, hearsay evidence is often not allowed. **Hearsay** is testimonial evidence about something that was not actually witnessed by the witness. For example, if a witness testifies that they heard something relevant about the case from another person (*e.g.* John told Bob that Bill did it). This type of evidence is second or third hand and its reliability is not acceptable.

46.2 The Conduct of a Criminal Trial

The conduct of a criminal trial is another area in which jurisdictions differ, but there is enough commonality to the process throughout the U.S. that the general scheme can be presented here. By examining the process of a criminal trial we can more clearly see where forensic scientists fit into the picture in their role as expert witnesses.

A general regularity in criminal trials is that the prosecution gets to speak first and the defense gets to speak last. This has evolved from the idea that the last words that the jury hears are likely to be the most influential. Thereby, the defense is perhaps given a slight advantage, which is a core principle upon which the American system of justice is based.

A trial formally begins with a reading of the charges against the defendant. This consists of a statement of what section of the criminal code the defendant is thought to have violated. The next step is opening statements by both sides. During opening statements no arguments, evidence, or objections are allowed. Instead, the attorneys present an outline or plan of how they intend to present their side of the case. The prosecution goes first, followed by the defense. At the end of opening statements there is usually a brief recess. Recesses are part of the trial process, and their purpose is to allow attorneys to gather their wits and paperwork, and to catch their breaths.

The next phase of a trial is the presentation of cases. It is in this phase that the witness stand becomes the center of attention as witnesses are presented in turn, each giving their testimony. Each witness is examined using a standard process that consists of direct examination, cross examination, redirect examination, and recross examination.

During direct examination the witness will be asked a series of questions by the attorney for their side. These questions are designed to introduce the witness, and their qualifications if they are an expert witness, to the jury; and to allow each witness to testify to what they saw, did, analyzed, found, etc.

During cross examination the attorney from the opposing side asks the questions. The job of the opposing attorney is to destroy the credibility of the witness. If the witness is an expert, the opposing attorney will probably question their qualifications as well as demanding that they justify the way in which they handled the evidence and drew conclusions from it. Another common tactic is to ask questions designed to "trick" the expert witness into drawing a conclusion that is not warranted by the nature of the evidence or to give an opinion in an area outside their area of expertise. If the expert witness allows themself to be led into this trap it is possible that their entire testimony may be disqualified, or that the jury may come to regard the expert as less than fully qualified.

During redirect examination, the attorney for the side that called the witness will attempt to undo any damage the opposing attorney might have done during cross examination. The same testimony presented during direct examination may not be covered again, and the questions must be confined to issues raised by the opposing attorney.

During recross examination, the opposing attorney gets a final chance to discredit the witness, but she or he must confine their questioning to issues raised during redirect examination.

The prosecution will present their case first, and it is during this part of the trial that most forensic scientists will testify as expert witnesses for the prosecution. When the prosecution has exhausted their list of witnesses, they will rest their case. A recess normally follows, as the defense does last minute preparation to present their case. The presentation of the defense case often begins with a motion for dismissal of the case on the grounds of lack of evidence. This motion is rarely granted, but it creates the impression among the members of the jury that the prosecution's case suffers from lack of evidence or credibility. The defense will then present their witnesses, including any experts they have worked with. After the defense has presented all its witnesses and evidence it rests.

The next phase of the trial is the prosecution rebuttal and the defense rejoinder. It is typical for both sides to in some way save their best evidence or witnesses for this part of the trial. However, the foundation for this evidence must have been laid during that side's presentation of their case, and completely new topics are usually not allowed. There is some exception to this rule, for example if a new witness comes forward during the trial. During this phase of the trial it may be possible to admit "hypotheticals" as evidence. **Hypotheticals** are statements or reports made by experts who do not actually testify at the trial.

As usual, the prosecution goes first with their rebuttal. The prosecution may call witnesses who testified previously, or witnesses that were held in reserve for this part of the trial. The goal is to reacquaint the jury with the main points of the prosecution's case, and to present evidence that makes the case appear airtight. After a recess, the defense rejoinder will follow. Like the prosecution, the defense may call previous witnesses or present additional witnesses or evidence. One common strategy is to call a prestigious expert witnesses who can summarize or comment on the testimony of expert witnesses that were previously heard.

As the trial begins to move toward its end, each side presents closing arguments. The purpose of closing arguments is to touch on the highlights of the case and present arguments based on the entirety of the evidence that has been presented. Each side tries to come up with sound bites that will stick in the jury's mind. The attorneys may become highly dramatic, and attempt to appeal emotionally to the jury. The other side's arguments are portrayed as unfair strategies that are not based in truth but just designed to win the case. The closing arguments usually consist of the prosecution summation and the defense summation. Some courts allow a prosecution rebuttal and a defense rebuttal. The prosecution makes their closing arguments first. The defense then follows with their closing arguments. The prosecution rebuttal is a reply and commentary on the summation by the defense, and the defense rebuttal is a reply to the prosecution rebuttal.

Before sending the jury into deliberations the judge will give them instructions. These instructions are not orders to vote in a certain way, but instead usually consist of what the law requires their verdict to be if they believe a certain set of events to have occurred.

The first part of jury deliberation is for them to elect from among themselves a foreperson, who will speak for them to the judge. It is typical for the jury to become polarized, with some favoring guilt and others favoring innocence. Eventually, in most cases, they arrive at a consensus. The phenomenon of a "hung" jury, in which a consensus can not be reached among the jury members is handled in a variety of ways. In some jurisdictions it is grounds for a declaration that the defendant is innocent. In others, it is grounds for a mistrial, which requires the trial to be replayed with a new jury. It is also possible under the Allen act for the judge to order dissenting jury members to vote with the majority under some conditions.

When the jury has reached a verdict, they will notify the courtroom attendants, who will reconvene the trial. The verdict is nothing more than a statement of guilty or innocent. As discussed in chapter 4, sentencing is done in a separate proceeding. The verdict is passed to a bailiff, who presents it to the judge. The judge reads the verdict, then asks the foreperson of the jury if this is indeed their verdict.

46.3 The Roles of Experts

Experts are people with special knowledge, skill, experience, training, and/or education that goes beyond that of ordinary citizens. This definition applies to forensic scientists, and when forensic scientists analyze evidence or testify in court they do so as experts. The role experts play varies from case to case. Let me list some of the more typical roles of experts.

- Educating attorneys and peace officers about things that lie within the expert's area of knowledge. This also includes the nature and results of analyses.
- Analysis of evidence is another obvious role of experts. As forensic scientists, our primary job is to analyze fingerprints, firearms evidence, blood, bones, bugs, or whatever evidence lies within our area of expertise.
- Do literature research or conduct experiments to find the answers to specific questions that arise in a case and which are within the expert's field.
- Locate and recruit other experts, both for help with specific questions and in those instances where the relevant question lies outside the expert's field.
- Prepare demonstrations to present in court.
- Prepare written reports, grant interviews to the opposing side during discovery proceedings, and testify in court. These duties have been discussed in previous chapters with respect to what is expected of forensic scientists.

46.4 Being an Expert Witness

When an expert testifies in court about something within their area of expertise, they are called, logically enough, an **expert witness**. Expert witnesses are to a certain extent treated differently from normal witnesses. For example, as discussed in chapter

8, an expert witness is expected to bring notes and refer to them during his or her testimony. Also, the attorneys from the opposing side are often more highly motivated to discredit an expert witness than other witnesses. This makes being an expert witness a far from pleasant experience. However, serving as an expert witness is part of the job of a forensic scientist – it would be unfair if the expert were not required to take the witness stand where their evidence and testimony can be scrutinized. In this section, I will discuss some of the issues involved with being an expert witness.

46.4.1 What Attorneys Look For

Except for the rare situation where a certain expert is the only one available, attorneys choose the expert they wish to work with. The things they look for in choosing an expert are straightforward, but are important to be aware of. Some of the things attorneys look for are as follows.

- Compatibility. If the attorney and the expert witness can not work together, due to differences in style or personality, then the attorney will not hire that expert.
- Presentation Skills. Presentation skills are important for an expert witness. Judges and juries will generally not be impressed with an expert who is nervous, speaks poorly, or mutters.
- Education, Training, Degrees, Certifications, and Licenses. The main criteria that a judge or jury has available for evaluating the qualifications of an expert witness is the various certificates, diplomas, and similar documents that attest to the expert's training and performance in their field.
- Specific Knowledge. Each case is different, and some cases require expertise in a very specific realm of knowledge or type of analysis.
- Practical Experience. The more experienced the expert is, including the more times they have served as an expert witness, the more acceptable he or she will be to a judge and jury.
- Publications, Teaching, and Special Recognition Within Their Field. These things all indicate that the expert's ideas and work are considered exemplary by their colleagues.
- Availability. Attorneys are much more likely to choose an expert witness from a nearby location than from a far away location. Similarly, they are better served by an expert who they can contact easily, rather than one who rarely answers their phone.

46.4.2 The Issue of Balance

An expert who always testifies for one side (prosecution/plaintiff or defense) is engaging in a pattern that is not well respected by ethical attorneys. Experts who always work for one side are sometimes referred to as "prosecution whores" or "defense whores" (not my terminology). An expert who fairly presents evidence, who works for both sides, and is trusted by both prosecutors and defense attorneys, is best.

This raises two issues, however. The first is that those forensic scientists who work at a crime lab, and many others as well, inevitably appear as witnesses for the prosecution. This only reflects the fact that it is a government that funds the facility they work at. Second, there do exist unethical attorneys who do not seek fair and balanced expert witnesses, but instead seek out those experts who will tell the jury what their side wants it to hear.

46.4.3 Slide: How Much Are Experts Paid?

Asking how much an expert is paid is like asking "How much does a house cost?" The answer is "It depends." First, experts are not paid for their testimony. This would be perceived as "buying someone's testimony" and is unacceptable. What experts are normally paid for is their time – the time they spend in analyzing evidence, meeting with attorneys, writing reports, and testifying. It is also acceptable for an expert to charge for their expenses, including materials used and travel expenses. Experts who are in high demand often have a minimum fee, which discourages attorneys from consulting them in small or minor cases (for better or worse).

Most experts are paid at rates comparable to the normal fees they earn from other assignments. For example, in the few rare instances I have been paid for my work as a forensic anthropologist, I have charged the hourly rate that my university calculates that they pay me. If the expert demands an amount that is abnormally large, the judge or jury may become suspicious that the expert is being paid extra in order to present their findings in a way more favorable for the side that hired them than the evidence actually warrants.

46.5 Ethics of Forensic Scientists

The American Academy of Forensic Sciences (AAFS) has a code of ethics that their members subscribe to. Since the vast majority of forensic scientists in the U.S. are members of the AAFS, this can be considered the basic code of ethics for forensic scientists. The only substantial tenet of the AAFS code is the matter of misrepresentation of one's expertise or the results of one's analysis. The AAFS code says: *"Every member of the AAFS shall refrain from providing any material misrepresentation of education, training, experience or area of expertise."* Elsewhere it says: *"Every member of the AAFS shall refrain from providing any material misrepresentation of data upon which an expert opinion or conclusion is based."*

Personally I think this code of ethics is weak. Note that there is no specified ethical obligation to seek and present the truth, nor is there an ethical obligation for fairness, balance, or impartiality. Also missing is an ethic of confidentiality. Perhaps this is unnecessary because most jurisdictions have laws that govern what information an expert witness (or any witness) may reveal to whom.

46.6 Wrongful Convictions

Jurisprudence is also concerned with **wrongful convictions**, which are convictions of innocent people. **Exoneration** is the reversal of a conviction when the person is proven to be innocent. Wrongful convictions are an important issue for two reasons, for reasons I have given several times previously. First, an innocent person is punished for something they didn't do. Second, the real criminal is still out on the street, and may offend again.

The American system of justice is a fairly good one, but it does have problems and flaws. Although many aspects of it are designed to err on the side of letting a guilty party go free rather than putting an innocent party in jail, mistakes are at least occasionally made. In the past it has been difficult to get a realistic estimate of how many wrongful convictions occur. Recently, new understandings of wrongful convictions have come from applying DNA evidence to older cases where a person is in prison for a crime, but the DNA evidence from the scene of the crime excludes them from being the perpetrator of the crime. One effort of this sort is the Innocence Project, based at the Cardozo School of Law in New York City. Nobody is willing to give exact statistics about the frequency of wrongful convictions at this point, but the number is being found to be surprisingly high.

We can profitably ask why wrongful convictions occur. The Innocence Project has compiled statistics for 70 cases in which a convicted person was exonerated based on DNA evidence. Although this is a relatively small sample, the results are informative, and in many ways predictable. What the Innocence Project found in their study is that the number of wrongful convictions based on forensic science is fairly small, except for cases where the main evidence against the person was traditional serological evidence or hair evidence. They found that serological and hair evidence supported the wrongful conviction in a fairly large number of cases, as may be predicted based on the fact that neither serology's genetic markers nor the characteristics of hair include individual characteristics. However, I think this observation is misleading because of the fact that DNA was the evidence used in the exoneration. All 70 cases are ones in which the original evidence included a source of DNA, and both blood and hairs are sources of DNA. This means that the sample is biased in favor of cases where the serological evidence or hair evidence was wrong. Also, we have to remember that the genetic markers examined by serologists are class characteristics and the way to narrow down the probability of anyone other than the suspect being the origin of the biological evidence is to use many markers. In the cases where serology gave the wrong answer we don't know how many genetic markers were used. In the worst case scenario it might just have been the suspect's ABO blood type. So, I don't think we need to condemn serology or hair analysis, but we do need to be aware of the limitation of these forms of evidence. We also need to be aware that since the Innocence Project only takes cases in which a potential DNA sample exists, there is potentially a large number of currently invisible cases in which someone was wrongfully convicted on the basis of other forms of forensic evidence that were incorrect.

What the Innocence project found was that the causes of wrongful convictions, in order from most common to least common were: mistaken identity, misapplication of serology, police misconduct, prosecutorial misconduct, defective or fraudulent science, defense attorney incompetence or misconduct, misapplication of hair evidence, false witness testimony, false testimony from informants, false confessions, misapplication of forensic evidence other than serology or hairs, and misapplication of DNA evidence. The finding that misapplication of forensic evidence (with the exception of serology and hairs) was the least common item on the list is encouraging, and suggests that the trend in the justice system toward greater use of forensic science is a positive one.

In addition to greater reliance on forensic evidence, there have been other changes in the justice system since the 1960's that should help to prevent wrongful convictions. One of the more important of these changes has been strengthening of rules of evidence, which have made it more difficult for police misconduct to affect a case. Yet, this still remains an issue. We don't like to believe that the police would be unprofessional or unethical, yet they remain as human as the rest of us. Police are often under enormous pressure, especially in high profile cases, to bring someone, anyone, to justice, even if it's the wrong person. Prosecutorial and defense misconduct can arise from the same pressures to put someone in prison. The situation isn't helped by the fact that government attorneys, especially public defenders, are overworked and underpaid. These problems can only be combated from within the justice system itself.

Despite these advances, much more needs to be done. Many people, including conservative Supreme Court Justice Sandra Day O'Connor, have called for reform of the justice system. However, such reform has to overcome the inertia of a justice system that has evolved over hundreds of years to its present form. Another problem is that it is difficult to envision what shape such a reform would take. Any reform that would make it more difficult to convict an innocent person would also make it more difficult to convict a guilty person. Society is already paying the price of the difficulty of convicting the guilty in the form of criminals who are free because the evidence is not strong enough to convict them. Clearly, something needs to be done, but exactly what remains to be worked out.

46.7 The 2009 National Academy of Sciences Report

In early 2009 the National Academy of Sciences (NAS) reported on a several-year long study of the forensic sciences, focusing on criminalistics. They concluded that all was not well in the forensic sciences, and called for several changes. First, they called for the establishment of a National Institute of Forensic Sciences to oversee and enforce standards both for forensic scientists and for the evidence and methods they uses. Such official bodies are found in a number of countries. Second, the report concluded that additional research was needed to establish the reliability of current methods and develop new ones for the future. Third, crime labs should be independent of law enforcement agencies and prosecutor's offices. Fourth, that accreditation and certification should be mandatory for more forensic scientists. Fifth, that several of the forensic sciences did not have a strong enough base of evidence on their reliability and that validation studies were needed to determine whether they were as reliable as we have previously assumed. Finally, they concluded that courtroom testimony needs to be more grounded in science and that the statistical probabilities of findings should be made part of expert testimony.

This is a significant report, and for the most part the forensic science community has taken it seriously. Some opposition to the report has emerged at local crime labs, mostly taking the position that it is not the forensic science that is faulty, but instead its use by unscrupulous prosecutors. There is a lot of merit to this argument, yet it seems difficult to deny that forensic science methods and the justice system could benefit from validation studies.

As of this time, the national forensic science community has begun moving forward with new validation studies. However, there seems to be little movement toward some of the other significant recommendations of the report, such as making crime labs independent from law enforcement and the prosecutorial side of the justice system or the establishment of a National Institute of Forensic Science.

46.8 Questions for Study and Review

1. Define jurisprudence as a branch of the forensic sciences.
2. What does the U.S. Constitution and the Constitution of most states demand of search and arrest warrants?
3. Describe why peace officers often do not need a search warrant in order to search a motor vehicle.
4. What does the term "probable cause" mean?
5. What is the difference between a sworn oath and an affirmation?
6. Describe how the terms "relevant" and "fair" apply to evidence.
7. What is "hearsay" evidence?
8. In a trial, which side gets to speak first and which side gets to speak last?
9. Summarize the stages of a criminal trial.
10. What are "hypotheticals" in a trial?
11. List all six of the common roles of forensic scientists and describe them.
12. How is an expert witness different from other witnesses?
13. Discuss the things attorneys look for in an expert witness.
14. In what way is the issue of balance unrealistic for forensic scientists who work at a crime lab?
15. In most cases, how much are expert witnesses paid?
16. What is the basic ethical issue that is addressed by the code of ethics of the AAFS?
17. What are wrongful convictions, and what are at least two ways in which they are bad for society?
18. What is an exoneration?
19. Describe the work of the Innocence Project.
20. What were the major points of the NAS report on forensic science, and what have forensic scientists done in response to this report?

46.9 Sources

Albritton, W. Harold, Thompson, Myron H., and De Ment, Ira, 1999. Standing Order on Criminal Discovery. Online at <http://www.almd.uscourts.gov/Web%20Orders/Criminal%20Discovery%20General%20Order.htm>.

[Anonymous]. n.d. General Information About Expert Witnesses and Consultants [Internet]. [cited 2007 Oct 30]. Available from: http://www.expertpages.com/news/new1.htm.

[Anonymous]. 2006. On the Stand: Expert Witness Share Their Strategies (Discussion). The Forensic Examiner 15(4): S8(2).

Babitsky S. n.d. The 10 Biggest Mistakes Experts Make During Depositions [Internet]. [cited 2007 Oct 30]. Available from: http://expertpages.com/news/ten_biggest_mistakes.htm.

Cantor BJ. n.d. Expert Witness Fees [Internet]. [cited 2007 Oct 30]. Available from: http://www.expertpages.com/news/expert_witness_fees.htm.

Colbridge TD. 1999. Search Incident to Arrest [Internet]. FBI Law Enforcement Bulletin 68(5). [cited 2007 Oct 19]. Available from: http://www.fbi.gov/publications/leb/1999/may99leb.pdf.

Committee on Identifying the Needs of the Forensic Science Community, Committee on Science, Technology, and Law Policy and Global Affairs, Committee on Applied and Theoretical Statistics, Division on Engineering and Physical Sciences, National Research Council of the National Academies. 2009. Strengthening Forensic Science in the United States: A Path Forward. Washington D.C.: National Academies Press.

Connors E, Lundregan T, Miller N, McEwen T. 1996. Convicted by Juries, Exonerated by Science: Case Studies in the Use of DNA Evidence to Establish Innocence After Trial [Internet]. IPT Journal 10(3). [cited 2007 Oct 30]. Available from: http://www.ipt-forensics.com/journal/volume10/j10_3.htm.

Giannelli PC. 1999. Expert Qualifications & Testimony [Internet]. Scientific Testimony: An Online Journal. [cited 2007 Oct 23]. Available from: http://www.scientific.org/distribution/law-review/giannelli.pdf.

Harrison DH. 1991. Bad Science [Internet]. [cited 2007 Oct 30]. Available from: http://www.crimeandclues.com/badscience.htm.

Hendrie E. 1998. Warrantless Entries to Arrest [Internet]. FBI Law Enforcement Bulletin 67(9). [cited 2007 Oct 19]. Available from: http://www.fbi.gov/publications/leb/1998/sept98.pdf.

Hendrie E. 2005. The Motor Vehicle Exception [Internet]. FBI Law Enforcement Bulletin 74(8). [cited 2007 Oct 19]. Available from: http://www.fbi.gov/publications/leb/2005/august2005/august05leb.htm#page22.

Holcomb JW. 2004. Consent Searches Scope [Internet]. FBI Law Enforcement Bulletin 73(2). [cited 2007 Oct 19]. Available from: http://www.fbi.gov/publications/leb/2004/feb2004/feb04leb.htm#page_23.

Holcomb JW. 2005. Revoking Consent to Search [Internet]. FBI Law Enforcement Bulletin 74(2). [cited 2007 Oct 19]. Available from: http://www.fbi.gov/publications/leb/2005/feb2005/feb2005.htm#page25.

Hoover LA. 2007. Getting Schooled in the Fourth Amendment [Internet]. FBI Law Enforcement Bulletin 76(3). [cited 2007 Oct 19]. Available from: http://www.fbi.gov/publications/leb/2007/march2007/march2007leb.htm.

Innocence Project. 2001. The Innocence Project [Internet]. [cited 2007 Oct 30]. Available from: http://www.innocenceproject.org/.

Matley M. 1999. The Expert Ambush [Internet]. [cited 2007 Oct 30]. Available from: http://www.expertpages.com/news/expert_ambush.htm.

Montana Code Annotated 2007. 2007. [Internet]. [cited 2007 Oct 30]. Available from: http://data.opi.state.mt.us/bills/mca_toc/index.htm.

O'Connor T. 2004. Complaints, Indictments, Arraignments, Notifications, and the Pretrial Right to Counsel [Internet]. [cited 2007 Oct 17]. Available from: http://faculty.ncwc.edu/toconnor/arraignments.htm.

O'Connor T. 2004. Pretrial Procedures and Police Testimony [Internet]. [cited 2007 Oct 17]. Available from: http://faculty.ncwc.edu/toconnor/315/315lect08.htm.

O'Connor T. 2006. Format of a Criminal Trial [Internet]. [cited 2007 Oct 17]. Available from: http://faculty.ncwc.edu/toconnor/trialfrm.htm.

O'Connor T. 2006. Affidavits and Warrants [Internet]. [cited 2007 Oct 17]. Available from: http://www.apsu.edu/oconnort/3000/3000lect03c.htm.

Petrowski TD. 2001. Miranda Revisted: Dickerson v. United States [Internet]. FBI Law Enforcement Bulletin 70(8). [cited 2007 Oct 19]. Available from: http://www.fbi.gov/publications/leb/2001/august2001/aug01p25.htm.

Ramage M. 1996. 96-02: Revised State Criminal Discovery Rules, Effective 10/1/96 [Internet]. [cited 2007 Oct 17]. Available from: http://www.fdle.state.fl.us/OGC/Legal_Bulletins/lb9602_9-24.html.

Silverman M. n.d. Litigation: Selecting and Preparing Expert Witnesses [Internet]. [cited 2007 Oct 30]. Available from: http://expertpages.com/news/selecting_and_preparing_expert_witnesses.htm.

The Constitution of the State of Montana [Internet]. [cited 2007 Oct 30]. Available from: http://data.opi.state.mt.us/bills/mca_toc/Constition.htm.

Chapter 47
OTHER FORENSIC SCIENCES

In a survey of the type this textbook undertakes there is inevitably a section devoted to those topics that don't fit well elsewhere. This is that chapter for this text. There are several reasons for being included in this chapter rather than in a more prominent location. Some of the forensic sciences and technologies discussed here have limited applicability (*e.g.* forensic knot analysis). Others are included here because their primary venue is in civil court (*e.g.* forensic economics). Others are included because they are helping professions that work with victims or criminals rather than being the sort of forensic science that analyzes evidence (*e.g.* forensic social work).

I apologize if your favorite forensic science ended up here. Actually, two of my favorites, forensic mathematics and forensic phylogenetics, are described in this chapter. One of them, forensic phylogenetics is mentioned nowhere else except for one sentence in chapter 2.

47.1 Forensic Sciences Used Most Often In Civil Court

Let's start with three forensic sciences that have not been described in detail before because their main application is in civil cases. These are forensic economics, forensic medicine, and forensic meteorology. All are common forensic sciences with large numbers of practitioners, and deserve chapters of their own. In fact, I will not describe them in detail, even here, since this text focuses more on the criminal arena.

47.1.1 Forensic Economics

Forensic economics was mentioned previously in chapter 43. It is the application of economic theory and statistics to questions of economic loss in a legal context. Forensic economists take the approach that people are assets, valued according to the income that they produce. This is not because they dislike people, or fail to appreciate things about people other than the money they make – it is simply that some questions concern the value of a person in terms of their income producing potential. That value is altered by injury or death, and it is the forensic economist's role to determine the resulting change in value.

Forensic economists are often involved in civil suits that arise from an injury or a death. For example, if a pianist loses her left hand in an accident, what is the value of that left hand in terms of her ability to earn money to support themself over her lifetime. In the case of a death, the deceased person may have contributed to the finances of their family. What is the impact on that family now that the person is deceased. These are tough questions that forensic economists can answer.

Other things besides people also need to be assigned a value on occasion. As discussed in chapter 43, economists can estimate the financial damage to a company that arises from some event such as a fraud or a computer system attack. Although a forensic accountant can give a dollar amount to the actual crime, the result of a high profile crime may have an additional impact on a company due to loss of customers and other factors. An economist can estimate the amount of this damage.

Other types of cases to which forensic economists contribute include divorces. For example, in a divorce, if one spouse claims that they help put the other spouse through college and are entitled to a part of the value of the college degree the other spouse earned, a forensic economist can estimate the value of the degree.

Forensic economists can also estimate the value of intangible items, such as a person's reputation, which can be damaged by a wrongful termination from their employment, a false accusation, or some poor judgement on the part of the media. I recently heard of an economist who placed a value of $10,000 per year on the wonderful scenery and favorable living conditions of the city in which I live. This is a cost to all of us who live here in terms of salaries that are better elsewhere, housing costs that are lower elsewhere, and similar considerations.

47.1.2 Forensic Medicine

Here I am using the term "medicine" in the broad sense, and including in the term "forensic medicine" not only conventional medicine as practiced by medical doctors, but also chiropractic, naturopathy, osteopathy, accupuncture, homeopathy and other alternative health systems. The history of forensic medicine is shared by forensic pathology and was discussed in chapter 25.

With the exception of forensic pathology, psychiatry, and forensic podiatry, forensic medicine is mostly seen in civil court, and principally in suits that involve an injury. Doctors of all sorts testify about the nature of a person's injuries, treatments required, prognosis (the expected outcome), the extent of their pain and suffering, and other matters that help in deciding the proper amount of damages to award to the injured person.

47.1.3 Forensic Meteorology

Most of us are aware that a **meteorologist** is someone who studies the weather. Forensic meteorology is the application of atmospheric sciences to investigations. The main application of forensic meteorology is in traffic accident cases, where it is believed that the weather may have played some part. Occasionally, forensic meteorologists may contribute to a criminal investigation that involves some aspect regarding the weather.

Primarily, forensic meteorologists consult weather records to reconstruct the weather conditions at a certain time and place. Most often, this is a time and place where a traffic accident has occurred. Some of the information that can be provided by forensic meteorologists includes the detailed weather conditions prior to and at the time of the incident, including times of sunrise and sunset, wind, rain, snow, ice, temperature, lightning, etc. They can also render an opinion on whether the people involved had notice of the hazardous conditions, and whether the type of hazard involved is common or rare.

47.2 Very Specialized Forensic Sciences

In this section I will discuss four forensic sciences that have a subject matter that is very specialized, and therefore utilized only occasionally in an investigation. The forensic sciences that fall into this category include forensic knot examination, forensic mathematics, forensic phylogenetics, and forensic zoology

47.2.1 Forensic Knot Examination

Forensic knot examiners specialize in knots and ligatures. A **ligature** is any item used to restrain, tie or compress something. Examples of ligatures are various types of rope, string, and twine; nylon panty hose; belts; wire; tape; and clothing; to name just a few. The variety of things that can be used as ligatures is nearly endless. Knots are tied in ligatures to fix the ligature at a certain length or configuration.

There are many cases in which a person is tied up, or some item is tied down or packaged. Most of these cases do not require the expertise of a forensic knot analyst, but if there is something unusual about the knot or ligature, or the circumstances of the case are complex, a forensic knot analyst has the skills to draw out any information possible from the knot or ligature. There are cases where knots and ligatures are actually the cause of death, as in a hanging.

There is an abundance of information that can be drawn from competent analysis of a ligture. Most of these are class characteristics, such as manufacturer, material, size, color, typical use, where and how it may be obtained, and its strength. Individual characteristics are confined primarily to torn ends of the ligature, but there may be subtle variations due to manufacture or use and abuse of the ligature that constitute individual characteristics.

Tying knots, however, is individual. Knot tying is similar to handwriting in that there is both natural variability in the knot tying of a given individual, and in that the knot tying of different individuals is unique. Elements of the theory of handwriting, such as motor programs, can profitably be applied to knots. If the knot is simple, however, it may be difficult to identify the individuality. As in any form of matching, suitable standards are necessary. Because of the uniqueness of each knot, any marks made by that knot on the body of a victim or on some item can be matched to the knot that made the marks.

A complicating factor is that knots are tied either **by habit**, as when we tie our shoe laces, or by **design**, as when I tie a bowline knot and have to recite the jingle "the little squirrel comes out of the hole, goes around the tree, and goes back into its hole". A knot that is tied following a set of instructions, such as from a book, is also considered to be tied by design. This distinction applies more to formal knots that to the types of informal knots that a criminal would normally make when tying up a person or item. Informal knots are often the result of a sort of ad-hoc process in which the person fiddles with the ligature until a knot is achieved. This adds to the uniqueness of the knot, of course.

Not all knots are tied by human hands. For many years I have blamed gnomes and various other mythical beings for the knots that invariably occur in my garden hose, my computer's power cord, or for that matter in any piece of rope or cord I handle that is longer than a meter or so. These occur during normal manipulation of ligatures and does not necessarily imply any intent on the part of someone to tie a knot. Although these accidental knots are usually simple, I run into some darn complicated ones that I am sure must have been tied by gremlins with a high level of expertise.

When knots are used during the commission of a crime, some of their features constitute behavioral evidence that can be used to profile the perpetrator. The knot tying skill of the perpetrator is usually obvious, as is the fact that the perpetrator has some familiarity with certain types of knots that are used by people in certain professions or who have certain hobbies. For example, I have never seen anyone other than my father tie a "trucker's hitch" who was not a present or former truck driver. It may also be possible to gauge the perpetrator's state of mind at the time, under the assumption that someone who was nervous or agitated would not be able to tie as precise a knot as someone who was calm and collected.

47.2.2 Forensic Mathematics

Having recently completed a second B.S. degree, in Computer Science and Mathematical Sciences, for no reason other than the simple enjoyment of it, I am fascinated with forensic mathematics. The TV show, Numbers, is the only forensic science show I have ever watched, and I enjoyed both episodes I saw, even though it is clearly not a reasonable portrayal of forensic mathematics.

Forensic mathematics, as the name implies, is the application of mathematics to generate evidence in an investigation. There are three main applications that I see at present. The first is in the realm of serology and DNA analysis, and involves generating the likelihood of a certain combination of genetic markers being encountered at random. The second is helping forensic scientists and attorneys avoid certain common mathematical fallacies. The third was introduced in chapter 45 and involves helping model terrorist cells so that they can be disrupted most effectively.

The subject of estimating probabilities of combinations of genetic markers was discussed in chapter 20 and need not be repeated here. The second concern of forensic mathematicians provides some object lessons for forensic scientists, and reinforces some of the ideas I have tried to make in previous chapters.

One common fallacy, which we have encountered in several previous chapters, is that forensic evidence stands on it's own. In reality, as we have seen before, many types of forensic evidence need to be validated by independent and competent investigation. Let me present as an example two common fallacies to which forensic scientists and attorneys are subject. In fact, these are known as the prosecutor's fallacy and the defense attorney's fallacy. Both arise from ignoring evidence, or perhaps in not seeking the evidence that independent investigation may provide.

The **prosecutor's fallacy** unfairly limits the range of suspects by ignoring data. Let's say that a suspect has been matched to semen at a crime scene using DNA and that the probability of finding another person with the same DNA profile as a suspect is 1 in 7,000. To commit the prosecutor's fallacy is to believe that this means there is only 1 chance in 7,000 that a person other than the suspect left the DNA. The fallacy lies in the idea that the probability of a certain suspect being the donor of the DNA can be computed from the DNA evidence alone. This implies, illogically, that other evidence in the case is irrelevant. For example, the suspect might be female, therefore incapable of leaving semen at the crime scene; or the suspect may have been filmed in the act, which makes it certain that he was the perpetrator. In neither of these examples is the probability that someone other than the suspect left the semen at the crime scene 1 in 7,000 – in the first case it is zero and in the second case it is 100%.

The **defense attorney's fallacy** overly expands the range of suspects, again by ignoring data. Using the same example as above, assume that 1 in 7,000 males could have left the evidence at the crime scene. Let's say that this crime occurred in Missoula, MT, which has a population of about 70,000 people. To commit the defense attorney's fallacy is to say that since half of the people in Missoula are male, 35,000/7,000 = 5 males exist in Missoula that could have left the evidence at the crime scene. While this is mathematically true, it is not likely that all 35,000 males, or even all five with the suspect DNA profile had access to the crime scene. Likely, if all five could even be identified, four of them likely have an alibi. What the defense attorney's fallacy ignores is the evidence from the investigation of the case, which is likely to have provided a small list of men who might have had access to the crime scene. If among this list one man's DNA profile matches and those of the others don't match, it's very highly likely that this man was the perpetrator.

Modeling terrorist cells is a new area for forensic mathematics. It relies upon a branch of mathematics known as graph theory. Let's take as an example the terrorist cell shown in figure 47.1, which is allied with a group seeking the downfall of country of Mordor and it's leader Sauron.

Figure 47.1: An Example Terrorist Cell

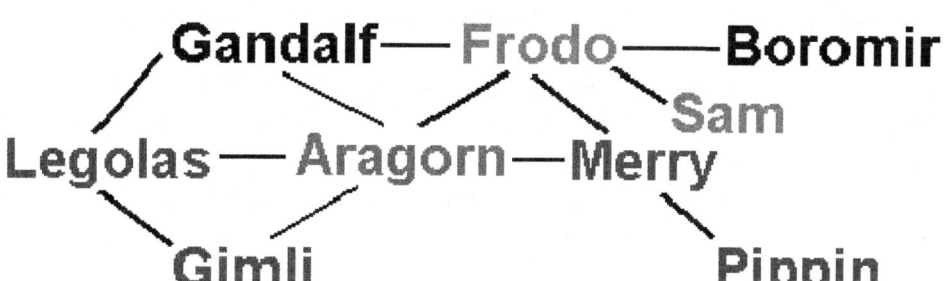

In figure 47.1, the lines connecting the members of the terrorist cell represent who communicates effectively with whom. This diagram, a graph, in mathematical terms, allows us to judge the degree of **connectedness** of each terrorist. Aragorn and Frodo tie for being the most highly connected, because both connect with 5 of the 9 terrorists. These highly connected individuals are normally the ones that the authorities want to take out of the picture first. Those of you familiar with the Lord of the Rings books or movies know that, unfortunately for Sauron and his allies, they were only able to take out Gandalf and Boromir (and Galdalf only temporarily). This resulted in splitting the cell into two smaller cells. Frodo and Sam formed an execution cell of their own with an independent mission, which left most of the cell intact with Aragorn as the focus.

Another feature of the terrorists, which is visible from the graph of figure 47.1 is betweenness. **Betweenness** is gauged by how many separate groups would be formed if this person were taken out of the picture. Frodo is the most "between" of the group, since without him the group would break into three parts: Boromir alone, Sam alone, and everyone else. Merry is the second most "between" person, because without him Pippin would be unconnected to the rest of the cell. The authorities definitely want to neutralize these people, and again Sauron's forces were unsuccessful in doing this.

The third feature of the relationships between these terrorists is closeness. **Closeness** is a bit more complex, and is measured by the number of times a person is on a shortest possible path between two others. These are the communicators of the group, and they are the best people to wiretap or surveil. Frodo is the person with the greatest closeness, followed by Aragorn.

A forensic mathematician might recommend the following optimal strategy for neutralizing this terrorist cell. First arrest or kill Frodo and Merry, then wiretap Aragorn's cell phone. The fact that Sauron apparently had no forensic mathematicians to consult with ultimately led to his downfall.

47.2.3 Forensic Phylogenetics

Phylogenetic analysis is the term used for a variety of techniques that are used to determine evolutionary relationships between living species, extinct species, or both. In actuality, it can be used to examine changes that occur through time between any type of items that have a historical relationship. I was a phylogeneticist, in a non-forensic arena, long before I was a forensic anthropologist, and therefore this is my favorite forensic science of all.

The applicability of forensic phylogenetics is very limited, however. The cases so far have involved tracing strains of contagious disease to determine who passed the disease on to whom. In most cases, this involves a disease, such as HIV, which mutates so rapidly that a direct comparison of the infectious microorganisms of two people would not produce a match even if they contracted the disease from the same source. A forensic phylogeneticist can make a match in a situation such as this by showing that the two infectious microorganisms both evolved from an ancestor that they shared recently.

For example, in one case, a doctor was accused of trying to kill his former lover by injecting her with blood from a patient infected with HIV. The woman did test HIV positive. In order to prove that she was infected by HIV from the doctor's patient, the lover's HIV was compared to the patient's HIV. An expert on HIV, who was also familiar with phylogenetic methods discovered that in fact the lover's HIV was closely enough related to the patient's HIV that it was likely that the patient was the source of the lover's HIV infection.

47.2.4 Forensic Zoology

Forensic zoology is the application of knowledge about non-human animals (hereafter **fauna**) to criminal investigations. There are two main applications, the first being the identification of the species of deceased fauna or their parts. The second is the application of what is known about the behavior of fauna in cases that involve an encounter between humans and fauna.

The species identification of fauna or their parts is useful in conjunction with forensic anthropology. If bones are submitted to a forensic anthropologist, who determines that they are not human bones, a forensic zoologist can determine which species they actually belong to. Species identification of fauna and their various parts is also important for the enforcement of wildlife protection laws, as will be described in the next chapter.

A knowledge of faunal behavior is sometimes important in figuring out what happened in cases where humans and fauna interact. For example, if a grizzly bear mauls a person, it is important to know why this event occurred. If the bear considers humans to be a food source, then authorities may want to destroy it. However, if it was a mother bear defending its cubs from humans who were too close, then another approach may be called for. Also, forensic anthropologists often encounter human bones that have been scavenged by fauna. The scavenging pattern of some species of carnivores produces characteristic and sometimes alarming distributions of the bones. For example, when a skeleton is found in a wooded area and the hands are missing, peace officers are likely to assume that it represents a homicide in which the perpetrator hacked off the victim's hands to prevent the victim from being identified by their fingerprints. In reality, members of the dog family, particularly coyotes, wolves, and foxes are likely to remove the hands in the course of scavenging. I have consulted with a forensic zoologist on occasion in connection with cases that involve faunal scavenging.

47.3 Forensic Helping Professions

The forensic helping professions, forensic nursing and forensic social work, are emerging as specialties within their respective fields. In both cases the focus is on providing essential services, either to the victims of crimes, or to criminals who are on probation, parole, or have served their sentences.

47.3.1 Forensic Nursing

Forensic nursing has emerged as a specialty within the field of nursing. One company that provides forensic nursing services describes forensic nursing as "the application of forensic aspects of health care combined with the bio-psychosocial education of the registered nurse in the scientific investigation and treatment of trauma and/or death of victims of abuse, violence, criminal activity, and traumatic accidents." This is a wordy way of saying that the primary concern of forensic nurses is with people who are the victims of interpersonal violence.

An epidemic of violence and its associated trauma is widely recognized as a critical health problem in North America and throughout the world. Forensic nursing is one strategy for providing an appropriate health care response to the results of criminal and interpersonal violence.

In most cases the forensic nurse provides care to a person, most often female, who has been the victim of a sexual assault or of domestic violence. We need to recognize, however, that providing care is only part of what forensic nurses do. In many cases they act as informal investigators to help identify victims or potential victims. Nor is the subject of a forensic nurses care or attention necessarily a woman, they also stand on the front lines of child abuse and elder abuse issues.

One of the most widely sought services of a forensic nurse is in gathering evidence in sexual assault cases. In many jurisdictions, the crime lab puts together "**rape kits**" that consist of specimen containers, swabs, and various other materials used in the collection of sexual assault evidence. A forensic nurse has the ability to help a victim through this terribly invasive ritual with the greatest amount of dignity. I am neither a female nor a sexual assault victim, but if I were, I would much rather have the necessary evidence collected by a nurse than by a peace officer.

Forensic nurses have also taken the lead in gathering evidence in cases of domestic abuse, child abuse, and elder abuse. This includes photographic documentation of bruises and other injuries, taking victim statements, and counseling the victim.

47.3.2 Forensic Social Work

Social workers are those people who strive to make sure that people have the necessities of life and fit into general society with the greatest facility of which they are capable. It is not uncommon for social workers to encounter victims of crime, especially victims of abuse or neglect. For example, a woman fleeing from an abusive husband may need housing, food, and protection for herself and her children. One of the slogans that have emerged from the crisis of violence is care for victims "from crime scene to court". Forensic nurses often provide immediate care, and the responsibility for care is taken up by a forensic social worker when the victim leaves the hospital.

Also, many social workers encounter criminals among their clientele. These are most often former criminals, who are on probation, parole, or have served their sentence, and are having problems fitting back in to normal society. They may, for example, need help in finding a job and housing.

Forensic social workers have emerged from within their field as specialists in working with people who fall into these categories. The special training they seek allows them to do the job they are hired for better. Like forensic nurses, forensic social workers also have considerable responsibility for investigation, such as identifying evidence of further abuse.

Another realm of forensic social workers is, often under the authority of a court order, follow up on certain clients to evaluate their life skills. For example, in the movie "Mrs. Doubtfire", a social worker appeared at the home of Robin Williams' character to check up on him, as ordered by a court in relation to a case that involved custody of children.

47.4 Questions for Study and Review

1. For each of the following forensic sciences, briefly describe it and discuss whether it is more commonly applied to civil cases or criminal cases: forensic economics, forensic medicine, forensic meteorology, forensic knot examination, forensic mathematics, forensic phylogenetics, forensic zoology, forensic nursing, forensic social work.
2. Define the following terms or concepts: value of intangible items, ligature, typing knots by habit or by design, prosecutor's fallacy, defense attorney's fallacy, connectedness, betweenness, closeness, fauna, rape kit.
3. Discuss the class characteristics and individual characteristics of knots.
4. Describe how forensic mathematicians model terrorist cells.
5. Discuss what the slogan "care from crime scene to court" refers to within the forensic helping professions.

47.5 Sources

[Anonymous]. 2007. Unravelling Crime: Knots Can Reveal All Kinds of Clues [Internet]. Gazette magazine 62(2). [cited 2007 Nov 1]. Available from: http://www.rcmp-grc.gc.ca/gazette/vol69no2/knot_e.htm.

Bellis C, Ashton KJ, Freney L, Blair B, Griffiths LR. 2003. A Molecular Genetic Approach for Forensic Animal Species Identification. Forensic Science International 134(2-3): 99-108.

Bernard EJ, Azad Y, Vandammec A, Weait M, Geretti AM. 2007. The Use of Phylogenetic Analysis as Evidence in Criminal Investigation of HIV Transmission [Internet]. [cited 2007 Nov 1]. Available from: http://www.aidsmap.com/files/file1001199.pdf.

Boykins AD. 2005. The Forensic Exam: Assessing Health Characteristics of Adult Female Victims of Recent Sexual Assault. Journal of Forensic Nursing 1(4): 166(6).

Brenner CH. n.d. Forensic Mathematics of DNA Matching [Internet]. [cited 2007 Nov 1]. Available from: http://dna-view.com/profile.htm.

Brownell P, Roberts AR. 2002. A Century of Social Work in Criminal Justice and Correctional Settings. Journal of Offender Rehabilitation 35(2): p1(17).

Burgess AW, Clements PT. 2006. Elder Abuse: a Call to Action for Forensic Nurses. Journal of Forensic Nursing 2(3): 110(2).

Chisnall R. 2000. The Forensic Analysis of Knots and Ligatures. Jacksonville (FL): Lightning Powder Company.

Climatologyical Consulting Corporation. 2007. Forensic Meteorology [Internet]. [cited 2007 Nov 1]. Available from: http://www.ccc-weather.com/.

Connor JM. 2007. Forensic Economics: An Introduction With Special Emphasis on Price Fixing [Internet]. [cited 2007 Nov 1]. Available from: http://www.agecon.purdue.edu/staff/connor/papers/ForensicEconomics.pdf.

Constantino RE, Crane P, Symonds H, Sutton LB. 2002. The Role of the Forensic Nurse in the Assessment of Abuse: among Female Suicide Survivors. The Forensic Examiner 11(5-6): 22(6).

Day SA. 1999. Use of Forensic Economists in Commercial Litigation: a Defense Perspective. Defense Counsel Journal 66(4): p552.

Economatrix Research Associated. n.d. Forensic Economics [Internet]. [cited 2007 Nov 1]. Available from: http://www.economatrix.com/forensic-economics.php.

Fish and Wildlife Research Institute. n.d. Species Identification in Fish and Wildlife Forensic Investigations [Internet]. [cited 2007 Nov 1]. Available from: http://research.myfwc.com/features/view_article.asp?id=26585.

Fitzgerald P. 2002. The Forensic Chiropractic Examiner: Duties and Professional Opportunities, Part I [Internet]. Dynamic Chiropractic 20(13). [cited 2007 Nov 1]. Available from: http://www.chiroweb.com/archives/20/13/03.html.

Guha SJ. 2002. Doctors and Lawyers: `Fixing' Medical Malpractice -- One Doctor's Perspective of a Non-system in Need of National Standardization. The Forensic Examiner 11(1-2): 38(7).

Hunter P. 2006. All the Evidence - the Combination of Molecular Biology, Zoology and Botany in Forensic Science Comes as a Great Help to Crime Investigators. Embo Reports 7(4): 352-354.

Krebs V. 2002. Uncloaking Terrorist Networks [Internet]. First Monday 7(3). [cited 2007 Nov 1]. Available from: http://www.uic.edu/htbin/cgiwrap/bin/ojs/index.php/fm/article/view/941/863.

Krueger KV, Albrecht GR, Ward JO. 1998. It's about Time: the Forensic Economic Evaluation. Journal of Forensic Economics 11(3): 203(1).

Lawson L. 2006. Evidence-based Forensic Nursing Practice: Benefits and Challenges. Journal of Forensic Nursing 2(1): 5(2).

Logan TD, Cole J, Capillo A. 2007. Sexual Assault Nurse Examiner Program Characteristics, Barriers, and Lessons Learned. Journal of Forensic Nursing 3(1): 24(11).

Potter S. 2004. Pieces of Evidence: the Practice of Forensic Meteorology. Weatherwise 57(3): p28(6).

Roberts AR, Springer DW (Editors). 2007. Social Work in Juvenile and Criminal Justice Settings. Springfield, IL: Charles C Thomas.

Salamone S. 1989. Forensic Meteorology. Technology Review 92(3): 7(2).

Steadman DW, Worne H. 2007. Canine scavenging of human remains in an indoor setting. Forensic Science International 173(1): 78(5).

Sugarman N, Glinka C. 1994. Explaining Pain: How You Do It, Who Can Help. Trial 30(11): 92(7).

Tarbell S, Haggard W. 2003. Forensic Meteorology. Bulletin of the American Meteorological Society 84 (1): 24-25.

Tsvetovat M, Carley KM. 2005. Structural Knowledge and Success of Anti-Terrorist Activity: The Downside of Structural Equivalence [Internet]. Journal of Social Structure 6(2). [cited 2007 Nov 1]. Available from: http://www.cmu.edu/joss/content/articles/volindex.html.

Turvey B. 1996. A Guide to the Physical Analysis of Ligature Patterns in Homicide Investigations [Internet]. [cited 2007 Nov 1]. Available from: http://www.corpus-delicti.com/ligature.html.

Vogel G. 1997. Phylogenetic analysis: getting its day in court. Science 275(5306): 1559(2).

Yost JR, Burke TW. 2006. Forensic Nursing: An Aid to Law Enforcement [Internet]. FBI Law Enforcement Bulletin 75(2). [cited 2007 Oct 19]. Available from: http://www.fbi.gov/publications/leb/2006/feb2006/feb2006leb.htm#page7.

Chapter 48
Advanced Topics
FORENSIC SCIENCE FOR WILDLIFE PROTECTION

In this chapter we will explore the application of the forensic sciences to enforcing regulations that protect or apply to wildlife. I have designated this an advanced topic since no new forensic sciences will be introduced. Instead, this topic illustrates an interesting and important application of the forensic sciences.

The chapter will start will a consideration of why animals and their parts are valuable. We will then take a look at wildlife protection laws. Next, the chapter will turn to a study of how forensic scientists become involved with the investigation of violations of wildlife related laws and the wildlife-related crime labs that have appeared to provide forensic science services to law enforcement agencies that enforce these laws. The chapter will conclude with some observations about non-human forensic science.

48.1 Economics and Uses of Animal Products

The economic value of wild animals is staggering. One authority estimates that it is at least a billion dollar industry. While that doesn't seem like much when compared to the budget of the U.S., it's still enough to fuel a large criminal industry. Many criminal activities that exploit wildlife are organized crime operations, with member involved in capture, transportation, and sales of animals and their parts.

Before the rise of civilization wild animals furnished our ancestors not only with food, but also with essential raw materials for daily life. Their skins were the raw materials of clothing and tents, their sinews were used for attaching spear and arrow heads to shafts, their bones and horns were used to make tools, their tusks were used in creating art. Though we now obtain most of these raw materials from plants or from synthetic sources, there is a considerable interest among many people in owning products that include animal parts in them. I own a few myself, including two hunting horns made of cow horn and a knife that has a handle made of deer horn. Of course, my belt and parts of my shoes are leather. These items were all made from legitimately acquired animal parts, and with the exception of the knife handle come from domestic rather than wild animals.

Some of the uses of wild animals or parts of their bodies in modern times include those in the following list.

- Use for food.
- Displaying them in zoos, circuses or other places that make a profit from this activity.
- Use in medical research, where they take the place of human subjects.
- Use as medicines, or supposed medicines. Many things, from bear gall bladders, to chimpanzee testicles have been thought to be cures for

impotency, aphrodisiacs, strength boosters, weight-loss aids, or other types of medicine. Few of them actually work.
- Displaying as trophies or other decorations. If you go to many places here in Montana, you will find trophies such as moose head mounts and bear skin rugs used as decoration.
- Use in various commercial products. Animal skins and parts have been used in numerous products. Some include: alligator shoes and suitcases, colobus monkey capes, fur coats of many sorts, whale oil, turtle shell buttons, gorilla hand ashtrays, and snake skin belts, to name but a few.
- The exotic pet trade uses a lot of animals. Many people want to have interesting and exotic animals as pets.

Another problem is the destruction of animals that are considered nuisances or dangerous. Animals destroyed for this reason range from baboons to prairie dogs, and from grizzly bears to rattlesnakes.

48.2 Wildlife Protection Laws

The problem that we experience in modern times is that there are just too many people, and the human population continues to grow. In order to accommodate our expanding population, we have unavoidably changed the environment to create places for us to grow our food, build our houses, and drive our cars. However, much of this environmental change in the past was not done with an eye toward sustaining faunal populations (or human populations), and this has led to the extinction and near extinction of many species. In modern times we are becoming aware of human caused environmental damage and in some cases taking steps to halt or reverse it. Yet, this is a battle that will continue to be fought.

The use of wild animals as food or as raw materials for items is natural and part of our human heritage, but in modern times it is problematical simply because of the extremely large number of humans that live on Earth. I am not an environmentalist, yet I too feel the urgency to protect wildlife before human greed and sheer stupidity drive many of our familiar species into extinction. The battle has begun. Part of this battle has involved several federal, state, and international laws that have been created to protect wildlife.

48.2.1 State Laws

The first level of protection for wildlife comes from state laws. In the United States, each state has laws that regulate the relationships between humans and non-human animals. Counties and cities may have additional laws that apply within their jurisdictions. These laws regulate treatment and use of domestic animals (animal control, agricultural regulations, etc.), wild animals in captivity (such as in zoos and circuses), animals in research labs, and the hunting or other capture of wild animals. Many states also have strict prohibitions against the introduction of non-native wildlife. While laws vary from state to state, they do a pretty good job of protecting wildlife, if they can be enforced.

The use of forensic science in wildlife protection applies mostly involves contributing to investigations of violations of those laws regulating hunting, trapping, fishing, and other economic, subsistence, or recreational uses of wild animals. These laws regulate when, where, by what manner, and how many of each species of animal can be taken. Capturing or killing wild animals other than as approved by law is known as **poaching**.

48.2.2 Federal Laws

There are many U.S. federal laws that apply to wildlife. Most of them focus on import or export of endangered animals into or out of the U.S., thereby leaving regulation of wildlife native to the U.S. in the hands of the various states.

Some of the most beautiful and interesting souvenirs offered for sale abroad are made from the furs, hides, shells, feathers, teeth, and other parts of creatures threatened with extinction. Although tourists may lawfully buy such souvenirs in a number of foreign countries, it is often illegal to bring them home to the United States. Many of the people selling these items make the argument that the animal is already dead so it doesn't matter if you buy something made from it. This argument is false. Should you buy items fashioned from endangered species, you'd be adding to the demand for such products and supporting a market for which more animals will be killed. When you consider purchasing a wildlife product during your travels, first make sure you can legally bring it home. Don't rely on assurances by the vendor, but instead check with U.S. Customs officials. There are no refunds if your purchase is seized by Customs or by wildlife inspectors, and you might also find yourself subject to penalties.

Federal restrictions on the import and export of protected wildlife also apply to hunters who take trophies, businesses that deal in the animal and animal product trade, and scientists or teachers who use animals for research or educational purposes. Some exceptions are allowed, but most imports or exports of wildlife require that a federal permit be obtained in advance.

The most widely applied of the federal wildlife protection laws is the **Endangered Species Act** of 1973. This law protects species that are thought to be in danger of extinction by prohibiting import or export of any part of them. More than 1,000 species of animals and plants are officially listed under U.S. law as endangered or threatened. Some examples of things that can't be brought into the U.S. includes the following:
- Anything made of sea turtles or their parts including shells, "tortoise" shell jewelry, soup, facial cremes, shoes, handbags, belts, wallets, luggage, and other manufacture items;
- Rugs, pelts, hunting trophies, handbags, compacts, coats, wallets, key cases, or any other type of manufacture item made from the skins and/or fur of endangered or threatened animals;
- Elephant ivory, or art or other products made from it;
- Crocodiles and certain other reptiles, or their parts;
- Marine Mammals, such as seals, whales, dugongs (sea cows), porpoises, walruses, sea otters, polar bears, and manatees; or their parts;
- Certain birds and their feathers, including most migratory birds and most psittacine birds (parrots, macaws, etc.); and

- Live animals classified as injurious wildlife, such as fruit bats, mongooses, walking catfish, and java sparrows.

The U.S. federal government also prohibits the hunting, capturing, or harassing of marine mammals, endangered or threatened species, or raptors (eagles, hawks, and owls) without a special permit. In addition, no hunting or trapping of non-game migratory birds (sea birds, songbirds, etc.) is allowed without a special permit. The federal government authorizes special hunting seasons for certain migratory game birds, such as ducks.

The U.S. Department of Agriculture (USDA) also regulates import and export of plants, birds, and certain animals that might be carriers of disease. Some animals are prohibited; others must be held in USDA Animal Import Centers or quarantine stations for 30 days after entry.

The **Lacey Act** helps foreign countries and our individual States enforce their wildlife protection laws. Under the Lacey Act, it is a violation of federal law to import, export, transport, sell, receive, acquire, or purchase in interstate or foreign commerce any wildlife, including fish, that was taken, transported, possessed, or sold in violation of any State or foreign law, or taken or possessed in violation of other Federal law or Indian tribal law. This essentially means that if an animal was captured, killed, or purchased anywhere in the world under circumstances that were illegal at the time, that this is a violation of U.S. federal laws and can be investigated in the U.S.

48.2.3 CITES

CITES stands for the Convention on International Trade in Endangered Species of Wild Fauna and Flora (pronounced SITE-ease), It is an international treaty designed to protect endangered and threatened wildlife. Nearly 150 nations are members of CITES, including the U.S. Like U.S. federal laws, CITES primarily focuses on restricting import and export of endangered and threatened species or their parts. Import/export of these species requires a permit from the wildlife management authorities of both countries involved.

48.3 Forensic Science and Wildlife Protection

The investigation of violations of wildlife protection laws has a great deal of similarity to death investigation. Most of the applicable state laws regulate what species may be killed or otherwise taken, and restrict the taking of the animal to certain times at certain places. Therefore, the main questions involved are these:
- What is the species of the animal that was killed;
- What is the sex and age of the animal killed.
- What was the cause of death of the animal;
- What was the manner of death of the animal;
- Who killed the animal;
- When was the animal killed; and
- Where was the animal killed.

Those laws that regulate import and export of certain animals and their parts are more strictly concerned with what species the animal belongs to. If the case involves a crime organization controlling the acquisition, transport, and sale of the animal or its parts, then other obvious questions will be important as well.

The most important and commonly asked question in all these investigations is what species the animal belongs to. This is a task for a forensic zoologist, who can often identify the species from even small parts or bones. In other cases the species of animal can be determined from the details of its hairs, as discussed in chapter 10. In cases where the species identification of an animal can not be determined using one of these lower tech methods DNA can be used. It is not possible to exactly identify the species in all cases, but usually at least the taxonomic family or subfamily can be identified. In any case, the taxonomy of the animal is identified as exactly as possible from the available material.

The sex of the animal is important because in some cases the laws apply differently to males and females of the species. For example, during some hunting seasons it may be legal to kill male deer, but not females. Sex determination from anything other than the whole animal is difficult for many species, but can be accomplished using DNA analysis by looking for the presence or absence of certain STR's present only on the Y chromosome, which is present only in males. The age of an animal that may be taken is also sometimes regulated, and so this occasionally becomes an important question in an investigation. Age determination is usually done using a method called **cementum annulation analysis** of the teeth. The roots of teeth are held in their sockets by a calcified substance composed of sugar and protein molecules. A layer of cementum is laid down over the roots of the teeth each year, to form a pattern similar to tree rings. By grinding a slice of the tooth thin enough, the rings can be counted using a microscope illuminated by the transmission method as discussed in chapter 9.

The cause of death of an animal is sometimes obvious. When it is not, the equivalent of an autopsy, called a **necropsy**, may be performed on its remains. Manner of death is a legal category that is based on all the circumstances of the animal's death, and as for human death investigation determines whether the case will be investigated further. The manners of death for fauna are similar to those for humans and usually classified as: killed by a human, natural, accidental, or undetermined. Suicide is not normally recognized as a manner of death for non-humans.

Establishing who killed the animal, when, and where, is done by traditional investigation by a peace officer, usually a game warden. Processing of the crime scene or scenes is similar to traditional crime scene processing, including the requirement for a warrant. The case against a suspect is built in the same manner as it would be in a homicide case, based on a combination of testimonial and physical evidence. The types of physical evidence that may be found include many of those discussed in previous chapters, including fingerprints, trace evidence, firearms evidence, toolmarks and impressions evidence, and DNA evidence. It is not hard to imagine cases in which a crime organization stores records on a computer, thereby producing digital evidence and financial evidence.

It is clear that many of the forensic sciences normally encountered at a crime lab apply in the enforcement of wildlife protection laws, and in some cases the physical evidence from a wildlife related crime is sent to a traditional crime lab. However, crime labs are busy places and wildlife related crimes are often considered low priority by human-oriented crime labs. Therefore, until recently, in order to convict someone of poaching, wildlife agents had to either catch the person in the act of poaching, or catch them immediately afterward with a whole animal in their possession. Once the animal had been reduced to parts, it was too difficult to identify.

48.4 Some State Wildlife Forensic Laboratories

Today the situation is very different, and wildlife related crimes are much easier to investigate because wildlife protection agencies in some states have found it advantageous to establish their own crime labs. One example of this is California, which has a crime lab dedicated to investigations of wildlife issues at California Department of Fish and Game Northern Regional Office number 2 in Rancho Cordova, CA. It may be that California can claim to have the oldest forensic wildlife lab in the U.S., but I have not been able to verify the details of when this lab was first opened because California Fish and Game officials have not responded to my emails inquiring into the history of the facility. As you surely remember from chapter 1, California, specifically the Los Angeles Police Department, opened the first human-oriented crime lab in the U.S.

The most recent wildlife crime lab of which I am aware opened in Idaho in the mid 1990's. The Idaho State Wildlife Forensics DNA Laboratory in Caldwell, Idaho, now known as the Wildlife Health Laboratory, was created by a unique collaboration of state and private organizations. The organizations involved were the Shikar Safari Club International, the Foundation for North American Wild Sheep, the Rocky Mountain Elk Foundation, and the Idaho Department of Fish and Game. It provides DNA services only. Its primary mission seems to be to do sex determinations and species identifications. The lab also plans to do genetics research, primarily on disease resistance in bighorn sheep.

It is possible that the next wildlife forensic laboratory to open will be here in my back yard, in Missoula, MT. There has been talk of this around town, and the conditions are right for the type of state and private collaboration as took place in Idaho. The Missoula area is home to the Montana State Laboratory of Criminal Investigation, the national headquarters of the Rocky Mountain Elk Foundation, and a private lab that does some analyses.

48.5 The Clark R. Bavin National Fish and Wildlife Forensics Laboratory

The premier wildlife related crime lab for the United States is the Clark R. Bavin National Fish and Wildlife Forensics Laboratory, located in Ashland Oregon. It was opened in 1989 by the U.S. Fish and Wildlife service. It offers a full range of forensic services for the investigation of wildlife related crimes.

48.5.1 The History of the Bavin Lab

Most people credit the idea for the Bavin Lab to Ken Goddard. Goddard is an interesting figure, with a strong background in criminalistics, who had apparently worked at the wildlife forensic facility in California earlier in his life. He is also a mystery writer. In 1979, he was hired to be the Federal Fish and Wildlife's Service's first chief of forensics. At that time, the Fish and Wildlife Service had no laboratory, and had no money allocated to build one. Goddard lobbied the government for funds to establish a national laboratory.

In 1987, Congress appropriated the money and Southern Oregon State College donated land on its Ashland campus. The facility opened in June 1989. It was named for Clark R. Bavin, who was the Chief of the Division of Law Enforcement of the U.S. Fish and Wildlife Service from 1972 to 1990. During Bavin's administration, the Fish and Wildlife Service's officers underwent a transformation from game management officials to criminal investigators. Today, the Clark R. Bavin National Fish and Wildlife Forensics Laboratory is a modern, 23,000 square-foot concrete and glass facility that employs more than 30 forensic scientists and a staff of support people.

48.5.2 The Organization of the Bavin Lab

The Bavin Lab is organized into 6 branches, which are broadly reminiscent of the sections of a traditional crime lab as described in chapter 10. These branches include the following.
- The Administration Branch includes the administrators and the evidence collection team.
- The Criminalistics Branch corresponds to the branches of a traditional crime lab that employ criminalists, forensic chemists, and forensic toxicologists. Some of the types of evidence it handles includes wildlife parts and products; chemicals, such as poisons; firearms evidence; fingerprints; and trace evidence.
- The Morphology Branch uses visual and microscopic methods to determine the species of an animal from parts and products that are recovered as evidence. If appropriate, the sex and age of the animal are also determined. In cases where the parts of several animals are recovered the minimum number of individuals present is determined.
- The Pathology Branch conducts necropsies in order to establish cause of death.
- The Serology Branch uses serology and DNA analysis identify species and sex of an animal. These methods can also be used to determine whether two or more samples are from the same individual, different individuals, or different species.
- The Technical Support Branch is primarily responsible for keeping the computers, networks, and miscellaneous high tech services running. Photography and video services are also handled by this branch.

48.5.3 What goes on at the Bavin Lab

In general, the activities at a wildlife forensic laboratory are similar to those at a traditional crime lab. The main difference is that either the suspect or the victim is not a human being. The activity revolves around the many types of evidence that are submitted for analysis. The forensic scientists who work at the Bavin Lab, like their crime lab counterparts, also spend a considerable amount of time preparing reports, and often testify in court.

48.5.4 Applying for Jobs and Internships at the Bavin Lab

Students often ask me whether they can work at the Bavin Lab. The answer, as will all jobs, is maybe. Since the Bavin Lab is a facility of the U.S. Fish and Wildlife Service, all position opening are advertised and filled following the procedures common to all U.S. government positions. This means that if and when a job exists it is advertised everyplace U.S. federal government jobs are advertised. I recommend checking the U.S. Office of Personnel Management website, at <http://www.usajobs.opm.gov/>. The job announcement will specify the qualifications for the job.

The Bavin Lab does have a limited number of volunteer and student project positions. The best way to find out about them is on their web site at <http://www.lab.fws.gov>. A modest amount of browsing around will reveal the process to use in applying for these types of opportunities.

In general, the Bavin Lab hires people with extensive experience in one of the forensic sciences that are used at the lab. This experience is typically gained through working at a traditional crime lab.

Idaho State Wildlife Forensics DNA Laboratory also has jobs and volunteer positions. This is a state laboratory and the application process follows Idaho state guidelines. The best way to find out about these opportunities is to browse their website at <http://wwwnt.state.id.us/fishgame/info.htm>.

48.6 Observations about Non-Human Forensic Science

Wildlife oriented forensic science seems to be a small segment of forensic science, yet it will surely grow as states find the task of protecting their wildlife becoming more of a challenge as a result of expanding human populations. I would bet that demand for forensic wildlife services will continue to grow.

So far, no laboratories exist that provide forensic services to those people who enforce laws related to the protection of plants, fungi, and other organisms that are not animals. There are certainly many endangered and threatened species of non-animal life that also need protection. I see some potential here, but I have not heard or read of any plans for such a facility.

48.7 Questions for Study and Review

1. Discuss how people use wild animals and their body parts.
2. What jurisdictions regulate hunting and which jurisdictions investigate cases of poaching.
3. What approach do most federal laws take in order to protect endangered species that is of relevance to the forensic sciences.
4. How does the Lacey Act help in the enforcement of laws that protect wildlife?
5. What is CITES? Is the U.S. a member of CITES?
6. What are the most important forensic questions in the investigation of wildlife related crimes? Which of these questions is asked most often?
7. What is cementum annulation analysis?
8. What is a necropsy?
9. Describe how the section organization of the Clark R. Bavin National Fish and Wildlife Forensics Laboratory compares with the organization of a typical human-oriented crime lab.
10. Is it most likely that in the future the demand for forensic science in the investigation of wildlife crimes will be less, be greater, or remain the same?

48.8 Sources

[Anonymous]. 1991. The "Scotland Yard" of Wildlife Crime. The Futurist 25(2): 48(1).

Culotta E. 1995. Ivory identity crisis still unsolved. Science 267(5202): 1264(1).

Drew M, Mamer P, Mulholland J, Moran S, Rudolph K. 2006. Wildlife Health Laboratory Study I, Job 1: July 1, 2005 to June 30, 2006 [Internet]. [cited 2007 Nov 1]. Available from: https://research.idfg.idaho.gov/wildlife/Wildlife%20Technical%20Reports/W-179-R-5%20PR06-Wildlife%20Lab.pdf.

Dunn T. 2003. Caught, After the Act: How Crime Solvers Use Scientific Sleuthing to Stay Hot on the Trail of Wildlife Criminals [Internet]. ZooGoer 32(6). [cited 2007 Nov 1]. Available from: http://nationalzoo.si.edu/Publications/ZooGoer/2003/6/Animal_forensics.cfm.

Endangered Species Act of 1973 [Internet]. [cited 2007 Nov 1]. Available from: http://www.nmfs.noaa.gov/pr/pdfs/laws/esa.pdf.

Freund HL. 1990. Getting the goods on poachers. Sports Illustrated 72(6): 10(2).

National Association of Home Builders. 1996. Developer's Guide to Endangered Species Regulation. Washington (DC): Home Builder Press.

Naylor RT. 2004. The Underworld of Ivory. Crime, Law and Social Change 42(4-5): p261(35).

Neill MJ. 1992. Hunting the Human Predator: Scientist Ken Goddard Tracks Criminals Who Prey on Wildlife. People 38(23): 151(4).

Repanshek K. 1995. Tracking Poachers with Forensic Science. Technology Review 98(6): 22(2).

Taylor A, Cooper D. 1998. "Wildlife forensics": New Zealand's pests are Australia's treasures [Internet]. Biology Bytes 1(3). [cited 2007 Nov 1]. Available from: http://www.bio.mq.edu.au/school/mag/intro/98bytes/may98/Bytes_May98.html.

U.S. Fish and Wildlife Service. 2000. Facts About Federal Wildlife Laws [Internet]. [cited 2007 Nov 1]. Available from: http://training.fws.gov/library/Pubs9/wildlife_laws.pdf.

U.S. National Fish & Wildlife Forensics Laboratory. n.d. National Fish and Wildlife Forensics Laboratory [Internet]. [cited 2007 Nov 1]. Available from: http://www.lab.fws.gov/.

Weir K. 2003. Forensic Park: in One of the World's Most Unusual Labs, Scientists Use the Latest Technology to Fight Crimes--against Bears, Fish, and Other Wildlife. Current Science, a Weekly Reader publication 89(2): p10(4).

Chapter 49
Advanced Topics
FORENSIC SCIENCE APPLIED TO HISTORY AND PREHISTORY

There are many questions from history and prehistory that have been answered to some level of satisfaction by the application of modern forensic science. On TV this field is known as "forensic history", which would not be a correct name for the field given contemporary naming schemes for the forensic sciences. Considering the names forensic accounting, forensic botany, forensic computer science, and so on down the alphabet to forensic zoology, would be the application of the knowledge and methods from the science of history to develop evidence that might be used in court. In reality there is no such field. History is not a science – it is one of the humanities, and although the fascinating and abundant body of knowledge that we know as history is important and interesting, it is not generated using the scientific method, and therefore can not be one of the forensic sciences. Further, the raw material for the study of history is not physical evidence, but testimonial evidence, primarily in the form of various kinds of documents.

Since the name "forensic history" is not available, we are left with "historical forensic science" which does not conflict with any known word usage rules within forensic science and seems to adequately capture the goal of the field. I have heard the term "historical forensics", and while this is almost acceptable, we must remember that "forensics" and "forensic science" are two different things, as explained in chapter 1. Forensics is a term that is already use for something else and may not legitimately be used as a substitute or abbreviation for forensic science. There we have it – **historical forensic science**, and that is the term I will use in this text.

49.1 How Can Forensic Science Answer Questions in History and Prehistory?

Most forensic sciences have histories that go back a century or more, but the best and most powerful techniques in almost all of the forensic sciences have only emerged in relatively recent times. Some techniques, such as DNA analysis and almost any form of analysis that involves computers have only been in use since the late 1980's or 1990's. Therefore, there are many cases and incidents that occurred at times in history during which the relevant forensic science or the techniques and technologies that would have shed light on the event, were not available. The term **prehistory** is applied to the times in a region's past before the invention of writing or before there were people living there who recorded events in writing.

Historical forensic science is interesting in that in most cases the goal is not to bring a suspect to justice. Although the goal, or part of it, might be to identify a perpetrator, many of these cases are so old that the perpetrator would have long since passed away. Therefore, historical forensic science is less of an applied field than many of the other topics we have examined in this text. The goal is often simply to satisfy curiosity or to know the truth for truth's sake.

The most well known person in the field of historical forensic science is James E. Starrs. Starrs is an attorney and professor of forensic science at George Washington University. He is the person who has done the most to popularize the application of forensic science to solving historical questions.

There are a variety of questions that have been investigated by historical forensic science. These tend to fall into six broad categories:
- What did this person look like?
- What was the true cause of death of this person?
- Was this person the ancestor of certain persons living today?
- Is the person buried here really who we think it is?
- Did a certain person really die at the time and place history records?
- Are these people victims of human rights violations?

The topic of investigating human rights violations was discussed sufficiently in chapter 28 and will not be repeated here.

49.2 Exhumation

Some of the cases investigated require a person's grave to be opened to give access to their body. The process of excavation of an established grave is known as **exhumation**. Exhumation is often a sensitive subject to the descendants of the person being exhumed, because they may not want their ancestor's body to be disturbed. Therefore, there is a legal process that has to be followed. In some jurisdictions a coroner can order an exhumation for the purposes of establishing the facts of a case, even if that case occurred long ago. In some jurisdictions, the team seeking to do the exhumation must appear before a judge and get a court order. In some jurisdictions, approval of the family or descendants of the individual being exhumed is needed. Some jurisdictions require some combination of these three things. The approval required depends on the case and obtaining this approval is often the most difficult step in doing historical forensic science.

The exhumation is usually performed using a backhoe or other type of ground-moving equipment. Unlike a more traditional forensic recovery, such as performed by a forensic archaeologist, for example, there is not likely to be any evidence inherent in the way the original burial was done. The evidence being sought is usually something inherent to the body or possessions of the buried individual. Greater care is taken once the casket or other burial container is opened. The placement of the body and any other items in the casket may be evidence, depending on the nature of the question being asked.

49.3 What Did the Person Look Like?

As odd as it may sound, some historical forensic science is conducted simply to determine what a historical figure really looked like. This may be true even if there are conventional images of the person in existence. If the person lived so long ago that their appearance has not been definitively recorded, say by photography, their flesh will have decayed long ago. Therefore, the goal is usually to obtain the skull for facial reconstruction, as described in chapter 7. Remember that the success and accuracy of facial reconstruction are quite low. Therefore, it is questionable how accurately the appearance of a person can be reconstructed using this method. Here are a couple of interesting examples.

49.3.1 Kennewick Man

Perhaps the best known facial reconstruction done just to find out what a person might have looked like was the reconstruction of Kennewick Man. Discovered in 1996 in Southeastern Washington, near the city of Kennewick, Kennewick Man probably died about 9,000 years ago. The initial discoverers believed that it had a "Caucasoid" (peoples of Europe, the Middle East and the subcontinent of India) appearance, a characterization that is still believed to be partly true, though misleading. Actually, statistical analyses of measurements of Kennewick Man's skull have repeated shown that he bears the most similarity to the Ainu peoples, a minority group that live primarily on the northern island of Japan. This makes some sense, because the scientific evidence at hand suggests that at least most of the ancestors of today's Native Americans migrated from northern Asia and that the ancestors of the Ainu also migrated to Japan from northern Asia. The Ainu have often in the past been regarded as having "Caucasoid" features. For example, Ainu males are often capable of growing beards that are heavier and bushier than those that most Asian men can produce. Regardless, a facial reconstruction was done on a cast of the skull and the result looked quite a bit like the actor Patrick Stewart (who played Captain Jean Luc Picard in Star Trek: TNG).

I have not included a picture of the reconstruction of Kennewick man for copyright reasons, but images of the reconstruction and details of how it was done are available from the website for the NOVA television program on PBS, at http://www.pbs.org/wgbh/nova/first/kennewick.html.

49.3.2 Early Cajuns

The early history of the Cajun peoples of Louisiana is relatively poorly known. It is known that they began arriving in Louisiana in the middle 1700's after fleeing persecution both in France and in Accadia (present day Nova Scotia, Canada). Cajun is a corruption of the name "Accadian". In the early 1990's two burial crypts were discovered near Thibodaux, Louisiana. They contained the bodies of 18 early Cajuns who were buried during the 1850's. These graves and the artifacts they contained have yielded some good historical information about the Accadian settlement of southern Louisiana. Facial reconstructions were done of a few of the graves' occupants in order to get an impression of what they may have looked like. One of the people whose face was reconstructed was known to have been named Celeste Leontine. The result was narrow faced and finely featured people, who we can certainly imagine as being of French ancestry.

Again, I have not included a picture of the reconstruction of Celeste Leontine for copyright reasons, but images of the reconstruction are available on the Historical Exhumation Project's web site at http://www.forensicartist.com/hep/leontine/3dreconstruction.html.

49.4 What Was the Real Cause of Death?

Many historical figures, ranging from Moses to Elvis Presley, have died under conditions that have left questions in the minds of some people. Sometimes these questions are unfounded. For example, Elvis really is dead. It is indisputable from the results of his autopsy that he died of a massive heart attack, probably brought on in part by his famous overindulgence in sex, drugs, and rock and roll. In case someone lacks a sense of humor, this last part is a sick attempt at a joke.

The list of people for whom the cause of death is questioned or simply unknown is long. In recent years the cause of death for some of these people has been investigated using modern forensic methods applied to old evidence.

49.4.1 Was it Poison?

One example of modern forensic science applied to historical cases is analysis of the hair of the deceased person for traces of arsenic or other poisons. Poisons have other effects besides causing death, and some of the most interesting questions from history involve the speculation that certain peoples' health problems may have been caused by sublethal doses of poisons. This type of investigation requires the expertise of a forensic toxicologist. Here are two famous cases involving a death, and one involving an non fatal but puzzling mystery.

49.4.1.1 Napoleon Bonaparte

Napoleon Bonaparte, the Emperor of France and near conqueror of the entire European continent, spent his final days exiled to the island of St. Helena in the Atlantic. Napoleon is famous for always having his hand inside his shirt pressed against his stomach. The mainstream theory, which goes back to Napoleon's autopsy in 1821 was that he died of stomach cancer. However, there is another theory – that he was poisoned by Count Charles de Montholon under orders from the British.

A few strands of Napoleon's hair, which were collected by bystanders at the time of his burial preparation have been preserved. Some of them were tested by the FBI lab and found to contain elevated levels of arsenic. This caused something of an international stir. The French tested a couple of other strands and found arsenic but concluded that Napoleon had been exposed to it over a long period of time, not just at the time of his death, and so dispute the murder theory. However, those who see Napoleon's death as a conspiracy point out that the most typical way to poison someone with arsenic was to give it to them gradually over several months, in doses low enough that it didn't kill them right away. Arsenic is a cumulative poison, and when administered this way the person can eventually be killed by a comparatively small dose.

So, the cause of Napoleon's death remains disputed, but arsenic poisoning seems possible, if not likely.

49.4.1.2 Zachary Taylor

Zachary Taylor, the 12th president of the United States, died suddenly, and very conveniently for his political opponents, in 1850. The 1840's through 1850's was an era of extreme political tension over the issue of slavery. This tension erupted in the American Civil War of the 1860's. During the Taylor administration, however the issue was over the admission of California to the United States, as a free state (i.e. a state in which slavery was illegal). At that time, the number of free and slave states was approximately balanced and the coalition of slave states was afraid that allowing more free states to join the Union would put them at a political disadvantage. A famous piece of legislation, called the Compromise of 1850, was drafted to allow California to join as a free state if one of the two next states to join (New Mexico and Utah) would be admitted as a slave state. Taylor opposed this legislation and would have exercised his power of veto, because he believed it was each state's right to choose for itself whether to be a slave state or a free state. This brought the nation to the brink of war. Taylor's sudden death in July, 1850, let the Compromise be passed and postponed the Civil War for another 10 years.

More than 140 years later, some scholars, notably the late William Maples, a forensic anthropologist at the University of Florida, thought that the description of Taylor's last few hours was consistent with being arsenic poisoned. So, with permission from Taylor's descendants, his body was exhumed and his hair tested for arsenic poisoning. The result was that no arsenic was detected. Therefore, Taylor wasn't murdered, and his death was one of those events that occur randomly to change the face of politics. However, conspiracy theorists still believe that Taylor was poisoned, and explain the fact that no arsenic was detected in his hair as due to a flawed laboratory procedure.

49.4.1.3 Ludwig van Beethoven's Hearing Loss

Ludwig van Beethoven, the brilliant German composer, died in 1827 of pneumonia at the age of 56 years. Although his death was unexpected it is not unexplained. What remains a mystery about Beethoven is his deafness, which began afflicting him when he was about 30 years old. Since it isn't recorded that he had an illness that led to his hearing impairment, people have always suspected metal poisoning as the cause. Metals, including mercury, lead, and arsenic, can cause a variety of problems if ingesting in large enough doses. The symptoms they cause can range from fatigue to mental illness. Mercury in particular frequently causes hearing impairment, but other metals can as well.

The prime suspect for causing Beethoven's hearing loss was mercury. Beethoven lived during a time when syphilis was rampant in Europe. The common treatment for syphilis involved rubbing compounds that contained mercury onto the syphilitic lesions, which caused some improvement in the lesion but did nothing to actually cure the disease, which is caused by a bacterium. Therefore, it was believed that Beethoven had syphilis, and was treated with mercury, which led to his hearing loss.

In order to test this theory, forensic scientists again turned to hair analysis. Beethoven was so well-liked that during his funeral so many people cut locks of his hair, that he was nearly bald when they buried him. So, a few of Beethoven's hairs were obtained and analyzed. The result: no lice, no syphilis, and no mercury, but they did find lead. Lead is also known to occasionally cause hearing impairment. Although the source of the lead poisoning is unknown, this seems currently to be the best explanation for why Beethoven lost his hearing.

49.4.2 Was Bad Medicine to Blame?

A version of the "what was the true cause of death?" question is the question of whether a person died due to mishandling by doctors. Good medicine, like good forensic science, is a product of the past century or so, and modern medical authorities believe that the practices of historical times could kill a person as often as save them. Here are a couple of cases.

49.4.2.1 Wolfgang Amadeus Mozart

Wolfgang Amadeus Mozart, another great German composer, died in December 1791 at the age of only 35 and at the height of his career. This is an interesting case, because not only was his cause of death an issue, but his skull had to be identified as his.

During Mozart's time, people who were not nobility were buried in common mass graves, and Mozart was no exception despite his genius. However, 10 years after his death, the gravedigger who buried him returned to the grave and produced a skull that he claimed was Mozart's. Eventually, the skull was confirmed to be Mozart's by the technique of **photographic superimposition** in which photographs of the skull were superimposed upon a painting of Mozart. The features matched well enough to convince most people that the skull was Mozart's, though without an actual photograph of Mozart it's difficult to be completely sure.

There are many theories for the cause of Mozart's early death, including poisoning by a rival, kidney disease, rheumatoid arthritis (which doesn't kill), streptococcal infection, or as the result of a conspiracy of freemasons. However, evidence on the skull (assuming it's Mozart's) gives us a more realistic picture. The skull had a partly healed fracture on the left forehead. This fracture had healed to an extent consistent with an injury occurring 6 months to a year previously. This injury would probably have torn some of the cerebral blood vessels, creating a slow leak of blood into Mozart's skull and putting pressure on the brain. The pressure would have caused headaches, mood swings, depression, and lack of coordination in his right hand, all of which are documented as affecting him during his last few months of life. It is recorded that Mozart's doctors suspected that he had a problem with his brain.

If Mozart had lived in Africa instead of Europe, he could have consulted a healer who would have performed a **trepanation**, which consists of removing part of the bone of the skull. This would have relieved the pressure and eventually allowed the torn blood vessel to heal. If done nothing at all had been done to him, he would have probably have survived and perhaps recovered in time. However, he lived in Europe, at a time when European doctors believed in bloodletting as the cure for everything. In his already weakened state from the constant loss of blood, Mozart was not able to survive the loss of more blood, so soon after his doctors drained a few pints out of him, he died.

49.4.3 Was it Homicide or an Accident

Knowing the true cause of death often forces a reevaluation of the true manner of death. Here is a case that involves a cause of death other than poisoning and which is also prehistoric in time frame.

49.4.3.1 Otzi the Ice Man

Otzi the ice man is an **ice mummy**, a term that applies to bodies that have been preserved by freezing climatic conditions. Otzi is named for the Otzal region in which he was discovered, in the high Italian Alps close to the border with Switzerland, in 1991. He had a few possessions, which have told archaeologists a lot about the way of life of people during the time Otzi was alive, which was about 5,300 years ago, and before people in that region learned to write.

For 10 years after the discovery of Otzi, it was assumed that he had died of exposure to the cold weather of that region after being caught in a disastrous storm high in the mountains. In 2001, however, a scientist doing some x-rays of Otzi spotted what could only be a flint arrowhead embedded in Otzi's left shoulder. After thawing the surface of the cadaver slightly, it could be shown that the entrance wound for the arrowhead was fresh -- not an old, partly healed wound. The arrowhead lay embedded in an area where there are a lot of blood vessels, and it would surely have cut at least one of them, causing Otzi to lose blood, weaken, and eventually die. The wound would not have been immediately fatal, and people now assume that Otzi was able to escape from whoever shot him, only to die later from loss of blood.

49.5 Is This Person My Ancestor?

Tracing family and genetic relationships through time, in situations where marriage, birth, and death documents are not available was impossible before the advent of DNA analysis. Modern DNA analysis, offers two tools, mitochondrial DNA analysis (MtDNA) and Y chromosome DNA analysis. Conventional DNA becomes useless for tracing ancestry after a few generations, because each child inherits DNA from both parents, all four grandparents, all eight great grandparents, and so on.

Although approximately one eighth of a person's genetic material is inherited from, say, great granny, this is generally not enough to make a match using STR technology. MtDNA, however, as described in chapter 21, is inherited entirely from one's mother. Thus, it is passed from mothers to their offspring of both sexes without mixing with the father's DNA. This allows ancestry to be traced through the maternal line for a considerable distance back in time. The DNA contained in the Y chromosome is similar. Because the Y chromosome is only present in males, it is passed from fathers to their sons, intact and unmixed, and therefore can be used to trace descent through male ancestors for a considerable distance back in time.

There are a few types of investigations that ask the question of whether a certain living person is, or a group of people are, the descendant of a deceased person. In this case the identity of the deceased person is usually known, at least to the extent that the question can be asked. Let's look at a couple of examples.

49.5.1 Thomas Jefferson's African-American Descendants

Thomas Jefferson, the third president of the U.S., and author of the Declaration of Independence, owned slaves. Although he spoke against slavery, he either would not or could not free his own slaves. Since Jefferson's time there have been questions about his relationship with one of his slaves, Sally Hemings. It has long been alleged that Jefferson was in love with her, or at least had sex with her, and that she bore him several children, including three sons: Thomas, Madison, and Eston. This is interesting because Jefferson had no surviving sons with his wife. It has been a family tradition in the Hemings family that they are descended from Jefferson.

In 1998, a group of scientists and historians set out to test whether any of Hemings's son's might have been fathered by Jefferson. They did this using the Y chromosome. This proved somewhat difficult, since Jefferson had no surviving unquestioned sons. However, some descendants of Jefferson's paternal uncle, Field Jefferson, were located. These men share a male ancestor (Thomas's grandfather) with Thomas Jefferson, and would have the same Y-chromosome as he had.

Male descendants of two of Sally Hemings' sons, Thomas and Eston, could be located, but Madison Hemings did not have any male descendants who survived the civil war. Comparing Y-chromosomes showed that Jefferson positively could not have been the father of Sally's son Thomas. However, his Y chromosomes matched those of the descendants of Eston Hemings. This is not definitive proof that Jefferson was Eston's father, but either he or one of his paternal relatives was. Some historians point out that Thomas Jefferson's brother Richard was a good candidate. However, we should give some credence to the Hemings family tradition that Thomas Jefferson was the father of at least one of Sally's sons. Historians now tend to accept this. What is not agreed upon is whether Jefferson exploited his vulnerable slave or whether theirs is a touching story of love overcoming the obstacles of different races and social classes.

49.5.2 Descendants of Cheddar Man

Cheddar Man is the name given to a skeleton discovered near the town of Cheddar in England around the beginning of the 20th century, and dating to about 9,000 years ago. One theoretical issue in population genetics studies is how long a genetic legacy can persist in a certain region. This is a good case in point. We do not know who the people living in England 9,000 years were, or what language they spoke. We do know that 500 years or so BCE, Celtic speaking peoples, relatives of today's Welsh and Cornish peoples, arrived in England bringing the Iron Age with them. After the Celts came the Angles and Saxons, and eventually the Normans. The Romans also controlled England for a short time. Given all this migration and undoubted genetic admixture, it is interesting to ask whether there is any legacy of Cheddar Man's people in modern English populations.

In order to test this question, scientists extracted MtDNA from the pulp cavity of one of Cheddar man's teeth in 1997. Cheddar Man's MtDNA was compared to that of people living in Cheddar, and some people who have the same mitochondrial DNA as Cheddar man were identified. One of them is Adrian Targett, a schoolteacher, whose many "greats" removed grandmother was an ancestor shared by Cheddar Man.

The lesson of Cheddar Man can be profitably applied to Kennewick Man, discussed previously, which dates to the same approximate time as Cheddar Man. One question in several legal battles over the custody of Kenniwick Man's remains is whether the Native Americans of the area in which he was found could have been his descendants. Unfortunately, no DNA was recoverable from Kennewick man, yet the example of Cheddar Man shows that it is possible for an ancestral lineage to persist in a region for 9,000 years even when there is considerable migration of peoples from outside the region.

49.6 Is The Person Buried Here Who We Think It Is?

Another question regarding the identity of a deceased person is whether a person buried at a certain location is really the person people think it is. On one hand, this seems as lame a question as "who's buried in Grant's tomb?" (answer: Ulysses S. Grant). On the other hand, there are many cases in which reasons for doubting the stated identity of a person buried at a certain place exist. People can be mistaken, they can forget, or they can even lie. Let's take a look at a couple of cases of this sort.

49.6.1 Giotto

Giotto was a pre-Renaissance painter of Italian frescoes -- a forerunner of Michaelangelo. Most of his work has been destroyed by earthquakes and time, but enough of it remains to establish him as one of the grand old masters. Giotto died in 1337 C.E., and was buried in the Church of Santa Maria Del Flore in Florence, Italy.

In the 1970's three skeleton's were unearthed during renovations. Twenty years later, one of them was identified as Giotto based on forensic anthropological and pathological evidence, which suggested that the skeleton was that of a person who was robust, ugly, talked a lot, kept gazing skywards, injured himself on a scaffold, and kept nibbling on a paintbrush containing supposed paint components of arsenic, manganese, iron, lead, aluminum, copper and zinc. This team also attempted to match the jaw recovered with an existing self-portrait of Giotto. They made a good case, but other scholars have presented compelling evidence that the identification of the bones as Giotto's is incorrect. First, a book written in the 1500's places Giotto's grave on the other side of the church from where the skeleton's were found. Second, some coins were found with the burials that date to about 40 years after Giotto's death. A general principle in archaeology is that a person can be buried with things from the past, but never with things from the future.

So, where are the real Giotto's bones? The side of the church where Giotto is said to have been buried underwent renovation in the 1960's. Ten graves were found, but the head of the Historic Monuments Commission of Florence ordered them destroyed. Giotto's bones are probably lying crushed in the Florence landfill.

49.6.2 The Vietnam Unknown Soldier: Michael Blassie

Every historic war since the Civil War has a memorial grave of the unknown soldier located in Arlington Cemetery, near Washington, D.C. The Vietnam war was no exception, and had an unknown soldier up until 1998. There had always been some evidence that the Vietnam unknown soldier was actually Air Force pilot Michael J. Blassie. The Blassie family had always believed the evidence and when DNA technology became good enough they began pressing for an exhumation. Exhumation and DNA analysis were done in 1998, and the analysis of the DNA showed that the unknown soldier was indeed Michael Blassie. So, at present, there is no Vietnam unknown soldier. Given DNA technology, it is unlikely that there will ever again be an unknown soldier for any of the wars of the present of future.

49.7 Did this Person Really Die as Claimed?

The final identity issue I will discuss is the question of whether a certain person is really dead. Like Elvis Presley, the death of several historical figures is questioned. This is especially true in cases where the presumed dead person is an outlaw or other notorious figure. Enough time has passed in most of these cases that there is no doubt that the person is actually dead, the question is whether they were killed at the time and place that history records. Let's look at a couple of cases of this sort.

49.7.1 Josef Mengele

Josef Mengele, the Nazi doctor, sent 400,000 people to their deaths in the gas chambers of Auschwitz during World War II. He also did "experiments" on countless others. After the war, Mengele escaped justice by fleeing to South America, where he lived for 34 more years before he accidentally drowned in 1979. Mengele spent at least most of those years in Brazil, near Sao Paolo, where he lived under a variety of aliases, including "Peter Hochbichled" and "Wolfgang Gerhard". He was buried in a grave near the town of Embu, Brazil, with a headstone that read Wolfgang Gerhard.

In the early 1980's the information the Mengele had died and was buried at Embu reached Nazi hunters, who organized a team of Brazilian experts to see if it really was Mengele. The skeleton was of a man of the right height and pathology, and the handwriting in some documents written by "Wolfgang Gerhard" matched that of Mengele, which was on record in Germany. Therefore, they concluded that the body was Mengele's. This wasn't good enough for the Nazi hunters, however, who very badly wanted him to still be alive so he could be captured and tried. So, another team was assembled, this time from international experts. This team also concluded that the body was Mengele's, and was also disbelieved by the Nazi hunters. What finally clinched the issue in the minds of everybody in the 1980's, except the most ardent Nazi hunters, was a photographic superimposition done by a German Scientist. He overlaid a picture of Mengele with the skull and got a very good match, down to the gap between Mengele's front teeth. The identification of the deceased person as Mengele was confirmed in 1992 by DNA analysis.

49.7.2 Jesse James

Jesse James was one of many former confederate soldiers who turned outlaw after the Civil War. According to the standard history books, James was killed in 1882 in St. Joseph, Missouri, where he was living under the alias Thomas Howard. History records that he was killed by a member of his gang, Bob Ford, as he turned around to dust off a picture. His body was buried near a family member's cabin in Kearney, Missouri, and was later moved to Kearney's Mount Olivet cemetery.

Almost immediately, rumors began circulating that Jesse James was not dead. There were some problems with the identification of the dead man as Jesse. First, Jesse James had red hair and the dead man supposedly had dark hair. Second, Jesse's mother initially said that it wasn't him, but later changed her story claiming that she had lied so that Ford wouldn't be able to collect the reward money for killing Jesse.

Several people claimed to be the famous outlaw, but their claims were easily dismissed. There is one claim, however, that is still believed by some people today. J. Frank Dalton, who lived in Granville Texas until his death in 1951, claimed for many years to be Jesse James, and "confessed" to this identity on his deathbed. The claim couldn't be dismissed out of hand, and there are many in Granville (and possibly elsewhere) who still believe it. Some investigation has been done to "prove" that Dalton was actually Jesse James including an age progression from a photograph of James in the 1800's, through various hypothetical intermediate steps to Dalton near the time he died. Granville's biggest industry is tourism, and J. Frank Dalton's/Jesse James' grave is a prime tourist attraction.

In 1995, the Kearney, Missouri grave of Jesse James was exhumed in order to do DNA testing to see whether the body within was really that of Jesse James or not. This investigation was headed by James Starrs. They found that the DNA recovered from the skeleton did indeed match that of family members of Jesse James. However, the Granville, Texas crowd are not convinced. In 2000, they exhumed the J. Frank Dalton grave with the idea of testing whether the DNA of the man within matched that of

Jesse's family members. Some bone was sent to my friend and mentor David Glenn Smith, at the University of California, Davis for testing. The test was negative -- no match. Now, however, the Granville crowd are claiming that they got the wrong grave, and that J. Frank Dalton's grave is actually the one next to the grave they exhumed. A new exhumation order is in the works, and we will see whether they get a match this time or not.

49.8 Questions for Study and Review

1. In your own words, define the following terms: prehistory, exhumation, Ainu, photographic superimposition, trepanation.
2. What is the goal of historical forensic science? Is it likely in most cases that a perpetrator will be brought to justice because of a historical forensic investigation?
3. Who authorizes exhumations?
4. What group of people does Kennewick Man most resemble?
5. Compare and contrast the cases of Napoleon and Zachary Taylor.
6. Compare and contrast mitochondrial DNA with Y-chromosome DNA.
7. Compare and contrast the cases of Kennewick man and Cheddar man in terms of the methods that were used to determine who their descendants are.
8. The case of Giotto's grave was solved partially by examining the dates of some coins found with it. Explain why the coins provided strong evidence that the grave was not Giotto's.
9. Explain why there will not likely be a "Tomb of the Unknown Soldier" for the Iraq war.
10. The case of Jesse James provides some examples of testimonial evidence. Apply the things you learned about testimonial evidence in previous chapters to the testimonial evidence in this case to draw a conclusion about whether the DNA results that substantiate the historical count of James' death should be believed.

49.9 Sources

[Anonymous]. n.d. President Zachary Taylor and the Laboratory: Presidential Visit from the Grave [Internet]. [cited 2007 Nov 1]. Available from: http://www.ornl.gov/info/ornlreview/rev27-12/text/ansside6.html.

[Anonymous]. 1994. Napoleon's Last Strands: a Lock of the Emperor's Hair May Solve a 19th Century Mystery. People Weekly 42(11): 107(1).

[Anonymous]. 1997. Caveman's Descendant. Maclean's 110(11): 27(1).

[Anonymous]. 1998. Unknown Soldier Confirmed As Michael Blassie [Internet]. [cited 2007 Nov 1]. Available from: http://www.cnn.com/ALLPOLITICS/1998/06/30/unknown.soldier/.

[Anonymous]. 2000. Lead poisoning seen as probable cause of Beethoven illnesses [Internet]. CNN October 17, 2000. [cited 2007 Nov 1]. Available from: http://archives.cnn.com/2000/HEALTH/10/17/beethoven.hair/.

[Anonymous]. 2001. In Search Of Jesse James: Did Famed Outlaw Die In 1882? [Internet]. [cited 2007 Nov 1]. Available from: http://www.cbsnews.com/stories/2001/02/01/tech/main268780.shtml.

[Anonymous]. 2002. British 'Cleared' of Napoleon's Murder [Internet]. BBC News 29 October, 2002. [cited 2007 Nov 1]. Available from: http://news.bbc.co.uk/1/hi/sci/tech/2371187.stm.

[Anonymous]. 2003. Death of the Iceman [Internet]. BBC Horizon 2nd June 2003. [cited 2007 Nov 1]. Available from: http://www.bbc.co.uk/science/horizon/2001/iceman.shtml.

[Anonymous]. 2005. Zachary Taylor Demands Justice [Internet]. [cited 2007 Nov 1]. Available from: http://www.shoutingground.com/~bigred/ZachTay.html,

Chambers C. 2007. Autopsy in Medical Education. Student BMJ 15: 36(2).

Chatters J. 2000. Meet Kennewick Man [Internet]. [cited 2007 Nov 1]. Available from: http://www.pbs.org/wgbh/nova/first/kennewick.html.

Friedrich O. 1985. Mengele [Internet]. Time 125(25). [cited 2007 Nov 1]. Available from: http://www.time.com/time/magazine/article/0,9171,959474,00.html.

Glausiusz J. 1994. The Banal Death of a Genius. Discover 15(3): 25(1).

Historical Exhumation Project. n.d. Celeste Leontine [Internet]. [cited 2007 Nov 1]. Available from: http://www.forensicartist.com/hep/leontine/3dreconstruction.html.

Historical Exhumation Project. n.d. 145 Years Ago in a Small Cemetery Lay a Family of French Acadians Whose Names Had Long Been Forgotten [Internet]. [cited 2007 Nov 1]. Available from: http://www.forensicartist.com/hep/.

Kubba AK, Young M. 1996. Ludwig van Beethoven: a Medical Biography. The Lancet 347(8995): 167(4).

Darrow S. 1997. DNA Sample Links 2 Men, 9,000 Years Apart [Internet]. CNN Interactive July 31, 1997. [cited 2007 Nov 1]. Available from: http://www.cnn.com/TECH/9707/31/cheddar.man/.

Flores G. 2007. What killed Napoleon? Natural History 116(3): 14(1).

Hall SS. 2007. Last Hours of the Iceman. National Geographic 212(1): 68(14).

Heard A. 1991. Exhumed Innocent: the Grave-digging Craze. The New Republic 205(6): 12(2).

Hebert T. 2002. Acadian-Cajun [Internet]. [cited 2007 Oct 12]. Available from: http://www.acadian-cajun.com/acadiancajun.htm.

Iyer P. 1985. Searches the Mengele Mystery Time 125(25): 38(7).

Lang JS. 1985. Why the Nazi Hunters Keep Pressing on. U.S. News & World Report 98: 31(3).

Lemonick MD. 1996. Hair Apparent: Beethoven's Locks Could Reveal Why He Went Deaf. Time 147(24): 69(1).

Lindelof B. 2000. DNA Tests May Solve Old West Mystery. *The Sacramento Bee* (CA), June 1, 2000 , p. A1.

Marriott M. 1991. President Zachary Taylor's Body To Be Tested for Signs of Arsenic [Internet]. NY Times June 15, 1991. [cited 2007 Nov 1]. Available from: http://query.nytimes.com/gst/fullpage.html?res=9D0CE2D81E31F936A25755C0A967958260&sec=&spon=&pagewanted=print.

Masterson S. 2000. Master Painter Buried Six Centuries Later [Internet]. ABC News January 8, 2000. [cited 2007 Nov 1]. Available from: http://abcnews.go.com/Technology/story?id=119709&page=1.

Mille, DA. 2003. Jesse James [Internet]. [cited 2007 Nov 1]. Available from: http://www.millersparanormalresearch.com/Pages/Jesse_James_Tombstone.htm.

Riley M. 1992. Tales from the Crypt. Time 140(11): 64(2).

Stover D. 1999. Was it Murder?. Popular Science 254(4): 78(4).

Wash WJ. 2000. Beethoven Project: Press Conference - October 17, 2000 [Internet]. [cited 2007 Nov 1]. Available from: http://www.sjsu.edu/depts/beethoven/hair/hairtestpc.html.

Wikipedia contributors. 2007. Compromise of 1850 [Internet]. Wikipedia, The Free Encyclopedia; 2007 Oct 30, 02:16 UTC [cited 2007 Nov 1]. Available from: http://en.wikipedia.org/w/index.php?title=Compromise_of_1850&oldid=167994816.

Wikipedia contributors. 2007. Death of Mozart [Internet]. Wikipedia, The Free Encyclopedia; 2007 Oct 22, 17:09 UTC [cited 2007 Nov 1]. Available from: http://en.wikipedia.org/w/index.php?title=Death_of_Mozart&oldid=166317292.

Wikipedia contributors. 2007. Jefferson DNA Data [Internet]. Wikipedia, The Free Encyclopedia; 2007 Oct 22, 04:36 UTC [cited 2007 Nov 1]. Available from: http://en.wikipedia.org/w/index.php?title=Jefferson_DNA_data&oldid=166212136.

Williams PN. 2000. England, a Narrative History. Part 1: The Prehistoric Period [Internet]. Encyclopdia Brittania. [cited 2007 Nov 1]. Available from: http://www.britannia.com/history/narprehist.html.

Chapter 50
Advanced Topics
PSEUDOSCIENCE

In this chapter we will consider the phenomenon of pseudoscience. I loathe pseudoscience, perhaps because I love science and the fact that pseudoscience masquerades as science angers me. This chapter is designated an advanced topic because no new forensic sciences will be presented. I can not even claim that you will see any science here at all. My goal for this chapter is to define pseudoscience, examine some examples of pseudoscience, explore how to tell the difference between a science and a pseudoscience, and consider how a scientist might become a pseudoscientist. Along the way I also hope to dispel the myth that while pseudoscience isn't helpful it isn't harmful either.

50.1 Introducing Pseudoscience

Pseudoscience is body of belief and/or practice that looks like science, and which may even use the language of science, but which isn't based in ordinary reality. Instead it is based in an extraordinary reality that is self-consistent, but inconsistent with facts and processes of ordinary reality. Ordinary and extraordinary realities were discussed in chapter 3. Pseudoscience is an increasingly common phenomenon. Many of the "science" programs on cable TV, even on those networks that claim to be science oriented, are actually pseudoscientific. This leads me to think that most people, perhaps even many scientists, don't know the difference.

Due to the Daubert decision, mentioned several times in previous chapters, pseudoscientific evidence may not be used in court, because it does not pass the test of having been generated using a scientific method. Therefore, we people who work as part of the justice system darn well better know the difference, and we need to maintain vigilance to make sure that the evidence that makes it to court and claims to be scientific, really is.

50.2 Psychic Detectives

The first pseudoscience we will examine is psychic detectives. There is a program that runs on the Court TV channel named Psychic Detectives. This program seeks to show cases in which psychic detectives have contributed to investigations. In reality, the cases they present are heavily edited with the facts and actual interview dialog distorted to make it seem as if the psychic detectives are contributing to the case, when in reality they are not. The fact that this runs on the Court TV network is disturbing, because it gives the impression that this pursuit is authentic and approved by the court system. Nothing could be farther from the truth.

Psychic phenomena are simply not detectable in ordinary reality, in spite of thousands of experiments searching for evidence of them. Either they do not exist at all, or if they exist they do not manifest themselves reliably enough that they can be used in ordinary reality.

No doubt many of you readers are thinking, no, you're wrong, I've experienced psychic phenomena. Well, so have I, and here we are back at the subject of anecdotes, as discussed in chapter 3. Personal experiences sell products in commercial advertisements, but they do not make up science. The problem is that psychic phenomena are not replicable. Many people have experienced what seems to be some form of psychic phenomenon (including myself), but the problem is replicability. Psychic phenomena may or may not be real, but they are indisputably not replicable. Not everybody experiences them, and those that do can't do so reliably. Therefore, psychic phenomena of all sorts, including psychic detective work, are not science and can not be allowable in court. If, however, you can reliably produce your psychic experience on demand, get in touch with me. We'll do the experiments, and we'll share the Nobel Prize money between us.

50.2.1 Psychic Detectives and Missing Bodies

A common argument runs like this. Even if psychic evidence can not be used in court, it might still be useful in investigations. One area that is widely regarded as one to which psychics can contribute is in locating the body of a murder victim. In fact, one of the nation's most respected detectives, Vernon J. Geberth, has a section on psychic detectives in his widely used book, *Practical Homicide Investigation* (pp. 718-719 in the 2006 edition). The theory is that if a body was discovered using psychic evidence, the judge and jury would still be willing to allow it as evidence in court as long as it was recovered according to the rules of evidence (search warrant, etc.). This theory has yet to be tested, because the number of cases in which psychic detective work has led to the discovery of a victim's body has been zero – despite what you see on TV.

Michael Corn, an investigative producer for "Inside Edition" (a television news magazine), who produced a story debunking psychic detectives, says, "These guys don't solve cases, and the media consistently gets it wrong". Moreover, the FBI and the National Center for Missing & Exploited Children maintain that to their knowledge, psychic detectives have never helped solve a single missing-person case. For example, FBI agent Chris Whitcomb says, "Zero. They go on TV and I see how things go and what they claim but no, zero".

50.2.2 My Experiences With Psychic Detectives

Let me share some anecdotes with you. As is true for all anecdotes they prove nothing, but may be interesting. I have a couple of experiences with psychic detectives myself. In one case I got a call from a county sheriff's deputy asking me to accompany them as they opened a closed mine entrance to search for human remains. The story was that a woman had been missing for a few years and they suspected her husband of murdering her, but couldn't find the body. The officer said that they had long suspected that her body was in this mine, which her husband had dynamited closed soon after she disappeared. However, reopening a mine shaft is difficult and expensive, and they were hesitant to do it without some evidence that the body was there. So, they consulted a psychic who reported that the body was in a "dark place". A sealed mine is certainly a dark place, and on the basis of this "evidence" the sheriff's department decided that the body must be in the mine. I declined to go with them to reopen the mine, but assured them that I would come right away if they found bones. I never heard back from them, but I heard through contacts at the crime lab that they found absolutely nothing.

In another case I received a bone from the crime lab that was sent to them from one of the larger counties in the state. The law enforcement agents in the case had shown it to a psychic who held it in their hands and declared that it was the bone of a woman who had died violently. It turned out to be the cannon bone (bone between the hock and hoof) of a deer. This deer may indeed have been female, and may have died by an hunter's bullet or a mountain lion's teeth, but certainly wasn't a woman.

50.2.3 Formal Studies of Psychic Detectives

What I have written so far is merely testimonial evidence, anecdotes, and we all know that such accounts are not reliable as evidence. Therefore, let's nail shut the coffin on psychic detectives by looking at a more formal scientific study of the results of psychic detectivess.

One of the most thorough studies was done by Arthur Lyons and Marcello Truzzi and published in their book, *The Blue Sense: Psychic Detectives and Crime.* Lyons and Truzzi divide psychic detectives into two groups: deluded, meaning that they actually believe that they have some powers; and frauds, who use common tricks of the magician's trade to seem to know something. Most are frauds and use a few standard approaches to seem to be giving accurate information. Some of these "tricks" are reading the client, shotgunning, and vagueness.

Reading the client involves paying attention to the client's language and body language in order to judge what they want to hear. Some "psychics" are very good at this. Shotgunning refers to providing an enormous number of statements, some of which are likely to be correct. Then, a phenomenon called confirmation bias takes over. **Confirmation bias** occurs when the person judging the accuracy of something remembers the successes and forgets or ignores the failures. Doing this, you can come up with a large list of successes, which looks good if you don't consider the even larger list of failures.

Vagueness is a strategy in which the psychic provides vague information, or information that has a large statistical probability of being correct. For example, if you say that a murder victim's body is "in a dark place", you are almost guaranteed to be correct. Even a body that is not hidden or buried will be in a dark place at night, so the statement is accurate at least part of the time no matter where the body is. Also, in places like Montana, if you say "I see trees close by" you can hardly be wrong. Another trick is to state something and then partially contradict it. For example, the psychic might say "I see wooden boards .... no, maybe it's branches or some other type of wood – I'm not sure". In this case, if there happen to be boards or wood or something similar nearby, this information is judged as accurate, and if not it's still judged as accurate because the psychic wasn't sure.

50.2.4 The Harm that Psychic Detectives Do

The bottom line is that psychic detectives are not helpful to police. In fact they are harmful to the process of an investigation in that they often give information that leads police in the wrong direction, wasting time and resources. The practice is also harmful to the families and loved ones of victims. I read one testimonial by a man who's child had disappeared. Frequently, psychics would call him suggesting a possible location for the child's body, and he would drop everything and go search that location. He never found the body, but in the meantime spent many agonizing hours searching, and missed a considerable amount of work. He reported that he know rationally that the psychics didn't know anything, but would always search anyway because he was tormented by the idea that maybe just this once, they were correct. Unfortunately, they never were.

50.3 Fingerprints and the Mind: Chirology

Let's look at another type of pseudoscience – one which doesn't rely on psychic phenomena and which appears at some level to actually be scientific. This pseudoscience is called **chirology**, known informally as palm reading or palmistry, and branches of it purport to be able to determine an individual's characteristics, or even their future, from examination of their fingers and palms. At first glance, this does not seem to be much different from dermatoglyphics, the study of skin ridge patterns; and palm print matching is certainly a core forensic science. It is tempting to think that if dermatoglyphics is a science, the chirology must also be a science, but this is a mistake.

50.3.1 A Tyranny of Pseudoscience?

Since chirology allows us to determine the characteristics of a person from analysis of their finger and palm prints, including whether or not they are criminals, why even wait for someone to commit a heinous crime. Why not have a national screening program to identify criminals and arrest them before they could commit their crime. The movie "Minority Report" was based on a premise similar to this and perfectly illustrated the problem with this "proactive" approach. The problem is that this seems wonderful, unless you happen to be someone with a finger or palm pattern that indicates criminality. Given that you have this criminal palm pattern, and given that you have perhaps committed a crime, can we allow simply the possession of an incriminating palm pattern to substitute for the investigative work of showing that a crime has been committed and that the perpetrator is you? If you have not committed a crime yet, and are destined to commit one, what should your punishment be? Should you be locked up in prison so that you can not commit the crime in the future? No! God save us from the tyranny of pseudoscience.

Actually, we don't need God to save us, because we already have the Daubert decision. Fortunately, we also have constitutional protections from such repression. Our legal system, for all its faults, seeks to protect us from this form of tyranny. However, there has been at least one case in which a government has adopted such a measure. In the late 1800's Cesare Lombroso, an Italian physician, believed that he could predict criminality based on the "science" of **phrenology**, which examines the shape and measurements of the head. Phrenology is not a science, but a pseudoscience. Lombroso was able to convince authorities that his method was valid and it was adopted for a short time. Fortunately, genuine scientists working with experts in the law were able to convince the authorities to give up this practice, but not before many people were persecuted because of the shape of their head. It is precisely because of this danger that the U.S. Supreme Court ruled in Daubert v. Merrell Dow Pharmaceuticals (1993) that evidence presented in court must be derived by the scientific method.

50.3.2 Chirology Explored

The pseudoscience of chirology has two branches: chiromancy and chironomy. **Chiromancy** seeks to predict a person's future from the details of their hands. **Chironomy** seeks to determine a person's character from the details of their hands. Chirology combines both chiromancy and chironomy and is a synonym for palmistry or palm reading. This pseudoscience bears the same relationship to the study of dermatoglyphics that phrenology does to skeletal anatomy and astrology does to astronomy.

How do we know that palmistry is a pseudoscience? There are abundant clues. First, common sense tells us that there is no way to predict the future with any certainty, even when using the best available 21st century science. Consider for a moment attempts to predict the weather or the stock market. There are two problems with such predictions. They are occasionally, perhaps often, inaccurate; and the more specific the prediction, such as the high temperature tomorrow will be 42 degrees, the less likely they are to be accurate. Also, these predictions focus on averages. The stock market may rise or fall in accordance with predictions, but the performance of any single stock is more difficult to predict. Therefore, any method that claims to predict the future of an individual with any accuracy must be a pseudoscience.

Second, chirology is recognized as a pseudoscience because it is not based on deduction from scientific experiments. The history of chirology shows that it developed from a collection of folk wisdom from a variety of cultures ranging from Hindu to Celtic. Indeed, a good indicator of a pseudoscience is that it is proud of its ancient origins. There are no truly scientific studies that support the validity of chirology by showing that there is an actual verifiable predictive link between a person's hands and their character or future. Studies that purport to do so suffer from clear experimenter's bias (the experimenter effect as discussed in chapter 3).

Third, chirology is recognized as a pseudoscience because there is no plausible mechanism for a link between a person's hands and their future or character. In the case of chirology any such link must be either metaphysical or genetic. A theory of metaphysical linkage takes the form of an assertion that those forces which shape a person's destiny also shape their hand. These metaphysical forces must belong to a special reality and not to ordinary reality, therefore can not be investigated by the scientific method. But, is there a genetic connection between the skin ridges and the mind. If so, chirology could be considered partially scientific.

The hypothesis of a genetic connection stems from the observation that there are some connections between details of the hand and some aspects of a person's nature. However, there are two problems with these connections. First, they are weak. The strongest such observation is the fact that 35% of people with Down's syndrome have ulnar loops on all ten of their fingers, as compared to only 5% of the general population. On the face of it this seems to offer some predictive power because the frequency of having ten ulnar loops is much higher in people with Down's syndrom. However, because Down's syndrome is rare, there are vastly more people without Down's syndrome who have 10 ulnar loops than people with Down's syndrome who have 10 ulnar loops. Actually, I have 10 ulnar loops, and do not have Down's syndrom. Therefore, if you take all of the people with ten ulnar loops , the vast majority of them would be normal, and not people with Down's syndrome. Therefore, although having ten ulnar loops offers some help in diagnosing a child who is experiencing developmental difficulties as having Down's syndrome, it has no predictive power on its own.

The second problem with connections between skin ridges and personal characteristics is that the nature of the connection is poorly known and likely to be indirect. There is nothing about Down's syndrome that causes ulnar loops, nor is there anything about ulnar loops that cause Down's syndrome. Instead, both are probably reflections of some genetic factor related to the 21st chromosome, of which people with Down's syndrome have three instead of the normal two.

Putting this all together, the connections between fingerprints and Down's syndrome do not allow one to predict or diagnose the personal characteristic of Down's syndrome simply by looking at the fingerprints. None of the other connections between characteristics of the hand and personal characteristics are nearly as well established as this one.

50.4 Cadaver Dogs and Other Scent-Trained Dogs

I love dogs and know them well. My constant companion is a blue heeler named Shadow who literally goes with me everywhere, including to class – much to the amusement of my students, I think. I am able to have her with me on a campus that is otherwise inexplicably hostile to canines, due to a disability accommodation. I am tempted to claim that Shadow is the best educated dog in the world, but in fact she sleeps through lectures, never takes notes, and refuses to take tests, so she ends up having to take the same classes over and over. In short, she's not one of my best students.

I have seen dogs trained to do incredible things, such as herding sheep and cattle and working with peace officers. "Police dogs" have been tremendous assets to peace officers in locating and apprehending criminals. In addition, they are of great service in search and rescue operations. In recent times dogs' ultra-sensitive noses have been recruited to help in searches for drugs, explosives, and accelerants; where the fact that a dog reacts to some item or container is now considered probable cause for conducting a warrantless search. Dogs have also been trained to sniff out dead bodies, including body parts and items such as weapons that have blood or tissue on them. These dogs are known as cadaver dogs and are the death investigators of the canine world.

As a forensic anthropologist I have worked directly or indirectly with cadaver dogs on many occasions. I have worked with cadaver dogs trained by a variety of people, and even made an unsuccessful effort to train a cadaver dog myself. I have seen and read many positive descriptions of their work. I have read the book on cadaver dogs coauthored by fellow forensic anthropologist Marcella H. Sorg.

As a result of this observation and research I must regretfully say that the use of cadaver dogs is a pseudoscience. This is not to imply that cadaver dogs can not find cadavers – they can, but they also find every squirrel hole, every place where a deer has urinated, and similar spots interesting to a dog, but irrelevant to the search for a body. I have wasted so much time on investigating, even excavating, locations where a cadaver dog made a "hit" (as the reaction associated with smelling desired evidence is called), that I now insist that the investigating officers find me an actual human bone before I will bother to respond to their request for assistance if cadaver dogs are involved. So far, I am still waiting for a call telling me that they have found such a bone.

Claims of the success of cadaver dogs are clear cases of confirmation bias. Doubtless, many advocates of cadaver dogs will say that I just haven't worked with the right cadaver dogs. Well, I suppose that I haven't worked with the right psychic detectives either. Perhaps the art of working with and training cadaver dogs has simply not progressed to the point where the dogs' ability to locate a cadaver is replicable. Should that become the case I will suspect that the replicability in locating the cadavers rests more in the trained brain of the human handling the dog than in the nose of the dog. Only when I see the scientific study showing that cadaver dogs are able to correctly locate human remains, while ignoring all other signals, at a rate of 95% of the time or better, under double blind studies in which the handlers can not detect signs of soil disturbance or other cues, will I accept that cadaver dogs are a worthwhile addition to the search for a buried body.

What disturbs me even more is the possibility that my characterization of the use of cadaver dogs as a pseudoscience may extend to dogs used to locate other forms of evidence, such as drugs, explosives, or accelerants. Especially disturbing is that if a one of these dogs makes a hit then it is regarded as probable cause for a search. Studies exist that question the scientific nature of trained dogs used in these ways. I think the reality is that the peace officers who handle these dogs have, through long experience, developed the ability to identify people and containers likely to contain some sort of contraband and communicate this information, perhaps unconsciously, to their dogs, who then signal a hit, giving the officer the opportunity to conduct a search. Confirmation bias takes over from there. The only double blind study I have been able to locate, which concerned the abilities of trained dogs to detect a type of explosive, produced these initial results.

> "Of the 46 searches conducted, the dogs responded to the target by indication or marked interest 34 times, of which 17 were full responses and 17 were considered marked

> interest. Thus, half of the indications were clear, whereas the other half were open to handler interpretation. The dogs completely missed 12 times, thus failing to indicate the target odor in 26.1 percent of the searches. Eight false indications, mostly on introduced distractors, were also noted." (Schoon and others 2006)

They authors of this study proceeded from this to revise their test procedure in several arbitrary ways until they got relatively acceptable results, but this tweaking with the procedure only demonstrates the unreliability of the use of trained dogs. Until correctly done double blind studies of scent dogs are produced, the use of dog alerts as justification for probable cause for searching should be disallowed by the court system because it can not be demonstrated to be scientific.

50.5 How Does a Scientist Become a Pseudoscientist?

How does a person with scientific training, such as Cesare Lombroso, the proponent of using phrenology in the justice system of Italy, make the slide from science to pseudoscience? Lombroso was a legitimate scientist – a physical anthropologist like I am. How was he, despite his training, fooled by something so clearly pseudoscientific to us today?

The progression from the proper application of science to improper application of science, to outright pseudoscience often follows a definite pathway, which is informative to examine. This is a different pathway from which a non-scientist may come to embrace a pseudoscience.

There are two types of fallacies that enable a body of valid scientific observation to mutate into a pseudoscience. The first, and probably most important is confirmation bias. The second is in using a non-representative sample to confirm one's findings.

Confirmation bias, discussed earlier in this chapter, occurs when a person has a hypothesis, and tends to notice and remember observations that support their hypothesis while ignoring or discounting observations that refute it. The most common way that refuting evidence is discounted is by explaining it away as an exception. For example, the fact that I have all ulnar loops but do not have Down's syndrome can be explained away (by someone who did not understand the underlying genetics of fingerprints and chromosomal abnormalities) by saying that I must have the genes for Down's syndrome which were not expressed because of some circumstance in my life, or even more insidiously by saying that I don't have the condition now but I must be destined to develop it later in life. By explaining away or otherwise discounting observations that refute their hypothesis a scientist may unintentionally cross over to the "dark side" and become a pseudoscientist. Confirmation bias is a natural human tendency and scientists must be ever vigilant to guard themselves against it.

It is also tempting for a scientist to gather their data using an unrepresentative sample. An unrepresentative sample is one that is different enough from the population the scientist is interested in drawing a conclusion about that the conclusion drawn can not be generalized to that population. The problem of using a non-representative sample occurs when we count the number of people with Down's syndrome who have all ulnar loops. The conclusion that a fairly high percentage of them have this fingerprint pattern is an interesting item of information about the population of people who have Down's syndrome, but it does not apply to the entire population of human beings to say that anyone with all ulnar loops has Down's syndrome. Some historians believe that this was one of Cesare Lombroso's mistakes. These historians believe that he looked at a non-representative sample, namely people who were in prison, and made observations about their skull shapes, which he then generalized to the entire population by saying that people with skull shapes similar to those commonly found among criminals must be criminals. Having formed this hypothesis, supporting it is simply a matter of confirmation bias. If someone with a skull shape common among criminals is convicted of a criminal act, then the hypothesis is affirmed. If such a person is not convicted, then this fact can be explained away by saying that they are a criminal who hasn't been caught yet, that they had a good lawyer who got them off, or that they are a criminal who hasn't yet broken the law but is destined to at some time in the future.

50.6 Why Do Non-Scientists Believe in Pseudoscience?

Non-scientists who believe in a pseudoscience often do so for different reasons than fallen scientists. Typically, they subscribe to the pseudoscientific belief because they accept testimonial evidence or because the belief follows from some aspect of an extraordinary reality in which they already believe. A scientist is not likely to be convinced by an acquaintance's testimonial concerning their belief that the powers of the universe have given us chirology to enable us to know and control our destiny and that this knowledge has improved the acquaintance's life. A non-scientist might be convinced by this type of evidence, however. Also, there is a substantial number of people who distrust science and scientists. Such people might subscribe to a pseudoscience out of spite.

50.7 How To Recognize a Pseudoscience

There are some distinctive differences between a science and a pseudoscience that will help us distinguish between them. These are presented in table 50.1.

Table 50.1: Distinctions Between Science and Pseudoscience

science	pseudoscience	comment
The primary goal of science is to achieve a more complete and more unified understanding of the physical world.	Pseudosciences are more likely to be driven by ideological, cultural, or commercial goals.	Some examples: astrology (from ancient Babylonian culture,) UFO-ology (popular culture and mistrust of government), "structure-altered" waters (commercial quackery.)
Most scientific fields are the subjects of intense research which result in the continual	The field has evolved very little since it was first established. The small amount of research	The search for new knowledge is the driving force behind the evolution of any scientific field.

expansion of knowledge in the discipline.	and experimentation that is carried out is generally done more to justify the belief than to extend it.	Nearly every new finding raises new questions that beg exploration. There is little evidence of this in the pseudosciences.
Workers in the sciences commonly seek out counterexamples or findings that appear to be inconsistent with accepted theories.	In the pseudosciences, a challenge to accepted dogma is often considered a hostile act if not heresy, and leads to bitter disputes or even schisms.	Sciences advance by accommodating themselves to change as new information is obtained. In science, the person who shows that a generally accepted belief is wrong or incomplete is more likely to be considered a hero than a heretic.
Observations or data that are not consistent with current scientific understanding, once shown to be credible, generate intense interest among scientists and stimulate additional studies.	Observations or data that are not consistent with established beliefs tend to be ignored or actively suppressed.	Have you noticed how self-styled psychics always seem eager to announce their predictions for the new year, but never like to talk about how many of last years' predictions were correct?
Science is a process in which each principle must be tested in the crucible of experience and remains subject to being questioned or rejected at any time.	The major tenets and principles of a pseudoscience are often not falsifiable, and are unlikely ever to be altered or shown to be wrong.	Enthusiasts incorrectly take the logical impossibility of disproving a pseudoscientific principle as evidence of its validity.
Scientific ideas and concepts must stand or fall on their own merits, based on existing knowledge and on evidence.	Pseudoscientific concepts tend to be shaped by individual egos and personalities, almost always by individuals who are not in contact with mainstream science. They often invoke authority (a famous name, for example) for support.	Have you ever noticed how proponents of pseudoscientific ideas are more likely to list all of the degrees they have?
Scientific explanations must be stated in clear, unambigous terms.	Pseudoscientific explanations tend to be vague and ambiguous, often invoking scientific terms in dubious contexts.	Phrases such as "energy vibrations" or "subtle energy fields" may sound impressive, but they are essentially meaningless.

50.8 Other Kinds of Defective Science

There are other types of defective science that we need to be aware of and to guard against. These include pathological science, junk science, and just plain bad science. A few other, less common, forms of defective science exist, but a full treatment of the subject is beyond the scope of this book.

50.8.1 Pathological Science

Pathological science is a term applied to cases in which someone thinks they have made a discovery and rushes to press with it, but it turns out that they didn't do the proper checks and their discovery is bogus. Most scientific journals insist on a process of review and replication before they will publish something, but the popular press does not. The remedy for this is to print a retraction, but some people only see the original announcement and not the retraction. Our response to this as forensic scientists might be to remain skeptical about the announcement of new discoveries until it becomes clear that there is no following retraction.

One example of pathological science is the "discovery" of cold fusion. Nuclear fusion is a nuclear reaction that normally occurs at temperatures of millions of degree, such as inside the sun. In 1989, scientists at the University of Utah rushed to press with the story that they had discovered how to achieve nuclear fusion at room temperatures. Everybody was very excited, because "cold fusion", *i.e.* nuclear fusion at room temperature, would be a cheap and inexhaustible form of energy. However, the results of their experiments hadn't been verified by other scientists. It quickly became apparent that not only could scientists at other lab not reproduce the cold fusion experiments, neither could the team from Utah. So, unfortunately, cold fusion still remains an elusive goal.

50.8.2 Junk Science

Another term, **junk science**, is often used to describe scientific theories or data which, while perhaps legitimate in themselves, are mistakenly or intentionally used to support an invalid conclusion. There is usually an element of political, ideological, or commercial interest in the use of junk science. For example, a huge amount of commercial advertising would fall into this category. One example is the advertisements for Ivory soap. Ivory uses the slogan "99 & 44/100% Pure: It Floats". This is junk science because the term "pure" is meaningless when applied to an undefined mixture such as a bar of soap. The assertion that it floats actually argues against the purity of the product, because soap is denser than water and should sink. They create its ability to float by beating air bubbles into it, actually reducing the "purity" of the product, and in a sense cheating the consumer who is buying a bar of soap that consists of a considerable portion of air.

50.8.3 Bad Science

Bad science is a term commonly used to describe well-intentioned but incorrect, obsolete, incomplete, or over-simplified expositions of scientific ideas. An example would be Bohr model of the atom that I used in chapter 11, which envisions electrons revolving in orbits around the atomic nucleus. This model of the atom was discredited in the 1920's, but is so much more vivid and easily grasped than the true relationship that it shows no sign of dying out.

50.9 Questions for Study and Review

1. Define pseudoscience.
2. Discuss why anecdotes do not constitute scientific evidence.
3. What conditions would need to occur for psychic evidence to be considered scientific evidence for purposes of use in court?
4. To the best of the knowledge of people in the law enforcement community, how many cases have been solved by psychic detective?
5. What two types of psychic detectives did Lyons and Truzzi find in their study?
6. Describe "reading the client", "shotgunning", and "vagueness" as techniques used by some psychic detectives.
7. What is confirmation bias, and how does it relate to a person's belief in a pseudoscience?
8. Describe how psychic detectives cause harm.
9. In your own words, define chirology, chiromancy, chironomy, and phrenology.
10. Explain how you would reconcile the fact that having ulnar loops on all ten fingers is used to diagnose Down syndrome with the fact that I have ulnar loops on all ten fingers but do not have Down syndrome.
11. How does confirmation bias explain the "success" of cadaver dogs and other scent trained dogs?
12. Compare and contrast the factors that might lead a scientist to become a pseudoscientist with the factors that might lead a non-scientist to adopt a belief in a pseudoscience.
13. "Intelligent Design", the most current version of "Creation Science" presents a body of evidence purporting to show life on Earth was created by God as described in the Bible and that the findings of evolutionary scientists over the past 150 years are wrong. Apply as many of the criteria in table 50.1 as you can to "Intelligent Design" and draw a conclusion as to whether it is a pseudoscience or not. If you don't know the details of "Intelligent Design" you can easily find them on the internet using your favorite search engine.
14. Compare and contrast "pathological science", "junk science", and "bad science". How do these differ from pseudoscience?

50.10 Sources

Bulzomi MJ. 2000. Drug Detection Dogs: Legal Considerations [Internet]. FBI Law Enforcement Bulletin 69(1). [cited 2007 Oct 19]. Available from: http://www.fbi.gov/publications/leb/2000/jan00leb.pdf.

Campbell ED. 1998. Fingerprint and Palmar Dermatoglyphics [Internet]. [cited 2007 Oct 19]. Available from: http://www.edcampbell.com/PalmD-History.htm.

Carroll RT. 2007. Confirmation Bias [Internet]. The Sceptic's Dictionary. [cited 2007 Nov 1]. Available from: http://skepdic.com/confirmbias.html.

Carroll RT. 2007. Palmistry [Internet]. The Sceptic's Dictionary. [cited 2007 Nov 1]. Available from: http://skepdic.com/palmist.html.

Carroll RT. 2007. Phrenology [Internet]. The Sceptic's Dictionary. [cited 2007 Nov 1]. Available from: http://skepdic.com/phren.html.

Carroll RT. 2007. Psychic Detective [Internet]. The Sceptic's Dictionary. [cited 2007 Nov 1]. Available from: http://skepdic.com/psychdet.html.

Derksen AA. 2001. The Seven Strategies of the Sophisticated Pseudo-scientist: a Look into Freud's Rhetorical Tool Box. Journal for General Philosophy of Science 31(2): 329(22).

Dodes JE. 2001. Junk Science and the Law. Skeptical Inquirer 25(4): 31.

Eve RA. 2007. Science Education and Belief in Pseudoscience: Good News--but the Glass Is Still Two-thirds Empty. Skeptic 13(3): p14(2).

Geberth, Vernon J. 2006. Practical Homicide Investigation: Tactics, Procedures, and Forensic Techniques. Fourth Edition. Boca Raton (FL): CRC Press.

Harrison DH. 1991. Bad Science [Internet]. Midwestern Association of Forensic Science Newsletter [cited 2007 Nov 1]. Available from: http://www.crimeandclues.com/badscience.htm.

Holtzman A. n.d. A Comprehensive Overview [Internet]. Psychodiagnostic Chirology Magazine. [cited 2007 Nov 1]. Available from: http://chirology.org/default.asp?SubId=220.

Hunt R. 1999. The Benefits of Scent Evidence [Internet]. FBI Law Enforcement Bulletin 68(11). [cited 2007 Oct 19]. Available from: http://www.fbi.gov/publications/leb/1999/nov99leb.pdf/.

Kruglyakov E. 2002. Why Is Pseudoscience Dangerous?..Skeptical Inquirer 26(4): 33(4).

Lower S. 2001. Pseudoscience: What is it? How can I recognize it? [Internet]. [cited 2007 Nov 1]. Available from: http://www.chem1.com/acad/sci/pseudosci.html.

Lowy A, McAlhany P. 2000. Human Remains Detection "Cadaver Dogs": The latest Police Canine Detector Specialty [Internet]. [cited 2007 Oct 23]. Available from: http://www.crime-scene-investigator.net/cadaverdogs.html.

Lyons A, Truzzi M. 1991. The Blue Sense: Psychic Detectives and Crime, New York: The Mysterious Press.

Nickell J. (Editor). 1994. Psychic Sleuths. Buffalo (NY): Prometheus Books.

Nickell J. 1998 How Psychic Sleuths Waste Police Resources [Internet]. Skeptical Inquirer Electronic Digest. [cited 2007 Nov 1]. Available from: http://www.csicop.org/list/archive/0093.html.

Patton MD. 2003. Frauds, Hoaxes and Pseudoscience: a Course in Argumentation. Academic Exchange Quarterly 7(4): 204(5).

Rebmann A, David E, Sorg M. 2000. Cadaver Dog Handbook: Forensic Training and Tactics for the Recovery of Human Remains. Boca Raton (FL): CRC Press.

Saeta, PN. 1999. What Is the Current Scientific Thinking on Cold Fusion? Is There Any Possible Validity to this Phenomenon? [Internet]. Ask the Experts: Physics. Scientific American. [cited 2007 Nov 1]. Available from: http://www.sciam.com/askexpert_question.cfm?articleID=0007CC4D-394F-1C71-84A9809EC588EF21&catID=3.

Schoon A, Gotz S, Heuven M, Vogel M, Karst U. 2006. Training and Testing Explosive Detection Dogs in Detecting Triacetone Triperoxide [Internet]. Forensic Science Communications 8(4). [cited 2007 Oct 21]. Available from: http://www.fbi.gov/hq/lab/fsc/backissu/oct2006/research/2006_10_research01.htm.

Stenger VJ. 1996. ESP and Cold Fusion: Parallels in Pseudoscience [Internet]. [cited 2007 Nov 1]. Available from: http://www.colorado.edu/philosophy/vstenger/ESPColdFusion.pdf.

van Mensvoort IMC. n.d. 10 Years- Handanalysis Research: An Overview [Internet]. [cited 2007 Nov 1]. Available from: http://www.handresearch.com/hand/Evolutie/overzichtEngels.htm.

Walker JS. 2001. Using Drug Detection Dogs: An Update [Internet]. FBI Law Enforcement Bulletin 70(4). [cited 2007 Oct 19]. Available from: http://www.fbi.gov/publications/leb/2001/apr01leb.pdf.

Wikipedia contributors. 2007. Cesare Lombroso [Internet]. Wikipedia, The Free Encyclopedia; 2007 Oct 14, 15:38 UTC [cited 2007 Nov 2]. Available from: http://en.wikipedia.org/w/index.php?title=Cesare_Lombroso&oldid=164509964.

Wikipedia contributors. 2007. Cold Fusion [Internet]. Wikipedia, The Free Encyclopedia; 2007 Oct 17, 16:03 UTC [cited 2007 Oct 17]. Available from: http://en.wikipedia.org/w/index.php?title=Cold_fusion&oldid=165204041.

Wojcikiewicz J. 1999. Dog Scent Lineup as Scientific Evidence [Internet]. A paper presented at the 1999 meeting of the International Academy of Forensic Sciences in Los Angeles, CA, USA. [cited 2007 Oct 14]. Available from: http://forensic-evidence.com/site/ID/ID_DogScent.html.

Zanoni MM, Morris A, Messer M, Martinez R. 1998. Forensic Evidence Canines: Status, Training, and Utilization [Internet]. Paper presented at the annual meeting of the American Academy of Forensic Sciences, February 1998 - San Francisco CA. [cited 2007 Nov 1]. Available from: http://www.prusik.com/K9Forensic/AAFS%20paper.html.

Chapter 51
CAREERS AND EDUCATION IN THE FORENSIC SCIENCES

In this chapter I would like to talk about the various careers in the forensic sciences and related fields and how to go about obtaining the education for them. I'm writing this chapter at least partly in self defense, because there are times when I receive up to ten inquiries per day, by email, phone, letter, or someone just walking into my office, about how to become a forensic scientist. At one time, I made up a handout, which my colleagues and I have passed out literally by the hundreds. I have this handout online as well, and point anyone who inquires by email to it (http://www.anthro.umt.edu/studguid/forensic.htm). For those of you patient enough to have waded through this text to this point, here it is – all I know about jobs and education in the forensic sciences and related areas.

51.1 Careers in Law Enforcement

Although not forensic scientists in the typical sense, peace officers use forensic science daily in pursuit of their duties. I believe that most students who enjoy the forensic science programs on TV should consider a job in law enforcement instead of forensic science. The actors in those TV shows may claim to be forensic scientists, but they act like peace officers. If you are attracted by the excitement, you should definitely consider a career in law enforcement – real jobs in forensic science are somewhat repetitive and tedious with little action.

All of you college students have the basic qualifications for being a city, county, or state level peace officer, which is a high school diploma or GED. Some of these local level agencies require an associate or bachelor's degree, but you are working toward that. In addition, you will need to go to police academy, which is usually a three to five month program. Some law enforcement agencies will hire you first then send you to police academy. In this case they will usually pay your academy fees and give you a small salary while you are completing your training. Other jurisdictions expect you to go to police academy first, at your expense, before they hire you. Almost everyone starts their state or local law enforcement career as a patrol officer and moves up with experience. Local and state level peace officer jobs usually have starting salaries in the range of $25,000 to $50,000 per year, depending on the location.

Federal level law enforcement careers usually require a four-year college degree, and for the most part they don't care very much what major your degree is in. In addition, they will provide specialized training, such as the program at the FBI academy. As discussed in chapter 4, there are many federal agencies that employ peace officers, and my advice to any of you who are interested in these positions is to browse each agency's web site. Federal level peace officer positions usually have starting salaries in the range of $30,000 to $55,000 per year.

51.2 Careers that Depend on Certification

In previous chapters I discussed a few forensic sciences for which certification is critically important. While certification is available and important in many of the forensic sciences, it is most important for these. For example, a person earns the right to call themself a Crime Scene Investigator, as opposed to an evidence technician or something similar, by earning Crime Scene Investigator certification from the International Association for Identification. Other branches of forensic science that rely heavily on certifications include forensic photography, forensic accounting, and forensic computer science.

Certifications in these fields are awarded only to people who have certain combinations of specialized training and experience. Therefore, expect that your first position will be as a lower level employee, a technician or assistant, and to work and study hard in that position to earn certification.

51.3 Careers in Criminalistics

By now any reader of this text should know who the criminalists are, and I will not repeat the list here. The basic requirement for an entry level criminalist is a Bachelor of Science degree in one of the physical or biological sciences. Although a degree in chemistry or biology is often preferred, the following major fields are generally considered acceptable for criminalists: chemistry (any type), biochemistry, biology, physics, geology, genetics, mineralogy, petrology, chemical engineering, forensic science, pharmacology, microbiology, molecular biology, biological sciences, immunology, and criminalistics. Regardless of the major you choose, be sure that you take at least a year of general chemistry with lab, plus quantitative analysis or instrumental analysis.

It's tougher to get a job as a criminalist if your major is not one of those listed above. This includes the horde of people with majors in anthropology and criminology/sociology. My best advice to those of you in this situation is to either change your major to chemistry or biology, do a double major with one of your majors being chemistry or biology, add a minor in chemistry or biology, or take a hefty dose of chemistry and biology classes, including those mentioned in the previous paragraph, along with microbiology and cellular biology classes. Also, take all the laboratory classes you can get. There are many classes in a variety of subjects that apply to a career in the forensic sciences. Your task as an applicant for a crime lab job is to package these and present them in a convincing manner to the people who might hire you. Non-science classes that are especially relevant include public speaking, technical writing, drawing, photography, archaeology, and criminology.

The entry positions for these jobs usually have the title criminalist or forensic technician. I have observed that the various specialties within the broad spectrum of criminalistics often operate on an apprentice system. For example, if the Firearms and Toolmarks section needs another person, they will often choose one of the forensic technicians, who they will then train in their specialty. Starting salaries for criminalist positions usually range between $30,000 and $50,000 per year depending on the location.

51.4 Careers that Require a Specialized College Degree

To be considered for a career in certain forensic sciences, one needs a more specialized degree. For a job as a forensic chemist or toxicologist a Bachelor of Science degree in chemistry, biochemistry, or toxicology is required. For a job in forensic serology or DNA Analysis a degree in biology with a concentration in cell and molecular biology is necessary. Positions in forensic accounting, computer science, nursing, social work, engineering, audio, video, and other technical forensic sciences the appropriate degree in that field is required.

Expect to be hired at a technician or assistant level, and to work under the supervision of a more experience forensic scientist for several years. For some of these fields expect to spend a year to several years working toward certification as a forensic specialist. Typical starting salaries for these profession are between $35,000 and $50,000 per year, depending on location.

51.5 Careers that Require a Doctoral Degree

Many careers in the forensic sciences require a doctoral degree of some sort. This certainly applies to all those in the medical or health fields, with the exception of nursing. Forensic pathologists, psychiatrists, and podiatrists need an M.D. degree; forensic odontologists need a D.D.S. or similar degree in dentistry; and forensic pharmacologists need a Pharm. D. Degree. Other types of forensic medical specialists need the doctoral degree that their specialty awards. Forensic pathologists may be hired by a crime lab or work in private practice. Those hired by a crime lab can expect starting salaries greater than $95,000 per year. Forensic psychiatrists who work for a government funded institution can expect salaries greater than $70,000 per year. Other forensic medical specialists will normally work in the private sector, with starting salaries that vary considerably but are usually greater than $70,000 per year.

Those forensic specialties that I have described as university forensic sciences normally require a Ph.D. degree. There are some fields, especially in the arts where a master's degree (*e.g.* the Master of Fine Arts in photography) is the highest degree awarded, and these are acceptable in place of a Ph.D. I have seen people who have been successful in the university forensic sciences with only a master's degree, but with the interest in forensic science since the late 1990's degree inflation is a reality. **Degree inflation** refers to the phenomenon where jobs that used to require a bachelor's degree now require a master's degree and jobs that used to require a master's degree now require a doctorate.

Fields that normally require a doctorate include forensic anthropology, forensic botany, forensic economics, forensic entomology, forensic geology, forensic linguistics, forensic mathematics, forensic psychology, and forensic zoology. These specialists will normally find employment at a college or university, starting their career as an assistant professor with a salary between $45,000 and $65,000 per year depending on a variety of factors.

51.6 Other Related Careers

There is a wide variety of government offices and institutions that are related to the justice systems. These include state departments of justice, local offices of probation and parole, state and local regulatory agencies, correctional institutions, and a large number of federal agencies. Many of these positions are filled by applicants with a bachelor's or master's degree in sociology with a concentration in criminology. People with degrees in other social sciences, such as anthropology, psychology, economics, and political science can also be successful in landing these positions. What seems to be required is a strong background in statistics and a familiarity with the justice system. Starting salaries for these positions range from $25,000 to $45,000 per year depending on many factors.

Another career closely related to the forensic sciences is as an attorney or judge. Attorneys need to earn a bachelor's degree in some discipline, then go to law school. Law school takes a minimum of three years. After graduating from law school, with a J.D. degree, the person must pass the bar exam for their state in order to be a practicing attorney. There is a need for both prosecuting attorneys and defense attorneys in the justice system. The starting salary for attorneys is enormously variable. I recently saw an advertisement by the State of Montana, looking for an attorney for one of their environmental agencies with an advertised starting salary of $23,000 per year. On the other hand, an attorney hired by a large, prestigious law firm can make $100,000 per year or more. Judges usually (though not always) begin as attorneys. Salaries for judges are also variable, ranging from $30,000 per year for a judge in a small city to more than $100,000 per year for some federal judges.

51.7 Where to Find a Good College Program

The quick answer to where to find a good college program in forensic science is at the college where you are taking the class that requires this textbook. Obviously, your professor is smart and competent enough to have chosen the absolutely best textbook in the world for you to use <just kidding>. More seriously, any four year college or university is likely to have a competent program in chemistry or biology, with the possible exception of some smaller liberal arts colleges. I would bet that you can prepare yourself adequately right where you are.

For those readers to which the previous paragraph does not apply, my best advice is to figure out exactly what it is that you want. The more specific you can be, the easier it will be to find an appropriate college or university that offers the program you need. Having done this, start up your computer and browse the website of the American Academy of Forensic Sciences at <http://www.aafs.org>. Look around the AAFS website for their link to educational programs, which they maddeningly keep moving around. Under this link you will find a list of programs in a wide range of forensic sciences. The AAFS has a branch of their organization, FEPAC, that oversees accreditation of certain types of forensic science education programs. In my opinion you can't do better than one of these AAFS accredited programs for a career in general forensic science. If you find yourself in a non-accredited program, you could maximize your chances of getting the best possible forensic science education by taking classes

that are consistent with FEPAC's curriculum guidelines. These guidelines include the following classes.
- Biology: One course for science majors, with lab;
- Physics: Two courses for science majors, with lab;
- Chemistry: Two courses in general chemistry, two in organic chemistry, all with lab;
- Mathematics: One course in differential and integral calculus, one course in statistics;
- 12 additional semester hours in advanced science, including two courses with lab;
- 15 semester hours in forensic science coursework that covers courtroom testimony; introduction to law; quality assurance; ethics, professional practice, background; evidence identification, collection, processing; and, a survey of forensic science;
- A minimum of 19 advanced semester hours in courses that provide greater depth beyond an introductory level in the program.

Other programs are listed on the AAFS web site because officials at those colleges and universities are proud of their programs and have advertised them through the AAFS. As I view this list I see the top programs in the country. Chances are, there is a good one in your home state.

Don't forget that there are huge numbers of programs that do not advertise through the AAFS, and that most of you will be well served by any college or university with a solid program in chemistry of biology.

If you have decided to focus on a more narrow specialty, then locate the website for the profession organization for that specialty using a search on Google or your favorite search engine. For example, the professional organization for firearm and toolmark examiners is the Association of Firearm and Toolmark Examiners (AFTE) which has a website. Browse around this website and see what they recommend. Sometimes, members will even post their email addresses so that people like yourself can contact them with questions.

51.8 Where to Find Job Postings

The best place to find job postings for positions in forensic science is, again, the AAFS website <http://www.aafs.org>. Most of the jobs in the country get posted there. Since most jobs in forensic science or related fields are state, local, or federal government jobs, they are required to be posted to the public job boards of the advertising government. All states have their job boards available online now, and they all support searching by key word. Finding these job boards should be easy, and searching the job board for the key word "forensic" should locate any appropriate jobs that are available. If a search for "forensic" doesn't produce anything, try the search words "criminalist" and "scientist" before giving up. For U.S. federal government jobs try the website of the U.S. Office of Personnel Management <http://www.usajobs.opm.gov/>.

Jobs in the private sector are more difficult to locate. I suggest this approach. As an example, let's say that you are interested in forensic accounting. First do an internet search (using Google or whatever) for "forensic accounting". The search should return a list of links, many of which are to firms that provide forensic accounting services. Browse to these firms and look for their jobs, employment, or human resources pages. This will turn up a surprising number of job leads, especially in a growing field such as forensic accounting.

Jobs in related fields are also best searched for online. Many of them are government jobs, and the previous comments apply. For jobs in law enforcement try <http://www.lawenforcementjobs.com/> and <http://www.policeemployment.com/>. For jobs, internships, and volunteer positions with the FBI try <http://www.fbijobs.gov/>. For jobs at colleges or universities try the job postings provided by the Chronicle of Higher Education, <http://chronicle.com/jobs/>. Don't panic when they ask for money, because (at least at this time) they will let you browse their postings that are more than a week old for free.

51.9 My Best Advice

The advice I find myself giving most often is this. If you are serious about wanting a job at a crime lab earn a degree in chemistry or biology. I teach in a department of anthropology, and we have a thriving program in forensic anthropology. Some of my former students with degrees in anthropology have found employment at a crime lab, but the reality is that the vast majority of them will end up selling insurance, or working at one of the other jobs in the community that require a college degree but don't care which field it's in.

We live in a mobile society. Therefore, be prepared to move to where the job you want is – it's not likely to come to you. Those people who get the best jobs are most often those who are willing to uproot themselves and take a great position that just happens to be in the least desirable part of the country. Utilize the concept of stepping stones by taking what is available now, then moving through a succession of jobs to eventually get to where you want to be. When one of my students asks me what their prospects for a job in their field are, I ask them where do they want to live. If the student replies that they couldn't possibly think of leaving Missoula, I have to sadly inform them that their job prospects are pretty poor. If, however, they reply that they don't care where they live, then I tell them that their prospects are pretty good. These considerations may not be so important if you live in a large city, but if you live in a small college town, plan to move on after earning your degree.

Although not required, having volunteer experience or an internship in your field of choice can often improve your chances of getting that great first job. Finding an internship or volunteer experience is like finding a job, in that it is up to you to beat the bushes and find it. Many students ask their professors if there are any of these opportunities available, but we professors are always the last to find out about such things. Instead, seek out the resources that your college or university provides. For example, the University of Montana – Missoula has a student job board that lists internships and volunteer opportunities in addition to actual employment opportunities. Browse the websites of agencies or companies with which you would be interesting in volunteering or interning and see what they have available. Be vigilant – the best internships are filled rapidly and you need to check your sources often. Be flexible –

many students with interest in forensic science might profit from a volunteer experience with the local city police of sheriff's office cleaning out the evidence locker or something similar. Be creative – identify a need that an agency has, that they may not even be aware of, and volunteer to fill the need. Be persistent – if one opportunity or agency doesn't work out, try another. Avoid badgering the same people or agency over and over, however, as this is unlikely to make you any friends.

Finally, find out all you can about the forensic science you are interested in, and be realistic about yourself in deciding whether you are really cut out for it or not. For example, if you are someone who likes to smoke the occasional joint, there is probably no place for you at a crime lab or in law enforcement. If you are the type of person who craves excitement, do you really want to be a trace evidence examiner and spend your days counting dust grains and examining hairs through a microscope? Can you be meticulous about keeping good records? Can you be depended upon to do your analyses thoroughly? Can you work well as a member of a team? Can you work well independently? Can you face the prospect of testifying regularly in court? Are you willing to drop what you are doing to go process a crime scene at a moment's notice? There is no embarrassment in choosing a different career that is more suitable to your personality and abilities.

51.10 Questions for Study and Review

1. Discuss why some people who are attracted to a career in forensic science because they like the forensic science programs on TV would probably enjoy a career in law enforcement more.
2. What careers in the forensic sciences and related areas are open to a person with a high school diploma (or equivalent)?
3. What careers in the forensic sciences and related areas are open to a person with a 4-year college degree in biology, chemistry, or another of the "hard" sciences?
4. What careers in the forensic sciences and related areas are open to a person with a 4-year college degree in most majors, such as anthropology, sociology, English, Art, etc.? What careers are open to a person with a 4-year degree or a graduate degree in your major?
5. What careers in the forensic sciences and related areas require education beyond a 4-year degree.
6. What is the role of certifications in some forensic science fields?
7. If your college or university does not have a major in forensic science, criminalistics, or something similar, how could you put together a selection of classes from those offered at your college or university to acquire the equivalent education?
8. Visit the AAFS website or one of the law enforcement job websites listed in this chapter and examine their job postings. How many jobs did you see that you might be interested in applying for? What educational background would you need in order to be considered for these jobs?
9. In general, are forensic job prospects better for those people willing to move to another city?

10. Can you think of any internship or volunteer opportunities that may improve your chances of getting a job in the forensic sciences?

51.11 Sources

American Academy of Forensic Sciences. n.d. So You Want to Be a Forensic Scientist! [Internet]. [cited 2007 Nov 1]. Available from: http://www.aafs.org/default.asp?section_id=resources&page_id=choosing_a_career.

[Anonymous]. 2005. Great Schools for Criminal Justice. Careers & Colleges 26(2): 49(1).

[Anonymous]. n.d. Missoula Police Department Employment Information [Internet]. [cited 2007 Nov 1]. Available from: http://www.ci.missoula.mt.us/police/pdemploy.htm.

Douglas JE. 1998. John Douglas's Guide to Careers in the FBI. New York: Kaplan Books.

Forensic DNA Consulting. n.d. Careers in Forensic Science [Internet]. [cited 2007 Nov 1]. Available from: http://www.forensicdna.com/careers.htm.

Gaensslen RE. 2003. How Do I Become a Forensic Scientist? Educational Pathways to Forensic Science Careers. Analytical and Bioanalytical Chemistry 376 (8): 1151-1155.

Northway W. 1997. Looking for a Few Good People, Fbi Broadens its Recruiting Reach [Internet]. Mississippi Business Journal August 11, 1997. [cited 2007 Nov 1]. Available from: http://www.msbusiness.com/archives/archives_article.cfm?ID=12143.

Nute D. n.d. Advice about a Career in Forensic Science [Internet]. [cited 2007 Nov 1]. Available from: http://www.criminology.fsu.edu/faculty/nute/FScareers.html.

Roberti JW. 2004. Personality Characteristics of Undergraduates with Career Interests in Forensic Identification. Journal of Employment Counseling 41(3): 117(9).

Tebbett IR, Wielbo D, Khey D. 2007. Beyond Boundaries: Future Trends in Forensic Education. The Forensic Examiner 16(2): 62(5).

Chapter 52
WHAT I WISH DEFENSE ATTORNEYS KNEW ABOUT FORENSIC SCIENCE

I write this final chapter knowing full well that I will anger and offend some of my friends and colleagues. While this is regrettable, it is unavoidable. As discussed previously my professional ethics forbid me to be silent when confronted with a situation where my speaking the truth as I see it may make a difference. So it is in this chapter. I am thankful, and you probably are too, that this is the shortest chapter in the book.

I have been guilty in this text of portraying the justice system as kinder and gentler than it is. I have portrayed it as a system that seeks fairness, which may occur occasionally but not as often as it should. In fact, it seems that in many ways justice has been lost from the justice system. Often, I am appalled that I myself contribute to the system. The problem begins at the grass roots level where underfunded law enforcement agencies set priorities for enforcement that are driven by revenue potential. It continues with peace officers who rate their personal value in terms of how many arrests they make rather than the number of incidents they resolve peacefully.

The problem continues with prosecutors and judges who are out to make their reputations based on the number of "bad guys" they send to prison. Judges, furthermore, often pride themselves on the length of the sentences they give these "bad guys". This is followed up by a correctional system that has been described as an "academy of crime", where if a person enters as a reasonably decent human being they are very unlikely to emerge that way. In the past month I have seen two documentaries on television about the abysmal state of affairs at our prisons. Journalists tell us that the United States now has more people in prison than any other country – more than any other country has ever had. We have more people in prison right now than the Soviet Union had under the dictatorship of Joseph Stalin. Something is desperately wrong here. At the top level are the lawmakers who seem to feel the need to define more and more things as criminal acts, and to mandate ever more harsh sentences in the name of deterrence.

No, the criminal justice system is not a system of justice. It is in fact multi-billion-dollar industry that thrives on dealing out punishments. It cries for reform, and I believe that everyone in the system knows this, with the possible exception of a few real bullies. How has this come to pass? The reason for the problems with the justice system stare back at us from our mirrors each time we comb our hair or brush out teeth. Yes, it's us. It's the citizenry that lacks any sense of community or connection with those fellow human beings outside their narrow circle of friends. It's the crusaders, who want really harsh penalties for certain crimes, drunk driving for example. It's the pathological apathy that considers this to be someone else's problem. It's the feeling that I'm a good guy and I'll never run afoul of the law, which lasts just until some trivial incident lands you in the hands of the justice system. It comes from viewing the process of the justice system as theater or as a sport, wherein we watch people's lives destroyed in front of our faces in a way reminiscent of the ancient Romans watching Christians being fed to the lions. It comes from a political system wherein lawmakers author and vote on bills in response to how much will be contributed to their campaign fund rather than following their consciences. Possibly most of all, it is our willingness to give up freedom for safety.

We no longer live in a free society. My personal test for a free society is one in which I can legally drive my car without wearing seatbelts. Here is a "crime" that harms nobody except the person who does not wish to wear a seatbelt. (By the way, I wear mine religiously – I simply don't believe that I should be required to by law.) The push to create and enforce seatbelt laws comes from the highest levels of government, in some misguided attempt to improve safety statistics. The most persuasive argument for it is that if someone is severely injured because they were not wearing a seatbelt, they may have to go on public medical assistance and be a financial burden to the taxpaying public. Who ever said that freedom doesn't have costs? Is this a trivial issue? Perhaps, but I find it to be a perfect illustration of how we are willing to let freedom be taken from us by paternalistic lawmakers and government agencies in the name of keeping us safe.

Further, the justice system indisputably favors the wealthy. It is also racist (non-Whites get harsher treatment), classist (lower socioeconomic classes get harsher treatment), and sexist (males get harsher treatment). Is this what we want our justice system to be like? I doubt that most of us do.

52.1 The Importance of Defense Attorneys

Well, what does this rambling have to do with defense attorneys? It is clear to me that, if unchecked, the criminal justice system would rapidly degenerate into a more streamlined system for routing citizens into prisons. We are already well on the way, through the practice of plea bargaining, to a system that skips the trial and moves a person fairly rapidly from arrest to sentencing. This is particularly true when physical evidence is present and used as leverage to obtain that plea bargain.

Only one thing stands in the way of this grim scenario – the thin and fragile line of defense attorneys. Only defense attorneys seek to prevent sending people to prison. Only defense attorneys stand up for us small and weak citizens. It is the possibility of a defense attorney's objections that motivates us forensic scientists to process crime scenes adequately and to keep good chains of custody. It is this same threat that motivates peace officers to seek search or arrest warrants, makes sure that prosecutors and grand juries fairly weigh the strength of the evidence against a suspect, and insures that our forensic examinations are done in a fair and thorough manner. Defense attorneys are the unsung heroes of the justice system. It is only because they exist that any semblance of justice remains.

Every defendant, whether guilty or innocent, has the right to the best defense possible. Anyone who doesn't believe this now, will have a change of heart the instant he or she is accused of some crime and has to stand trial. This includes all the technicalities of evidence admission. If a guilty person goes free on a technicality it is not the fault of the defense attorney, it is the fault of the prosecution, the law enforcement agency, or the forensic scientists that failed to do their job well. Each such case is a victory for justice that slows the slide of the justice system into tyranny.

Those forensic scientists who work at a crime lab, or who for other reasons tend to most often appear as a witness for the prosecution hate defense attorneys, and I believe that the feeling is mutual. This probably stems from the traditional hard time that defense attorneys give these expert witnesses during cross examination. But consider this, if we forensic scientists were more fair and cooperative, perhaps defense attorneys would not question us so harshly. When I talk with defense attorneys, what I hear from them is that forensic scientists are so difficult to get information out of during a interview, that they are forced to "beat it out of them" on the witness stand. Come on, let's call a truce in the name of justice for all.

52.2 A Forensic Scientist's Advice for Defense Attorneys

The remainder of this chapter is devoted to advice for defense attorneys. I do not have extensive courtroom experience, but in my limited experience I have not yet seen a defense attorney ask what I considered to be appropriate questions about the evidence I was presenting. I do not know whether this is true for other forensic experts or not, because in all cases except one I was sequestered and did not hear any other part of the trial, and in that one exception I only heard one other scientist's testimony. I believe that in some cases, defense attorneys do not do as good a job as they are capable of, simply because they don't understand forensic science well. Perhaps this advice will help.

52.3 Making a Moral Argument

Forensic scientists see themselves as highly moral people, and I agree. However, they also see themselves as rational, dispassionate scientists who are offering their testimony in a way that they consider neutral and unbiased. In this they are deluding themselves. Further, the court and the jury tend to also perceive forensic scientists as only interested in the truth and disinterested in the outcome of a trial. Hogwash! As an analogy, consider watching a sports event. Is it possible to watch a football game, for example, dispassionately and without caring which side wins? Is it possible to be associated with one or the other of the teams, even in the most trivial way (say the person who brings water to the players) without becoming emotionally invested in the team? These things may be possible for exceptional people, but not for very many. Many forensic scientists often find themselves as part of the prosecution team, and it is difficult to remain dispassionate and neutral under these conditions.

Now, let's add a social pressure to this mixture. Society tells us that people are getting away with heinous crimes all the time. Now, let's say that someone stands accused of such a crime. We "know" that crimes happen all the time with devastating consequences and that the detection of these crimes is fairly rare. Under these conditions, forensic scientists, as human beings and members of their society, may feel the moral obligation to try to prove guilt. They may further feel that they are making a moral stand on the issue and going beyond actually punishing this particular offender to send a message to other offenders that they too can be found out and punished.

Putting the team mentality together with the perceived moral obligation to stand up for victims produces bias. It's probably unavoidable. Your job, as the defense attorney, is to combat this bias. I believe that the best way to do this is to make a subtle appeal to the forensic scientist's inherently moral character. Perhaps just a simple question prefaced with the phrase "if this person were innocent" can be a call to clarity. During your interview with the forensic scientist try asking them how they would reinterpret their evidence if you showed them incontrovertible proof of the defendant's innocence. Try reminding them of the tragedy of wrongful convictions, especially the point that if the wrong person is convicted then the true guilty party remains free to commit similar offenses. You are trained for this – you can do it. Maybe it won't win your case for you, because the defendant might really be guilty, but even so justice is served by a more fair interpretation of evidence.

52.4 Forensic Evidence is Often Not as Strong as it Seems

When I was in 7th grade, my mathematics teacher, Mrs. Renfro, once told me that if things do not add up correctly, then there must be an error in my calculations. In Mrs. Renfro's class close enough just wouldn't do. There are some prosecution cases that add up precisely, but others only add up if we accept that $2 + 2 = 5$ at this certain point in the calculation. This is your opportunity as a defense attorney. The assertion that $2 + 2 = 5$, just this once, is probably more common when the evidence is testimonial, but occasionally you will find it applied to physical/forensic evidence as well. Don't be intimidated just because it's forensic evidence. With forensic evidence the assertion that $2 + 2 = 5$ is even more alarming, and probably even more obviously bogus. Since expert witnesses are often sequestered during a trial, it may be that the forensic scientists themselves don't realize that the evidence they developed and presented is being used to justify $2 + 2 = 5$. You, the defense attorney are likely to be the only one who realizes this. You must point out that $2 + 2$ never equals 5, and if necessary you need to ask for a recess so you can find and call an expert witness who can back you up.

It is possible for forensic evidence to be in error. Here is where you need to do some research, and I don't mean just a review of the law literature. I mean getting on the internet and doing some searches for cases where forensic evidence was in error or used incorrectly. Read everything written by the Innocence Project. Also, nearly every issue of the journal The Forensic Examiner, <http://www.acfei.com/forensics.php>, has a review of a case in which someone was exonerated. If the prosecution case hinges on a certain type of evidence, find cases where this type of evidence was in error. A good place to start is at the Forensic-Evidence.com web site, <http://forensic-evidence.com/site/MasterIndex.html>.

As I understand it, perhaps poorly, your job is to introduce cause for reasonable doubt about the correctness of the prosecution case. Fingerprint evidence and firearms evidence in particular have questionable scientific standing in the minds of many scientists (see discussions in previous chapters) because they focus on similarities between evidence and standard items while either ignoring differences or not checking for them at all. This in not a fair application of the comparative method and even the most scientifically untrained people will understand the principle that even one difference yields an exclusion no matter how many similarities are present. If differences were not looked for, then the possibility of an exclusion was overlooked. Even within a crime lab, scientists who work in other sections often refer to the fingerprint section and firearms section as "soft science". It may be that the forensic scientists working in these sections do not even have a degree in a science and may be untrained in the scientific method. Similar questions can be directed at other forms of forensic evidence. Your ammunition is out there on the internet and in libraries waiting for you to use it.

52.5 Interviewing Experts

What would you say if I told you that I have testified at two murder trials, both of which rested on my testimony that certain small fragments of bones were human remains, and that I was not interviewed by the defense team? This is true. Assuming that the penalties for their clients would have been much less severe if the bones were actually deer bones or cow bones, this seems like an omission that borders on misconduct. For Heaven's sake, don't waive your right to interview the prosecution's forensic witnesses. If they are too far away to conveniently interview in person, there are many high tech solutions that will bring you face to face with someone via the miracle of computers and the internet. Ask the information technology person at your local court or law school for some ideas, such as net conferencing. At the very least you could give the expert a phone call.

When interviewing an expert, you can easily determine whether this person is a fair and balanced forensic scientist or not by asking them what are the strengths and weaknesses of their analysis and evidence. I have never generated evidence that did not have at least some weaknesses. If the expert won't share the weaknesses of their evidence with you, then it's time to consult another expert. It may be expensive to buy a couple of hours of an independent expert's time to review the evidence against your client, but I think it's worth the expense just to be sure that those weaknesses don't exist. If they don't exist, then you know what you are faced with, and if they do exist, then you know what to do.

52.6 In the Courtroom

I believe that hammering at the qualifications and competence of an expert is probably pointless, unless it really is the case that the expert is underqualified or incompetent. In this case, hammer away. Otherwise, you are probably wasting your time and making enemies of the expert witnesses. Instead, direct your questioning toward those weaknesses that you found out about during the interview with the expert, or from information obtained from another expert. In one interview, I all but handed the case on a platter to the defense team, but they failed to follow up in court by asking me a single question directed toward the weaknesses in my evidence that I had informed them of. Perhaps these weaknesses were nullified by other evidence presented in the case – I don't know because I was sequestered and not present for other testimony. I suppose I should have asked the defense attorney afterward, but I was so flabbergasted by this omission that I wasn't thinking clearly. In any case, I would have thought that a few of the proper questions would have raised some reasonable doubt in the minds of the jurors.

If no weaknesses in the forensic evidence emerged from the discovery process, then educate yourself about that type of evidence and use your court time to probe for weaknesses. When you find a weak spot poke at it until either something gives or the judge tells you to leave off that line of questioning.

If it's your forensic witness be sure you know what they are going to say. At one murder trial for which I testified as one of the prosecution's key witnesses, the defense team brought in a regionally prominent forensic pathologist to refute my testimony about the number of blows to the head the victim had received. Not only did this expert not even convincingly try to refute my testimony, when asked if he could determine the cause of death he said that in his opinion the person had died of "homicidal violence". This was the defense's star witness! It is possible that this witness tricked the defense team, but I doubt it. I think it is much more likely that they never even bothered to ask him what his testimony was going to be. Then, they asked entirely the wrong questions when he was on the witness stand. I really do hope that the defendant in that case was not innocent.

Two words: jury nullification. OK, I know that we're all supposed to keep this under the table, and that juries are instructed not to do it, but as a citizen I think it's good for justice and for society. Jury nullification is the first step toward recognizing the unfairness of existing laws and a step toward rehabilitation of the justice system. Heck, it's good enough for Denny Crane on Boston Legal.

52.7 Final Encouragements

The task you have undertaken as a defense attorney is often thankless, therefore let me humbly and sincerely thank you. It is often underappreciated, therefore let me tell you that there is at least one screwball in Montana who appreciates you. It often makes enemies for you, and therefore I wish you many friends and colleagues with which to share your victories and defeats. Your task is often difficult, and nothing I have said in this chapter makes it any easier. Indeed, taking my advice perhaps makes it more difficult. The task is endless, and in the words of Tim Allen's character from the movie "Space Quest", you must "never give up, never surrender." You are the often underpaid heroes of the justice system, and I hope that your job repays you by being both interesting and morally satisfying.

52.8 Questions for Study and Review

1. Describe the problems with the American system of justice as it exists at present.
2. ~~In what respect can it be said that defense attorneys are the "heroes" of the justice system?~~
3. How does the fact that there are defense attorneys make forensic science better?

4. How would you describe the nature of the relationship that has historically existed between forensic scientists and defense attorneys?
5. What are some factors that may cause a forensic scientist to become biased?
6. Where is the best place to find out about possible problems with a certain type of forensic evidence?
7. Is it true that forensic evidence may sometimes be in error?
8. Is it true that forensic evidence may sometimes be misinterpreted?
9. In what situations might a defense attorney strengthen her or his case by consulting an independent forensic expert?

52.9 Sources

[Anonymous]. 2003. U.S. Correctional Population [Internet]. FBI Law Enforcement Bulletin 72(5). [cited 2007 Oct 19]. Available from: http://www.fbi.gov/publications/leb/2003/may2003/may03leb.htm#page_6.

Burnett H. 2007. Experts Aren't Always Right: the Problems Unearthed with One Ontario Pediatric Forensic Pathologist's Work Should Be Wake-up Call for Lawyers and Judges. Canadian Lawyer 31(9): 47(4).

Cole SA. 2007. The fingerprint controversy. Skeptical Inquirer 31.4: 41(6).

Decker SH, Alarid LF, Katz CM. 2003. Controversies in Criminal Justice: Contemporary Readings. Los Angeles: Roxbury Publishing.

Dror IE, Charlton D, Peron AE. 2006. Contextual Information Renders Experts Vulnerable to Making Erroneous Identifications. Forensic Science International 156 (1): 74(5).

Giannelli PC. 2003. Scientific Evidence [Internet]. Criminal Justice Magazine 18 (1). [cited 2007 Oct 18]. Available from: http://www.abanet.org/crimjust/spring2003/scientific_evidence.html.

Giannelli PC. 2007. Daubert challenges to firearms ("ballistics") identifications. Criminal Law Bulletin 43(4): 548-568.

Koppl R. 2007. Breaking up the Forensics Monopoly: Eight Ways to Fix a Broken System. Reason 39(6): 44(2).

Kramar K. 2006. Coroners' Interested Advocacy: Understanding Wrongful Accusations and Convictions. Canadian Journal of Criminology and Criminal Justice 48(5): 803(19).

Kruglick K. n.d. A Beginner's Primer on the Investigation of Forensic Evidence [Internet]. Scientific Testimony: An Online Journal. [cited 2007 Oct 23]. Available from: http://www.scientific.org/tutorials/articles/kruglick/kruglick.html/

Masterton RP. 2006. A View from the Bench: Defense Requested Experts. Army Lawyer (Sept 2006): 39(4).

Modisett J. 2004. Shifting the Emphasis from Prison to Education: How Indiana Saved over $40 Million. Black Issues in Higher Education 21(3): p40(2).

Moenssens AA. n.d. Is Fingerprint Identification a Science? [Internet]. [cited 2007 Nov 1]. Available from: http://forensic-evidence.com/site/ID/ID00004_2.html.

Murphy E. 2007. The New Forensics: Criminal Justice, False Certainty, and the Second Generation of Scientific Evidence. California Law Review 95(3): 721-797.

Pyrek KM. 2007. Forensic Science under Siege: The Challenges of Forensic Laboratories and the Medico-Legal Investigation System. Burlington (MA): Elsevier Science & Technology Books.

Risinger DM, Saks MJ. 2003. A House with No Foundation: Forensic Science Needs to Build a Base Rigorous Research to Establish its Reliability. Issues in Science and Technology 20(1): 35(5).

Steele LJ. 2002. All we want you to do is confirm what we already know" A Daubert Challenge to Firearms Identifications. Criminal Law Bulletin 38 (4): 466-483.

Glossary

AAFS: The American Academy of Forensic Sciences. The AAFS is the largest association of forensic scientists in the United States. It holds an annual convention, and fosters education, research, and improvement of practices in the forensic sciences. The AAFS has a web site at http://www.aafs.org.

Abnormal result (in forensic accounting): The result of an examination of financial records in which discrepancies were found that seem to be deliberate and therefore evidence of a crime.

Absorption: A type of interaction between electromagnetic radiation and matter in which the matter absorbs the EMR and its energy.

Absorption-elution method: An indirect method for determining blood type. This method is commonly used with dried blood. In the case of secreters, whose red cell antigens also appear in other body fluids such as semen and saliva, this method must be used if blood type is to be determined from these non-blood fluids. In this method an antiserum is allowed the opportunity to react with the questioned sample. Then, the antiserum is recovered and allowed to react with its normal target red cells. If the recovered antiserum does not react with its normal target red cells, then the antibodies present in it must have previously reacted with antigens in the questioned sample. Thus, it can be inferred that the questioned sample contained antigens of the type the antiserum targets.

Absorption spectrum: The pattern of wavelengths of light absorbed by a substance. See spectrophotometry.

Accelerant: A substance, such as gasoline, used to make a fire burn faster or hotter.

Acid phosphatase color test: A test used by serologists to determine whether a suspect fluid or stain is semen. In this test the chemical reagents change color in the presence of semen.

Action: The action of a firearm is the mechanism by which the firearm is prepared for firing. This may involve moving a cartridge or shell into the firing chamber and/or moving the firing hammer or pin to the position where it is ready to strike the cartridge or shell..

Affirmation: A solemn statement that testimony given is true that is made without reference to God.

AFIS: An acronym for automated fingerprint identification system. An AFIS is a state or local level fingerprint database.

AFIS Method: An approach to biometric verification based on examination of fingerprint characteristics similar to those used by AFIS systems.

Age progression and updating: The process in which a forensic artist makes or modifies a drawing or photograph to show known changes in the appearance of the subject of the drawing or photograph. In age progression the forensic artist makes changes designed to show how the subject's appearance is likely to have changed due to normal aging since the previous photograph or drawing was made. In updating the forensic artist makes changes to reflect recently acquired information about how the subject has changed their appearance (growing a beard, for example).

Age regression: A form of memory refreshment (via hypnosis) in which the subject is guided to return to an earlier age.

Agglutination: An antigen-antibody reaction that causes cells to be clumped together, usually by the action of antibodies attaching to antigens on two different cells. Agglutinated red cells are referred to as precipitin.

Aggravated assault and battery: A type of battery that resulted in serious injury or was accomplished using a deadly weapon.

Aggression: A behavior that includes the threat to cause harm to another person.

Alcohol in possession of a minor: Possessing or drinking of alcohol by a person who is under the age of 21.

Algor mortis: Cooling of the body after death.

Algorithm: A method for accomplishing a desired task.

Allele: The actual genetic instructions found at a locus.

Amino acids: The building blocks of proteins. Proteins are composed of combinations of amino acids, which define the protein's physical shape and properties. There are approximately 20 different amino acids.

Amphetamines: A type of stimulant.

Amplifying DNA: See PCR.

Anatomical position: A positioning of the body that anatomists consider standard. In anatomical position the person is standing or lying supine (on the back) with the palms of the hands facing forward.

Anecdote: A story a person tells about their personal experience. Anecdotes are a form of testimonial evidence, and are not considered reliable by scientists.

ANFO: An acronym for ammonium nitrate and fuel oil -- a combination that produces a high explosive.

Anterior: Also referred to as ventral (toward the belly), this direction on the body is toward the front of the body. The opposite of anterior is posterior.

Anthropometry: The science of measuring the human body and interpreting these measurements.

Antibodies: Proteins of the globulin category found in the plasma or serum of blood. Antibodies comprise the humoral part of the immune system and function by attaching to an antigen, thereby killing or inactivating the cell carrying the antigen. Antibodies have a physical shape similar to the letter 'Y' with an area on the tip of both forks of the 'Y' enabling them to attach to two antigens simultaneously.

Antigens: Molecules attached to the surfaces of cells. The function of antigens is not known in all cases, but some of them appear to function to identify cells as being of a certain type and others appear to function in communication between cells and tissues. In immune reactions antibodies bind to antigens to inactivate the cell carrying the antigen.

Antiserum: A serum (or synthetic serum-like substance) that contains antibodies specific to a certain antigen or set of antigens. For example anti-A serum reacts with the antigen associated with blood type A to agglutinate type A red cells.

Antisocial personality disorder: A set of related psychological disorders in which the affected person is less capable than normal of understanding the feelings of others.

Appellate courts: Courts that hear or try cases that have been appealed after having been heard or tried at a court of origination. The U.S. Courts of Appeals and the U.S. Supreme Court are appellate courts.

Appendicular skeleton: The division of the skeleton that includes the bones of the limbs.

Armed Robbery: A form of robbery in which a weapon is used.

Army System of Tooth Designation: A system of referring to the 32 human teeth by number. The numbering starts with tooth 1, which is the upper right third molar. The numbering continues in sequence along the upper tooth row to tooth 16, which is the upper left third molar. Tooth 17 is the lower left third molar, and the numbering continues sequentially along the lower tooth row to tooth 32, which is the lower right third molar.

Arson: Intentional burning of an item or structure in an illegal manner.

Assault: An unlawful attempt to touch or strike a person, or the threat to do so.

Assault with a deadly weapon: Assault that was committed with a weapon capable of causing death, such as a gun, knife, or club.

Asset misappropriation: Stealing of an organization's assets by an employee.

Atom: The smallest particle of an element. Atoms contain a nucleus containing protons and neutrons, which is surrounded by a probability cloud of electrons. The number of protons in the nucleus determines the element of which the atom is a sample.

Atomic weight: The atomic weight of a an atom is a function of the number of protons and neutrons in its nucleus. For example, normal hydrogen has one proton and has an atomic weight of 1. Normal oxygen has 8 protons and 8 neutrons in its nucleus and has an atomic weight of 16.

Audibility Analysis: An investigation of the circumstances surrounding a sound event (a gunshot, for example) to determine whether the sound could have been heard at certain locations or by certain individuals.

Authentic image: An image that has been determined to have been made and to show what is claimed for it by testimony.

Authentication: Testing some item to see whether it was produced at the time and place, by the person, and under the circumstances that are claimed for it.

Autopsy: Also called a post-mortem examination, this is an examination of a body after death in order to determine the cause of death. An autopsy can be external, in which the characteristics of the body are examined without opening the body cavities. Many autopsies also include an internal examination in which the body cavities are opened to examine the brain and internal organs.

Axial skeleton: The division of the skeleton that includes the skull, the jaw, the hyoid, the vertebrae, the ribs, and the sternum.

Backdoor: In the realm of computers this is a method of accessing a computer or network that was intentionally programmed into software by its creator.

Bad science: Well intentioned but incorrect, obsolete, or oversimplified explanations or presentations of scientific matters.

Ballistic Curve: The path of a projectile during the external phase of ballistics.

Ballistics: The scientific study of the motion of projectiles, including projectiles fired from firearms.

Bandwidth: The amount of data that a network is capable of handling.

Barbiturates: A type of depressant drug.

Base: In genetics "base" is a synonym for nucleotide. The nucleotides are the components of DNA and RNA – argenine, cytosine, guanine, and thymine in DNA. In RNA uracil replaces thymine.

Basic units of the metric system: The basic units of the metric system are the meter for length, the gram for weight, the liter for volume, the Newton for energy, etc.

Battery: The result of a successful assault – the person is actually struck.

Beetles: Insects belonging to the order coleoptera. Some beetles eat the tissues of dead bodies and are therefore of interest in forensic entomology.

Behavioral evidence: In the realm of psychological profiling behavioral evidence refers to the details of how the crime was committed (e.g. a murder by gunshot vs a murder by strangling) as well as details of other actions the criminal took in addition to the crime itself.

Behavioral sciences: Several social sciences, including at least psychology, sociology and sociocultural anthropology.

Bertillonage: The method developed by Alphonse Bertillon (1853–1914) for the purpose of identifying criminals based on measurements of several parts of their bodies. Bertillonage is widely recognized as the first reliable method for determining the true identity of individuals.

Betweenness: A concept from forensic mathematics that describes how many subgroups of a group would result from removing a certain member of the group.

Bias (in science): Error introduced into data by the person doing an experiment. Bias can be introduced in several ways, most commonly by experiments that are designed poorly and do not account for all extraneous factors that may alter the results.

Binary: In general, the condition of having exactly two states. In the realm of computers it refers to the coding system, consisting of 0's and 1's, used internally by the computer.

Biometric Identification: The process of scanning the characteristics of a large number of people in order to identify wanted individuals or terrorists.

Biometrics: The field that focuses on identifying a person immediately or within a very short time (seconds or minutes) by matching characteristics of the person's body to a database of characteristics of known individuals. Current applications of biometrics generally use characteristics of the hands (fingerprints, palm prints) or eyes (iris pattern, retina pattern).

Bit: A bit is a single 0 or 1 that is part of the information used or stored by a computer.

Blended Inheritance: A form of inheritance in which the offspring seem to be intermediate in some feature between the conditions found in the parents. Blended inheritance is actually an illusion seen in complex traits due to the action of several genes plus the environment.

Blood group: A system of red blood cell antigens, defined by alleles at a locus. For example, the ABO (transfusion) blood group includes antigen A and antigen B.

Blood spatter analysis: Analysis of the shapes and patterns of blood droplets from a bleeding person that contact a surface (wall, floor, etc.). Blood spatter analysis can yield information about the location and movements of a bleeding person.

Blood type: The characterization of an individual's phenotype with respect to a blood group. For example, a person may be blood type A with respect to the transfusion (ABO) blood group. Most blood types result from the presence of certain antigens on the surface of red blood cells.

Blow flies: A taxonomic group of flies that lay eggs on the bodies of dead organisms. The eggs hatch into larvae (maggots) that consume the flesh of the dead organism.

Bonding: The joining together of atoms or molecules to form other molecules during a chemical reaction.

Bookmaking: A type of organized illegal gambling on the outcomes of sporting events.

Breaking and entering: Breaking into a building in order to commit a crime that is not a felony.

Buccal: The direction within the mouth that is toward the cheek. For those teeth and portions of the mouth that are not within the cheek, the equivalent direction is labial. The opposite of buccal is lingual.

Buffer (in chemistry): See electrophoresis.

Buffer (in computers): An area of memory that is used to collect bytes arriving at at computer from an input device.

Buffer overrun: A condition that occurs when data arriving at a computer from an input device fills the buffer designated for it and overwrites data or program code that follows the buffer.

Bug: In the field of electronic surveillance a bug is a device that can be hidden at a location in order to capture an audio or video record of the activities that occur there. The information obtained by the bug is usually transmitted to a person or computer that records it.

Burglary: Breaking into a building in order to commit a felony theft.

Byte: A set of eight consecutive bits.

Cache: In the realm of computers this is an area of non-volatile memory that is used to store browsed material (graphics, etc.) so that they can be loaded more rapidly the next time the web page containing them is accessed.

Cadaver dog: A dog purported to be trained to locate buried bodies by smell.

Caliber: The diameter of a rifle or handgun barrel, and hence the diameter of the bullet it fires.

Cancellous bone: Also called trabecular bone or spongy bone, this is a bone material that is formed of bony plates that intersect at approximately right angles to form air spaces. Cancellous bone forms the inner portions of most bones, where it provides considerable strength while being lighter than compact bone.

Cartridge: The type of ammunition utilized by a rifle or handgun.

Cast: A replica of an item or impression.

Casting: The replication of the characteristics of an item using a casting material. The details of impressions are often recovered by casting.

Casting material: The material of which a cast is made. There are many casting materials including several forms of plaster, plastic, and resin.

Cause of death: The actual phenomenon or mechanism that caused a person to die. Examples are drowning, gunshot, electrocution, heart attack, drug overdose, and thousands of others.

Cavitation: The action of tissue or other matter moving aside as a projectile moves through it. Cavitation is one of the mechanisms by which a projectile causes damage to tissue.

CBN: See unconventional weapons.

CCH: An acronym for computerized criminal history. An CCH is a state or local level database of criminal histories.

Cells (in terrorism): Most terrorist groups are organized as cells, consisting of groups of a few to several individuals that operate independently from other such groups. In general, members of a cell do not know of the existence or activities of other cells or their members.

Cellulose: A tough, fibrous plant material formed of sugar molecules linked together in a specific manner.

Cementum annulation analysis: A method for estimating the age of an animal by counting the number of annually formed layers of cementum on the roots of its teeth.

Central processing unit: This term is used in two contexts. In the first context it refers to the main computer "box" (the container and components within it). In the second context it refers to the integrated circuit (processor chip) that carries out instructions.

Certification: The process of undergoing training and testing in order to pass the requirements for a certificate administered by an association of professionals that attests to the competency of the person in an area of forensic science.

Chain of custody: A detailed record of evidence recovered from a crime scene and what happened to it between the time it was recovered and the time it was presented at a trial.

Change of venue: In the realm of legal proceedings this refers to moving a trial to a court in a different place in order to be able to find a judge and jury that are less prejudiced.

Chemical symbol: A short abbreviation or representation of the name of an element. For example, O for oxygen, Li for lithium, or Pb for lead.

Chemical explosion: An explosion that occurs due to an exothermic chemical reaction that occurs extremely rapidly.

Chief Justice: The Chief Justice of the U.S. Supreme Court is the leader of the 12 U.S. Supreme Court justices (judges). The Chief Justice has powers and responsibilities beyond those of the other justices.

Chief Deputy Coroner: Since Coroners are elected officials and often the Sheriff, it is usually impossible for the actual Coroner to conduct all inquests. Therefore, the Coroner will appoint one or more Deputy Coroners to conduct these investigations. The Chief Deputy Coroner is a Deputy Coroner that supervises other Deputy Coroners and who often speaks for and as the Coroner. In some jurisdictions the Chief Deputy Coroner is referred to as the "Coroner", leading to confusion about who exactly is the Coroner.

Chief Judge: The senior judge of a U.S. Court of Appeal, who has additional administrative duties beyond hearing cases.

Chirology: The pseudoscience that attempts to derive information about a person, the person's character, or the person's future by examination of the person's fingerprints palm prints.

Chiromancy: The attempt to find out about a person's future using principles of chirology.

Chironomy: The attempt to find out about a person's character using principles of chirology.

Chromatography: A method for separating the chemical components of a mixture based on difference in the solubility and molecular weight of each component. Several types of chromatography exist, including paper, liquid, thin layer, gas, and others. In each form of chromatography there is a substance called the "mobile phase" and a substance called the "stationary phase". The mobile phase moves past the stationary phase, carrying the components of a mixture varying distances.

Chromosomes: Bodies made of double stranded DNA found, in the nucleus of a cell, that store an individual's genetic information and which are replicated and passed to the next generation as part of inheritance.

Circumstances of death: The place where a death occurred, whether other people were present at the time of death, the behavior of the deceased around the time of death, and many other types of facts surrounding the death. Circumstances of death are important for determining manner of death. For example, if a person dies of a gunshot in a locked room while alone, the manner of death is most likely accidental or suicide. However, if someone else was present in the room, then there is an increased probability that the manner of death was homicide.

CITES: An acronym for Convention on International Trade in Endangered Species, an international treaty that seeks to protect endangered species by restricting import or export of their parts or products among nations who are signatories to the treaty.

Civil court: A court that handles cases of civil law.

Civil law: A body of law that defines and sets procedures for handling civil suits. A civil suit involves some action that is not defined as criminal but which may have caused some loss or injury to one or more persons.

Clark R. Bavin National Fish and Wildlife Forensics Laboratory: A facility established at Ashland Oregon in 1989 to provide crime lab services and analyses for agencies that protect wildlife.

Class characteristics: Characteristics that are shared by a group of items of a certain type. For example, a person's sex, race, and age are class characteristics of a human being.

Closeness: A concept from forensic mathematics that describes how many individuals a message from one member of a group to another must pass through.

CODIS: An acronym for combined DNA index system. A CODIS is a DNA database maintained by a state or a local jurisdiction.

Codon: A group of three consecutive DNA or RNA bases that are the code for a certain amino acid.

Color tests: A test for the presence of a drug, blood, semen, or some other substance in which the chemical reagents change color in the presence of the substance being tested for.

Combustion: Also known as burning. A type of oxidation reaction in which fuel is consumed to release light and heat in a form that is usually described as fire.

Commodity item: An item that has inherent value, but which is not in itself money. Gold is a good example. One money laundering strategy is to buy commodity items with illegitimate money.

Compact bone: Dense bone tissue that forms outer surface of bones.

Comparative method: The basic scientific process wherein two items are compared in order to find their similarities and differences.

Comparison microscope: A type of microscope that allows two items to be magnified and viewed side-by-side at the same time. Typically, it consists of two compound microscopes united by an optical bridge that brings their images together for viewing.

Competency: In the realm of courts or forensic psychology competency refers to the ability of a defendant, by virtue of the mental or psychological state, to understand the actions being taken against them. There are other contexts in which the term may be applied, such as a person being competent to write their will or being competent to serve as their own attourney.

Composite drawing: A likeness of a person (suspect, victim, etc.) made by a forensic artist from descriptions of the person given by witnesses.

Compound microscope: A microscope that contains more than one lens.

Concave lens: A type of lens in which at least one surface curves inward. A item viewed through a concave lens often appears to be smaller or farther away.

Conditional release: A type of release of an arrested person until the start of their trial under a set of conditions that the arrested person must obey. Common conditions include monitoring, restriction from the use of alcohol, or restriction from certain locations.

Confirmation bias: The tendency for people to remember evidence that supports their view or hypothesis and overlook or forget evidence that refutes it. It is thought that many people who believe in a pseudoscience do so because of confirmation bias.

Connectedness: A concept from forensic mathematics that describes how many other members of a group a particular individual is connected to or communicates with.

Consistent match: The result of a matching wherein the conclusion is that the evidence item and the standard item have the same class characteristics, but the individual characteristics are unavailable or not yet examined. This implies that the evidence and standard items might have the same origin, but it can not be proven that they do. Consistent identification and consistent match mean the same thing. This result is expressed in a variety of alternative ways, including "presumptive match/identification", "circumstantial match/identification", "inconclusive", "undetermined", and (simply) "identification".

Control experiments: Experiments designed to establish baseline parameters under multiple conditions. For example, if we want to understand something about a person's physiology while they are asleep we also need to determine what their physiology is like while the are awake, so that an accurate comparison can be made between the physiology under these two conditions. In this example, determining the physiology while awake is a control experiment.

Control questions: In the field of lie detection these are questions designed so that the person being tested will give an untrue response. This gives the polygraph operator an idea of what a person's response is when lying.

Controllers: Small computers that control the functions of the components of a larger computer system. For example, a typical desktop computer has a data bus controller, a hard drive controller, and a graphical display controller, among many others.

Convex lens: A type of lens in which at least one surface curves outward. Convex lenses usually cause an image view through them to be magnified.

Coronal plane: A plane of the body running in such a direction that it divides the body into upper (superior) and lower (inferior) portions.

Coroner: A type of detective who conducts investigations of deaths, called inquests, in which the cause, manner, time, and circumstances of death are determined to the greatest extent possible. The Coroner is normally an elected official of a county and in many jurisdictions the office of Coroner is combined with the office of Sheriff. Nationally, the trend is for Coroners to be replaced by Medical Examiners, and therefore coroners do not exist in all states.

Corporate Crime: The type of occupational fraud in which the perpetrator is the company or other organization and the victims are the organization's customers or employees.

Correctional psychologist: A psychologist or related forensic scientist who is employed by a correctional facility or work primarily with convicts.

Corrections: The portion of the justice system that handles incarcerations and other punishments that were given as sentences.

Correlation-based Approach: An approach to biometric verification or identification that samples characteristics of the body part being examined at several locations (hundreds). The characteristics of the sampled areas are then compared to databased characteristics of these areas for known individuals.

Corruption: A large set of crimes related to occupational fraud in which someone received money that they did not earn via legitimate means. An example is bribery.

Counterfeiting: Making illegitimate or unauthorized copies of an item of value and selling or passing it as if it were genuine.

Court psychologist: A forensic psychologist whose major focus is on witnesses, jurors, and others who participate in trials.

Courts of Appeal: The United States Courts of Appeal are the 13 courts that form the middle level of the U.S. federal court system.

Courts of origination: Courts that hear or try cases for the first time. The U.S. District Courts are courts of origination for the U.S. federal court system.

CPU: See central processing unit.

Cracking: Defeating the copy protection present in commercial software.

Cranial skeleton: The division of the skeleton that includes the bones of the cranium, the mandible, and the hyoid bone.

Credit card front-end loading: A money laundering strategy in which illegitimate money is used to prepay a credit card.

Crime lab: A facility dedicated to forensic analysis of items that are thought to be evidence of crimes. A crime lab houses a variety of scientific instruments for use by forensic scientists.

Crime lab photographer: A forensic photographer who works at a crime lab, primarily documenting the steps in forensic analysis of evidence.

Crime organization: An organization that engages in organized crime.

Crime scene: The location where a crime, or activities related to a crime, occurred.

Crime scene illustration: A redone version of a crime scene sketch made by a forensic artist using professional methods and conventional symbols, which may be more effective than the original rough crime scene sketch when presented in court.

Crime scene investigator: A person who processes a crime scene, particularly doing the documentation and evidence collection parts of the process. Depending on the case this can be a patrol officer, a detective, an evidence technician, or a team of forensic scientists from the crime lab. There is a certification process through which a person can be certified crime scene investigator. The people you see on the CSI television series may claim to be crime scene investigators, but actually they act like detectives.

Crime scene layout: The crime scenes for most cases can be conceptualized as having two or more areas, arranged like the circles of a target. The primary area of the crime scene is conceptualized as the bullseye of the target and is the area in which the most important events related to the crime occurred and where the most evidence exists. The secondary area is conceptualized as surrounding the primary area. Logically, the perpetrator had to move through the secondary area to get to the primary area. Therefore, evidence probably exists in the secondary area, though not as much of it as in the primary area. For some crime scenes it's helpful to think of a tertiary area that surrounds the secondary area, and which might contain some evidence.

Crime scene reconstruction: An animation or other hypothetical artistic reconstruction of the events of a crime. Crime scene reconstructions are helpful tools in an investigation, but should not be shown in court because jurors get a false impression that the details of the crime are known accurately.

Criminal court: A court that handles cases of criminal law.

Criminal history: The record of a criminal and his or her crimes.

Criminal law: A body of law that defines and sets procedures for handling criminal acts, which are those acts investigated, prosecuted, and punished by some level of government.

Criminal profiling: The process of making inferences about the characteristics of a person who committed a crime from the nature of the crime, how the crime was committed, and/or where the crime was committed.

Criminalist: A forensic scientist whose expertise is in one of the forensic sciences that focus on using the comparative method to do matching. This is a poorly defined category. Fingerprint examiners, firearms examiners, trace evidence examiners, toolmark examiners, and impressions examiners are universally considered to be criminalists. Some forensic scientists, such as serologists, forensic chemists, and questioned document examiners are considered to be criminalists by some authorities and not by others. Some types of forensic scientists, such as evidence technicians, are often considered criminalists as well, even though they do not do matching.

Criminologist: A social scientist, usually a person with a college degree in sociology, who is interested in theories for why some people become criminals, the workings of the justice system, corrections, and the impact of crime on society.

Cross examination: The part of the testimony of a witness in court during which the questions are asked by an attorney representing the other side in a case. For example, a witness for the prosecution will be questioned by the defense attorney during cross examination.

Crown: The portion of a tooth that is covered by enamel and includes the occlusal surface.

Cryptography: The field that studies codes. Also refers to the processes of encoding or decoding data in order to keep it secret.

Crystallization tests: A family of tests used by serologists to determine whether a suspicious red liquid is blood. In this type of test, the suspect liquid is treated in such a way that hemoglobin crystals will form if it is blood.

Currency smuggling: A money laundering strategy in which money is physically moved from one location to another to disguise its source.

Cusps: Projections from the occlusal surface of a tooth.

Cybercrime: Traditional types of crime, such as theft, fraud, or stalking, that are done or facilitated by the use of a computer or network.

Cyberstalking: The crime of stalking that is facilitated or enabled by the use of a computer or information obtained over the internet.

Data: Facts, observations, information, or other evidence collected during an experiment.

Data bus: The wires and other components used to carry information from one component of a computer to another. For example, data is moved from RAM to the processor chip over the data bus.

Data mining: The application of statistical and related methods to large data sets in order to detect patterns of behavior, consumer buying, or similar things. This method is sometimes used by forensic accountants to detect fraud.

Database: A set of related files that are managed together.

Daubert v. Merrell Dow Pharmaceuticals: A U.S. Supreme Court decision, made in 1993, which requires evidence purporting to be scientific to have been acquired using the scientific method. It empowers judges to determine the scientific credibility and relevance of evidence.

DBMS: Software that manages the files in a database.

Decompaction: The process of reducing the compaction of soil that naturally occurs over time. Digging into the ground decompacts the soil that is removed from the hole.

Decriminalization: Lessening the penalties associated with a certain crime.

Decryption: The process of decoding encrypted data so that it can be read and understood.

Defense attorney's fallacy: A logic fallacy (false understanding) in which evidence that might argue for the guilt of a suspect is ignored. In most cases this involves improperly expanding the range of possible suspect.

Degree inflation: The phenomenon whereby the education requirements for a particular job increase over time.

Denial of service attack: A type of system attack, often targeted at a server, that works by sending so many requests for service that the computer system or network is not able to handle them all.

Dental charts: Formal, ad hoc, graphical diagrams used by dentists to record the locations of tooth fillings and other dental issues in a patient. Usually these are made by shading in the location and shape of a filling on a standard graphic of the teeth.

Deposition: A formal interview of a witness before trial, outside the courtroom, normally by attorneys, and often under oath. During a deposition attorneys may ask questions and these questions and their responses are reduced to writing in proper legal form. The information in a deposition can be used as evidence and under certain conditions can be presented in court.

Depressants: A category of drugs that cause sleepiness without relieving pain. Depressants lower the activity of (depress) the central nervous system.

Dermatoglyphics: Fingerprint patterns and minutia. Dermatoglyphics can also refer to the study of fingerprints.

Detective: A type of peace officer whose primary duty is to do thorough investigations of crimes.

Detector: A device for detecting the presence of substances as they emerge from a chromatography device. A detector is often a form of spectrophotometer.

Developing fingerprints: The process of making a latent fingerprint visible. Usually this involves treating the fingerprint with a powder or a chemical substance such as superglue fumes.

Dialog decoding: The process of distinguishing between similar words (say "gun" versus "gum") in an audio recording by using a spectrograph to distinguish between the speech sounds in question.

Digital evidence: Evidence recovered from a computer, network, or other digital device or system. Also refers to the collection of this evidence.

Diphodont: Having two sets of dentition over an individual's lifetime. Mammals are diphodont, having a set of deciduous teeth (milk teeth, baby teeth) early in life and a set of permanent teeth (adult teeth) later in life.

Direct examination: The part of the testimony of a witness in court during which the questions are asked by the attorney who subpoenaed the witness. For example, a witness for the prosecution will be questioned by the prosecuting attorney during direct examination.

Direction indicator: This is some mechanism for locating the direction north, which should be included in all photographs of evidence collected at a crime scene.

Dirty bomb: A type of nuclear weapon designed to kill people while leaving buildings and other infrastructure relatively undamaged.

Discovery proceedings: A process that occurs before the start of a trial in which the prosecution and defense teams gather evidence. Under most condition, both teams must share the nature of their evidence with the other.

Disorganized serial killer: Serial killers who do not dress well and who have trouble keeping jobs or relationships. These serial killers often behave oddly, are not careful to minimize the evidence they leave at crime scenes, and are easier to identify than are organized serial killers.

Distal: The direction on a limb of the body that is toward the tips of the fingers or toes. The opposite of distal is proximal. When applied to the dentition, distal refers to the direction along the curving tooth row toward the posterior surface of the third molars. In the context of the teeth, the opposite of distal is mesial.

District Courts of the United States: The 94 District Courts form the lowest level of the U.S. federal court system.

Dizygotic twins: Also called "fraternal" twins, dizygotic twins are the product of two separate conceptions and no more genetically similar than normal siblings.

DMORT: An acronym for Disaster Mortuary Operational Response Team. DMORT's are organizations of volunteer forensic scientists that respond upon request to scenes of mass disasters.

DNA: Deoxyribose nucleic acid. The material in which genetic information is encoded and which forms the chromosomes.

DNA fingerprint: See RFLP.

Documenting the evidence: Making an unambiguous record of the nature and location of evidence at a crime scene. This can be done by photography, videorecording, and/or sketching.

Domestic terrorists: Terrorists who are citizens of a country and whose terrorist acts revolve around issues internal to that country.

Dominant alleles: Alleles that are always expressed if present.

Dorsal: The direction on the hand or foot that is toward the back of the hand or the upper portion of the foot. The opposite of dorsal is palmar on the hand and plantar on the foot. When applied to the body in general, dorsal has the same meaning as posterior.

Driving under the influence: The operation of a motor vehicle with a blood alcohol level over the limit defined as legal in that jurisdiction.

Dry stick fracture: A type of fracture of bone in which the bone snaps cleanly like a dry stick. Dry stick fracture normally indicates that the trauma involved was postmortem.

Duress: Threat or stress. For example, a signature made under threat of bodily harm is described as having been produced under duress.

Dynamite: An initiating explosive made of nigroglycerin mixed with wood pulp or another stabilizing substance; sodium nitrate, which serves to furnish oxygen for complete combustion; and a small percentage of a stabilizer such as calcium carbonate.

EAAF: An acronym for Equipo Argentino de Antropología Forense, which translates to English as Argentine Forensic Anthropology Team. This organization is the first and most prominent group to investigate human rights violations around the world by recovery and analysis of remains of victims.

EGIS: An acronym for environmental graphical information system. An EGIS system is essentially a portable GCMS unit that is used in a variety of ways, including investigation of bombing incidents.

Electromagnetic radiation (EMR): Traditionally defined as self-propagating waves in a vacuum or in matter, electromagnetic radiation can be just as validly thought of as particles (photons) moving through a vacuum or matter. EMR has three properties. Intensity is, depending on conceptualization as waves or particles, the height of the wave crests or the number of photons striking a unit area of a surface. The energy of ERM is the energy inherent in the wave or photon, and in the case of a wave is expressed as the frequency or wavelength of the wave. The direction of EMR is the direction in which the waves or particles are moving.

Electronic Surveillance: Monitoring the location, activities, or communication of an individual using electronic equipment. As a field, electronic surveillance is considered to include the use of bugs, cameras, satellite images, wiretapping, and similar activities.

Electrons: Negatively charged subatomic particles that are normally found encircling the nucleus of an atom. According to the principles of quantum physics the location of an electron is not known until it is measured. Therefore, electrons are described as existing in a "probability cloud" surrounding the nucleus.

Electrophoresis: A method of separating the components of a mixture based on molecular weights, solubilities, and electrical charges of the components. Electrophoresis is similar to chromatography in that there is a stationary phase and a mobile phase, and in addition an electrical field is also applied. The stationary phase is most often a gel (starch or a synthetic material) and the mobile phase (usually called a "buffer") is a liquid. Electrophoresis is most commonly used to separated biological molecules such as proteins, peptides, and DNA fragments.

Elevation sketch: A sketch made using elevation perspective, meaning that it is drawn as it would be seen by someone standing beside the area or item.

Email flood: A system attack carried out by sending more emails to a user or computer than the system can handle, thus degrading the performance of the system.

Embezzlement: A form of theft in which the perpetrator legally receives the property of another for a certain purpose, but converts the property to their own use. A charge of embezzlement does not require the intent to permanently deprive the victim of their property.

Encryption: The process of encoding data in such a way that it becomes more difficult for anyone other than the intended recipient of the data to read it.

Endangered Species Act: U.S. federal legislation passed in 1973 that seeks to protect species that are thought to be in danger of extinction. The provisions of the Endangered Species Act primarily restrict import and export of endangered species.

Endothermic: A type of chemical reaction that consumes energy.

Environment: There are several uses of this word. In ecological studies it refers to the resources, climate, and geography of the world around us. In genetics it refers to the causes of variability that are not genetic. In sociology it refers to the social and economic conditions under which a person was raise or in which they now live.

Ethnicity: A cultural group to which individuals are born into by virtue of their physical characteristics or the national origin of their parents. The modern American concept of races and nationalities really refers to ethnicities.

Event Sequence Analysis: The process of producing a timeline of events using recorded sound information.

Evidence items: Items of physical evidence for which it is not known who left them at a certain location of interest, such as a crime scene. During matching these items are compared with standard items in an attempt to determine who left them at the location.

Exclusion (in matching): The result of a matching wherein the conclusion is that the evidence item and the standard item differ in at least one characteristic such that they can not be from the same origin.

Execution cell: A terrorist cell that carries out an actual act of terrorism.

Exemplars: Standard items. The term "exemplars" is commonly used in document examination.

Exhumation: The excavation of a body that was previously legitimately interred, for the purposes of continued investigation of a case or for historical forensic science.

Exoneration: In general, exoneration is the finding that a person is innocent of a crime for which they have been charged. It may also refer to the overturning of the conviction of a person who is later found to be innocent due to new evidence or reevaluation of evidence used at their trial.

Exothermic: A type of chemical reaction that produces more energy than it consumes.

Experiment: A procedure that includes observing the effects that occur when items and events are manipulated or naturally occur in a specified way.

Experimenter effect: A form of bias that results from the natural human tendency to expect certain results to occur during an experiment. Operating under this expectation a scientist may imagine results, or may (consciously or unconsciously) misinterpret or manipulate the results.

Expert witness: A witness with special knowledge, skill, experience, training, and/or education that goes beyond that of ordinary citizens. Expert witnesses often testify about the evidence in the case rather than presenting details of the crime that they themselves witnessed. When forensic scientists testify at a trial they do so as expert witnesses.

Expressivity: Factors arising from an individual's unique biochemistry that have an effect on the expression of a genetic trait.

External ballistics: The phase of ballistics in which the projectile is traveling through the air between the firearm and the target.

Extortion: A form of robbery in which there is a threat of force or retaliation in the future.

Extraordinary realities: The many and varied realities thought to exist by some individuals and groups of people, but not all. Usually, extraordinary realities include deities and other supernatural elements that are at least partly responsible for events that happen in that reality. Most extraordinary realities require some element of belief in order to be perceived.

Facial reconstruction: Reconstructing the facial features of an individual from their skull or an x-ray radiograph of their skull using scientific data about tissue depths at certain places on the face, and the placement of facial elements (noses, ears, etc.) in relation to features of the skull. Historically, facial reconstruction was done by forensic anthropologists and the nearly identical process of postmortem reconstruction was done by forensic pathologists. In recent times, it is typical for forensic artists to do facial reconstructions.

False Accept Rate: In the field of biometrics this refers to the percentage of times that the biometric procedure recognizes a person not in the biometric database in use at the time as someone who is in the database.

False positive results: In drug testing this is a result in which the test suggests that a drug is present in the test sample when it in fact is not present.

False Pretenses: A form of theft in which the property of another is taken through trickery or misinformation.

False Reject Rate: In the field of biometrics this refers to the percentage of times that the biometric procedure fails to identify a person who is known and is present in the biometric database used at the time.

Falsification (in the scientific method): See hypothesis testing.

Fauna: Non-human animals.

Fear-induced aggression: A type of aggression in which a person feels cornered and strikes out in order to protect themself.

Felonies: Serious crimes. Most crimes that involve more than a minor injury or more than a small amount of money are felonies. Each state decides which crimes are felonies.

Field: In database management a field is an item of information that is part of a record. For example, a record of an employee may include the employee's name, their address, and their social security number, all of which are different fields of the record.

Financial release: Also known as "bail", this is a type of release of an arrested person until the start of their trial in which they provide a sum of money as an assurance that they will appear at their trial. If the defendant appears at their trial the money is returned, but if the defendant fails to appear the money is forfeited.

Fingerprint patterns: The class characteristics of fingerprints, categorized as various types of arches, loops, and whorls.

Firearm function examination: A type of firearm examination in which the examiner seeks to determine whether the firearm is in safe operating condition. Particularly, the examiner often seeks to determine whether it could have been accidentally discharged in the manner described by a suspect or witness.

First appearance: A type of hearing before a judge, which is required in some jurisdictions when a suspect is arrested without a warrant. At this hearing the judge will hear evidence that the suspect should really have been arrested and if the evidence is insufficient the suspect may be released.

First degree murder: A type of murder in which the death was deliberate and premeditatied.

Fissures: When applied to a tooth, this applies to the grooves between the cusps.

Flesh flies: A taxonomic group of flies that deposit live larvae on the bodies of dead organisms. The larvae consume the flesh of the dead organism.

Forensic accounting: The application of accounting to investigations.

Forensic anthropology: The application of anthropology, particularly archaeology and skeletal analysis, to investigations.

Forensic archaeology: The subfield of forensic anthropology that includes searching for and recovering buried (occasionally unburied) evidence using archaeological methods.

Forensic art: The application of art to investigations or to documenting evidence.

Forensic audio analysis: The application of audio technology and knowledge to investigations. The most common task of a forensic audio analyst is to enhance the listenability or intelligibility of recorded audio.

Forensic botany: The application of the science of botany (the study of plants) to investigations.

Forensic chemistry: Broadly, the application of chemistry to investigations. In practice, most forensic chemistry involves testing suspect substances to see whether they are drugs.

Forensic computer science: The application of computer science to investigations.

Forensic drug testing: See forensic toxicology.

Forensic economics: The application of the theory, statistics, and knowledge of economics to establish the value or cost of a loss or injury.

Forensic engineering: The application of the several branches of engineering to investigations.

Forensic entomology: The application of the science of entomology (the study of insects) to investigations..

Forensic geology: The application of the science of geology (the study of the processes shaping the features of the Earth) to investigations.

Forensic hypnosis: The application of hypnosis to investigations, usually in an attempt to enhance a witness's or victim's memory of an event.

Forensic image analysis: The analysis of a photograph, video, or other visual information in order to enhance details of the image or authenticate the image.

Forensic knot examination: The application of specialized knowledge of knots and ligatures to investigation.

Forensic linguistics: The application of the science of linguistics to investigations.

Forensic mathematics: The application of the methods and knowledge of mathematics to investigations.

Forensic medicine: A broad category of forensic specialties in which medical knowledge (conventional medicine, chiropractic, naturopathy, etc.) is used to determine the extent of an injury and the cost of treating it. According to some authorities forensic pathology and forensic podiatry are also branches of forensic medicine, though these specialties are usually involved in criminal issues while other branches of the field are usually involved in civil suits.

Forensic meteorology: The application of the science of meteorology (study of weather) to investigations.

Forensic nursing: The application and provision of nursing services to victims of violent crimes.

Forensic odontology: The application of the science of dentistry to investigations.

Forensic osteology: The subfield of forensic anthropology the includes the analysis and interpretation of skeletal remains.

Forensic pathology: The application of the medical field of pathology to investigation of deaths and related issues. Forensic pathologists are often trained both in anatomic and clinical pathology plus have specialized training in the forensic aspects of these fields.

Forensic pharmacology: The application of the science of pharmacology to investigations. In contrast to forensic toxicologists, forensic pharmacists are more concerned with prescription drugs, prescription fraud, and drug interactions.

Forensic photography: The application of photography to investigations or to documenting evidence.

Forensic phylogenetics: The application of the science of phylogenetic analysis (the study of evolutionary change through time) to investigations.

Forensic podiatry: The application of knowledge from the medical field of podiatry to investigations. Podiatry is the branch of medicine concerned with feet.

Forensic psychology: The application of the psychology and related sciences (psychiatry, behavioral science) to investigations.

Forensic serology: The application of the scientific principles of serology to investigations. See serology.

Forensic social work: The application and provision of social services to victims of crimes and to convicts who have been released.

Forensic toxicology: The branch of the forensic sciences that tests for the presence of drugs, alcohol, or poisons within an individual's body. There are several branches, including post-mortem forensic toxicology in which sample collected at an autopsy are tested, human performance forensic toxicology that is concerned with the effects of substances on a person's ability to function well, and forensic drug testing that focuses on screening people for use of drugs.

Forensic zoology: The application of the science of zoology (the study of non-human animals) to investigations. There are two main contexts in which forensic zoology is important. The first is the identification of animals from their remains or body parts. The second is the interpretation of an encounter between a human and a non-human animal using knowledge of the non-human animal's behavior.

Forensics: Competitive debate between teams. The teams usually consists of college students majoring in Rhetoric, Communications, or English. Despite popular usage "forensics" is not a good substitute for "forensic science".

Fossa: A depression in the surface of a bone.

Fraudulent statements: The crime of intentionally misrepresenting an organization or company's financial status on a required financial statement.

Frequency: In the context of electromagnetic radiation frequency is the number of cycles per second of a wave – a measure of it's inherent energy. In the context of serology and probability the frequency is the proportion of individuals with a certain allele or other characteristic.

Front companies: Companies that carry on legitimate business but also have a component of illegitimate business, such as money laundering or organized crime.

Fuel: A substance that is capable of undergoing an oxidation reaction such as combustion. Most fuels contain carbon.

Fuel cell: In general, a device in which a fuel undergoes a chemical reaction that releases electrons thus producing a small electric current. The number of electrons released, measured as a voltage or current, is proportional to the amount of fuel. Newer breathalyzers utilize this principle to determine the amount of alcohol in a person's breath.

Fuming with superglue: Also called cyanoacrylate fuming, this is a very common method of developing fingerprints.

Gas chromatography: A type of chromatography in which the stationary phase is a liquid and the mobile phase is a gas.

Gauge: The diameter of a shotgun barrel.

GCMS: See mass spectrometer.

Gel: See electrophoresis.

Genetic markers: Characteristics of a person that are determined by the person's genes. Common genetic markers used in forensic serology include blood groups, blood proteins, and DNA sequences.

Genetic profiling: The process of estimating characteristics of an individual, such as their race or sex, from the presence of certain DNA markers. Estimation of sex from DNA is probably reliable, but estimation of race is only a statistical guess.

Genotype: An abbreviation or code for the two alleles an individual possessed for a certain locus. For example, a person who inherited an A allele and an O allele for the transfusion (ABO) blood group system would be genotype AO.

Geographic profiling: A type of criminal profiling in which the home or workplace of a serial criminal is inferred from the locations of where the crimes were committed.

Governmental Crime: The type of occupational fraud in which the perpetrator is a government agency and the victims are the people the agency is supposed to serve.

Grand: A term applied to a form of theft in which the value of the stolen item is large. Grand thefts are usually felonies.

Grand jury: A group of citizens who examine the evidence in felony cases to determine whether it is sufficient to justify prosecuting a defendant. In most jurisdictions grand juries test the strength of evidence in felony cases and a judge tests the strength of evidence in misdemeanor cases. See preliminary hearing.

Green stick fracture: A type of fracture of bone in which the bone does not snap cleanly and some component of bending of the bone can be seen. Green stick fracture normally indicates that the trauma involved was premortem or perimortem.

Gross: In anatomical terms gross refers to the body as a whole, as opposed to its constituent cells. Therefore, gross anatomy is the study of the features of the whole body, its muscles, bones, blood vessels, nerves, etc.

Hacking: The act or process of gaining unauthorized entry to or use of a computer or network.

Hallucinogen: A category of drugs that alter the state of consciousness and/or alter perception. In rare cases an actual hallucination can be produced.

Handgun: A firearm that is designed to be small enough to be held and fired with one hand. Pistols are a common form of handgun.

Haplotype: An abbreviation or code for a single allele. In the case of nuclear (chromosomal) DNA a haplotype is simply half of a genotype.

Hashing: The process of generating a relatively unique number that characterizes the information in a file for the purpose of indexing the file so it can be searched for easily in the future.

Hearsay: Testimonial evidence that consists of what someone told the witness but the witness did not witness themself.

Hedonistic motive: The motive for serial killing in which the killer kills for fun, excitement, or thrill.

Heritability: The proportion of the variability in a characteristic that is caused by genetics.

Heterodont: Having different types of teeth in the dentition. Mammals are heterodont, with dentition consisting of incisors, canines, premolars, and molars.

Heterozygous: The condition of having two different alleles for a locus.

High explosive: An explosive substance that causes a shock wave that is faster than the speed of sound.

High performance liquid chromatography (HPLC): A type of chromatography in which the stationary phase is (usually) resin beads and the mobile phase is a liquid.

Historical forensic science: The application of forensic science to cases that happened long enough in the past that the perpetrator(s) of the crime are presumed to be dead. The goal is to simply know the truth of an event rather than to bring a suspect to justice.

History: In the realm of computers this is a list maintained by a web browser of the web sites that have been accessed.

HLA: Human leukocyte antigens. Antigens on the surfaces of white blood cells and other tissues. This is the same system as the MHC system.

Homeland security: The U.S. national effort to prevent terrorism, detect and arrest terrorists, and recover from terrorist attacks.

Homicide: See murder.

Homozygous: The condition of having two identical alleles for a locus.

Hub: In the realm of computers a hub is a now obsolete type of network device that receives packets from one computer and re-broadcasts them to all computers connected to the hub.

Human-performance forensic toxicology: See forensic toxicology.

Human produced fibers: Also known as "man-made" fibers, these are fibers that are produced in a laboratory rather than by an organism. Examples are rayon, nylon, and polyester.

Hydrocarbon: A chemical molecule formed entirely of carbon and hydrogen. Common hydrocarbons are methane, propane, and octane.

Hydrocarbon fingerprint: The unique pattern of trace substances in a sample of hydrocarbon. For example, gasoline refineries combine petroleum from a variety of sources in each batch of gasoline produced. The petroleum from each source has a distinct combination of trace substances, leading to each batch of gasoline having a unique combination and quantity of these trace substances.

Hypercompliance: A mental state induced by hypnosis in which the hypnotized person seeks to please authorities by answering questions even if the answer is not known. Often the answers given are erroneous.

Hypnosis: The process by which a person is led into a hypnotic state.

Hypothesis: A tentative explanation of a phenomenon based on the evidence available at the moment. One of the core processes of the scientific method is hypothesis testing.

Hypothesis testing: A hypothesis is tested by evaluating whether can explain the results obtained during a set of experiments. If the hypothesis can explain these results then it is considered to be "supported". If it can not explain these results then it is considered to be "falsified" or "rejected" and must be revised to accommodate the new evidence.

Hypotheticals: Statements or reports made by experts who do not actually testify at the trial during which their evidence is used.

I/O system: See input/output system.

IAFIS: An acronym for integrated automated fingerprint identification system. The is the U.S. national level fingerprint database.

IBIS: An acronym for integrated ballistics identification system. An IBIS is a regional level firearms evidence database.

Ice mummy: Essentially a freeze dried corpse. A cadaver may become an ice mummy in a cold, dry environment.

Identification: Identification often refers to the same process as matching. In some contests it refers to a matching process, or part of a matching process, in which class characteristics are examined. Identification can also be used as an alternative term to express the result of consistent match in matching.

Identification and Individualization: Alternate terms for consistent match (identification) and positive match (individualization).

If...then statements: Statements of the form "if <a hypothesis> is true then <a certain result> should occur when <a certain set of conditions exist>. These statements are used to express predictions used in hypothesis testing.

Ignition energy: The activation energy required to start a combustion. Activation energy is the energy required by an exothermic reaction in order for the reaction to begin. After the exothermic reaction begins energy is produced, normally more than was required for the activation energy.

Ignitor: A device used to ignite a fuel.

Illegal distribution of alcohol: Making alcohol available outside of the jurisdiction's normal regulations and taxes.

Image enhancement: The process of improving the quality of an image so that its details can be more clearly seen.

Immunity: In the realm of criminal prosecution immunity refers to a promise made by the prosecution that a person will not be tried for a crime in exchange for testimony.

Immunoassay: A type of test for the presence of drugs or to test for drug use via testing urine or blood. Drug-specific antibodies, which are labeled in some manner that they can later be detected, are introduced to the test sample. If the antibodies react with the test sample it is inferred that the target drug is present in the sample.

Impression: A representation of an item impressed into a relatively soft surface.

Inconclusive (in matching): See consistent match.

indeterminate (in matching): The result of a matching wherein it is concluded that no result of exclusion, consistent match, or positive match is possible. Usually this results when evidence items are too fragmentary or damaged to allow reliable class or individual characteristics to be observed.

Individual characteristics: Characteristics that are unique to a certain individual item. For example, details of fingerprint minutia and pigment distribution in the iris of the eye are unique to every human being.

Individualization (in matching): A matching, or the part of a matching, in which individual characteristics are examined. Individualization can also be used as an alternative term to express the result of positive match in matching.

Inferior: The direction on the body that is toward the bottoms of the feet. This direction is used roughly synonymously with caudal (toward the tail), especially when describing direction on the bodies of non-human animals. The opposite of inferior is superior.

Infrastructure: In the field of homeland security this refers to those services that enable our normal way of life, such as electrical power, water, and transportation.

Inhalants: Substances that are inhaled (huffed, snuffed) in order to produce a "high". Many inhalants work by simply displacing oxygen. Others cause physiological changes independent of displacing oxygen.

Initiating high explosive: A high explosive that needs only a small input of energy in order to detonate.

Input/output system: The set of input devices and output devices used by a computer. Input devices send data to the computer and output devices receive data sent to them by a computer.

Inquest: See Coroner.

Insanity defense: A strategy by the defense team for a criminal trial in which the idea is to demonstrate that the defendant is innocent or has reduced responsibility because of a mental illness or other unusual psychological state.

Instars: A stage within the larval stage of insect development. The instars are defined by molts (shedding of the exoskeleton) of the larva.

Instrumental aggression: The type of aggression that a person uses because they have learned that they can get what they want if they are aggressive.

Intelligibility: A characteristic of an audio recording that refers to how easy it is to understand or interpret the recorded sounds or speech.

Intermale aggression: A type of aggression between two or more males. The conflict is often over the affections of a female.

Internal ballistics: The phase of ballistics that occurs inside a firearm or other device that propels projectiles.

International terrorists: Terrorists whose activities occur within a country of which they are not citizens and which revolve around issues that are international in scope.

Internet: In general, a large scale, non-private WAN. "The internet" is a good example.

Intranet: A large private WAN, that restricts access to members of a certain company or other organization.

Investigative accounting: A context of forensic accounting in which financial records are examined to determine whether a crime has been committed.

Involuntary manslaughter: A type of manslaughter in which the death was not deliberate but occurred due to negligence.

Irrelevant questions: In the field of lie detection these are question unrelated to the subject of the investigation that the person being tested should normally answer truthfully. This gives the polygraph operator and idea of what a person's response is when telling the truth.

Irritable aggression: The type of aggression that a person who is stressed by circumstances may display.

JPAC: An acronym for Joint Prisoners of War, Missing in Action Accounting Command. JPAC is a U.S. military organization that seeks to account for service personnel lost or missing during military conflicts. Some JPAC operations include recovery of the remains of lost servicepeople.

Junk science: The phenomenon in which legitimate scientific theories or data are used improperly to support an idea or conclusion that is false.

Jurisdiction: A set of laws enforced by a law enforcement agency and a geographic area in which the agency is expected to enforce them.

Jurisprudence: The forensic science concerned with how evidence is collected, preserved, admitted, and presented in court.

Justifiable homicide: A situation in which it is recognized as allowable to cause a death. For example, killing someone in defense of one's own life or the life of another is normally considered justifiable homicide.

Kastle Meyer test: A test used by serologists to determine whether a suspicious red liquid is blood. In this test, the test reagents change color in the presence of blood.

Key (in cryptography): In the realm of cryptography a key is the translation table or method that is used to encrypt (and later decrypt) a file. For example, the rot1 key replaces each letter in the alphabet with the next letter in sequence (a = b, b = c, etc.) so that dog is encrypted as eph.

Key (in database management): In the realm of database management systems the key is the information field that is used to designate each separate record. For example, The University of Montana uses an internal ID number to designate each employee and students. In this scheme my key is 790217986.

Kinetic energy: The energy inherent in a moving object by virtue of its mass and velocity. The amount of kinetic energy is calculable from the object's mass (m) and its velocity (v) as $(mv^2)/2$.

Knot tying by habit: Tying a knot in the case where the method of tying it has been internalized as a motor program and does not need to be specifically thought about. For example, tying one's shoes is by habit for most of us.

Knot tying by design: Tying a know when the method of tying it has not been internalized as a motor program and the steps involved must be deliberately remembered or thought about.

Labial: The direction within the mouth that is toward the lips. This direction applies to those teeth and portions of the mouth that lie behind the lips, and the equivalent direction for more distal teeth and portions of the mouth is buccal. The opposite of labial is lingual.

Laceration and crushing: Laceration is creating a wound by cutting. Crushing is creating a wound by application of pressure. Laceration and crushing is one of the mechanisms by which a projectile causes damage to tissue.

Lacey Act: U.S. federal legislation that makes it illegal in the U.S. to import, export, or otherwise deal with wildlife that were taken in violation of the laws of any state or nation.

LAN: See local area network.

Lands: The part of rifling that is the firearm barrel surface between grooves.

Larceny: A form of theft that involves intentional taking of another's property, with the knowledge that the property belongs to another. A charge of larceny requires that there is an intent to permanently deprive the victim of their property.

Larva: In general, the immature form of an invertebrate organism. When applied to insects it refers to the maggot (worm-like) stage of life.

Latent fingerprints: Fingerprints that are not visible to the naked eye and must be developed by some method in order to be detected and studied.

Lateral: The direction on the body that is toward the right or left side – away from the median sagittal plane (see sagittal plane). The opposite of lateral is medial.

Law enforcement agency: A government agency that employs peace officers.

Law of association: The rule from the field of archaeology which says that the closer two items are in time and space the more likely it is that there is a conceptual or behavioral relationship between them. This law also applies to at least most evidence collected at a crime scene.

Layout of a crime scene: The application of the theoretical perspective on a crime scene that divides it into a primary area and a secondary area. Alternatively, the designation of primary and secondary areas of a particular crime scene.

Leadered cells: A type of terrorist cell organization in which there is a leader of all the cells who communicates regularly with the cells and coordinates their activities.

Leaderless cells: A type of terrorist cell organization in which the leader of the organization (if such exists) does not communicate regularly with the cells. The cells therefore operate autonomously, follow a predetermined master plan, or wait for a coded communication from the organization in order to carry out an operation.

Leading question: A question that implies the desired answer.

Left wing domestic terrorists: Domestic terrorists in the U.S. who espouse philosophies of opposition to capitalism and American business and industrial practices. They can be communist oriented, anarchist oriented, or have a poorly defined orientation.

Lens: A device made of glass or other material, which is designed to have a certain shape that controls the refraction of electromagnetic radiation (usually light) for some desired effect such as magnification.

Ligature: The material (string, wire, rope, cord, etc.) in which a knot is tied.

Lignin: A protein found in the cell walls of plants.

Lingual: The direction within the mouth that is toward the tongue. The opposite of lingual is labial or buccal, depending on the exact location within the mouth.

Listenability: A characteristic of an audio recording that refers to how comfortable it is to listen to. Clicks, pops, and other background sounds detract from listenability.

Litigation support: In forensic accounting this refers to the work of a forensic accountant in an ongoing case in which the goal is to determine the nature or amount of money involved in a crime.

Livor mortis: Pooling of blood in a person's body after death, resulting in a purplish color change in the parts of the body closest to the ground.

Local Area Network: A small scale network, usually connecting the computers within a room, set of rooms, or a building.

Locard's exchange principle: The principle first stated by Edmond Locard (1877 – 1966) as ""with contact between two items, there will be an exchange". This principle is generally understood to mean that when a person visits a location they both leave something there and take something with them. Thus, evidence that may link a suspect with a crime scene will always exist.

Locus: A location on a chromosome that where the genetic blueprint for some biochemical product (usually a protein) is stored.

Log files: Files that record the time and nature of use of a computer system or network.

Low explosive: An explosive substance that causes a shock wave that is slower than the speed of sound.

Luminescence: A type of interaction between electromagnetic radiation and matter in which the EMR stimulated the matter to release additional EMR, usually at a different wavelength from the stimulating EMR.

Luminol: A chemical substance used by serologists to detect the presence of blood. Luminol reacts with blood in such a way that is will fluoresce under ultraviolet light. Luminol is extremely sensitive to even trace amounts of blood.

Lust motive: The motive for serial killing in which the the killer seeks to satisfy sexual lust.

Lysis: Bursting of cells. In the laboratory, lysis is often done to release the contents of cells for serological testing. Lysis is also one of the ways in which antibodies react with cells carrying target antigens. Lysis kills the cell.

Magistrate judges: A judge appointed by a U.S. District Court Judge for an eight-year term, who assists the regular judges by presiding over lesser cases.

Manner of death: The legal category into which a death falls, which determines how the justice system will handle it. Most jurisdictions recognize five manners of death: homicide, suicide, accident, natural causes, and undetermined. Some jurisdictions recognize a sixth manner of death– therapeutic complication.

Manslaughter: A type of murder in which the death was either not intentional, was without malice, or was provoked by the victim.

Masking noise: Background noise in a location that makes it more difficult to hear a sound.

Mass murderer: A murderer who kills several people in one event.

Mass spectrometer: A method of identifying and quantifying substances. The substance in question is ionized, then subjected to an electrical or magnetic field, resulting in the substance moving in a specific manner depending on its mass to charge ratio. The mass to charge ration of many substances is unique and known, enabling rapid identification of the substance. Often, a mass spectrometer is coupled with a gas chromatograph in such as way that the substances exiting the gas chromatograph are identified by the mass spectrometer. Such as device is call a gas chromatograph mass spectrometer or GCMS.

Matching: The process of comparing evidence items with standard items in order to determine the origin of the evidence items. Matching can result in an exclusion, a consistent match, a positive match, or the conclusion that none of these three results is possible (indeterminate). Various forms of matching are also referred to as identification and as individualization.

Materials science: The study of the properties of materials of all sorts, particularly those materials used to manufacture items.

Maternal aggression: The type of aggression in which a female seeks to protect her offspring.

MD5: Short for message digest version 5, this is a method for generating a hash code (number) for a file.

Mechanical explosion: An explosion that occurs due to increasing the pressure within a container to the point that the container ruptures explosively. An example is popping a balloon.

Mechanical fingerprint: The individual characteristics of a firearm, including features of the barrel, firing chamber, and firing hammer/pin, that are transferred to the bullet or casing.

Medial: The direction on the body that is toward the median sagittal plane (see sagittal plane). The opposite of medial is lateral.

Medical Examiner: A Medical Examiner is a medical doctor, often a specialist in forensic pathology, who is charged with investigating deaths. Some states have Medical Examiners and no Coroner. Other states have both.

Medicolegal forensic entomology: The branch of forensic entomology that includes investigation of crimes. One of the more common applications is in estimating time since death of an individual from insect evidence associated with the remains

Memory distortion: A phenomenon that occurs when the memories of a hypnotized person become confused or distorted with respect to what actually happened.

Memory refreshment: The process of enhancing the memory of a witness or victim by having them remember an event under hypnosis.

Mental template: A picture in the mind of how something should look when produced by the body. In theory, handwriting begins as a mental template.

Mesial: A direction within the mouth that is toward the midline of the tooth row – i.e. between the two central incisors. The opposite of mesial is distal.

Metric system: A system of measurements of distance/length, weight, volume, and other measured variables in which units, subunits, and superunits, are related by powers of ten. The metric system is used in nearly every country in the world except the United States.

MHC: Major histocompatibility complex. Antigens on the surfaces of many types of tissues including white blood cells. This is the same system as the HLA system.

Microscopy: The science of microscopes, their operation, and their use.

Mineralogy: The chemical composition of the particles in a sample of soil.

Minutia: The individual characteristics of fingerprints, including ridge bifurcations, dots, ridge ending, short ridges, and similar features.

Misdemeanors: Less serious crimes that result in only minor injury or loss of a small amount of money.

Missionary-oriented motive: The motive for serial killing in which the killer's goal is to rid the world of something bad or impure.

Mistakes result (in forensic accounting): The result of an examination of financial records in which some discrepancies were found but it is not clear that they are the result of an intentional crime (they might just be mistakes).

Mitochondria: Small bodies (organelles) that are found within the cells of eukaryotes (animals, plants, and fungi). Mitochondria are closely involved with regulation of metabolic energy within the cell.

Mitochondrial DNA: Small strands of DNA found in the mitochondria of cells. Often abbreviated as MtDNA. Mitochondra are small bodies (organelles) found within cells that are involved with energy production and manipulation for the cell. MtDNA is separate from the conventional (nuclear) DNA that is found in the nucleus of a cell. MtDNA has an advantage in some types of analyses because it evolves rapidly, is maternally inherited (all of an individual's mitochondria come from the mother), and is resistant to degradation. Also, while each cell has at most one nucleus, it often has thousands of mitochondria, so a given sample of blood or tissue will have thousands of times more mitochondrial DNA than nuclear DNA.

Mixed System of Death Investigation: A system in which both Coroners and Medical Examiners exist. In these systems it is typical for the Medical Examiner to determine the medical facts surrounding the death, while the Coroner investigates the time and circumstances of death. Usually, it is the Coroner who will combine all this information to determine the manner of death.

Mobile phase: See chromatography.

Modifying prefixes of the metric system: The modifying prefixes of the metric system are deci for one tenth, centi for one hundredth, milli for one thousandth, micro for one millionth, nano for one billionth, deka for ten, hecto for hundred, kilo for thousand, mega for million, and giga for billion of a basic metric system unit.

Mole: The amount (weight) of Avogadro's number of molecules of a certain substance. Avogadro's number (in scientific notation) is 6.0221415×10^{23}.

Molecular weight: The molecular weight (in more modern terms, molecular mass) of a molecule is the sum of the atomic weights of the atoms that comprise it. For example, water has one oxygen with an atomic weight of 16 and two hydrogens, each with an atomic weight of 1. Therefore, the molecular weight of water is about 18.

Molecules: Substances formed from the combination of atoms. For example, a water molecule is formed by the combination of a hydrogen atom and two oxygen atoms.

Molting: Shedding of the exoskeleton or armored skin of an organism in order to allow the organism to grow.

Money or value transfer systems: Systems by which money or goods are transferred from on person or form to another person or form. For example, wiring money via Western Union.

Money laundering: The process of attempting to make money acquired illegitimately or illegally look as if it has been acquired legitimately.

Monozygotic twins: Also called "identical" twins, monozygotic twins are the product of a single conception and are genetically identical.

Motor program: A condition in which the brain, nerves, and muscles have been trained to do something (to write, for example) without the individual having to think about the discrete steps involved.

Munsell color chart: A standard chart that allows the identification of a large number of distinct colors.

Murder: The crime of causing the death of another person. Also known as homicide.

Mutation: A change in the DNA base sequence at some location.

Narcissist: A person who sees only themself and their desires as having meaning.

Narcotic: Strictly, a category of drugs that both relieve pain and cause sleepiness. In many jurisdictions the term is expanded to include all types illegal drugs, whether or not they are actually narcotics.

Natural fibers: Fibers produced by a plant, animal, or other organism. Examples are cotton, silk, and wool.

Natural variability of handwriting: The variation in the handwriting of an individual from one time to another.

NCIC: An acronym for National Crime Information Center. NCIC is the U.S. national database of crime and criminals. Unlike other U.S. database systems it contains a very wide variety of information.

NDIS: An acronym for National DNA Index System. This is the U.S. national level DNA database.

Neck: When applied to a tooth, this is the line where the crown meets the root.

Necropsy: The autopsy of a non-human animal.

Network: A system that interconnects two or more computers so that they can share information and other resources.

NIBIN: An acronym for National Integrated Ballistic Information Network. NIBIN is the U.S. national level firearms evidence database.

NICS: An acronym for National Instant Criminal Background Check System. NICS is the U.S. national level database of criminal histories.

No contest: A type of plea in which the defendant is admitting criminal guilt but not civil liability for a crime.

Nominees: In the realms of terrorism, organized crime, and money laundering, these are individuals who are respected members of a community and who speak for, negotiate for, or do tasks for the terrorist or crime organization.

Non-initiating high explosives: A high explosive that needs a substantial input of energy in order to detonate. This detonation energy is often provided by a primer containing a small amount of initiating high explosive.

Non-porous surfaces: Surfaces that, due to their construction or the material of which they are made, can not absorb a liquid.

Non-volatile memory: A type of computer memory that retains data stored in it even when the power to the computer is turned off.

Normal result (in forensic accounting): The result of an examination of financial records in which no crimes or errors were detected.

NSOR: An acronym for National Sex Offender Registry. NSOR is the U.S. national level database of sexual offenders.

Nuclear DNA: The DNA found within the nucleus of a cell, packaged into chromosomes.

Nuclear explosion: An explosion that occurs due to a nuclear reaction.

Numbers: A type of organized illegal gambling that resembles a lottery. People bet on whether certain numbers will be drawn.

Occlusal surface: The surface of a tooth that meets with (occludes with) a tooth from the opposite jaw during chewing.

Occupational fraud: A large set of crimes that involve stealing or misusing a company's or organization's resources for personal gain.

Occupational markers: Changes in the skeleton, usually at joints of muscle attachment sites, that result from repeated specific motions or actions. The repeated specific motions or actions can occur due to the activities of an occupation, a habit, an athletic activity, or any repeated activity at all.

Ockham's razor: The idea (not originally stated exactly this way) that the simplest explanation for a phenomenon has the highest probability of being true. This does not imply that the simplest explanation is always true, but it does give scientists a guide for which hypotheses are more likely to be true and should therefore be tested first.

Operating system: The software (computer program) that handles the basic operations of a computer, such as reading keyboard presses and saving files.

Opiates: Narcotic drugs that are derived from the opium plant.

Opioids: Synthetic narcotic drugs that mimic the effects of opiates.

Optimal foraging models: Part of the theory underlying geographic profiling, based on a large family of models used in ecological studies in which an animal is seen as balancing costs and benefits in choosing food sources.

Ordinary reality: The reality that is shared by all people, and in which the laws and principles of science apply. It is assumed that all things about ordinary reality are discoverable and knowable, though much knowledge has not yet been discovered.

Organized crime: The activities of an organization whose aim is to make money illegally.

Organized serial killers: Serial killers who are difficult to distinguish from ordinary people because they dress well, may have a good job, may have a family, and in other ways appear normal. Organized serial killers are often careful to minimize the evidence they leave at a crime scene and are often difficult to identify and arrest.

Ossification: The process of bone formation.

Overhead sketch: A sketch made using overhead perspective, meaning that it is drawn as it would be seen by someone located above the area or item sketched.

Oxidation: A type of chemical reaction in which oxygen is added to some molecule.

Oxidizing agents: Chemical substances that supply oxygen for an oxidation reaction.

Packets: Standard sized chunks of information sent over a network. Files or other information to be sent over the network are divided into packets before being sent and the receiving computer reassembles the file from the individual packets..

Palmar: The direction on the hand that is toward the palm. The opposite of palmar is dorsal.

Palynology: The study of plant pollen.

Paper chromatography: A type of chromatography in which the stationary phase is a type of paper and the mobile phase is a water or solvent-based solution.

Parole: The release of a confict before they have served their entire sentence.

Paternity testing: Comparing genetic markers of a man with those of a child to determine the likelihood that he is the father of the child.

Pathological science: The phenomenon where a scientist believes they have made a major discovery and rushed to press with an announcement of the discover, but in actuality they are mistaken and haven't made a major discovery.

Pathology: Illness of any form, including illness caused by infectious agents or to genetic diseases.

Pathology: The branch of medicine that focuses on determining the cause and nature of illness and death. There are two main branches, anatomic pathology and clinical pathology. Anatomic pathologist work primarily with the overall body, organs, and tissues. Clinical pathologists work primarily with blood, blood tests, and identifying microrganisms that may cause disease.

Patrol officer: A type of peace officer whose primary duty is to travel through some portion of a community seeking to prevent crimes, detect crimes, and respond to citizen reports of crimes.

PCR: Polymerase chain reaction. A method of copying (amplifying, in the jargon of DNA analysis) a DNA sequence several times so as to have more of it to work with.

Peace officer: A person authorized by some level of government to carry a weapon and make arrests.

Pedestal a skeleton: A method of excavating a skeleton such that the soil around the bones is removed, but the soil under the bones is not removed. Thus, the bones remain in place as the soil around them is removed giving the appearance that the skeleton is lying on a pedestal.

Peer-to-peer: A way of constructing a network so that the connected computers interact with each other as peers. Interacting as peers means that the computers have relatively equivalent roles – i.e. one is not designated a server and others designated clients.

Perimortem: A descriptive term for events that occur around the time of death.

Petit (or petty): A term applied to a form of theft in which the value of the stolen item is small. Petit thefts are usually misdemeanors.

Petty crimes: The least serious types of misdemeanors. For example, violating parking regulations is a petty crime.

PGM: Phosphoglucomutase, a serum protein.

Phenomenon: A general term for any item or occurrence in the universe.

Phenotype: The anatomical or biochemical expression of a genotype. For example, genotypes AA and AO both cause the expression of the phenotype blood type A.

Phenotypic variance: The variability in a phenotypic trait.

Photographic superimposition:

Phrenology: The pseudoscience that attempts to derive information about a person's character from examination of the features of their skull. Phrenology is distinct from the science of forensic osteology, which attempts to determine biological characteristics of a person (such as their sex) that are reflected in the skeleton.

Phylogenetic analysis: see forensic phylogenetics.

Physical evidence: Evidence that consists of actual physical items (bullets, fingerprints, hair, blood, etc.) or modifications to items (tool marks, wounds, alterations to computer files, etc.).

Pinciple of individuality: The idea in fingerprint studies that no two individuals have exactly the same fingerprints.

Pixels: A contraction of "picture element" that refers to the smallest component that makes up a digital image.

Planning cell: A type of terrorist cell that does not actually carry out an operation but which plans the operation, supplies information, and/or supplies resources for carrying out the operation.

Plant community: A set of plants found in a certain location.

Plantar: The direction on the foot that is toward the sole of the foot. The opposite of plantar is dorsal.

Plasma: The liquid (non-cellular) fraction of blood, comprising about 40% of the volume of blood. The plasma carries several proteins, including transport proteins (which transport some product), hormones, enzymes, the antibodies, the blood clotting factors, and more.

Plastic explosive: A type of high explosive, usually non-initiating, with a texture like putty and which is pliable and elastic enough to be shaped with the hands.

Plastic fingerprints: Fingerprints that have been pressed into a soft substance, thus forming and impression.

Plea: A response to the judge's question, "how do you plead?". Possible responses by a defendant are "innocent", "guilty", and "no contest".

Plea bargaining: The process in which the prosecution team offers a defendant a lighter sentence or lesser offense in exchange for a plea of guilty.

PMI: See post mortem interval.

Poaching: Illegal killing or capture of a non-human animal in violation of local, state, or federal laws.

Police photographer: A forensic photographer who works for a law enforcement agency doing crime scene documentation, photographic suspects, and other tasks.

Police psychologist: A forensic psychologist employed by a law enforcement agency.

Pollen profile: The quantities of specific types of pollen found at a certain location.

Polygraph: A lie detector device. A polygraph detects and records several physiological variables of the subject including pulse rate and skin resistance.

Polymorphism: The condition in which two or more alleles exist at a certain genetic locus, each with a frequency of at least 1%.

Porous surfaces: Surfaces that, due to their construction or the material of which they are made, can absorb at least small amounts of a liquid.

Port: An address at which a computer interacts with data arriving over a network.

Positive match: The result of a matching wherein the conclusion is that the evidence item and the standard item have the same class and individual characteristics. This implies that the evidence and standard items have the same origin. Positive identification and positive match mean the same thing. This result can also be expressed using the term "individualized"

Possession of a controlled substance: The general term for possession of drugs except as directed by a physician.

Postcranial skeleton: The division of the skeleton that includes all of the skeleton that is not part of the cranial skeleton.

Posterior: Also referred to as dorsal (toward the back), this direction on the body is toward the back of the body. The opposite of posterior is anterior.

Postmortem: A descriptive term for events that occur significantly after death.

Postmortem forensic toxicology: See forensic toxicology.

Postmortem interval: The time between death of an individual and discovery of their remains. Nearly synonymous with "time since death".

Postmortem reconstruction: See facial reconstruction.

Powder stippling: Burns in the surface material or tissue of a target caused by unburned powder and other residues discharged from a firearm at close or medium range.

Power and control motive: The motive for serial killing in which the killer seeks to exercise control or power over his victims.

Precipitin: A darkly colored substance formed by the agglutination of red blood cells. Once formed, precipitin will precipitate from (sink to the bottom of) the liquid in which the red cells were suspended.

Precipitin tests: A family of tests used by serologists to determine whether a sample of blood is from a human being. These tests involve exposing an antiserum specific to human blood to the suspect sample in such a way that an agglutination reaction occurs and precipitin is formed. The formation of precipitin is evidence that the blood is from a human and not some other species.

Predatory aggression: A type of aggression in which the aggressive person sees their victim as prey.

Prehistory: The period of time before the invention or introduction of writing in a particular region.

Prejudiced: The condition of having formed an opinion about an issue without having heard the evidence.

Preliminary hearing: A type of hearing before a judge, which is required in most jurisdictions for misdemeanor cases. During this hearing the judge decides whether the evidence against a defendant is sufficient to proceed with a trial. In most juridictions the strength of the evidence in a felony case is determined by a grand jury.

Premortem: Also referred to as antemortem, this descriptive term refers to events that occur significantly prior to death.

Pretrial: The portion of the justice process after a suspect has been arrested but before the person's trial begins.

Primary area: See crime scene layout.

Principle of permanence: The idea in fingerprint studies that once the fingerprints are formed they remain the same for an individual's entire life.

Print: A representation of an item transferred to a surface, but usually not actually impressed into the surface.

Probable cause: Apparent evidence that a crime has been committed or that evidence is present in a certain location. In general, a peace officer must show probable cause in order for a judge to issue a search or arrest warrant. Under some condition, if the probable cause is immediate and obvious a search or arrest may be made without a warrant.

Process (of a bone): A projection of the bone outward from the surrounding surface.

Processor chip: See central processing unit.

Prosecutor's fallacy: A logical fallacy (false understanding) in which evidence that might argue for the innocence of a suspect is ignored. In most cases this involves improperly restricting the range of possible suspects.

Protecting the crime scene: Taking steps so that the evidence at a crime scene is not damaged or tampered with. This is often accomplished by cordoning off the crime scene and controlling the entrance so that only a small number of authorized individuals can access the crime scene.

Protection racket: A tactic of organized criminals in which people are forced to pay money to the crime organization in order to avoid being victims of crime.

Protocol: In the realm of computers a protocol is a set of agreed-upon procedures for transfer of information over a network.

Protons: Positively charged subatomic particles that are normally found in the nucleus of an atom.

Proximal: The direction on a limb of the body that is toward the point where the limb attaches to the torso (shoulder or hip). The opposite of proximal is distal.

Pseudoscience: A system of purported knowledge that claims to be based on scientific principles but actually is not. Common examples include astrology, palmistry, and phrenology.

Psychic detective: A person who believes that they have the ability to find knowledge (such as the location of a person or item) via extrasensory perception or a similar psychic principle. Psychic abilities are a pseudoscience.

Psychological profiling: A type of criminal profiling in which the characteristics of the criminal are inferred from the manner in which the crime was committed and other behavioral evidence.

Psychopath: A person with antisocial personality disorder.

Psychophysiological detection of deception: A fancy name for a lie detector test.

Public intoxication: Being visibly and obviously intoxicated outside of a private dwelling.

Public key encryption: An encryption/decryption scheme in which each individual has two keys, that are related to each other by a complex mathematical relationship and known as the public and private keys. Each individual makes their public key known to anybody but keeps their private key secret. An individual sending a message will use their own private key and the public key of the recipient. The recipient can then decrypt the message using their private key and the sender's public key.

Pupa: A stage in the insect life cycle between the larval stage and the adult stage. The chrysalis of a butterfly is a pupa. In the case of most insects the pupa does not move, but remains in place until the pupal case (puparium) splits open to release the adult.

Questioned documents examination: The examination of documents for which the authenticity, authorship, or circumstances of production have been questioned.

Racketeering: A term for organized crime that comes from the federal Organized Crime Control Act. A racketeer is a person who engages in two or more of the activities listed in this act.

Radiographs: The "pictures" produced using x-rays.

RAM: An acronym for random access memory. RAM is the computers's main or working memory. RAM is a type volatile memory.

RAM space: Information originally present in a computer's RAM memory that is saved along with another file due to the fact that the CPU sends RAM data over the data bus in fixed size "pages". If the information to be saved does not fill a full page, the information that follows it in RAM will also be sent in order to fill up a whole page.

Range of Fire: The distance between a firearm and a target when the firearm discharges. Usually categorized as close or point blank range (firearm contacts the target), medium range (close enough that powder stippling is present), or long range (far enough that no powder stippling is present).

Rape: Sexual intercourse achieve by force or threat of force, without the consent of the victim.

Rape kits: A pre-packaged set of specialized equipment and supplies for collecting evidence from the body of a rape victim.

Recessive alleles: Alleles that are only expressed in the absence of a dominant allele.

Recognizance release: A type of release of an arrested person until the start of their trial that is without conditions.

Record: In the realm of databases a record is a set of information about a specific individual or item.

Recorder: A device for recording the presence and amount of a substance detected by a detector. Traditionally, these devices use a pen to draw a curve on a moving strip of paper. The height of the curve drawn reflects the amount of a substance detected by a detector. In recent times recording is often done digitally by a computer.

Red blood cells: Formally known as erythrocytes, red cells are one of the main types of cells in blood, comprising about 40% of the volume of blood. The function of red blood cells is to transport oxygen from the lungs to the tissues of the body. Red blood cells have antigens attached to their surfaces and have several proteins inside the cell membrane, which are known as red cell proteins.

Red cell proteins: These are proteins found inside red blood cells.

Red flags (in Forensic Accounting): Unusual features of financial records which suggest that a fraud may be occurring.

Reflection: A type of interaction between electromagnetic radiation and matter in which the EMR is bounced off the surface of the matter.

Reflective illumination: A way of illuminating an item that is being viewed under a microscope such that the light reflects from the item and is then captured by the microscope.

Refraction: A type of interaction between electromagnetic radiation and matter in which the direction of the EMR is changed, usually slightly, as it passes through the matter.

Regenerated fibers: Human produced fibers that are created from a product produced by a plant or animal. For example, rayon is a regenerated fiber produced from plant cellulose.

Regulatory law: Administrative rules developed and enforced by a regulatory agency at some level of government. These rules are designed to regulate the products or activities the regulatory agency oversees.

Relevant questions: In the field of lie detection these are questions related to the subject of the investigation and for which the truthfulness of the suspect's answer is being tested.

Replicable: Able to be repeated. In science, replicable refers to the ability of scientists to obtain similar results when an experiment is repeated.

Restriction enzymes: See RFLP.

Restriction fragments: See RFLP.

Restriction sequence: See RFLP.

Revivification: The re-experiencing of an event under hypnosis with the full range of emotions experienced again.

Revolver: A handgun in which the cartridges are held in a revolving cylinder.

RFLP: Restriction fragment length polymorphism. A method for comparing DNA samples. RFLP is rarely used in the 21st century, but several of the concepts surrounding it still have relevance. In this method a section of DNA is treated with restriction enzymes, which cut the DNA strand when a certain sequence of bases called a restriction sequence is encountered. There are several restriction enzymes, each of which targets a different restriction sequence. Subjecting the DNA to restriction enzymes causes it to be cut into several fragments called restriction fragments. The restriction fragments of two individuals will differ due to differences between them in their DNA base sequences. The differences in lengths is, therefore, a form of polymorphism. The restrictions fragments are separated into bands representing fragments of the same length using electrophoresis. The resulting pattern of bands is often called a DNA fingerprint. If it is sufficiently complex, the probability of finding two people with the same DNA fingerprint may be narrowed to a very small number.

RICO: An acronym for racketeer influenced and corrupt organization, which is the title of section IX of the Organized Crime Control Act.

Rifle: A long firearm with a rifled barrel. Rifles are intended for hunting and military applications.

Rifling: Grooves on the interior of a firearm barrel that are designed to impart spin to the bullet, thus increasing accuracy.

Right wing domestic terrorists: Domestic terrorists in the U.S. who are opposed to the authority of federal, state, and local governments and suspicious of government activities. They may focus on issues such as taxation and firearms control.

Rigor mortis: Stiffening of the body after death due to the action of enzymes within the muscles. Eventually, as these enzymes become inactive the body will become limp again.

RNA: Ribose nucleic acid. RNA is similar to DNA and in humans is used for transcripts of alleles (messenger RNA) and in the process of translation of these transcripts into proteins or other products.

Robbery: A form of larceny in which the owner of the property is present when it is taken and the taking involves force or the immediate threat of force.

Root: The portion of a tooth that fits into the tooth sockets and is not covered with enamel.

Router: A type of network device, usually used in a WAN, that received packets from a computer or another router and sends them on to another router that can forward the packets on so that they eventually reach the intended recipient computer.

Sagittal plane: A plane of the body running in such a direction that it divides the body into right and left portions. The median sagittal plane divides the body into equal right and left halves.

Sampling: A method of converting an analog signal (music, for example) into a digital signal. The level of the signal is measured several thousand times a second, with each measurement consisting of a number. The stream of numbers is then stored and can be used to recreate the original signal.

Satellite surveillance: The use of images captured by satellite systems to monitor activity at a location.

Scale: This is an object of known size, or upon which a known size scale is printed, which should be included in all photographs of evidence collected at a crime scene.

Scent-trained dog: A dog purported to be able to locate or identify an item or substance by scent. Although scent-trained dogs are widely used in law enforcement and investigations there is considerable evidence that their use is a pseudoscience.

Scientific method: A method for finding knowledge about ordinary reality. It is based on the use of results from replicable experiments to test hypotheses.

Scientific notation: A method for expressing numbers as a root value multiplied by a power of ten. Scientific notation is most commonly used to express very small or very large numbers.

Script kiddie: A type of hacker that uses scripts to exploit specific vulnerabilities.

Scripts: In general, a set of instructions that a computer will run. Scripts are often written in a high level programming language, sometimes called a scripting language. In the realm of hacking scripts are used to exploit specific vulnerabilities.

Second degree murder: A type of murder in which the death was deliberate but not premeditated.

Secondary area: See crime scene layout.

Secondary crime scenes: Locations where the less serious activities related to a crime occurred. For example, if a person is murdered then their body is transported to another location and buried, the vehicle in which the body was transported and the location where it was buried are secondary crime scenes.

Secondary depression: A smaller depression that occurs within the depression of a grave over the abdomen of the body within the grave. The secondary depression forms when the buried body bloats, leading to the swollen abdomen compacting the soil above it. When the bloated abdomen eventually ruptures, soil falls into the resulting cavity. This void in the soil eventually is transferred to the surface of the ground, forming the secondary depression.

Secret key encryption: A type of encryption/decryption scheme in which the key is only known by the sender and the recipient of a message.

Secreter locus: Also known as the Lewis blood group. People who have a positive allele for this blood group secrete their ABO blood group antigens in non-blood body fluids such as semen, saliva, gastric fluids, and vaginal fluids.

Semiautomatic pistol: A handgun that uses some of the force of the firing of a cartridge to eject the spent casing and load another cartridge into the firing chamber. Semiautomatic weapons can be fired as fast as the user can pull the trigger.

Sentencing: The formal handing out of a sentence to a defendant by a judge.

Serial crime: A series of crimes of the same type committed by the same criminal over time.

Serial killer: A murderer who kills several people over a long time span.

Serology: The study of genetic markers (blood groups, serum proteins, DNA, etc.) present in blood or other body fluids. It can also be defined as the application of knowledge of blood and other body fluids to investigations. Serology includes detecting blood, determination that the blood is of human origin, and matching genetic markers in the blood between samples. Serology is often considered to include DNA matching and analysis of blood spatter.

Serology:

Serum: Blood plasma from which the clotting factors have been removed. In an older usage a serum is a body fluid.

Serum proteins: Also known as plasma proteins, these are proteins carried in the plasma of blood.

Server-client: A way of constructing a network so that one computer is designated a servers and acts as a central repository for information or software. The other computers on the network are designated as clients or workstations which obtain data from the server, allow users to work with the data, then save any changes on the server.

Sex Related aggression: The type of aggression motivated primarily by lust or the desire to dominate a person one might have sex with.

Sexual dimorphism: Differences between females and males of a species. In humans sexual dimorphism manifests as differences in genital form, body size, pelvis shape, pigmentation, and fat distribution, among other things.

Shell: The type of ammunition utilized by a shotgun.

Shell companies: Companies that do not carry on legitimate business, through which illegally acquired money is passed as a form of money laundering.

Shock waves: Waves caused by the impact of a projectile with air, tissue, or other matter. The shock waves ripple through the matter, often causing additional effect.

Shotgun: A long firearm designed to shoot a cluster many small projectiles (shot) or less commonly a single projectile (slug).

Signal to Noise Ratio: In the realm of audio analysis this refers to the loudness of a sound compared to the loudness of the background or masking noise.

Slack space: Portions of the storage area of a non-volatile memory device (e.g. a hard drive or CD) that contains data from a previous file that was not completely overwritten.

Sleeper cell: A type of terrorist cell that remains inactive, perhaps for years, waiting for orders to carry out an operation.

Smurfing: The strategy used in money laundering in which a sum of money is deposited in several small increments instead of as one large lump sum in order to avoid triggering automatic notification of authorities when a sum over a certain amount is deposited.

Social engineering: The process of tricking the user of a computer or network into revealing their password.

Software audit: An investigation of the software used by the employees of a company or other organization to make sure that all software used was obtained legally (i.e. not by software piracy or cracking).

Software counterfeiting: Illegally copying commercially marketed software and selling the copy as if it were a genuine authorized copy.

Software piracy: Illegal copying of copyrighted software.

Soil grain size distribution: The composition of a soil in terms of percentages of particles of specific size classes.

Solubility: The solubility of a substance in a certain solvent is a function of the ease with which it is dissolved in that solvent.

SOR: An acronym for sex offender registry. A SOR is a state or local level database of sexual offenders.

Sound Identification: Determining the exact nature of a sound and how it was produced.

Speaker Identification: Identification of the speaker of a voice sample by a forensic linguist based on pronunciation, pitch, and speech pattern rather than on matching voiceprints.

Special interest domestic terrorists: Domestic terrorists in the U.S. whose activities focus on a single issue such as abortion, the environment, animal rights, etc.

Specific vulnerabilities: Programming errors in a piece of software that allow a hacker to compromise the computer running it using a certain strategy targeted specifically at the error.

Spectrogram: A recording of the output of a spectrograph.

Spectrograph: In the realm of audio analysis a spectrograph is a device that analyzes how the time, frequency, and/or amplitude of a sound changes over time.

Spectrophotometry: A method for determining the identity of a substance and in some cases the amount of it in a solution. As light passes through a substance, the substance absorbs certain wavelengths of the light. The patter of absorbed wavelengths is often unique to a certain substance. A device used to perform this type of analysis is known as a spectrophotometer.

Spin: Movement of a projectile around it's long axis.

Spontaneous combustion: An object undergoing combustion without ignition energy being supplied. True spontaneous combustion events are rare, and even then it is likely that the source of ignition energy is simply not obvious rather than not present.

Spousal immunity: An old idea, now not applicable in most jurisdictions, that a man has the right to sexual intercourse with his wife at any time.

Spree killer: A murderer who kills several people over a short time span.

Staged Image: An image that purports to be of an actual event, but which was actually staged purposefully in order to deceive.

Standard items: Items of physical evidence for which the origin is known because they were collected from a known individual (e.g. a person's fingerprints) or through a process that produces them from a known item or procedure (e.g. bullets fired from a known firearm). During matching, evidence items are compared with these item in an attempt to determine the origin of the evidence items.

Statements: In forensic accounting these are descriptions of financial gains, losses, etc. that must be made to govermental or administrative agencies.

Stationary phase: See chromatography.

Statistical profiling: A type of criminal profiling in which the characteristics of the criminal are inferred from statistical patterns among people who commit the type of crime involved.

Statutory rape: Sexual intercourse with someone who is a minor and therefore considered to young to give consent.

Steganography: Hiding a secret message within another message, picture, or other medium. For example, the data of a secret message can be hidden within the data of a picture and even if the picture is viewed by many people, only a person who knows where the secret message is hidden will be able to see the secret message.

Stimulants: A category of drugs that raise the activity of (stimulate) the central nervous system.

Stored products forensic entomology: The branch of forensic entomology that includes investigation of insect infestations of stored products such as grain. The results of urban entomology are most often used in civil or regulatory cases.

STR: Short tandem repeat. A DNA region that consists of several tandem (like boxcars on a train) repeats of a short sequence of bases. The variable factor is how many times the sequence is repeated. At present, most forensic DNA testing is based on STR's.

Strong consistent match: A category of matching or identification recognized by some authorities. A strong consistent match uses several class characteristics, the probabilities or frequencies of which are multiplied to yield an estimate of the probability of identification of a specific individual or item. Matches made using serology or DNA are of this type.

Subchondral bone: The type of bone tissue normally found where bone contacts cartilage. It is similar to compact bone, but has a more immature appearance.

Suggestibility: A mental state induced by hypnosis in which the hypnotized person becomes more likely to believe or do what someone else tells them

Superior: The direction on the body that is toward the top of the head. Also referred to as cranial, especially when describing directions on the bodies of non-human animals. Its opposite is inferior.

Supreme court: The United States Supreme Court is the highest court in the U.S. federal court system. Each state also has a supreme court that is the highest court in that states court system.

Switch: In the realm of computers, a switch is a type of network device used in a LAN that receives packets from one computer and sends them to the exact computer that is intended to be the recipient of the packets.

Sworn oath: A solemn statement that testimony given is true, which makes reference to God.

Synthetic fibers: Human produced fibers that are created from a non plant or animal based product, usually petroleum.

System attack: An action designed to degrade the performance of a computer or network.

Terminal ballistics: The phase of ballistics in which the projectile interacts with its target.

Territorial aggression: A type of aggression motivated by a conflict over territory, status, or possessions.

Terrorism: Any act done with the intention of causing fear within a population, thereby possibly reducing the population's confidence in their government.

Testimonial evidence: Evidence that consists of something a person says. Witness statements, suspect confessions, and victim complaints are all forms of testimonial evidence. Testimonial evidence is much less reliable than physical evidence.

Theft: The general term for taking another's property. Theft has several forms in most jurisdictions, including larceny, robbery, burglary, etc.

Theory: A hypothesis or set of hypotheses that has undergone a large amount of hypothesis testing without being falsified.

Theory of the case: In forensic science the theory of the case is a hypothesis (in the sense of the scientific method) about what suspect did which crime to what victim at which time and place, by some method. As is true of any hypothesis, it should be tested against the evidence of the case and modified to fit the evidence. It is a common practice to pervert the scientific method by trying to force the interpretation of the evidence to fit the theory of the case. This should be avoided.

Thin layer chromatography: A type of chromatography in which the stationary phase is a thin layer of a solid deposited on a stiff medium and the mobile phase is a liquid.

Time-distance study: A second by second accounting of the events that occurred during a traffic accident.

Time distortion: A phenomenon that occurs when the memories of a hypnotized person become confused or distorted with respect to when events happened or the order in which they happened.

Time of death: The time at which a death occurred.

Title card: Also known as a menu card or menu board, this is some device upon which the case number and evidence item number is printed, and which is included in all photographs of evidence collected at a crime scene.

TNT: An acronym for trinitrotoluene – a high explosive.

Toolmarks: Impressions or other damage transferred to some material when the tool was used on it.

Torn and broken edge matching: The examination of torn or broken edges of two objects to determine whether they are pieces of an original larger object. The pattern of a tear or break is a unique individual characteristics because an item will never tear or break in exactly the same way twice.

Trace evidence: Evidence that is small in size, such as hairs, fibers, dust, and paint or glass chips.

Trace evidence examination: The forensic science that examines small items (hairs, fibers, dusts, paint chips, glass chips, etc.) using the comparative method.

Traffic: In the realm of computers traffic refers to the amount of data handled by a network.

Tranquilizers: A type of depressant drug that has a lesser likelihood of causing dependence or addiction if used over a long period of time.

Transactional immunity: The form of immunity in which a witness is given complete protection from being prosecuted for the crimes about which he or she testifies.

Transactions: In forensic accounting these are events that result in the movement of money or goods from one account or status to another.

Transmission: A type of interaction between electromagnetic radiation and matter in which the EMR passes through the matter without being absorbed.

Transmissive illumination: A way of illuminating an item that is being viewed under a microscope such that the light passes through the item and is then captured by the microscope.

Transverse plane: A plane of the body running in such a direction that it divides the body into front (anterior or ventral) and back (posterior or dorsal) portions.

Trauma: Injury to the body.

Trepanation: A form of surgery in which a hole is made in the skull. In ancient times and some contemporary cultures in undeveloped countries this was performed to treat headaches, which were seen as the result of bad spirits or energy trapped inside the head. In modern medicine trepanation is performed to relieve pressure on the brain due to bleeding inside the skull.

Trojan Horse: In the realm of computers this is a program that seems useful, but when run opens a backdoor that allows a hacker access to the computer.

Tube and bag breathalyser: A type of device for measuring blood alcohol from the breath. A tube contains a substance that changes color when breath containing alcohol is blown across or through it. The amount of the substance that changes color is roughly proportional to the amount of alcohol in the breath. This is older technology and not longer widely used.

Unattended death: A death in which a physician or other medical professional was not present and had not seen the deceased recently.

Unconventional Weapons: These are also known CBN's (Chemical, Biological, Nuclear) and WMD's (Weapons of Mass Destruction). They refer to chemical weapons (such as poisons), biological weapons (such as disease organisms), or nuclear weapons that are designed to kill a large number of people at once.

Undetermined (in matching): See consistent match.

Uniqueness of construction: A source of individual characteristics. Individual characteristics that arise from uniqueness of construction are produced either by humans or by machines, neither of which can produce identical items. Humans are incapable of producing identical items consecutively. For example, writing your signature five times will produce five slightly different signatures. Machines are in a constant state of wear, so that consecutive items differ in slight ways.

Uniqueness of random occurrences: A source of individual characteristics. Random events always produce slightly different outcomes. For example, two similar sticks can not be broken so as to produce the same fracture pattern, nor can two windows be broken so as to produce the same shards of glass.

Uniqueness of usage: A source of individual characteristics. No two similar items can be used in exactly the same way to exactly the same extent. Therefore, difference that arise from use of the item are unique.

Unknown firearm examination: A type of firearm examination in which the examiner attempts to determine the brand, model, and caliber of the firearm that produced some recovered evidence.

Urban forensic entomology: The branch of forensic entomology that includes investigation of insect infestations of buildings or businesses. The results of urban entomology are most often used in civil or regulatory cases.

Use immunity: The form of immunity in which a witness may not be prosecuted on the basis of evidence they give as testimony but may be prosecuted on the basis of other evidence.

Verification: In the field of biometrics verification refers to the process of quickly determining a person's identity in order to grant them access to something.

Victimology: A technique used in criminal profiling in which the characteristics of a criminal's victims are studied in order to infer some information about the criminal.

Virtual memory: Slower, non-volatile memory, such as space on a hard drive, that is used to simulate faster, volatile memory such as RAM.

Virus: In the realm of computers a virus is a program that is designed to replicate itself and spread to other computers. A virus may harm a computer system in other ways, such as erasing data, but does not need to be malicious in order to be termed a virus.

Visible fingerprints: Fingerprints that are visible to the naked eye.

Visionary motive: The motive for serial killing in which the killer perceives that he is being told to kill by an outside person or entity. Sometimes the killer may hear voices that tell him to kill.

Voice Identification: Matching the characteristics of a recorded voice evidence to the characteristics of a standard voice sample of a known speaker to determine whether the same person produced both voice samples. This process often involves the matching of voiceprints..

Voiceprint: A spectrogram of the characteristics of a voice for use in voice identification.

Volatile memory: ,A type of computer memory that loses any data stored in it when the power to the computer is turned off.

Voluntary manslaughter: A type of manslaughter in which the death was intentional but occurred as a result of passion brought on by provocation.

WAN: See wide area network.

Warez: Software that has been cracked (see cracking) and often made available on the internet.

Wavelength: The length of a wave of electromagnetic radiation, which is a way of expressing it's inherent energy. See electromagnetic radiation.

Weapons of mass destruction: See unconventional weapons.

White blood cells: Formally known as leukocytes, white cells comprise about 1% of the volume of blood and comprise the cellular portion of the immune system. There are several types of white cells. White cells have antigens that are the same as those on most types of tissue cells, known as the human leukocyte antigens or the major histocompatibility complex antigens.

White Collar Crime: The type of occupational fraud in which the perpetrator is an employee (or other member of an organization) and the victim is the organization the perpetrator works for.

Wide area network: A network that is larger in scale than a LAN, in that it connects computers that are widely separated in space. Often a WAN interconnects LANS rather than individual computers.

Wireless Networks: A network wherein the data is transmitted over radio waves instead of over wires or fiber optic cables.

Wobbly felonies: A type of felony that is less serious and which may later be reduced to a misdemeanor if the convict serves their sentence and exhibits good behavior.

Wound ballistics: The application of terminal ballistics in the case where the projectile's target is living tissue.

Write-overs: A type of alteration of a document accomplished by writing over older writing with newer writing.

Wrongful Convictions: Cases of the conviction of a person for a crime when he or she was actually innocent.

Yaw: The angle between the direction a projectile's long axis is pointing and the direction in which it is traveling.

Index

AAFS (American Academy of Forensic Sciences)	11, 12, 248, 298, 363, 510, 570-571, 583
Absorption spectrum	204, 583
Absorption-elution method	4, 217-218, 221, 583
Accelerant	337-339, 583
Acid phosphatase color test	222, 583
Affirmation	503-504, 583
AFIS	173, 180, 358, 405, 412, 414-419, 583, 603
Age progression and updating	85, 583
Aggravated assault and battery	56, 584
Aggression	441, 443-444, 455, 584, 597, 604-605, 608, 614, 619, 622
Alcohol in possession of a minor	58, 584
Algor mortis	267, 584
Algorithm	357-358, 415, 584
Allele	161-168, 170, 214, 216, 221, 223-224, 235-236, 584
Amino acid	163, 202, 584
Ammunition	121, 127, 131, 141, 145-151, 301, 337, 343-344, 413, 416, 579, 587, 620
Amphetamine	59, 192, 243, 25- 258, 584
Amplifying DNA	235, 584
Anatomic pathology	276, 612
Anatomical position	283, 290, 584
ANFO	344-347, 496, 584
Anthropometry	402, 584
Antibodies	211, 213-218, 223, 259, 583, 584
Antigen	210-211, 213-218, 221, 223, 237, 239, 584
Antiserum	215-218, 231, 584
Antisocial personality disorder	424, 584
Appellate court	61-63, 584
Appendicular skeleton	285, 290, 584
Army system of tooth designation	295, 585
Arson	43, 55, 68, 126, 328, 335-339, 346, 484, 486, 504, 585
Assault	56, 57, 68, 79, 222, 438, 446, 448, 524-526, 584-586
Assault with a deadly weapon	56, 585
Asset misappropriation	471-473, 585
Atom	28, 122-125, 198, 563, 564, 585
Atomic weight	198, 585
Audibility analysis	390, 393, 585
Authentic image	389, 585
Authentication of documents	190
Autopsy	14, 24, 245, 264-265, 276-282, 292, 533, 542, 585, 610
Axial skeleton	285, 289, 585
Bad science	514, 563, 585
Ballistics	129, 130, 134, 136, 138, 139, 146, 415, 585
Bandwidth	355, 357, 382, 585
Barbiturates	59, 256, 586
Battery	56, 81, 586
Beetles	322, 586
Behavioral evidence	422, 441, 447-448, 453, 520, 586
Behavioral sciences	441, 586
Bertillon	3, 80, 83, 402, 586
Bertillonage	3, 586
Betweenness	522, 586
Binary number	360
Biometrics	3, 180, 349, 358, 363, 387, 396, 399, 402-410, 499, 586
Bite mark	311-314, 319

Blended Inheritance	167-168, 586
Blood group	162-166, 210, 214-218, 221, 225, 586
Blood spatter analysis	227, 586
Blood type	4, 162-165, 216-218, 221-225, 231, 511, 587
Blow flies	322, 587
Bookmaking	487, 587
Breaking and entering	58, 587
Buccal	293, 587
Buffer (Chemistry)	202, 587
Buffer (Computer Science)	377-378, 587
Buffer overrun	377, 587
Bug	399-400, 487, 508, 587
Burglary	58, 456, 587
Cadaver dog	301, 558-559, 587
Caliber	136-137, 145, 147, 150, 266, 345, 587
Cancellous bone	284, 587
Cartridge	126, 127, 129, 141-147, 150-151, 587
Casting	153, 157-160, 587
Casting material	158, 587
Cause of death	14, 24, 264, 265, 272, 275, 276, 278, 280, 306, 307, 323, 335, 519, 532, 533, 535, 540-545, 580, 588
Cavitation	136, 137, 588
CCH	416, 588
Cellulose	117, 324, 343, 588
Cementum annulation analysis	533, 588
Central processing unit	349-350, 588
Certification	68, 81-83, 96, 435, 466, 512, 568, 569, 588
Chain of custody	45, 49, 70, 72, 75, 91, 114, 503, 504, 588
Change of venue	423, 588
Chemical explosion	341, 588
Chemical symbol	122, 588
Chief deputy coroner	268, 588
Chief Justice	60, 61, 588
Chirology, chiromancy, chironomy	556-558, 588-589
Chromatography	117, 198-202, 206, 245, 247, 260, 328, 391, 589
Chromosomes	161-164, 236-237, 413, 546, 589
Circumstances of death	266, 271, 589
CITES	532, 589
Civil court	226-227, 246, 270, 275, 311, 370, 517, 518, 589
Civil law	2, 44, 226, 245, 248, 589
Class characteristic	7-9, 96, 115-119, 146, 147-151, 154-157, 164, 169-170, 175, 180, 185, 187, 190-191, 224, 227, 232, 303, 316, 325-326, 511, 519, 589
Clinical pathology	276-277, 612
Closeness	522, 589
CODIS	239, 414-416, 589
Codon	163, 589
Color test	212-213, 222, 244, 583, 589
Combustion	123-128, 337, 342, 344, 589
Commodity item	477, 589
Compact bone	284, 587, 589
Comparative method	12, 91, 94, 95, 153, 578, 589
Comparison microscope	4, 108-110, 115, 117, 141, 149, 154-155, 590
Competency	17, 249, 421, 423, 590
Composite drawing	79, 83-85, 590
Compound microscope	108-110, 590
Concave lens	106-107, 590
Conditional release	48, 50, 590

Confirmation bias	555, 559-561, 590
Connectedness	521, 590
Consistent match	8, 30, 149, 150, 154, 157, 175, 180, 187, 232, 326, 590
Control experiments	29, 590
Convex lens	106, 590
Coronal plane	283, 287, 293, 590
Coroner	9, 263-264, 267-273, 278, 540, 591
Corporate crime	471, 591
Correctional psychologist	421, 424, 591
Corrections	41, 50, 189, 421, 424, 591
Corruption	43, 341, 471-472, 482, 486, 541, 591
Counterfeiting	58, 370-372, 471, 474-476, 486, 591
Court psychologist	423, 591
Courts of Appeals	60-61, 591
Courts of origination	61, 591
Cracking	370, 375, 378, 591
Cranial skeleton	285-289, 591
Crime lab	4-5, 11-15, 22-23, 31, 37, 45-46, 49, 67, 70, 75, 79, 81, 92-93, 111-115, 121-122, 153, 159, 178-179, 199-201, 222, 231-232, 236-239, 243, 246, 248-249, 251, 259, 27- 278, 347, 389, 510, 512, 524, 529, 533-537, 555, 568-569, 572-573, 577, 579, 591
Crime lab photographer	81, 591
Crime Lab Sections	111, 113
Crime organization	446, 465, 481-485, 487, 488, 533, 591
Crime scene	2, 6-8, 11-15, 22-24, 30, 41, 44-46, 49, 67-77, 79-81, 83-84, 87-89, 92-95, 111-113, 115, 117-118, 130, 151-155, 159, 209-212, 226, 232, 234, 238-239, 301-302, 322-323, 325, 332, 338, 346, 367, 414, 416, 422, 448, 520-521, 524, 533, 568, 589, 591
Crime scene illustration	83, 84, 592
Crime scene investigator	6, 13, 45, 67-75, 92-93, 159, 238, 568, 592
Crime scene reconstruction	84, 592
Criminal court	64, 226, 275, 592
Criminal law	2, 13, 55, 245, 248, 371, 592
Criminal profiling	17, 441-442, 449-451, 453, 592
Criminalist	11-13, 97, 121, 568, 592
Criminologist	13, 592
Cross examination	49, 506-507, 577, 592
Cryptography	359, 369, 496, 592
Crystallization test	213, 222, 593
Currency smuggling	477, 593
Cusps	293, 593
Cybercrime	363-364, 368, 369, 593
Cyberstalking	375, 382-385, 593
Data bus	350, 591, 593
Data mining	466-468, 593
Database	150, 312, 317, 325, 354, 375, 402, 403, 411-419, 593
DBMS	411, 593
Decriminalization	251, 484, 593
Decryption	359, 593
Defense attorney	18, 75, 222, 423, 504, 512, 576-581, 593
Defense attorney's fallacy	16, 521, 593
Degree inflation	569, 593
Denial of service attack	364, 382, 384, 593
Dental charts	312, 593
Deposition	49, 51, 304, 594
Depressants	252, 255-258, 594
Dermatoglyphics	170, 174, 556-557, 594
Detective	2, 6, 42, 44-45, 47, 67, 263, 269, 554, 591, 592, 594
Dialog decoding	392, 594

Digital evidence	361, 363-364, 366-368, 372-374, 380-383, 466, 474, 487, 533, 594
Diphodont	294, 594
Direct examination	49, 506, 594
Dirty bomb	498, 594
Discovery proceedings	49, 508, 594
Disorganized serial killer	447, 450, 594
Distal	283, 290-294, 594
District Courts of the United States	60, 594
Disturbance of the ground and vegetation	300
Dizygotic	169, 594
DMORT Teams	316-317, 595
DNA	3, 13, 16, 86, 111-113, 115, 121, 161-163, 165, 202, 209-210, 223-224, 227, 231-237, 239-242, 278, 281, 311, 313-314, 316-317, 413-416, 441, 465, 499, 511-512, 520-521, 533-536, 539, 545-552, 569, 595
DNA fingerprint	234, 235, 595
Documenting Evidence	14, 69-75, 79-81, 92-93, 106, 115, 159, 117, 178-179, 238, 301, 303, 316, 332, 338, 524, 595
Domestic Terrorists	491, 495-496, 500, 595
Dominant alleles	164, 595
Dorsal	283-284, 595
Driving under the influence	58, 245-246, 595
Dry stick fracture	307, 595
Duress	188, 595
Dynamite	343-347, 595
EAAF	318, 595
EGIS	347, 595
Electrical charge	122, 202, 205, 346
Electromagnetic radiation (EMR)	101, 595
Electronic Surveillance	349, 363, 387, 399-402, 484, 587, 595
Electrons	109, 122, 205, 563, 585, 595
Electrophoresis	198, 202-203, 223-224, 233-234, 236, 587, 596
Elevation sketch	74, 596
Email Flood	364, 380-382, 596
Embezzlement	15, 57, 473, 596
Encryption	357, 359-361, 369, 377, 499, 596
Endangered Species Act	531, 596
Environment	97, 163, 166-171, 212, 224, 248, 254, 261, 307, 324, 328, 393, 424, 442, 443, 464, 496, 498, 530, 570, 586, 596
Ethnicity	88, 482, 485-486, 596
Event sequence analysis	393, 394, 596
Evidence item	7-8, 12, 23, 30, 68, 73-75, 79, 85, 97, 148-149, 303, 326, 337, 367, 395, 403, 596
Execution	55, 318, 492, 493, 522, 596
Execution cell	493, 500, 522, 596
Exemplars	7, 186, 192, 193, 597
Exhumation	540, 542, 548, 597
Exoneration	511, 597
Exothermic reaction	125, 597
Experiment	24, 28-29, 31-36, 97, 115, 174, 185, 203, 227, 267, 280, 508, 548, 554, 557, 586, 591, 563, 593, 597
Experimenter effect	35-37, 557, 597
Expert witness	469, 503, 506, 508-510, 578, 597
Expressivity	167-168, 170, 597
External ballistics	129-130, 597
Extortion	57, 465, 472, 481-483, 485-487, 498, 597
Extraordinary realities	23, 25, 553, 597, 597
False accept rate	404, 597
False positive result	246, 259-261, 597

False Pretenses	57, 597
False reject rate	404, 597
Fauna	523, 530, 532, 533, 597
Felony	47, 55, 56, 58, 375, 598
Fingerprint Examination	96, 153, 161, 173-180, 193
Fingerprint patterns	166, 170, 173, 175, 180, 358, 594, 598
Fingerprints	3, 7-8, 12, 44-45, 68, 70, 72-73, 79, 86, 93, 96, 113, 161, 164, 169-170, 173-182, 184, 187, 234, 235, 278, 281, 336, 402-405, 412, 414, 415, 508, 523, 533, 535, 556, 558, 560, 598
Firearm function examination	150-151, 598
Firearms Examination	4, 129, 141-151, 153, 155
First degree murder	56, 598
Flesh flies	322, 598
Forensic accounting	15, 461-467, 471, 474, 476-477, 487, 499, 518, 539, 568-569, 571-572, 583, 598
Forensic anthropology	11, 15-16, 74, 86, 91, 112, 158, 240, 263, 267, 281, 283, 286, 297-308, 312, 312, 314-318, 402, 436, 465, 510, 522-523, 543, 547, 559, 569, 572, 598
Forensic archaeology	297, 299, 301-302, 315, 540, 598
Forensic art	79, 83-87, 96, 598
Forensic audio analysis	363, 387, 390-396, 402, 598
Forensic Botany	16, 31, 321, 323-325, 539, 569, 598
Forensic chemistry	12-13, 16, 112, 195, 206, 243-244, 249, 336, 347, 535, 569, 598
Forensic computer science	14, 349, 351-353, 358, 363-372, 375, 380-383, 411, 465-466, 474, 476-477, 487, 499, 539, 568, 598
Forensic DNA analysis	231, 236, 242, 316
Forensic drug testing	245, 246, 248, 259, 598
Forensic economics	2, 15, 51, 474, 517-518, 525, 569, 598
Forensic engineering	15, 55, 331-338, 341, 599
Forensic entomology	16, 263, 267, 321-323, 569, 599
Forensic geology	16, 321, 325-328, 336, 569, 599
Forensic hypnosis	421, 433, 435-438, 464, 599
Forensic image analysis	14-15, 387-390, 402, 487, 499, 599
Forensic knot examination	15, 517, 519-520, 599
Forensic linguistics	16, 387, 396, 487, 499, 569, 599
Forensic mathematics	16, 499, 517, 519-522, 569, 599
Forensic medicine	4, 15, 156-157, 263, 269-272, 275, 334, 517- 518, 569, 599
Forensic meteorology	15, 517-519, 599
Forensic nursing	17, 523-524, 569, 599
Forensic odontology	15, 263, 281, 283, 286, 298, 303, 311-316, 569, 599
Forensic osteology	297, 302-308, 315, 599
Forensic palynology	324-325, 612
Forensic pathology	14, 68, 86, 112, 137, 139, 238, 245, 263, 269-272, 275-281, 286, 303, 306, 308, 311, 314, 316, 337, 421, 499, 518, 569, 580, 599
Forensic pharmacology	15, 243, 249, 252, 569, 599
Forensic photography	14, 79-84, 388, 568, 599
Forensic phylogenetics	16, 517, 519, 522, 599
Forensic podiatry	15, 153, 156-157, 275, 518, 599
Forensic psychology	11, 17, 51, 421-429, 433-438, 441-449, 454, 518, 569, 599, 615
Forensic serology	3-4, 6, 11, 13, 111-114, 161, 195, 202, 209-218, 220-228, 231, 239, 278, 313, 467, 511-512, 520, 535, 569, 600
Forensic social work	17, 517, 523-524, 600
Forensic toxicology	3, 11, 13-15, 111-114, 121, 195, 206, 243-249, 252, 256, 259, 263, 270, 278-279, 499, 535, 542, 569, 600
Forensic zoology	16, 286, 519, 523, 533, 539, 569, 600
Forensics	1, 426, 539, 600
Fossa	284, 600
Fraudulent statements	471-472, 600
Front companies	476, 498, 600
Fuel	124-127, 328, 337, 341-345, 486, 600

Fuel cell	247, 600
Fuming with superglue	179, 600
Gas chromatography	200-201, 206, 247, 260, 328, 600
GCMS	114, 206, 243-244, 246-247, 260, 600
Genes	161-170, 214, 240, 442, 560, 586, 600
Genetic markers	94, 112, 157, 209-214, 221-227, 231-237, 511, 520, 600
Genetic profiling	231, 239-240, 303, 601
Genotype	164-168, 216, 223-224, 601
Geographic profiling	441, 443, 453, 457-459, 601
Governmental crime	471, 601
Grand jury	47, 55, 484, 576, 601
Green stick fracture	307, 601
Hacking	14, 356-357, 363-364, 375-380, 497, 499, 601
Hallucinogen	253-255, 601
Handgun	136, 142-145, 148-149, 601
Haplotype	164-165, 236, 239-240, 413, 601
Hashing	349, 358, 367, 396, 405-407, 413, 415-417, 601
Hearsay evidence	505, 601
Heritability	168-170, 601
Heterodont	294, 601
Heterozygous	165, 601
High explosive	341-347, 584, 602
High Performance Liquid Chromatography	202, 204, 260, 602
Historical forensic science	539-550, 602
HLA	221, 227, 602, 608
Homeland Security	341, 399, 461, 471, 491, 499, 602
Homozygous	165, 602
Human produced fibers	117, 602
Hydrocarbon fingerprint	328, 602
Hydrocarbons	124-125, 127, 325, 327-328, 337, 341, 343, 602
Hypercompliance	437, 602
Hypnosis	91, 421, 433, 435-438, 442, 464, 583, 599, 602
Hypothesis	25, 27-29, 31-36, 71, 75, 84, 558, 560, 561, 591, 598, 602
Hypotheticals	507, 602
IAFIS	180, 412, 414, 415, 417-419, 603
IBIS	151, 414-416, 603
Ice mummy	545, 603
Identification and Individualization	97, 603
Ignition Energy	125-127, 343, 603
Ignitor	125-127, 337-338, 603
Illegal distribution of alcohol	58, 603
Image enhancement	387-390, 603
Immunoassay	259, 603
Impressions	12, 15, 44, 79, 111-113, 121, 153, 155-159, 170, 178, 301-302, 336, 533, 603
Individual characteristics	7-9, 22, 23, 96-98, 113, 115, 118, 130, 146, 148-149, 153-157, 164, 169-170, 174-175, 177, 180, 185, 187, 191, 224, 297, 303, 306, 312, 325-326, 406, 511, 519, 591, 603
Inferior	283, 604
Infrastructure attack	497, 500, 604
Inhalants	256, 604
Initiating high explosives	343-344, 604
Inquest	268-269, 271, 604
Insanity defense	17, 423-427, 604
Instars	322, 604
Instrumental aggression	444, 604
Intelligibility	392, 393, 599, 604
Intermale aggression	443, 604
Internal ballistics	129, 604

Internet	14, 114, 280, 354, 357, 359-360, 363, 365-366, 368-372, 375-376, 378, 380-383, 496, 500, 572, 578-579, 604
Internets	354
Intranets	354
Investigative accounting	462, 604
Involuntary manslaughter	56, 604
Irrelevant questions	434, 604
Irritable aggression	444, 604
JPAC	317, 605
Junk science	562-563, 605
Jurisdiction	16, 42-45, 49, 50, 55, 57, 61-64, 81, 111, 247, 265, 267-269, 271, 278, 303, 400, 412-414, 422, 483, 503-506, 508, 510, 524, 530, 540, 567, 605
Jurisprudence	11, 13, 45, 49, 84, 503, 511, 605
Justifiable homicide	56, 605
Kastle Meyer test	212, 605
Kinetic energy	135-136, 333, 605
Labial	293, 587, 605
Laceration and crushing	136, 605
Lacey Act	532, 605
Larceny	57, 456, 472, 606
LAN (see Local Area Network)	
Larva	322-323, 606
Latent fingerprints	111-113, 121, 170, 178, 179, 181, 606
Lateral	283, 286-288, 291, 292, 606
Law enforcement agency	17, 42, 44, 71, 79, 81, 252, 271, 303, 357, 363, 401, 412, 422, 484, 576, 605, 606
Law of association	80, 606
Layout of a Crime Scene	68, 606
Leading question	427-429, 437, 606
Left wing domestic terrorists	495, 606
Length polymorphism	232, 233, 618
Lens	105-109, 590, 591, 606
Lie Detector (see polygraph)	
Ligature	519-520, 606
Lignin	324, 606
Lingual	293, 587, 606, 606
Listenability	391, 392, 599, 606
Livor mortis	267, 606
Local Area Network	353-357, 606
Locard	4, 111, 115, 607
Locard's principle	7, 44, 68, 146, 607
Locus	162, 164-168, 170, 216, 221, 223, 235-236, 239, 607
Log files	380, 607
Low explosive	341-343, 607
Luminescence	103, 607
Luminol	212, 607
Lust motive	446, 448, 607
Lysis	215-216, 607
Magistrate judges	60-61, 607
Manner of death	112, 264-266, 272, 276-278, 332, 335, 532, 533, 545, 589, 607
Manslaughter	56, 607
Masking noise	393, 607
Mass murderer	444, 607
Mass spectrometer	205, 245, 601, 607
Matching	4, 7-8, 12-15, 22, 30, 37, 44, 85, 87, 94-95, 97, 106, 109, 112-115, 117-118, 125, 130, 141, 145, 148-150, 153-159, 165, 173, 175, 177, 180, 186-188, 209-210, 212, 214, 224, 231-232, 234, 237, 281, 298, 303, 311-314, 323, 325-328, 336-337, 367, 395-396, 403-405, 407-408, 414-416, 519-522, 544, 546, 548-550, 556, 607

Materials science	331, 608
Maternal aggression	443, 608
Mechanical explosion	341, 608
Mechanical fingerprint	149, 608
Medial	283, 287, 293, 606, 608
Medical examiner	6, 112, 263-264, 268-272, 278, 302, 306, 608
Medicolegal entomology	321, 608
Memory distortion	437, 608
Memory refreshment	437, 583, 608
Mental template	184-185, 395, 608
Mesial	293-294, 608
Meteorologist	518
Metric system	147, 195-196, 586, 608
MHC	221, 227, 602, 608
Microscope	4, 101, 105-109, 115-117, 141, 149, 154-155, 161, 189, 195, 222, 276, 284, 323, 347, 533, 573, 590, 608
Mineralogy	326-328, 568, 608
Minutia	177, 180, 389, 405, 415, 594, 604, 608
Misdemeanors	47-49, 55-56, 61-62, 415, 608
Missionary-oriented motive	446, 449, 608
Mitochondrial DNA	233, 237, 281, 413, 545, 547, 609
Mixed system of death investigation	263-264, 270-272, 609
Molecular weight	198, 589, 609
Molecule	121-125-127, 131, 161-163, 171, 198-199, 202, 204-205, 212, 216, 233, 327, 345, 533, 609
Molting	322, 609
Money laundering	15, 465, 471, 476-477, 481, 482, 486, 488, 498, 609
Monozygotic	163, 169, 609
Motive	265, 444, 446, 448-449, 601, 607, 608, 614, 624
Motor program	184, 606, 609
Munsell color chart	326, 609
Mutation	234-235, 237, 610
Narcissists	443, 610
Narcotics	43, 59, 252, 253, 255, 257, 258, 610
Natural fibers	117, 610
Natural variability of handwriting	184, 610
NCIC	412, 417-419, 610
NDIS	239, 414-415, 417, 610
Necropsy	533, 610
NIBIN	151, 414-416, 610
NICS	156, 157, 416, 610
No contest plea	48, 610
Nominees	498, 610
Non-initiating high explosives	344, 610
Non-volatile memory	351, 587, 610
NSOR	413, 416, 417, 610
Nuclear DNA	237, 281, 609, 611
Nuclear explosion	341, 611
Occlusal surface	293, 304
Occupational fraud	461, 464-465, 471-477, 611
Occupational markers	306, 611
Ockham's Razor	35, 611
Operating system	351, 365-366, 376, 378, 380, 611
Opiates	59, 253, 611
Opioids	253, 611
Optimal foraging models	457, 458, 611
Ordinary reality	23-27, 35, 553-554, 558, 561, 611
Organized crime	43, 145, 251, 252, 341, 345, 446, 461, 465, 471, 476, 481-488, 498, 529, 591, 611

Organized serial killers	447, 611
Ossification	304, 611
Overhead sketch	74, 611
Oxidation	121, 123-127, 137, 212, 341-342, 611
Oxidizing agents	126, 343-344, 611
Palmar	283, 612
Palynology	324-325, 612
Paper chromatography	199, 200, 202, 612
Parole	50, 376, 424, 435, 523-524, 570, 612
Paternity Testing	221, 226-227, 237, 612
Pathological Science	562, 563, 612
Pathologist	14, 86, 112, 137, 139, 238, 245, 263, 269, 275-279, 281, 286, 303, 306, 308, 311, 314, 316-317, 337, 421, 465, 499, 569, 580, 612
Pathology (the medical specialty)	276-278, 612
Pathology (in forensic anthropology)	308
Patrol officer	42, 44, 45, 67, 92, 332, 567, 592, 612
PCR	235, 236, 584, 612
Peace officer	42-46, 49, 55, 67-68, 70-71, 79, 82, 136, 239, 247, 251-252, 263, 281, 297, 300-302, 332, 334, 336, 346, 388, 400, 411, 422, 435, 482-483, 485, 504, 508, 513, 523-524, 533, 558-559, 567, 575-576, 612
Peer-to-peer	354, 612
Perimortem	306-307, 612
Petty crimes	55, 612
PGM	223-225, 612
Phenotype	164-168, 216, 223-224, 612
Phenotypic variance	168, 612
Photographic superimposition	544, 549, 613
Phrenology	557, 560, 613
Phylogenetics	16, 517, 519, 522, 600, 613
Physical evidence	13, 45, 46, 51, 55, 67, 71, 92, 111, 332, 383, 533, 539, 576, 597, 613
Pixels	389, 390, 613
Plantar	283, 284, 595, 613
Plasma	202, 209-211, 215, 216, 223, 235, 584, 613
Plastic explosive	345, 613
Plastic fingerprints	178, 613
Plea bargaining	47-48, 376, 400, 483, 576, 613
Pleas	48, 255, 268, 509
Poaching	43, 531, 534, 613
Police photographer	81, 613
Police psychologist	17, 421-422, 435, 613
Pollen profile	324-325, 613
Polygraph	17, 433-435, 438, 613
Polymorphism	232-233, 614
Porous surfaces	178-179, 614
Port	347, 351, 353, 382, 614
Positive match	8, 23, 30, 94-95, 118, 148-149, 154, 157, 180, 187, 224, 232, 313-314, 326, 614
Possession of a controlled substance	59, 614
Post mortem interval	322, 614
Postcranial skeleton	285, 289, 614
Posterior	283, 288-289, 614
Postmortem	86, 245, 263, 306-307, 316, 614
~~Postmortem forensic toxicology~~	~~245, 263, 614~~
Postmortem reconstruction	86, 614
Powder stippling	137-138, 614
Power and control motive	446, 614
Precipitin	213-215, 614
Precipitin test	213-214, 614

Predatory aggression	443, 614
Prehistory	245, 539, 614
Prejudiced	423, 614
Preliminary hearing	47, 614
Premortem	306-307, 615
Pretrial	41, 47-48, 50-51, 61, 428, 615
Printer	96, 183-184, 187, 191, 193, 351, 414, 475
Probable cause	45-46, 503-504, 559, 560, 615
Processor chip	350, 588, 593, 615
Product failure	15, 331, 335,-336, 338
Projectile	129-131, 133-139, 144-145, 343
Prosecuting Attorney	16, 18, 47-50, 75, 369-371, 376, 380, 383, 433, 437-438, 483, 505-507, 509-513, 520-521, 570, 575-580
Prosecutor's fallacy	16, 520, 521, 615
Protecting the Crime Scene	69-71, 301, 332, 338, 615
Protection racket	486-487, 498, 615
Protocols (in computer science)	355-356, 382, 615
Proton	122-123, 615
Proximal	283, 290, 615
Pseudoscience	301, 553-564, 615
Psychiatrist	421-422, 427, 429, 569
Psychic detectives	553-556, 559, 565, 615
Psychological profiling	421-422, 441-442, 444-445, 447-449, 615
Psychological tests	422, 424, 442
Psychopaths	424, 443
Psychophysiological detection of deception (PDD)	434-435, 615
Public intoxication	58, 616
Public key encryption	359-361, 377, 616
Pupa	322, 323, 616
Questioned documents examination	4, 6, 11, 14, 113, 183-193, 243, 474, 476, 499, 616
Racketeering	483-484, 486, 616
Radiograph	87, 103, 312, 387, 616
RAM	27, 179, 303, 351-353, 365, 367, 411, 576, 593, 616
RAM space	352, 353, 365, 367, 616
Range of Fire	137-139, 141, 616
Rape	57, 383, 616
Recessive alleles	164, 616
Recognizance release	48, 616
Recorder (analytical instrument)	201, 204, 328, 393, 616
Red blood cells	210-217, 221, 223, 235, 237, 587, 615, 616
Red cell proteins	223, 616
Red flags (in forensic accounting)	463, 466, 617
Reflection	104, 178, 617
Reflective illumination	108, 617
Refraction	104-106, 617
Regenerated fibers	117, 617
Regulatory law	2, 617
Relevant and fair evidence	505
Relevant questions (in lie detection)	434, 617
Replicable	22, 23, 298, 554, 559, 617
Restriction enzymes	233-234, 617
Restriction fragments	233-234, 617
Restriction sequence	233-234, 617
Revivification	282, 437, 617
Revolvers	70, 142-142, 150-151, 266, 617
RFLP	232-233, 235-236, 617
RICO	138, 483-484, 617

Rifle	126, 136-137, 143-145, 147-148, 151, 447, 617
Rifling	130, 134, 141, 143, 147-148, 150-151, 617
Right wing domestic terrorists	495, 618
Rigor mortis	267, 618
Robbery	57, 80, 456, 485-486, 585, 597, 618
Routers	355-357, 361, 618
Sagittal plane	283, 288, 293-294, 618
Sampling (of sounds)	360, 618
Satellite Surveillance	399, 401, 618
Scent-Trained Dogs	558-560, 618
Scientific method	21-25, 27, 31, 33-35, 453, 505, 539, 553, 557-558, 579, 618
Scientific notation	197-198, 618
Script kiddie	378, 618
Second degree murder	56, 618
Secondary crime scenes	44-45, 618
Secondary depression	300, 619
Secret key encryption	359, 619
Secreter locus	221, 619
Semiautomatic pistols	142-143, 147, 150-151, 619
Sentencing	41, 50, 505, 508, 576, 619
Serial crime	442, 619
Serial killer	374, 445-449, 453, 456, 458, 619
Serum	211, 213, 215, 223, 231-232, 619
Serum proteins	223, 231-232, 619
Server-client relationship (in computer science)	354, 619
Sex related aggression	444, 619
Sexual dimorphism	304, 619
Shell companies	476-477, 498, 619
Shells (ammunition)	144-145, 147, 619
Shock waves	133-134, 136, 342, 619
Short term memory	351, 436
Shotgun	143-145, 147, 151, 620
Signal to noise ratio	393, 620
Slack space	352-353, 366-367, 620
Sleeper cell	493, 620
Smurfing	476-477, 620
Social engineering	378, 620
Software audit	371, 620
Software counterfeiting	371-372, 474, 620
Software piracy	364, 369-372, 485, 620
Soil grain size distribution	326, 620
Soils	16, 118, 254, 300, 321, 325-327, 346, 413, 599
Solubility	198-200, 202, 589, 620
SOR	413, 416-417, 620
Sound identification	390, 394, 620
Speaker identification	396-398, 620
Special Interest Domestic Terrorists	496, 620
Specific vulnerabilities	378, 620
Spectrogram	391-396, 620
Spectrophotometry	105, 117, 198, 203-206, 583, 620
Spontaneous Combustion	126, 621
Spousal immunity	57, 621
Spree killer	444-445, 621
Staged image	390, 621
Standard items	7-8, 12, 22-23, 30, 44, 75, 97, 149, 154, 156, 192, 303, 314, 326, 338, 395, 403, 578, 621
Stationary phase	200-202, 621
Statistical profiling	441, 453-455, 621

Statutory rape	57, 621
Steganography	359-362, 496-497, 499, 621
Stimulants	252, 257-258, 621
Stored products entomology	321, 621
STR	233, 235-237, 239-240, 533, 546, 621
Structure failure	331, 335, 338
Subchondral bone	284, 621
Suggestibility	437, 621
Sworn oath	504, 622
Synthetic fibers	117, 622
System attack	364, 375, 380-382, 497, 518, 622
Terminal ballistics	129, 134-139, 622
Territorial aggression	443, 622
Terrorism	16, 341, 344-345, 359-360, 364, 369, 375, 404, 417, 461, 465, 476, 491-500, 520-522, 622
Testimonial evidence	31, 45-47, 71, 86, 92, 265-266, 281, 332, 505, 539, 555, 561, 622
Theft	43, 57-58, 364, 368-369, 417, 436, 455-456, 485-486, 622
Theory of the case	69-72, 75, 622
Thin layer chromatography	117, 202, 622
Time-distance study	334, 622
Title card	73-74, 622
TNT	344-345, 622
Toolmarks	12, 68, 111-114, 121, 153-158, 302, 325, 336, 533, 568, 571, 623
Torn and Broken Edge Matching	118, 623
Trace evidence	4, 6, 12, 16, 111-118, 243, 302, 321, 325, 336, 499, 533, 535, 573, 623
Traffic (on a network)	355-357, 360, 623
Traffic Accident Investigation	2, 15, 246, 265-266, 331-335, 336, 518
Tranquilizers	59, 249, 256, 623
Transactional immunity	483, 623
Transmissive illumination	108, 623
Transverse plane	283, 623
Trauma	17, 86, 265, 306-308, 317, 423, 436, 523-524, 623
Trepanation	545, 623
Trojan horse	378, 623
Tube and bag breathalyser	247, 623
Unattended death	278, 623
Unconventional weapons	497, 499, 623
Uniqueness of construction	96, 624
Uniqueness of random occurrences	97, 624
Uniqueness of usage	96-97, 624
Unknown firearm examination	150, 624
Urban entomology	321, 624
Use immunity	483, 624
Value transfer systems	476, 488, 610
Victimology	448-449, 453, 456-457, 624
Virtual memory	365, 624
Visible fingerprints	178, 624
Visionary motive	446, 624
Voice identification	394-395, 407, 624
Voiceprint	395-396, 624
Volatile memory	351, 365, 624
Voluntary manslaughter	56, 624
Warez	371, 624
Wavelength	102-103, 105, 203-204, 625
White blood cells	210, 214, 221, 235, 625
White collar crime	43, 471, 625
Whorls	170, 173, 175-176, 598

Wide area network (WAN)	354, 357, 625
Wireless Networks	357, 625
Wobbly felonies	55, 625
Wound ballistics	136-139, 625
Write-overs	189, 625
Wrongful convictions	503, 511-513, 578, 625
Yaw	134, 136, 625

CPSIA information can be obtained
at www.ICGtesting.com
Printed in the USA
FSOW03n2158170816
23918FS